T0259352

ENZYMES IN FOOD PROCESSING

Third Edition

FOOD SCIENCE AND TECHNOLOGY

International Series

SERIES EDITOR

Steve L. Taylor
University of Nebraska

ADVISORY BOARD

A complete list of all books in this series appears at the end of the volume.

ENZYMES IN
FOOD PROCESSING

Third Edition

EDITED BY

Tilak Nagodawithana

Research and Development
Universal Foods Corporation
Red Star Specialty Division
Milwaukee, Wisconsin

Gerald Reed

Milwaukee, Wisconsin

ACADEMIC PRESS, INC.
Harcourt Brace & Company
San Diego   New York   Boston
London   Sydney   Tokyo   Toronto

Academic Press, Inc.

1250 Sixth Avenue, San Diego, California 92101-4311

United Kingdom Edition published by
Academic Press Limited
24–28 Oval Road, London NW1 7DX

Library of Congress Cataloging-in-Publication Data

Enzymes in food processing / edited by Tilak Nagodawithana, Gerald
  Reed. — 3rd ed.
      p.   cm. — (Food science and technology)
    Includes bibliographical references and index.
    ISBN 0-12-513630-7
      1. Enzymes–Industrial applications. 2. Food industry and trade.
    I. Nagodawithana, Tilak W. II. Reed, Gerald. III. Series.
    TP456.E58E6   1993
      664'.024–dc20   92-41628
                                                                    CIP

Transferred to digital printing 2005.

Contents

CHAPTER 1

Introduction 1
GERALD REED

CHAPTER 2

General Characteristics of Enzymes
KIRK L. PARKIN

CHAPTER 3

Environmental Effects on Enzyme Activity
KIRK L. PARKIN

CHAPTER 4

Modern Methods of Enzyme Expression and Design
JO WEGSTEIN and HENRY HEINSOHN

CHAPTER 5

Immobilized Enzymes

PATRICK ADLERCREUTZ

CHAPTER 10

Applications of Oxidoreductases
THOMAS SZALKUCKI

CHAPTER 11

Milling and Baking
BRUNO G. SPROESSLER

CHAPTER 12

Starches, Sugars, and Syrups
RONALD E. HEBEDA

CHAPTER 13

Dairy Products
RODNEY J. BROWN

CHAPTER 14

Pectic Enzymes in Fruit and Vegetable Juice Manufacture
WALTER PILNIK and ALPHONS G. J. VORAGEN

CHAPTER 18

Fish Processing
GUDMUNDUR STEFANSSON

FOOD SCIENCE AND TECHNOLOGY

Contributors

Numbers in parentheses indicate the pages on which the authors' contributions begin.

Jens Adler-Nissen (159), Department of Biotechnology, Technical University of Denmark, DK 2800 Lyngby, Denmark

Patrick Adlercreutz (103), Department of Biotechnology, Chemical Center, Lund University, S-221 00 Lund, Sweden

Ramunas Bigelis (121), Biotechnology Division, Amoco Technology Company, Naperville, Illinois 60566

Rodney J. Brown (347), Department of Nutrition and Food Sciences, College of Agriculture, Utah State University, Logan, Utah 84322

Sven Erik Godtfredsen (205), Novo Nordisk, DK 2880 Bagsvaerd, Denmark

Frank E. Hammer (221), Biotechnovation, Inc., Oak Forest, Illinois 60452

Ronald E. Hebeda[1] (321), Enzyme Bio-Systems, Ltd., Arlington Heights, Illinois 60005

Henry Heinsohn (71), Genencor, Inc., South San Francisco, California 94080

Tilak Nagodawithana (401), Red Star Specialty Products, Universal Foods Corporation, Milwaukee, Wisconsin 53218

Kirk L. Parkin (7, 39), Department of Food Science, University of Wisconsin-Madison, Madison, Wisconsin 53706

Walter Pilnik (363), Agricultural University, Department of Food Science, 6700 EV Wageningen, The Netherlands

Joseph Power (439), Siebel Institute of Technology, Chicago, Illinois 60645

Gerald Reed (1), 2131 North Summit Avenue, Apartment #304, Milwaukee, Wisconsin 53202

Bruno G. Sproessler (293), Kirschenallee, 6100 Darmstadt, Germany

Gudmundur Stefansson (459), Department of Food Technology, Icelandic Fisheries Laboratories, 121 Reykjavik, Iceland

Thomas Szalkucki (279), Center for Dairy Research, University of Wisconsin-Madison, Madison, Wisconsin 53706

Jean-Claude Villettaz (423), Ecole d'Ingènieurs du Valais, Food and Biotechnology Department, 1950 Sion, Switzerland

Alphons G. J. Voragen (363), Agricultural University, Department of Food Science, 6703 WD Wageningen, The Netherlands

Jo Wegstein[2] (71), Genencor, Inc., South San Francisco, California 94080

---

[1] *Present address:* Corn Products, Moffett Technical Center, 6500 South Archer Road, Summit-Argo, Illinois 60501-0345.

[2] *Present address:* 6346 Escallonia Drive, Newark, California 94560.

Preface to the Third Edition

The 1967 monograph, *Enzymes in Food Processing*, and the 1975 edited second edition covered the uses of enzymes in the food industry in a comprehensive manner. The present and third edited edition has been completely rewritten because of the extensive changes in the way enzymes are used and the availability of new enzymes.

We believe that the third edition will be more useful to readers because it emphasizes basic information on enzymes, newly discovered uses, and uses that have not been adequately described in the literature. Thus, chapters on enzyme functionality and the effect of environmental parameters have been expanded, a chapter on the genetic modifications of enzymes has been added, and a chapter on the use of enzymes in fish processing has been included.

Therefore, less emphasis is devoted to the routine uses of enzymes, and subjects in which relatively few changes have occurred were omitted. The chapter on the production of microbial enzymes has been replaced by an extended introductory chapter that deals with practical aspects of the formulation, standardization, and assay of microbial enzymes as they are sold to the food industry.

We express our gratitude to Academic Press for their continued support of *Enzymes in Food Processing* and to Red Star Specialty Products Division, Universal Foods Corporation, Milwaukee, Wisconsin for their encouragement, and, most of all, to the authors who have so competently and patiently contributed to the book.

We hope that the third edition will be as well received and as widely used as the two earlier editions.

Tilak Nagodawithana
Gerald Reed

Preface to the Second Edition

The purpose and scope of *Enzymes in Food Processing* have been adequately described in the Preface to the first edition, which follows. A deeper understanding of the action of enzymes, some changes in enzyme technology, and the introduction of new enzyme processes into the food industry have made it desirable to publish a second edition.

This edition, unlike the first one, is an edited work. The authors of individual chapters have contributed a deeper knowledge of their field and a greater expertise than the editor could muster for the writing of the first edition. Hence, the chapters dealing with the properties of specific enzymes and the chapters dealing with enzyme applications should be authoritative and up-to-date.

I am greatly indebted to the contributors who have given freely of their time to share their expert knowledge with their colleagues. I am equally indebted to readers of the first edition who have contributed encouragement and criticism and who have made publication of a second edition worthwhile. I am grateful to Academic Press and to the Board of Editors of the "Food Science and Technology" series and, in particular, to Dr. George F. Stewart for assistance in editing. Finally, I wish to thank Universal Foods for permission to undertake this work.

Gerald Reed

Preface to the First Edition

The manufacture of foods has rapidly changed from an art to a highly specialized technology based on discoveries in the natural sciences. However, the translation of scientific knowledge from the fields of microbiology and biochemistry into useful food technology has been rather slow. Art and tradition still play an important part in the fermentation industries and in various uses of enzymes in food processing. It is, therefore, important to bridge the gap between available scientific knowledge and food technology in these particular areas.

During the past 25 years the use of commercial enzymes has grown from an insignificant role to an important aspect of food processing. However, no comprehensive treatise on the use of enzymes in food processing has been published in the past 15 years; this monograph was written to fill that gap. Primarily, it is directed to food technologists. They will find in it a description of the properties of those enzymes which are important in food processing as well as a description of the many practical applications of enzymes in their industry. It will also be of value to the microbiologist and enzyme chemist who may wish to acquire some knowledge of the fields in which their discoveries are put to practical use. This volume will acquaint them with present applications of enzymes in the food industry and will perhaps suggest new uses for enzymes.

The subject is treated in two parts. Part I describes the properties of enzymes in general and the properties of enzymes used specifically in food processing. Part II describes the practical application of these enzymes to various phases of the food industry with cross references to the basic properties of the enzymes described in Part I. It is hoped that this method will foster a clearer understanding of the relationship between the basic properties of enzymes and their application.

I am greatly indebted to the following people who have reviewed one or several of the chapters and who have provided extensive assistance: Dr. M. L. Anson, Mr. W. G. Bechtel, Dr. T. Cayle, Dr. S. L. Chen, Dr. G. I. de Becze, Mr. F. Hammer, Dr. K. Konigsbacher, Dr. E. R. Kooi, Dr. J. H. Nelson, Mr. M. C. Reed, Dr. D. Scott, Dr. E. Segel, Dr. C. V. Smythe, Dr. G. F. Stewart, and Dr. L. A. Underkofler. Dr. Underkofler has written Chapter 10, "Production of Commercial Enzymes." Without his help this subject could not have been treated adequately. He also reviewed a considerable portion of the manuscript and made many helpful suggestions. I am deeply grateful to Dr. M. L. Anson, one of the editors of the Food Science and Technology series. Without his advice, counsel, and criticism it would have been difficult to write this book.

I want to thank Mrs. F. W. Chen and Mr. R. Liu for the illustrations, and Mrs. M. Ziesch for typing the manuscript. My wife has encouraged me and helped me with the preparation of the manuscript.

April, 1966
Gerald Reed

CHAPTER 1

Introduction

GERALD REED

Enzymes used in food processing are often called "commercial enzymes." They may be derived from plant, animal, or microbial sources. These products are sold by the pound or the ton, in contrast to the highly refined enzymes produced for pharmaceutical or research purposes. The latter are often crystalline enzymes sold in milligram and gram quantities by firms that specialize in "fine chemicals."

Commercial food grade enzymes are not highly refined. Generally, they are sold and used in bulk. The most widely used animal-derived enzyme is chymosin, a milk clotting enzyme used in the production of cheeses. This enzyme is extracted from the stomach of calves. The plant-derived enzymes are principally cereal malt, papain, and bromelin. Barley malt and, to a lesser extent, wheat malt are produced by maltsters and by some brewers. Papain and bromelin, both vegetable proteases, are sold by import houses.

Commercial microbial enzymes were introduced into United States food processing by three immigrants: the Japanese Takamine, the German Wallerstein, and the German Haas, all of whom founded companies that bore their name. Since that time, these firms have been sold or their enzyme facilities taken over by other individuals.

The earliest uses of microbial enzymes were in cereal processing and fruit juice clarification. Beginning in the 1950s and 1960s, microbial enzymes began to replace animal- and plant-derived enzymes in many food applications. Chymosin became in short supply because the availability of calf stomachs did not increase with the large increase in cheese production. Chymosin of animal origin is now replaced often with microbial "rennet" derived from *Mucor* species.

Malt diastase is still the predominant enzyme used in the production of worts used for beer fermentation and of cereal mashes used for the production of distilled alcoholic beverages. However, fungal amylases have made considerable inroads into the fields of flour supplementation and baking. The production of industrial alcohol relies heavily on fungal preparations for the hydrolysis of cereals such as corn, rice, and wheat to fermentable sugars.

Papain has not been replaced yet by microbial enzymes because of its wide spectrum of catalytic activity and because of its relatively high temperature stability. The principal applications of this enzyme are in meat tenderizing and in beer chillproofing. The enzyme is not inactivated fully by beer pasteurization.

Microbial enzymes are produced from a number of species that have been recognized as safe, including some fungi, such as *Aspergillus niger, A. oryzae, A. awamori,* and *Mucor miehei;* some bacilli, such as *Bacillus subtilis* and *B. licheniformis;* some yeasts, such as *Saccharomyces cerevisiae;* and others. The microbes can be grown in surface culture on solid media, such as mold bran, or preferably in submerged culture in batch, fed-batch, or continuous culture. An enzyme preparation in the crudest form can be obtained by simply drying the mold bran on which a mold has been grown. Such preparations are often suitable for the production of industrial alcohol. Alternatively, the mold bran can be extracted with water and the enzyme can be precipitated from the liquid with an alcohol or with ammonium sulfate. The enzyme may be isolated similarly, from the liquid portion of a submerged culture. Recovery of intracellular enzymes, such as the lactase and the sucrase (invertase) of yeasts, requires rupturing the cell wall and separating the solid cell residue before the enzyme can be precipitated. The enzyme fraction can be dried at a low temperature or sold in liquid form in drums. The final enzyme preparations usually contain a large number of enzymes, that is, all the extracellular enzymes of an organism, as well as other proteins and soluble media constituents. The amount of the enzyme of choice in the final enzyme preparation may be quite small. For example, a fungal amylase standardized at 10,000 Sandstedt, Kneen, and Blish (SKB) alpha amylase units per gram contains only 0.2% alpha amylase by weight. This percentage can be calculated readily from the known molecular weight of alpha amylase and the known activity of about 5,000,000 SKB alpha amylase units per gram of pure enzyme. Such calculations can be made only if the enzyme has been obtained in crystalline form and assayed by the same method as the commercial enzyme preparation.

Typically, a fungal alpha amylase preparation contains various other amylases, mainly alpha 1,4-glucosidase, as well as proteases, lipases, and other enzymes. The preparation is sold and used satisfactorily as an alpha amylase by virtue of its standardization on the basis of alpha amylase activity. The same enzyme preparation can, on occasion, also be standardized on the basis of its proteolytic activity and sold as a protease.

Enzyme preparations produced on a plant scale show some batch to batch variation in activity. This variation requires standardization to a given activity by dilution of the preparation. This dilution may be carried out with any inert material, for example, water, salt, sugars, starch, diatomaceous earth, provided the diluent is compatible with the final usage of the enzyme in food processing.

Analytic methods to determine enzyme activity usually are chosen to reflect end usage wherever possible. For instance, alpha amylase activity of fungal preparations to be used in baking is determined by the SKB method at 30°C. Activity of bacterial alpha amylases for starch thinning at much higher temperatures is determined by viscosimetric methods. This pragmatic approach is useful to the industry and requires thorough knowledge of industrial practice.

In some instances, suitable analytic methods are lacking. For instance, there is no suitable method for determining the effect of proteases on bread doughs. The substrate is insoluble and small changes in the viscoelastic properties of doughs are difficult to determine with common bench methods. Fungal proteases for use in baking usually are standardized by the Anson hemoglobin method, which shows no correlation with the effect in bread baking. In such cases, users must rely on the uniformity of the production methods used by the enzyme supplier or, better still, they must determine the effect in their own plant by measuring the effect of the protease on the mixing time of the doughs.

A similar problem exists in the use of pectic enzymes for the clarification of various fruit juices. The crushed fruits or the juices contain pectins of various degrees of polymerization and methylation. The preparations offered for sale contain a variety of enzymes, such as various polygalacturonases, pectin methyl esterases, and pectin lyases. Hence, it is impossible to determine the effectiveness of a preparation for clarification of a given juice by analysis of individual enzyme activities. Fortunately, the effect on a given juice can be checked readily by the rate of precipitate formation and settling.

Enzyme producers usually supply good information on optimum pH and optimum temperature of enzymatic activity, as well as on stability of the enzyme at higher temperatures. This information can be very helpful to industrial users. For instance, the pH optimum of yeast-derived lactases is near neutral, which makes them suitable for treating milk. On the other hand, the fungal lactase from *A. niger* has a much lower pH optimum, which suggests use of this enzyme in cheese whey. However, pH activity values cannot be accepted without reservation. These values depend on the choice of substrate, temperature, stability of the enzyme at a given pH, and other environmental factors. Temperature optima also depend on various environmental factors, particularly on the duration of the reaction. For shorter time periods, the optimum temperature of an enzyme is higher than for longer reaction periods, reflecting the differential rate of inactivation at the two temperatures.

The presence of multiple enzymes in commercial preparations is generally inconsequential. However, their presence must be considered in each case, since examples of unfavorable reactions are well known. The presence of proteases in fungal alpha amylase preparations is detrimental if doughs have been made with weak flour. The presence of proteases in yeast lactase preparations leads to the undesirable hydrolysis of milk proteins. Proteases (other than chymosin) in microbial rennets lead to weakening of the cheese curd. The presence of transglucosidases in amylase preparations used for the production of corn syrups lowers the yield of glucose. Predicting the effect of such adventitious enzymes is rarely possible. Hence, it is imperative to test the enzymes under conditions of actual commercial use.

Enzymes can be immobilized on solid supports, arranged in packed beds, and used for the continuous hydrolysis of fluid substrates that are percolated through the bed. Such a process is used commercially for the conversion of 98 DE (dextrose equivalents) glucose syrups by xylose isomerase to a mixture of glucose and fructose. Other applications have been investigated with some success. The use of lactose hydrolyzed whey as a substrate for the growth of baker's yeast has been considered. Unfortunately, almost complete hydrolysis of the lactose is required since residual lactose causes sewage disposal problems and enzyme costs rise steeply with extent of conversion (see Chapter 5).

Hydrolysis of polymeric carbohydrates and of proteins and lipids can, of course, be carried out using inorganic acids. Although hydrolysis with acids is not specific, it is applicable if a uniform substrate, such as corn starch or soy protein, is available. Until the 1950s, corn syrup generally was prepared by acid hydrolysis which is fast and efficient. However, yields of glucose (dextrose) are lower and reversion products are formed if acid hydrolysis is continued to give syrups with high DE values. In past decades, enzymatic conversion of corn starch by bacterial alpha amylase and fungal glucosidase has resulted in the production of high DE syrups that also provide a good yield of commercial dextrose. Such high DE syrups are the substrate for xylose (glucose) isomerase in the production of high fructose corn syrups. This last product has replaced sucrose in major applications, such as the production of soft drinks. Soy protein is hydrolyzed commercially using concentrated hydrochloric acid at elevated temperatures in pressurized vessels. The resultant product, known as hydrolyzed vegetable protein (HVP), is a widely used savory flavoring. Enzymatically hydrolyzed soy and yeast fractions are available but have not replaced acid hydrolyzed proteins.

One important economic aspect of enzyme processing, the effect of enzyme use on processing costs, often is neglected. Two examples illustrate this point. Apple juice is prepared by pressing macerated apples. The resultant juice contains much particulate matter that can be removed partially by filtration, resulting in a cloudy product. The particles can be removed completely by pectic enzyme treatment, resulting in a clear product. Many consumers prefer a cloudy, "more natural" apple juice, but filtration without pectic enzyme treatment is so slow that it requires a much larger investment in filters and increases processing cost. The second example is the use of fungal proteases in bread doughs. Use of the enzyme, under certain circumstances, produces a somewhat better grain and texture, but the principal reason for the introduction of proteases into the industry is the reduction in mixing time, for example, from 12 min to 8 min, that can be achieved. Again, this effect reduces capital costs for expensive mixers.

The foregoing discussion of some practical aspects of enzyme usage addressed only hydrolytic enzymes. Traditionally and currently, hydrolases represent the majority of enzymes used for food processing. Some interesting uses of glucose oxidases are discussed in Chapter 9. However, neither oxido-reductases nor synthetases have played a major role in the food processing industry. This situation is likely to change in the coming decade as advances in science prepare the ground for new applications.

Catalytic activity as it is related to the molecular structure of enzymes is treated in Chapter 2; the effect of environmental variables on enzyme activity is discussed in Chapter 3. The genetic modification of microorganisms is likely to permit the construction of enzymes with vastly greater potential for food applications. This topic is treated in Chapter 4. These three chapters present basic concepts required to understand enzyme usage in food processing. These ideas ultimately will change the use of enzymes in food processing from a merely pragmatic approach to a scientific one.

CHAPTER 2

General Characteristics
of Enzymes

KIRK L. PARKIN

I. Introduction

The term "enzyme" is derived from the Latin words meaning "in yeast." As implied, once enzymes were thought to exist at an organismal level, until Sumner (1926) crystallized jack bean urease, demonstrating unequivocally that enzymes were proteins. An enzyme can be defined as a polypeptide(s) that catalyzes a reaction with a certain degree of specificity. The three key words in this definition are polypeptide, catalyze, and specificity.

II. Primary Features of Enzymes and Enzyme Reactions

A. Enzymes as Polypeptides

1. Protein Structure

All enzymes are proteins. The functionality of enzymes is a direct consequence of the protein's amino acid sequence (primary structure), the folding of the polypeptide into α-helical, β-pleated sheet, or aperiodic (random coil) conformations (secondary structure), the interactions of the amino acid side chains by electrostatic and hydrophobic forces and disulfide bridges (tertiary structure), and the nature of the assemblage of the protein subunits (quaternary structure). Enzymes range in molecular size from about 13,000 to several million daltons, although most enzymes exhibit a molecular size in the range of 30,000 to 50,000 daltons (Page, 1987). Extracellular enzymes are generally low molecular weight proteins and often contain several disulfide bridges to assist the maintenance of protein structure in response to the environment. Regulatory enzymes are generally of higher molecular weights and often are composed of protein subunits, allowing for strict metabolic control in the intracellular environment. However, within a class of enzymes, for example, the serine proteases, molecular sizes can vary significantly, for example, from 185 to 800 amino acid residues (Srere, 1984).

Although enzymes are proteins containing several hundred amino acid residues, only a few residues are involved directly in substrate binding or reaction catalysis. The nature of the reaction that is catalyzed by a given enzyme is dependent on the identity of the amino acids that constitute the active site. These residues, by virtue of their ionization potential and polarity, determine the reaction specificity and mechanism that are characteristic of a particular enzyme. The other amino acid residues of the polypeptide chain provide the proper spatial orientation between the active site and the sub-strate(s) and cofactors necessary for reaction catalysis. In addition, the struc-ture, shape, and topography of the folded polypeptide may assist in "trap-ping" substrates at the surface and may facilitate diffusion of substrate to the transforming locus (Srere, 1984; Sharp *et al.*, 1987). The flexibility of the polypeptide structure and its interaction with the surrounding (micro)envi-ronment are also responsible for control and regulation of enzyme activity. The once-held view that enzymes and their active sites are rigid structures is no longer tenable. Plasticity of enzyme structure also appears to play a role in substrate specificity (Bone *et al.*, 1989).

2. Thermal Lability

The property of thermal lability is often used to document whether or not a reaction is enzymatic (proteinaceous). Although it is true that enzymes, being proteins, are denatured by the application of thermal energy, it should be noted that some enzymes are stable to treatments of 100°C for up to 1 hr or more in aqueous (Russ *et al.*, 1988) and nonaqueous (Zaks and Klibanov, 1984) solutions. This property is imperative for the survival of organisms that exist in high temperature environments, for example, bacteria indigenous to hot springs (Brock, 1985). Therefore, other means are often necessary to document the enzymatic nature of a particular reaction; lability of an active fraction to protease treatments, use in the reaction of cofactors that typically are bound to proteins (flavins, nicotinamides, cobalamine, pyridoxal phos-phate, tetrahydrofolate, pantothenate, biotin, and lipoate), activity staining of proteins separated by electrophoresis under nondenaturing conditions, crys-tallization of a protein with maintenance of reactivity, and recovery of activity in the retentate after dialysis provide additional evidence that a particular reaction is enzyme catalyzed.

B. Catalysis

1. Transition State Theory and Thermodynamics

Another salient feature of enzyme action is catalysis of a chemical reaction. Catalysis can be defined as the acceleration of a process that would proceed at an otherwise slower rate under a defined set of conditions in the presence of an agent that remains unmodified during the course of the reaction. Reaction catalysis is a specific function of the mechanism of enzyme action (Page, 1987). This property, along with specificity, is a criterion governing the choice of enzymes in selected food processing applications. An example is the use of invertase in the production of candies with liquid centers. Given enough time, the acid-catalyzed hydrolysis of sucrose would reach the same equilibrium as that achieved by invertase. However, the rate at which this process initially takes place is about 2×10^{10} times faster in the presence of invertase (Schwimmer, 1981). Enzyme catalysis has been explained on the basis of the transition state theory, first advanced by Eyring (1935). This concept is illustrated in Fig. 1.

Although the change in free energy (ΔG) and, therefore, the K_{eq} ($\Delta G = -2.3RT \log K_{eq}$) of the reaction are identical for the enzyme-catalyzed and uncatalyzed reactions, the activation energy (E_a or $\Delta H\ddagger$) in the former case is reduced, thereby reducing the magnitude of the thermodynamic energy barrier to the reaction proceeding through a transition state to completion

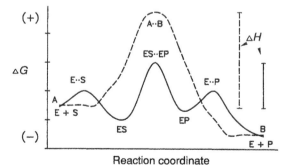

Reaction coordinate

Figure I Transition state theory for reaction catalysis. Relative changes in free energy (ΔG) for enzymatic (solid line) and nonenzymatic (dashed line) reactions are shown. The transition states are represented by the components separated by two dots. For the enzymatic reaction; E, enzyme; S, substrate; P, product; ES, enzyme–substrate complex; EP, enzyme–product complex. For the nonenzymatic reactions: A, reactant; B, product.

(equilibrium). Consequently, the rates of enzyme-catalyzed reactions are greater than those of uncatalyzed reactions. The power of catalysis also can be illustrated by the van't Hoff relationship (Eq. 1) for reaction equilibria and its transformation to the Arrhenius equation (Eq. 2) for reaction rate catalysis:

$$K_{eq} = C \, e^{(-\Delta H/2.3RT)} \tag{1}$$

$$k = Z \, e^{(-E_a/2.3RT)} \tag{2}$$

where K_{eq} is the equilibrium constant, ΔH is the heat of reaction, R is the gas constant, T is the temperature (K), C is an internal constant, Z is the frequency of favorable collisions, and E_a is the activation energy. If an enzyme is assumed, at 25°C, to reduce the activation energy (E_a) of an uncatalyzed reaction by 10 kcal/mol, the relative rates of the catalyzed and uncatalyzed reactions can be predicted by Eqs. 3 and 4:

$$\frac{k_{cat}}{k_{uncat}} = \frac{e^{(-E_{a_{cat}}/RT)}}{e^{(-E_{a_{uncat}}/RT)}} \tag{3}$$

$$\frac{k_{cat}}{k_{uncat}} = e^{[(E_{a_{uncat}}-E_{a_{cat}})/RT]} = e^{(10/0.592)} = 2.17 \times 10^7 \tag{4}$$

As can be seen in this example, in which the presence of a hypothetical enzyme reduces the E_a of the reaction by 10 kcal/mol, the rate of reaction is accelerated 7 orders of magnitude in the presence of the enzyme. To put this result into perspective, thermodynamically, an uncatalyzed reaction that would achieve equilibrium in 1 month would achieve equilibrium in 1 second in the presence of the catalyst. However, other constraints can prevent enzymatic reactions from progressing this rapidly. This extent of reduction in activation energy (10 kcal/mol) is realistic, since in one case (succinyl-CoA acetoacetate transferase) an enzymatic reaction is known to reduce the activation energy of a nonenzymatic reaction by as much as 18 kcal/mol, even though both reactions exhibit the same mechanism (Page, 1987).

2. Nature of Catalysis

Mechanistically, over 20 explanations have been advanced to account for enzyme catalysis (Page, 1987), some of which are more popular than others. The induced-fit hypothesis (Koshland, 1958) is based on the concept that binding of substrate causes a conformational change in the enzyme, resulting in a favorable alignment of reactive groups in the enzyme–substrate complex. A parallel concept is the distortion or rack mechanism of catalysis (Fano, 1947). In this model, substrate binding to enzyme induces torsional strain on the substrate, elevating it to a transition state and reducing the energy barrier for the reaction. Both these models implicitly rely on the concept that enzymes and their active sites are flexible structures. Orbital steering is suggested to align the proximity of substrate and active site reactive groups. Freezing or anchoring of substrate at the active site is believed to enhance the residence time of susceptible groups at the enzyme's transforming locus. To account for the catalytic nature of enzymes, all these models appear to favor a productive collision between enzyme and substrate.

Catalysis also can be attributed to an increase in the concentrations of reactants localized at the active sites of enzymes. Depending on the strength of the attractive forces between enzyme and substrates, the concentration of reactants localized at the active site can be increased effectively several orders of magnitude over that afforded by diffusion. Correspondingly, enzyme catalysis has been modeled as an intramolecular reaction to account for this proximity effect. Quantitatively, this effect translates into up to a 10^{12}-fold increase in the effective molarities of the reactants compared to intermolecular (nonenzymatic) catalysis (Whitaker, 1972; Page, 1987).

3. Reaction Mechanisms

When enzymatic and nonenzymatic reactions do not share the same mechanism, the mechanism of the enzymatic reaction has been held accountable for the acceleration in reaction rate (Page, 1987). The most common mechanisms of enzymatic catalysis include general acid–base behavior, charge neutralization, nucleophilic and electrophilic attack, and torsional strain (Table I).

The mechanism of an enzymatic reaction can involve many of these processes; the mechanisms of hydrolytic reactions are known to involve the formation of a covalent enzyme–substrate intermediate. The complexity of an enzyme mechanism can be illustrated in the case of subtilisin (Carter and Wells, 1988; Wells and Estell, 1988), a serine protease that can be considered representative of a hydrolytic enzymatic reaction (Fig. 2).

The susceptible peptide bond is brought in proximity of the transforming locus to form an enzyme–substrate complex. The first step in catalysis is the nucleophilic attack of Ser-221 on the peptide carbonyl. The nucleophilicity of this serine residue is enhanced by proton withdrawal by His-64, acting as a general acid–base. The acyl-enzyme intermediate is stabilized by hydrogen bonding of Asn-155 to the resultant electronegative oxyanion group. Electrostatic stabilization (charge neutralization) of the transition state complex is afforded by Asp-32, which also provides favorable electronic properties for the imidazole group of the His-64 residue. This charge relay feature is common to many enzyme mechanisms. Splitting the peptide bond yields the acyl-Ser-221 enzyme intermediate and a reduced length peptide as a leaving group. The catalytic cycle is completed when water, acting as a nucleophile, displaces the other peptide fragment from the Ser-221 residue. The spatial and electronic integration of these specific amino acid residues is critical to catalysis; replacement of Ser-221, Asp-32, or His-64 with an alanine residue reduces k_{cat} by a factor of $10^4 - 10^6$ without significantly affecting K_m.

TABLE I
Common Mechanisms in Enzyme Catalysis[a]

Process	Description	Amino acid residues, cofactors, or protein features commonly involved
Approximation	Binding of reactants in an orientation conducive to chemical transformation	Charge density, polarity, and solvation; active site configuration
General acid–base behavior	Lowering of transition state energy of substrate or substrate–product intermediate by proton addition or abstraction	Asp, Glu, His, Lys, Tyr, Arg, Cys; metal ions
Nucleophilicity	Formation of covalent enzyme–substrate intermediate by electron donation	Asp, Ser, Cys, His, Glu
Electrophilicity	Formation of enzyme–substrate intermediate by electron withdrawal due to protonation, a proximal charge, or an electron sink	Lys, Arg, His; thiamine pyrophosphate, pyridoxal phosphate, metal ions
Distortion or torsional strain	Use of binding forces to lower the transition state energy of the enzyme–substrate intermediate	Charge density, polarity, and solvation of active site; active site configuration and flexibility

[a] Compiled from Palmer (1981), Page (1987), Price and Stevens (1989), Saier (1987), and Whitaker (1972).

C. Specificity

1. Types of Specificity

Specificity of enzyme reactions is also a direct consequence of substrate and active site structures. Structural or shape (Fischer's lock and key concept) and electrical (Page, 1987) complementarities are factors important to substrate recognition by enzymes. Amino acid residues near the active site, but not directly involved in catalysis, also play a role in providing the substrate with access to the enzyme transforming locus; examples include the presence of a hydrophobic pocket near the active site of carboxypeptidase A to assist in recognition of peptide bonds with a nonpolar amino acid C terminus (Fig. 3) and the presence of a positive electrical field near the active site of superoxide dismutase to enhance diffusion of superoxide anion to the active site by a factor of 30 over the rate expected by diffusion in an aqueous environment (Sharp *et al.*, 1987).

Enzymes exhibit various types of specificity including group, bond,

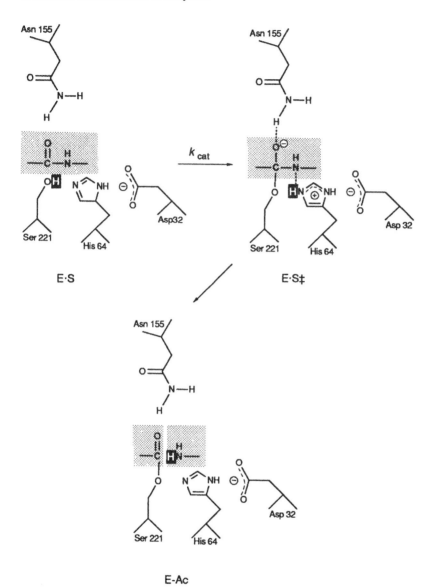

Figure 2 Reaction mechanism for the serine protease subtilisin. ES, Enzyme–substrate complex; E-Ac, acyl enzyme. (Reprinted from Carter and Wells, 1988, with permission.)

stereo-, and absolute specificities. Alcohol dehydrogenase exhibits group specificity because it oxidizes several primary alcohols. However, it also exhibits stereospecificity because a particular hydrogen is exchanged (as determined by isotopic studies) between the alcohol and NAD(H), although the substrate is not asymmetric. D- and L-amino acid oxidases are specific for

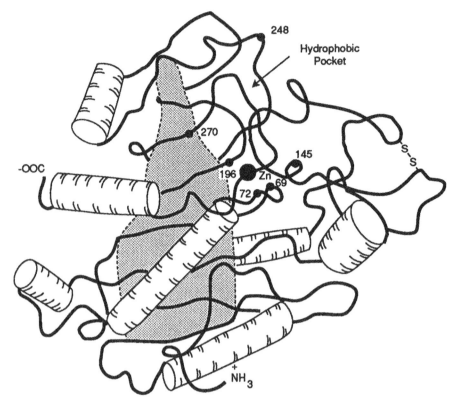

Figure 3 Representation of the three-dimensional structure of carboxypeptidase A, a metalloprotease. (Reprinted from Palmer, 1981, with permission.)

the D- and L-isomers of amino acids, respectively; glucose oxidase is 164 times more reactive on the β-anomer than the α-anomer of D-glucose.

Another interesting case of enzyme stereospecificity is glycerol kinase, which uses ATP to phosphorylate glycerol only at the *sn*-3 position and not at the *sn*-1 position. This example illustrates another, often disregarded aspect of enzyme specificity: product specificity. Another example of this feature is trimethylamine-*N*-oxide demethylase, which forms only dimethylamine and formaldehyde from trimethylamine-*N*-oxide, whereas the chemically catalyzed reaction forms a mixture of trimethylamine, dimethylamine, and formaldehyde (Parkin and Hultin, 1986).

Glucose kinase exhibits absolute specificity in exchanging phosphate only between ATP and glucose. Lipoxygenase is an example of an enzyme recognizing a functional group, namely, the 1,4-pentadiene structure. Lipoxygenases from various sources exhibit other types of specificity with respect to whether or not the fatty acid is esterified and to the stereochemistry of the initial products formed, the fatty acid hydroperoxides (Vick and Zimmerman, 1987).

Esterases, hexokinases, nonspecific phosphatases, and proteases exhibit bond specificity. Proteases and peptidases are also capable of acting on esters of amino acids. Within the class of serine proteases (trypsin, chymotrypsin, elastase), added reaction specificity is conferred for the residue on the N-terminal side of the peptide bond to be hydrolyzed, dictated in part by the presence of amino acid residues in proximity to the transforming locus (Fig. 4).

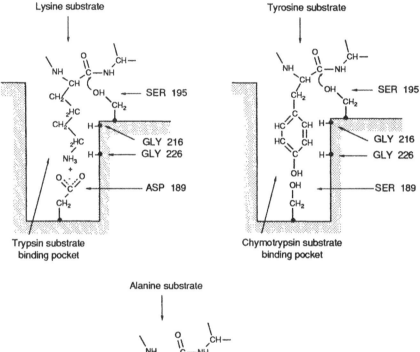

Figure 4 Representation of the binding pockets of the serine proteases trypsin, chymotrypsin, and elastase. (Reprinted from Shotton, 1971, with permission.)

2. Enzyme – Substrate Interactions

Reaction specificity is imparted by the nature of the enzyme–substrate interaction. Both long- and short-range interactions can play a role in substrate recognition and catalysis. This characteristic is illustrated by the reactivities of κ-casein analogs as substrates for chymosin (rennet) listed in Table II.

All the substrate analogs are characterized by the Phe–Met peptide bond (Phe-105–Met-106), the primary site of attack of chymosin on κ-casein. A peptide chain length of at least 5 residues is required for activity, provided that the Phe–Met bond is a nonterminal unit (compare substrates 1–7 and 8–31 in Table II). Efficiency in substrate utilization (K_m) and catalysis (k_{cat}) are both improved by at least an order of magnitude as the peptide analog is extended at the C terminus, resulting in an increase in catalytic efficiency (k_{cat}/K_m) of 2–3 orders of magnitude (compare substrates 12 and 13–17). As the peptide analog is extended toward the N terminus (compare substrates 16

TABLE II
Chymosin Reactivity with κ-Casein Substrate Analogs[a]

Sub-strate	Sequence[b]	k_{cat} (s^{-1})	K_m (nM)	k_{cat}/K_m (s^{-1} mM^{-1})
	100 105 106 110			
κ-casein	His-Pro-His-Pro-His-Leu-Ser-Phe-Met-Ala-Ile-Pro-Pro-Lys-Lys	—	—	—
1	Phe-MetOMe	No cleavage detectable		
2	Phe-Met-AlaOMe	No cleavage detectable		
3	Phe-Met-Ala-Ile-OMe	No cleavage detectable		
4	Phe-Met-Ala-Ile-Pro-Pro-LysOH	No cleavage detectable		
5	Ser-Phe-MetOMe	No cleavage detectable		
6	Leu-Ser-Phe-MetOMe	No cleavage detectable		
7	Ser-Phe-Met-AlaOMe	No cleavage detectable		
8	Ser-Phe-Met-Ala-IleOMe	0.33	8.5	0.04
9	Ser-Phe-Met-Ala-Ile-ProOMe	1.05	9.2	0.11
10	Ser-Phe-Met-Ala-Ile-Pro-ProOMe	1.57	6.8	0.23
11	Ser-Phe-Met-Ala-Ile-Pro-Pro-LysOH	0.75	3.2	0.24
12	Leu-Ser-Phe-Met-AlaOMe	0.58	6.9	0.08
13	Leu-Ser-Phe-Met-Ala-IleOMe	18.3	0.85	21.6
14	Leu-Ser-Phe-Met-Ala-Ile-ProOMe	38.1	0.69	55.2
15	Leu-Ser-Phe-Met-Ala-Ile-Pro-ProOMe	43.3	0.41	105.1
16	Leu-Ser-Phe-Met-Ala-Ile-Pro -Pro-LysOH	33.6	0.43	78.3
17	Leu-Ser-Phe-Met-Ala-Ile-Pro-Pro-Lys-LysOH	29.0	0.43	66.9
18	His-Pro-His-Pro-His-Leu-Ser-Phe-Met-Ala-Ile-Pro-Pro-LysOH	66.2	0.026	2509
19	His-Pro-His-Pro-His-Leu-Ser-Phe-Met-Ala-Ile-Pro-Pro-Lys-LysOH	61.9	0.028	2208
20	His-Leu-Ser-Phe-Met-AlaOMe	0.96	6.2	0.16
21	His-Leu-Ser-Phe-Met-Ala-IleOMe	16.0	0.52	30.8
22	Pro-His-Leu-Ser-Phe-Met-Ala-IleOMe	33.6	0.34	100.2
23	Leu-Ser-Phe-Met(O)-Ala-IleOMe	7.0	3.5	2.0
24	Leu-Ser-Phe-Leu-Ala-IleOMe	4.6	0.53	8.7
25	Leu-Ser-Phe-Nle-Ala-IleOMe	24.9	0.36	69.2
26	Leu-Ser-Cha-Nle-Ala-IleOMe	4.5	3.1	1.4
27	Leu-Ser-Phe(NO$_2$)-Nle-Ala-IleOMe	7.6	0.32	23.9
28	Leu-Ser-Phe(NO$_2$)-Nle-Ala-LeuOMe	12.1	0.50	24.2
29	Leu-Ser-Phe-Nle-Ala-GlyOMe	9.9	1.4	7.2
30	Gly-Ser-Phe-Nle-Ala-IleOMe	18.3	1.7	10.9
31	Leu-Gly-Phe-Nle-Ala-ILeOMe	1.8	1.0	1.8

[a] Reprinted from Visser (1981), with permission of the Netherlands Milk and Dairy Journal.
[b] Cha, Cyclohexylalanine; Phe(NO$_2$), 4-nitrophenylalanine.

and 18–19 in Table II), efficiency in substrate utilization (K_m) is greatly improved and the resultant catalytic efficiency (k_{cat}/K_m) approaches that of chymosin action on native casein. The segment His–Pro–His–Pro–His (residues 98–102) appears to be particularly important in enzyme–substrate recognition. The high positive charge density of this segment (at neutral or slightly acidic pH) implies a specific electrostatic interaction, or "anchoring," of the substrate near the active site of chymosin. The steric or polar properties of residues 105–106 are also important for substrate transformation (compare substrates 13 and 23–27 in Table II). Finally, the role of Ser-104 and its probable interaction with the enzyme through hydrogen bonding appears to be critical to reaction specificity (compare substrates 25 and 31 in Table II).

In a similar study, incorporation of cationic residues (Lys and Arg) into model peptide substrates for pepsin revealed that secondary electrostatic interactions between substrate and enzyme can increase efficiency of substrate binding several orders of magnitude, although pepsin action is highly specific for peptide bonds formed by aromatic and nonpolar residues (Pohl and Dunn, 1988).

3. Importance of Specificity

The unique pattern of substrate recognition by chymosin provides the rationale for its use as a milk-clotting enzyme. Another reason for the choice of chymosin over other proteases is its ratio of milk clotting and proteolytic activities (Visser, 1981). Chymosin has a high ratio; most other proteases have a low ratio. As a result, most other proteolytic enzymes are unsuitable for milk clotting because they produce a curd with inferior physical characteristics and often give rise to bitter peptides during cheese aging by continued proteolytic activity. Alternative sources of milk-clotting proteases continue to be evaluated. Microbial rennets have been used successfully as milk-clotting enzymes, especially the enzyme isolated from *Mucor miehei,* and now account for an estimated one-third of the worldwide use of enzymes in cheesemaking (Adler-Nissen, 1987). Advances in genetic engineering almost certainly will provide food biotechnologists with the means to design enzymes, including proteases, with a specificity optimized for a particular processing application (see Chapter 4). The application of biotechnology is based on the development of a fundamental understanding of the role of specific amino acid residues in enzymatic catalysis. Point mutations currently provide the means to study the effect of specific amino acid substitutions on enzymatic activity, as in the case of β-galactosidase (Ring *et al.,* 1988) and subtilisin (Wells and Estell, 1988). Recombinant DNA technology already has given scientists the ability to express a gene encoding chymosin in *Escherichia coli* (Kawaguchi *et al.,* 1987).

Perhaps one of the most intriguing areas of development in this field is the design of catalytic antibodies or "abzymes" (Lerner and Tramatano, 1987). This area of research represents the merging of the concepts of binding specificity (antibodies) and transition state catalysis (enzymes). This fertile, and still novel, area of research represents the potential to engineer or design "smart enzymes" that can catalyze very specific reactions. Although the most immediate developments are likely to occur in the biotechnological and biomedical fields, the subsequent extension of these discoveries into areas of food technology and food safety is inevitable.

III. Nomenclature

At this time, it is prudent to describe how enzymes are defined and cataloged. With over 2300 enzymes now cataloged (International Union of Biochemistry, Enzyme Commission (IUB), 1984), it is not always meaningful to refer to enzymes by their common names. "Diastase," "fumarase," "allinase," and "lactase" are examples that serve as points of discussion. Diastase, the name given to an isolated fraction of malted barley that is capable of hydrolyzing starch, gives no coherent indication of the nature of its action or the substrate transformed. However, in the brewing industry, it is common to refer to amylolytic activity as diastatic power. Fumarase is an enzyme that does not hydrolyze fumarate (as is implied), but catalyzes the reversible hydration of fumarate to form malate. Allinase is the enzyme responsible for the generation of pungent aromas in *Allium* species (e.g., garlic, onion, and leeks) by its action on the flavor precursors, the S-alk(en)yl-cysteine sulfoxides. However, it is believed that the enzyme also exists in brassicas, such as cabbage. The name lactase leads the naive scientist to believe that the enzyme acts on lactate. Although these names remain a mainstay of the food technologist's vernacular, another concern is the continuing promulgation of trivial or trade names for enzymes, or more specifically, enzyme preparations. Enzyme preparations available to the food industry often have trade names that describe their role or effect when used in food systems. Hence, the specific enzyme in the preparation that causes the desired change may not be obvious to the user. In spite of the confusion that may be caused by this type of nomenclature, trivial names and marketing-driven naming of enzymes undoubtedly will persist.

The nomenclature proposed by the IUB (1984) still retains the traditional use of the ending "ase" in the systematic naming of enzymes. (Proteolytic enzymes have been named trivially using the ending "in".) A summary of the classification of enzymes based on a four number sequence appears in Table III.

The first number represents the class of enzyme. The nature of the reaction catalyzed by each class of enzyme appears in the second column of Table III. Examples of enzymes in each class that are important in food systems are: 1, phenol oxidase, catalase, and lipoxygenase; 2, transpeptidase and starch synthase; 3, invertase, amylases, and lipase; 4, pectate lyase and S-alk(en)yl-cysteine sulfoxide lyase; 5, glucose isomerase; and 6, fatty acid

TABLE III
Systematic Basis of Enzyme Nomenclature[a]

First number (class)	Nature of reaction	Second number (first subclass)	Third number (second subclass)
1 Oxidoreductases	Electron transfer	Group in donor, D, oxidized	Acceptor, A, reduced
2 Transferases (synthases)	Group transfer	Group transferred from D to A	Group transferred (further delineated)
3 Hydrolases	Hydrolysis	Bond hydrolyzed: ester, peptide, etc.	Substrate class: glycoside, peptide, etc.
4 Lyases	Bond-splitting	Bond cleaved: C–S, C–N, etc.	Group eliminated
5 Isomerases	Isomerization	Type of reaction	Mix of S,[b] reaction type chiral position involved in iso-merization[c]
6 Ligases (synthetases)	Bond formation	Bond synthesized: C–C, C–O, C–N, etc.	Substrate S_1, cosubstrate S_2, third cosubstrate is almost always nucleoside triphosphate

[a] Adapted from Schwimmer (1981), with permission of Van Nostrand Reinhold.
[b] S refers to substrate.
[c] Isomerases, racemases, epimerases, mutases, ligases (decyclizing, isomerizing).

synthetase. These examples illustrate that synthases are actually transferase enzymes. Although they "synthesize" compounds, they do so using a group transfer mechanism. On the other hand, true synthetic enzymes (ligases) are named synthetases.

The second number in the systematic cataloging of enzymes represents the first subclass and refers to the group or bond undergoing the transformation or being donated. This list varies for each of the six basic classes of enzyme. The third number represents the second subclass and identifies an acceptor compound, a secondary characteristic of the group being transformed, a reaction type, or a positional characteristic. In this subclass, the number 99 is reserved for those enzymes that are ill-defined, for example, some phenol oxidases. The fourth number or third subclass is merely a bookkeeping number. All enzymes sharing the first three numbers are listed and assigned a final number arbitrarily. For example, the enzymes invertase (EC 3.2.1.26) and β-galactosidase (EC 3.2.1.23) share the first three numbers in this system. The numbers 3, 2, and 1 indicate, respectively, that the enzymes are hydrolases, act on a glycosidic bond, and recognize the O-glycosyl bond. The fourth number is reserved for the exact nature of the carbohydrate recognized: sucrose and galactose, respectively. (Although β-galactosidase is known principally for its action on lactose, it also can hydrolyze o-nitrophenyl esters of D-galactose).

Although knowledge of the systematic nomenclature of enzymes is likely to be of little direct value to the practicing food technologist, individuals involved in research should be aware of how enzymes are named. Literature accounts of studies on specific enzymes almost invariably provide the EC number to allow for an unambiguous identification of the enzyme evaluated.

IV. Enzyme Kinetics

A. Basic Considerations

One of the features that distinguish enzymatic from nonenzymatic reactions is the property of substrate saturation and the corresponding attainment of a maximal velocity (V_{max}). Knowledge of the factors that control the rates of enzymatic reactions is central to controlling processing steps that rely on enzymatic modification.

One of the first quantitative evaluations of the relationship between enzyme activity and substrate concentration was provided by Brown (1902) in a study of invertase activity on sucrose. A rectangular hyperbola (Eq. 5),

$$v_0 = \frac{A\,[S_0]}{B + [S_0]} \tag{5}$$

where v_0 is the initial velocity, A and B are constants, and $[S_0]$ is the initial substrate concentration, described the relationship between reaction velocity and sucrose concentration and accounted for the observed reaction order with respect to sucrose concentration (first and zero order at low and high sucrose concentration, respectively). Equation 5 has the same form as the conventional Michaelis–Menten expression (Eq. 6),

$$v = \frac{V_{max}\,[S_0]}{K_s + [S_0]} \tag{6}$$

where V_{max} is substituted for constant A and K_s is substituted for constant B (Michaelis and Menten, 1913). This relationship describes the typical velocity vs substrate concentration curve (Fig. 5).

Although hyperbolic kinetics fit many enzymatic reactions, this feature should not be interpreted in a mechanistic sense. The Michaelis–Menten model—

$$\text{Model 1} \qquad E + S \underset{k_{-1}}{\overset{k_1}{\rightleftharpoons}} ES \underset{k_{-2}}{\overset{k_2}{\rightleftharpoons}} E + P$$

where E is free enzyme, S is substrate, P is product, ES is enzyme–substrate complex, and each step is characterized by a rate constant—is based on several assumptions:

- [S] is $\gg$ [E];
- the reverse reaction (k_{-2}) is negligible, [P] = 0;
- k_2 is negligible compared with k_1 and k_{-1}, the formation of ES from E and S occurs under equilibrium conditions and is characterized by the dissociation constant K_s (k_{-1}/k_1);

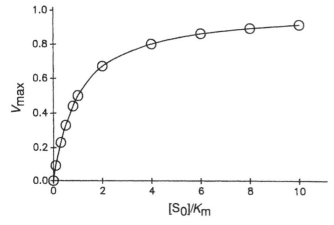

Figure 5   Relationship between reaction velocity and substrate concentration for an enzyme reaction that adheres to hyperbolic kinetics.

- the reaction is unimolecular in terms of substrate transformed and product generated; and
- the enzyme–substrate complex (ES) is the only reactive species.

Clearly many enzymes, including those used in food processing applications, do not adhere to these mechanistic parameters. All hydrolases require water as a cosubstrate of the primary substrate being transformed; these reactions are reversible, especially under conditions of low water contents (or water activities). Further, hydrolytic reactions by nature yield two products that result from the cleavage of the parent substrate molecule. In addition, enzymes that require cofactors for activity are multisubstrate reactions. Oxidoreductases, such as lactate dehydrogenase, exhibit reaction equilibria that are dependent on the pH of the system. Many multisubstrate enzymatic reactions display Michaelis–Menten kinetics, usually because one of the substrates is rate limiting to the reaction, yielding pseudo-first-order kinetic behavior.

Another restriction of the Michaelis–Menten relationship is that it was based originally on a rapid equilibrium model in which association and dissociation of enzyme and substrate was essentially instantaneous and reached equilibrium, and product formation (k_2 step) was negligible over the period of time required for measurement. Thus, E and S were assumed to interact and form ES in a manner dictated by the affinity (K_s or dissociation constant) of the substrate and enzyme. An assumption that is pervasive in the literature is that an experimentally determined K_m, based on the experimentally determined relationship between reaction velocity and [S_0] (apparent K_m or K_m^{app}), is a measure of affinity. This condition is satisfied only if the relative values of the rate constants k_1, k_{-1}, and k_2 are such that k_2 is negligible compared with k_{-1}. To make this assumption is scientifically tenuous; the mechanism of chymotrypsin activity on p-nitrophenyl esters illustrates this.

$$\text{Model 2} \qquad \text{E} + \text{S} \underset{k_{-1}}{\overset{k_1}{\rightleftharpoons}} \text{ES} \underset{k_{-2}}{\overset{k_2}{\rightleftharpoons}} \overset{\text{P}_1}{\text{EP}_2} \underset{k_{-3}}{\overset{k_3}{\rightleftharpoons}} \text{E} + \text{P}_2$$

The rate limiting step (k_3 step) of this reaction is the release of the acyl moiety (P_2) from the acyl-enzyme intermediate (EP_2). Thus, the dissociation step of the acyl group, $EP_2 \rightarrow E + P_2$, which is mediated by nucleophilic attack by water, must be considered in the determination of K_m, although the progress of the reaction is monitored spectrophotometrically by measuring the formation of nitrophenol (P_1). Clearly in this example, K_m does not represent a measure of affinity. When the apparent K_m is determined experimentally from the relationship between reaction velocity and substrate concentration (K_m^{app}), the author favors use of the term "reaction efficiency" (Segel, 1975, uses the term "substrate suitability") in lieu of "affinity" to compare the K_m values for enzyme action on various substrates or on one substrate under various conditions. This term is not to be confused with catalytic efficiency, which is defined as the relationship of k_{cat} and K_m under a defined set of experimental conditions.

Note that another model for enzyme kinetics developed in the early 1900s is based on a steady-state model, in which the rate of product formation is comparable to the rate of formation of ES complex (Briggs and Haldane, 1925). Referring back to Model 1, the steady-state model assumes that [ES] does not change during the early course of the reaction and product formation takes place at a rate that prevents [ES] from reaching its equilibrium concentration. Accordingly, the dissociation constant for [ES] is derived from the following expression:

$$d[\text{ES}]/dt = 0 = k_2[\text{ES}] + k_{-1}[\text{ES}] = k_1[\text{E}][\text{S}] \tag{7}$$

K_m in this case is defined as $(k_2 + k_{-1})/k_1$ and is not strictly a measure of substrate affinity.

B. Estimating Kinetic Constants

To control enzyme modification of foods or food ingredients, one should know the kinetic characteristics of the enzymatic reaction. The key kinetic constants are K_m and V_{max}. The most common way to estimate these parameters is from studies of the dependency of reaction velocity on substrate concentration (cf. Fig. 5) and transformation of the data to a double-reciprocal (Lineweaver–Burke) plot (see subsequent text). In such an analysis, whether the velocity vs [S_0] plot conforms to hyperbolic kinetics often is disregarded. If the data do not conform, then use of the double-reciprocal plot is not mathematically valid. Of equal importance is the choice in range of [S_0]/K_m used. The minimum range is about $0.3K_m$ to $3K_m$; broader ranges of [S_0]/K_m are more desirable. This consideration is important in eliminating any bias or improper weighting of data that could result if the range of [S_0] used was only higher or lower than K_m (Segel, 1975).

A dilemma exists because it is not always possible to generate an accurate estimation (within 5%) of V_{max}, due to limits of solubility of the substrate. Therefore, alternative means may be required to estimate V_{max}. One technique involves the direct linear (Eisenthal–Cornish-Bowden) plot (Cornish-Bowden, 1979; Fig. 6). The median intersection of the plots of velocity vs [S_0] yields a reasonable estimate of V_{max} and K_m. Another technique is transformation of the velocity vs [S_0] data to a logarithmic plot (Fig. 7; Segel, 1975). K_m can be identified as the [S_0] associated with the inflection point of the curve, in a manner analogous to the determination of pKs of ionizable groups. Once K_m is estimated, an estimation of V_{max} can be obtained from the first-order region (low [S_0]) of the velocity vs [S_0] curve, the slope of which is V_{max}/K_m (Fig. 5).

Using either of these techniques to estimate V_{max} and K_m allows an evaluation of whether the original velocity vs [S_0] data conform to hyperbolic kinetics. If K_m is known or initially estimated, then the velocity as a function of [S_0] expressed as multiples of K_m can be estimated using the Michaelis–Menten expression. For example, substituting $0.1K_m$, $0.3K_m$ $0.5K_m$, $0.8K_m$,

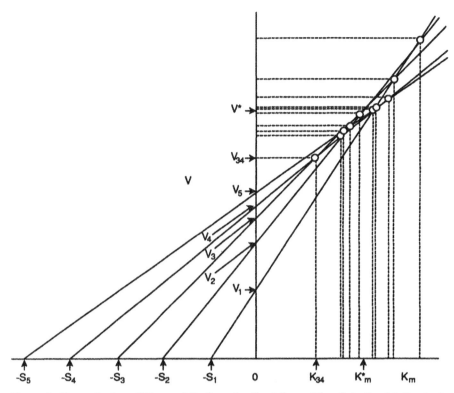

Figure 6 Determination of V_{max} and K_m from the direct linear (Eisenthal–Cornish-Bowden) plot. Experimental data for velocity (v) as a function of substrate concentration (S; plotted as negative values) are plotted to yield a series of intersecting lines. The estimates of V_{max} and K_m are determined from the median intersect. (Reprinted from Cornish-Bowden, 1979, with permission.)

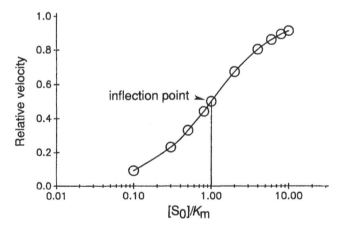

Figure 7 Plot of the relationship of velocity as a function of the logarithm of substrate concentration for an enzyme displaying hyperbolic kinetics. The data points are from Fig. 5. The inflection point indicates the K_m.

24

$2K_m$, $4K_m$, $6K_m$, $8K_m$, and $10K_m$ for $[S_0]$ predicts reaction velocities of $0.091 V_{max}$, $0.23 V_{max}$, $0.33 V_{max}$, $0.44 V_{max}$, $0.67 V_{max}$, $0.80 V_{max}$, $0.86 V_{max}$, $0.89 V_{max}$, and $0.91 V_{max}$, respectively. (These points appear in Fig. 5.) Then one can see easily how well the actual data fit Michaelis–Menten kinetics.

C. Use of Linear Plots

The transformations of velocity vs $[S_0]$ plots approximating hyperbolic kinetics to a Lineweaver–Burke plot (Eq. 8)

$$\frac{1}{v} = \frac{K_m}{V_{max}} \frac{1}{[S]} + \frac{1}{V_{max}} \tag{8}$$

an Eadie–Scatchard plot (Eq. 9)

$$v = V_{max} - \frac{K_m v}{[S]} \tag{9}$$

and a Hanes–Woolf plot (Eq. 10)

$$\frac{[S]}{v} = \frac{K_m}{V_{max}} + \frac{[S]}{V_{max}} \tag{10}$$

are all mathematically equivalent. However, differences in the plots lie in the portion of data that is emphasized in the determination of K_m and V_{max}; this difference has been demonstrated thoroughly in the example used by Cornish-Bowden (1979; Fig. 8). Taking into account the experimental errors in the measurement of reaction velocity (especially at low $[S_0]$), the effects of these errors have considerable impact on the graphical estimation of V_{max} and K_m. In the Lineweaver–Burke plot, maximal weight is placed on the data acquired at low $[S_0]$, whereas in the Eadie–Scatchard plot the data are weighted more equally. Although the Hanes–Woolf plot places the greatest emphasis on the velocities obtained at high $[S_0]$ (Fukagawa et al., 1985), these data are generally more reliable than those obtained at low $[S_0]$.

The Hanes–Woolf and Eadie–Scatchard plots also facilitate the identification of outlying data points. This advantage can be appreciated using the example of Fukagawa et al. (1985), who evaluated the use of these three plots and linear regression to estimate the K_m and V_{max} for the activity of o-diphenol oxidase on 3,4-dihydroxyphenylalanine (DOPA) (Fig. 9). One can easily ascertain the suspect points on the Hanes–Woolf and Eadie–Scatchard plots, whereas it is difficult to conclude that any of the points are errant on the Lineweaver–Burke plot. Omission of this suspect point yields good agreement between these three linear plots in the graphical determination of V_{max} and K_m. These data also exemplify the greater degree of error in the measurement of reaction velocities at low $[S_0]$ compared with high $[S_0]$. An added advantage of the Hanes–Woolf plot over the Eadie–Scatchard plot is that the dependent variable (velocity) is used on only one axis, whereas it is used on both axes in the Eadie–Scatchard plot. Calculation of the best fit

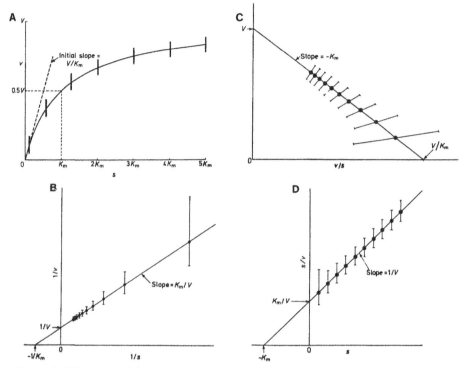

Figure 8 Effect of errors in velocity measurements on the estimation of V_{max} and K_m by linear plots. (A) Original data for velocity as a function of substrate concentration assuming an error in measurement of 5% of V_{max}. (B) Linear transformation of the original data to Lineweaver–Burke plot. (C) Eadie–Scatchard plot. (D) Hanes–Woolf plot. (Reprinted from Cornish-Bowden, 1979, with permission.)

line for the original data by nonlinear regression yields estimates of K_m and V_{max} of 10.0 and 4.3, respectively.

The Eadie–Scatchard plot often is used to analyze binding phenomena, such as numbers of receptor sites or "n" (analogous to V_{max} in enzyme kinetic analyses) and the presence of multiple enzyme forms (or the presence of isoenzymes). However, direct analysis of the data on an Eadie–Scatchard plot, without adequate review of the data examined as ligand bound vs log free ligand, can lead to gross misinterpretations, as illustrated by Klotz (1982).

In conclusion, the most important and revealing data are those presented as velocity vs $[S_0]$. Perhaps we all can benefit from a lesson briefly summarized by T. H. Huxley:

> Mathematics may be compared to a mill of exquisite workmanship, which grinds you stuff of any degree of fineness; but, nevertheless, what you get out depends on what you put in; as the grandest mill in the world will not extract flour from peascods, so pages of formulae [or graphs] will not get a definite result out of loose data.

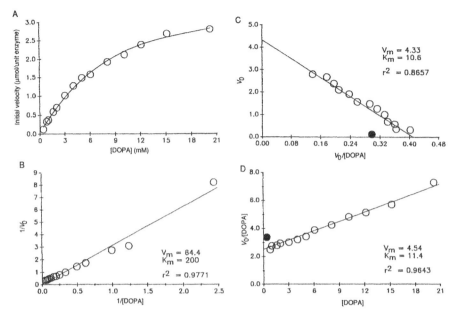

Figure 9 Estimation of V_{max} and K_m of o-diphenol oxidase action on 3,4-dihydroxyphenylalanine (DOPA). (A) Velocity as a function of DOPA concentration. (B), (C), and (D) Linear transformation of the original data to Lineweaver–Burke, Eadie–Scatchard, and Hanes–Woolf plots, respectively. The lines were fitted by linear regression to yield estimates of V_{max} and K_m. Filled circles indicate outlying points. (Data from Fukagawa *et al.*, 1985.)

D. Allosteric Kinetics

Regulatory enzymes tend to display allosteric kinetics and not simple hyperbolic kinetics. This feature is important to metabolic control in living organisms and respiring tissues. Allosteric enzymes typically display a sigmoidal plot of reaction velocity as a function of $[S_0]$ (Fig. 10). The feature central to allosteric enzymes is the presence of multiple binding or active sites that can display dissimilar affinities for substrate(s), modulator(s), or inhibitor(s). The behavior of allosteric enzymes has been modeled on the basis of symmetrical (Monod–Wyman–Changeux model) or hybrid (Koshland–Némethy–Filmer model) subunits. The important properties of allosteric enzymes are modulation of activity by a variety of chemical compounds and the marked influence of substrate concentration, especially in the range near K_s, on enzyme activity.

The scope of this chapter does not permit an analysis of the kinetics of allosteric enzymes and mechanisms of their control (see Segel, 1975, for a detailed presentation). Allosteric behavior can be determined from the nature of the velocity and $[S_0]$ relationship (Fig. 10). For an enzyme that exhibits hyperbolic kinetics, the ratio of substrate concentrations at $0.9V_{max}$

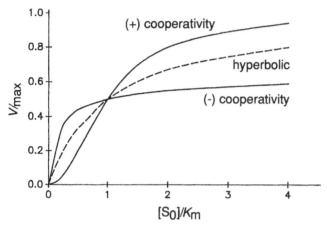

Figure 10   Kinetic behavior of allosteric enzymes. Solid lines indicate allosteric enzymes displaying positive and negative cooperativity. For comparison, dashed line represents an enzyme displaying hyperbolic kinetics.

and $0.1 V_{max}$ is defined mathematically as 81 (see previous discussion on hyperbolic kinetics, where $[S_0]$ is expressed as multiples of K_s). An allosteric enzyme will display ratios above or below this value. Ratios above 81 indicate negative cooperativity, which can be described mechanistically as initial binding of substrate making subsequent binding of substrate to other subunits less favorable. Positive cooperativity, with a ratio less than 81, reflects a greater affinity for substrate at other unoccupied sites as a result of the initial binding of substrate.

Although allosteric enzymes rarely are exploited in food processing applications, some cases are worthy of mention because they are critical to food quality. Phosphofructokinase (PFK), a key regulatory enzyme in glycolysis, appears to play a role in the quality of foods derived from plant and animal tissues. PFK is believed to be the enzyme that controls postmortem glycolysis during the proper conditioning of red meats (Hultin, 1985) and also appears to be involved in the regulation of ripening of tomato fruit (Isaac and Rhodes, 1986).

In the latter case, allosteric behavior is dependent on the enzyme being in the oligomeric form (Fig. 11). Hyperbolic kinetics are observed for the oligomer acting on substrate. However, the presence of citrate induces sigmoidal behavior of the PFK oligomer acting on substrate. Citrate binding to the monomer (inhibition) displays hyperbolic kinetics, whereas binding to the oligomer is cooperative. PFK also has been suggested to play a role in low temperature sweetening during the storage of potato tubers (ap Rees *et al.*, 1981), due to its cold lability.

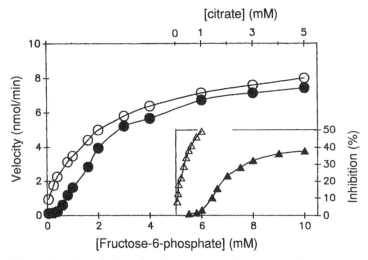

Figure 11 Allosteric behavior of phosphofructokinase (PFK) isolated from tomato fruit. Open and filled circles represent the behavior of the PFK oligomer in the absence and presence of 2 mM citrate, respectively. Inset displays kinetics of inhibition of the monomer (open triangles) and oligomer (filled triangles) by citrate. (Redrawn from Isaac and Rhodes, 1986, with permission.)

V. Inhibition

The foregoing discussion (Section IV, C) should not be misinterpreted as a call to abandon the use of the Lineweaver–Burke plot, but merely as a caution to the user of its potential pitfalls and limitations. Despite its limitations, the Lineweaver–Burke plot remains the method of choice for analyzing the nature of enzyme inhibition. The four basic types of singular inhibition are competitive, noncompetitive, uncompetitive, and irreversible. (There are also cases of mixed-type inhibition, a discussion of which is beyond the scope of this chapter. See Segel, 1975, for a detailed presentation.) The most common forms of reversible inhibition can be modeled as follows:

Model 3 (competitive inhibition) $\quad EI \underset{}{\overset{K_i}{\rightleftharpoons}} I + E + S \underset{k_{-1}}{\overset{k_1}{\rightleftharpoons}} ES \underset{k_{-2}}{\overset{k_2}{\rightleftharpoons}} E + P$

Model 4 (noncompetitive inhibition) $\quad E + S \underset{k_{-1}}{\overset{k_1}{\rightleftharpoons}} ES \underset{k_{-2}}{\overset{k_2}{\rightleftharpoons}} E + P$

$$+I \qquad\qquad +I$$
$$\Big\Updownarrow K_i \qquad\quad \Big\Updownarrow K_i$$
$$EI \qquad\qquad ESI$$

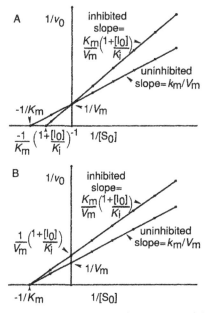

Figure 12 The common forms of reversible enzyme inhibition as illustrated on a Lineweaver–Burke plot. (A) Competitive inhibition. (B) Noncompetitive inhibition.

where I is the inhibitor and K_i is the dissociation constant for the enzyme–inhibitor complex (EI).

Development of the velocity equations from these models yields the following kinetic relationships for competitive inhibition

$$v = \frac{V_{max}\,[S_0]}{K_m\left(1 + \dfrac{[I_0]}{K_i}\right) + [S_0]} \tag{11}$$

and noncompetitive inhibition

$$v = \frac{V_{max}\,[S_0]}{K_m\left(1 + \dfrac{[I_0]}{K_i}\right) + [S_0]\left(1 + \dfrac{[I_0]}{K_i}\right)} \tag{12}$$

An enzyme reaction subject to inhibition through either of these mechanisms would yield relationships on a Lineweaver–Burke plot as shown in Fig. 12.

Although uncompetitive inhibition is exceedingly rare, probably for evolutionary reasons (Cornish-Bowden, 1986), at least two cases of its existence are reported. This mechanism is reported for L-phenylalanine inhibition of alkaline phosphatase (Ghosh and Fishman, 1966) and phytate inhibition of β-galactosidase activity on lactose (Inagawa *et al.*, 1987). In the latter case, experimental data were generated only for phytate concentrations of 0, 2, and 3 mM at lactose concentrations of 5–10 mM. The small windows of inhibitor and substrate concentrations used to characterize this inhibition suggest the need for a confirmatory study.

Irreversible inhibition or "poisoning" of enzymes usually results from a covalent interaction between the inhibitor and an amino acid residue essential for enzyme activity. A good example of this mode of inhibition is the action of iodoacetamide, which reacts with free sulfhydryl groups (cysteine residues). The various types of inhibition can be diagnosed by measuring enzyme activity as a function of [E] in the presence of fixed inhibitor concentration (Fig. 13). At a saturating level of substrate and a fixed inhibitor concentration, a noncompetitive reversible inhibitor will inhibit enzyme activity proportional to the uninhibited enzyme. For an irreversible inhibitor, added enzyme will be poisoned until the amount of enzyme added is in excess of the inhibitor. Further addition of enzyme will yield activity representative of the uninhibited system.

A form of inhibition similar to irreversible inhibition is reaction inactivation. Enzyme action on "suicide" substrates transforms these substrates into products that poison the active site of the enzyme, usually after a covalent reaction (Walsh, 1983). At least three examples of this behavior have been identified for enzymes related to food quality. Perhaps the best known example is the reaction inactivation caused by phenol oxidase action on o-dihydroxyphenols (Golan-Goldhirsch and Whitaker, 1985). Other examples include the inactivation of fish tissue lipoxygenase after its action on polyunsaturated fatty acids and the formation of the resultant hydroperoxides (German et al., 1986) and the action of xanthine oxidase on allopurinol (Walsh, 1983). In the latter case, the transformed substrate yields alloxanthine, which poisons the enzyme not by covalent modification but by formation of an alloxanthine–Mo(IV)–enzyme complex. This complex has a half-life of about 5 hr.

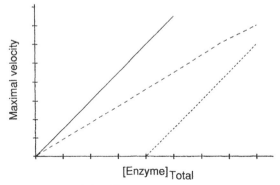

Figure 13 Relationship between enzyme concentration alone (——) and the presence of a noncompetitive reversible inhibitor (— —) and an irreversible inhibitor (- - -). (Redrawn from Segel, 1975, with permission.)

VI. Cofactors

Many enzymes require cofactors for activity. Many of these cofactors are derived from vitamins provided by the diet. Cofactors can be classified as prosthetic groups, coenzymes, and metal ions. Prosthetic groups are distinguished from coenzymes by their tighter binding to the enzyme, although there is considerable overlap in binding affinities of prosthetic groups and coenzymes (Whitaker, 1972). Enzymes with and without their prosthetic group(s) attached are referred to as holoenzymes and apoenzymes, respectively.

The mechanisms of action of the various types of cofactors are different; some cofactors exhibit multiple effects on the reaction (Saier, 1987). Cofactors can participate in the catalyzed reaction directly and, therefore, impact on the reaction mechanism, they can assist in the binding of substrate, or both. The binding of cofactors at or near the active site normally follows simple hyperbolic kinetics. However, since many of the regulatory enzymes rely on cofactors for activity, binding of cofactors to these enzymes often displays allosteric behavior.

The flavin (FMN, FAD) and nicotinamide (NAD, NADP) cofactors invariably participate in electron transfer reactions and interact with class 1 enzymes, the oxidoreductases. Generally, flavins are considered prosthetic groups whereas nicotinamides are considered coenzymes or cosubstrates. Electron transfer reactions facilitated by flavins often use oxygen as the electron acceptor. Accordingly, the flavoenzymes glucose oxidase and xanthine oxidase yield hydrogen peroxide by transfer of electrons from glucose and xanthine, respectively, to oxygen. However, xanthine oxidase is known to produce superoxide anion as a product also; this characteristic has implied a role for this enzyme in the oxidation of milk lipids. Xanthine oxidase, like many flavoenzymes, is also a metalloenzyme; molybdenum and iron exist as prosthetic groups and assist in the internal electron transfer reactions.

Oxidoreductases (e.g., lactate and alcohol dehydrogenase) that use nicotinamide cofactors also catalyze two electron reduction/oxidations of substrates. However, during the course of the reaction, a proton (H^+) is either liberated or consumed. Thus, not only does pH affect the reactivity of these enzymes by its effect on enzyme stability or catalytic activity, but it also affects the reaction equilibrium:

$$\text{Model 5} \qquad NADH + H^+ + A \underset{\text{enzyme}}{\overset{\text{enzyme}}{\rightleftharpoons}} NAD^+ + AH_2$$

and

$$K_{eq} = \frac{[NAD^+][AH_2]}{[NADH][H^+][A]} \tag{13}$$

Pyridoxal phosphate is another prosthetic group that can play a role in many types of reactions. Its action as a cofactor is expressed most often after the formation of a Schiff base between pyridoxal phosphate and the enzyme (usually with a lysine residue). The specific nature of the reaction catalyzed depends on the substrate. S-alk(en)yl-cysteine sulfoxide lyase (allinase) is an example of a food-related enzyme that uses pyridoxal phosphate as a cofactor.

Tetrahydrofolate is another cofactor that is involved in group transfers, principally those involving the methyl group. Similarly, cobalamine is involved in group transfer reactions, especially those involving protons or alkyl groups. Because of the presence of the transition metal, cobalt, this cofactor can take part in internal oxidation/reduction reactions also.

Biotin is a covalently bound prosthetic group that assists in carboxylation reactions. This group's primary importance is in fatty acid synthesis, in which it becomes activated by bicarbonate and transfers the carboxyl group to a receptor such as acetyl coenzyme A. Two other covalently bound cofactors are lipoamide and pantothenate (coenzyme A). These coenzymes are involved in oxidation/reduction and acyl transfer reactions and thioester formation, respectively. Coenzyme A (CoA), like biotin, is involved principally in fatty acid synthesis, since acetyl-CoA and malonyl-CoA are primers for the fatty acid elongation cycle.

Thiamine pyrophosphate is a prosthetic group involved in decarboxylation and transketolation reactions. Its principal role is in carbohydrate metabolism.

Metal ions constitute another class of cofactors. They can be reactive individually or as constituents of more complex structures, for example, in heme or iron–sulfur prosthetic groups. Heme enzymes, such as catalase and peroxidase, catalyze oxidation/reduction reactions and use the transition metal iron to facilitate these activities. Transition metals do not have to be coordinated as constituents of complex structures to participate in oxidation/reduction reactions. A good example is phenol oxidase, which contains the transition metal copper as a prosthetic group. The oxidation/reduction behavior of copper is directly responsible for phenol oxidase action on phenolic acids (Robb, 1984). Iron–sulfur prosthetic groups also are involved primarily in electron transfer reactions.

"Free" metal ion cofactors can participate in oxidation/reduction reactions, substrate activation, or substrate binding to the enzyme. Zinc is coordinated to His-69, Glu-72, and His-196 at the active site of carboxypeptidase A (Fig. 3). Its role appears to be in alignment of the carbonyl group of the susceptible peptide bond at the active site and in electron withdrawal from the carbonyl group. Alkaline earth metal cations such as calcium and magnesium often play a role in maintenance of enzyme structure. In addition, magnesium often acts as a cosubstrate for reactions involving the hydrolysis of ATP, since the substrate recognized by the enzyme is the magnesium–ATP complex.

Although ascorbic acid, glutathione, and ATP (and other purines and pyrimidines) often are classified as cofactors, the author favors regarding these components as cosubstrates. The rationale is that these molecules are not necessarily recycled during the course of a singular enzyme reaction and, thus, often are left in a transformed state as a result of a particular reaction.

VII. Enzyme Assays and Units

A. Assays

Perhaps the most important piece of information about an enzyme preparation to be used in food processing is its activity, expressed on a weight or volume basis, and most commonly expressed as Units. Before one can interpret fully the significance of the Units of activity reported for an enzyme preparation, it is prudent to provide a brief review of how enzymes are assayed. Obviously, if the Units (catalytic capacity) of enzyme activity are to be assayed, conditions must be present that allow the maximal expression of activity (i.e., V_{max}). Typically, conditions under which $[S_0]$ is saturating to the enzyme (or is at an optimal concentration) are used. If the limits of substrate solubility do not permit this condition to be used in practice, then the V_{max} can be estimated by other means (see Section IV, B). Obviously, V_{max} will depend on environmental factors such as pH, temperature, and ionic strength (see Chapter 3 for detailed discussions). Generally, conditions that are optimal for enzyme activity are used (optimal pH, excess substrate(s) and cofactors) to determine the Units of activity. However, in some cases, Units also may be defined under a set of conditions expected to prevail in a certain processing application. In either case, a full description of the conditions used should be made available to the potential user of the product.

Units often are expressed in terms of product formation per unit quantity of an enzyme preparation per unit of time. In some cases, enzymes are not easily assayed using their native or intended substrate. An example is β-galactosidase, which is most commonly (and easily) assayed for activity using o-nitrophenyl-D-galactopyranoside (ONPG). Accordingly, Units of β-galactosidase activity are often expressed as ONPG Units per quantity of enzyme preparation. The same enzyme from various sources can exhibit different dependencies on environmental conditions. Using β-galactosidase as an example again, most fungal enzymes exhibit slightly acidic pH optima whereas enzymes from yeasts exhibit pH optima near neutral.

B. Units

Depending on the profession of the enzymologist, the nature of the enzyme assayed, and its potential application, standard enzyme assays for catalytic activity or capacity (Units) usually are done at 25, 30, or 37°C. Although this practice can be somewhat confusing, the general temperature coefficient (Q_{10}) of 2 for most enzyme reactions permits a reasonable estimation of the activity that can be expected at a temperature other than the one used for assay (provided the enzyme is not inactivated). Enzyme Units (U) or International Units (IU) are defined as the amount of enzyme that transforms 1 micromole of substrate in 1 minute under a specified set of conditions. A more recent nomenclature developed for expressing units of enzyme activity is based on the *katal* (kat), which is the amount of enzyme capable of transforming 1 millimole of substrate per second under a specified set of conditions. The Units of activity per weight of enzyme preparation is a function of the purity of the preparation and the "turnover number" of the enzyme. The turnover number is the number of moles of substrate transformed into product by 1 mole of enzyme active site per second. Turnover numbers range widely among enzymes (from 10^1-10^4 for hydrolases to 10^3-10^7 for oxidoreductases; Whitaker, 1972; Schwimmer, 1981) and are a reflection of enzyme efficiency. Another related term is "catalytic efficiency," which is k_{cat}/K_m for the enzyme acting on a particular substrate.

The Units of activity describe the maximal rate at which a specified amount of enzyme can catalyze a reaction. In food processing settings, in which reactions often are required to go to completion (or equilibrium), conditions that support maximal activity certainly will not prevail toward the end of the process, due to the depletion of substrate. In this case, the parameter of K_m for the enzymatic reaction can be valuable. Enzyme activity can be expected to proceed at near maximal rates until [S] falls to a value near its K_m (at $2K_m$ the expected reaction rate would be 67% V_{max}). Therefore, a reasonable estimation can be obtained for the required duration of an enzyme modification based on the values of V_{max} (Units) and K_m, provided the enzyme is not subject to product inhibition, cofactor or cosubstrate depletion, or inactivation. A more precise estimation of the process (incubation) time required to achieve a desired degree of completion of the reaction can be obtained from yet one more transformation of the Michaelis–Menten expression (Eqs. 14–17; Segel, 1975):

$$v = \frac{-d[S]}{dt} = \frac{V_{max}[S]}{K_m + [S]} \tag{14}$$

by rearrangement,

$$V_{max}dt = \frac{-K_m + [S]}{[S]} d[S] \tag{15}$$

Integration yields

$$V_{max}t = 2.3K_m\log\frac{[S_0]}{[S]} + ([S_0] - [S]) \tag{16}$$

or

$$t = \frac{2.3K_m\log\frac{[S_0]}{[S]} + ([S_0] - [S])}{V_{max}} \tag{17}$$

Thus, the time (t) required for a desired conversion of substrate to product can be calculated if the K_m, V_{max}, and $[S_0]$ are known ($[product] = [S_0] - [S]$). The restrictions that were identified earlier for the estimation of reaction time also apply to this analysis.

Acknowledgments

This work was supported by the College of Agricultural and Life Sciences of the University of Wisconsin, Madison. The author is grateful for the assistance of B. Scullion in preparing this manuscript.

References

Adler-Nissen, J. (1987). Newer uses of microbial enzymes in food processing. *Trends Biotech.* **5,** 170–174.

ap Rees, T., Dixon, W. L., Pollock, C. J., and Franks, F. (1981). Low temperature sweetening of higher plants. *In* "Recent Advances in the Biochemistry of Fruits and Vegetables" (J. Friend and M. J. C. Rhodes, eds.), pp. 41–61. Academic Press, New York.

Bone, R., Silen, J. L., and Agard, D. A. (1989). Structural plasticity broadens the specificity of an engineered protease. *Nature (London)* **339,** 191–195.

Briggs, G. E., and Haldane, J. B. S. (1925). A note on the kinetics of enzyme action. *Biochem. J.* **19,** 338–339.

Brock, T. D. (1985). Life at high temperatures. *Science* **230,** 132–138.

Brown, A. J. (1902). Enzyme action. *Trans. Chem. Soc.* **81,** 373–379.

Carter, P., and Wells, J. A. (1988). Dissecting the catalytic triad of a serine protease. *Nature (London)* **332,** 564–568.

Cornish-Bowden, A. (1979). "Fundamentals of Enzyme Kinetics." Butterworth, London.

Cornish-Bowden, A. (1986). Why is uncompetitive inhibition so rare? A possible explanation, with implications for the design of drugs and pesticides. *FEBS Lett.* **203,** 3–6.

Eyring, H. (1935). The activated complex in chemical reactions. *J. Chem. Phys.* **3,** 107–115.

Fano, U. (1947). A possible contributing mechanism of catalysis. *J. Chem. Phys.* **15,** 845.

Fukagawa, Y., Sakamoto, M., and Ishikura, T. (1985). Micro-computer analysis of enzyme-catalyzed reactions by the Michaelis–Menten equation. *Agric. Biol. Chem.* **49,** 835–837.

German, J. B., Bruckner, G. G., and Kinsella, J. E. (1986). Lipoxygenase in trout gill tissue acting on arachidonic, eicosopentaenoic and docosohexaenoic acids. *Biochim. Biophys. Acta* **875,** 12–20.

Ghosh, N. K., and Fishman, W. H. (1966). On the mechanism of inhibition of intestinal alkaline phosphatase by L-phenylalanine. *J. Biol. Chem.* **241,** 2516–2522.

Golan-Goldhirsch, A., and Whitaker, J. R. (1985). k_{cat} inactivation of mushroom polyphenol oxidase. *J. Mol. Catalysis* **32,** 141–147.

Hultin, H. O. (1985). Characteristics of muscle tissue. *In* "Food Chemistry" (O. Fennema, ed.), 2d Ed., pp. 725–789. Marcel Dekker, New York.

Inagawa, J., Kiyosawa, I., and Nagasawa, T. (1987). Effect of phytic acid on the hydrolysis of lactose with β-galactosidase. *Agric. Biol. Chem.* **51,** 3027–3032.

International Union of Biochemistry, Enzyme Commission (1984). "Enzyme Nomenclature 1984." Academic Press, New York.

Isaac, J. E., and Rhodes, M. J. C. (1986). Phosphofructokinase from *Lycopersicon esculentum* fruits—II. Changes in the regulatory properties with dissociation. *Phytochemistry* **25,** 345–349.

Kawaguchi, Y., Kosugi, S., Sasaki, K., Uozumi, T., and Beppu, T. (1987). Production of chymosin in *Escherichia coli* cells and its enzymatic properties. *Agric. Biol. Chem.* **51,** 1871–1877.

Klotz, I. M. (1982). Numbers of receptor sites from scatchard graphs: Facts and fantasies. *Science* **217,** 1247–1249.

Koshland, D. E. (1958). Application of a theory of enzyme specificity to protein synthesis. *Proc. Natl. Acad. Sci. U.S.A.* **44,** 98–104.

Lerner, R. A., and Tramontano, A. (1987). Antibodies as enzymes. *Trends Biochem. Sci.* **12,** 427–430.

Michaelis, L., and Menten, M. L. (1913). The kinetics of invertase action. *Biochem. Z.* **49,** 333–369 *(in German).*

Page, M. I. (1987). Theories of enzyme catalysis. *In* "Enzyme Mechanisms" (M. I. Page and A. Williams, eds.), pp. 1–13. Royal Society of Chemistry, London.

Palmer, T. (1981). "Understanding Enzymes." Ellis Horwood, Chichester.

Parkin, K. L., and Hultin, H. O. (1986). Characterization of trimethylamine-*N*-oxide (TMAO) demethylase activity from fish muscle microsomes. *J. Biochem.* **100,** 77–86.

Pohl, J., and Dunn, B. M. (1988). Secondary enzyme–substrate interactions: Kinetic evidence for ionic interactions between substrate side chains and pepsin active site. *Biochemistry* **27,** 4827–4834.

Price, N. C., and Stevens, L. (1989). "Fundamentals of Enzymology," 2d Ed. Oxford, New York.

Ring, M., Bader, D. E., and Huber, R. E. (1988). Site-directed mutagenesis of β-galactosidase *(E. coli)* reveals that tyr-503 is essential for activity. *Biochem. Biophys. Res. Commun.* **152,** 1050–1055.

Robb, D. A. (1984). Tyrosinase. *In* "Copper Proteins and Copper Enzymes" (R. Lontie, ed.), Vol. II, pp. 207–240. CRC Press, Boca Raton, Florida.

Russ, E., Kaiser, U., and Sandermann, H. (1988). Lipid-dependent membrane enzymes. Purification to homogeneity and further characterization of diacylglycerol kinase from *Escherichia coli. Eur. J. Biochem.* **171,** 335–342.

Saier, M. H. (1987). "Enzymes in Metabolic Pathways. A Comparative Study of Mechanism, Structure, Evolution, and Control." Harper and Row, New York.

Schwimmer, S. (1981). "Source Book of Food Enzymology." AVI, Westport.

Segel, I. H. (1975). "Enzyme Kinetics. Behavior and Analysis of Rapid Equilibrium and Steady-State Enzyme Systems." Wiley-Interscience, New York.

Sharp, K., Fine, R., and Honig, B. (1987). Computer simulations of the diffusion of a substrate to an active site of an enzyme. *Science* **236,** 1460–1463.

Shotton, D. (1971). The molecular architecture of the serine proteases. *In* "Proceedings of the International Research Conference on Proteinase Inhibitors" (H. Fritz and H. Tschesche, eds.), pp. 47–55. de Gruyter, New York.

Srere, P. A. (1984). Why are enzymes so big? *Trends Biochem. Sci.* **9,** 387–390.

Sumner, J. B. (1926). The recrystallization of urease. *J. Biol. Chem.* **70,** 97–98.

Vick, B. A., and Zimmerman, D. C. (1987). Oxidative systems for modification of fatty acids: The lipoxygenase pathway. *In* "The Biochemistry of Plants" (P. K. Stumpf, ed.), Vol. 9, pp. 53–90. Academic Press, New York.

Visser, S. (1981). Proteolytic enzymes and their action on milk proteins. A review. *Neth. Milk Dairy J.* **35,** 65–88.

Walsh, C. T. (1983). Suicide substrates: Mechanism-based enzyme inactivators with therapeutic potential. *Trends Biochem. Sci.* **8,** 254–257.

Wells, J. A., and Estell, D. A. (1988). Subtilisin—An enzyme designed to be engineered. *Trends Biochem. Sci.* **13,** 291–297.

Whitaker, J. R. (1972). "Principles of Enzymology for the Food Sciences." Marcel Dekker, New York.

Zaks, A., and Klibanov, A. M. (1984). Enzymatic catalysis in organic media at 100°C. *Science* **224,** 1249–1251.

CHAPTER 3

Environmental Effects on Enzyme Activity

KIRK L. PARKIN

I. Introduction

Chemical reactions, including those catalyzed by enzymes, depend on environmental conditions. This characteristic of reactions is particularly important in food systems and processing applications, in which a broad spectrum of conditions is encountered. Knowledge of how a specific enzyme reaction is affected by its environment, whether the conditions are imposed or can be manipulated, is required to allow optimization or control of any enzyme-mediated process. Enzyme activity is affected by long-range and short-range, or macro- and microenvironmental, conditions. Emphasis in this chapter will be on macroenvironmental effects. However, important microenvironmental effects will be addressed also.

The dominant environmental factors affecting enzyme action are pH, water activity, and temperature. The individual effects of these factors are often interdependent. Therefore, an attempt will be made to identify the primary influences of each environmental factor. Effects of environmental conditions other than those just mentioned will be discussed briefly.

II. Effect of pH

A. General Considerations

The effects of pH on enzymatic reactions are caused largely by the reversible ionization of substrate or amino acid residues of the enzyme. These effects are manifested as changes in enzyme activity (catalysis), stability, or interaction (binding) with ligands, or a change in reaction equilibrium. The effects of pH on enzymes also depend on the presence or absence of specific ions or ligands, temperature, dielectric constant, and ionic strength. The pH of foods can range from very acidic (pH 3 in fruit juices or purees) to near neutral pH (milk, vegetables, grains, and some muscle foods) to mildly alkaline (hominy and ripe olives). The pH optima of enzymes can vary similarly, from pH 2 (pepsin) to pH 10 (alkaline phosphatase), most enzymes exhibiting optima near neutral pH (Whitaker, 1972). The same enzyme isolated from various sources can exhibit different pH optima, probably reflecting the physiological requirements of the organism from which the enzyme was isolated. Fungal enzymes often have slightly acidic pH optima, whereas those from yeasts often display pH optima near neutral pH. In fruit tissues, several forms of invertase with various pH optima are known to exist.

B. Mechanistic Considerations

1. Effect of pH on Enzyme Activity

The effect of pH on enzyme catalysis is caused by the ionization of substrate or enzyme, which can affect substrate binding or transformation to product directly or affect enzyme stability. To explain the effect of pH on enzyme activity, a simple model will be used. Although the pH effects are generally more complex for "real" enzymes, the example chosen is realistic since it yields a typical bell-shaped dependency of enzyme activity on pH. Consider a hypothetical enzyme the substrate of which is nonionizable, that has two ionizable groups essential to activity. An enzyme capable of binding substrate in all three ionization states (E^-, E°, E^+), but capable of transforming substrate (S) into product (P) in only the monoprotic form (E°), can be described by Model 1:

$$
\begin{array}{ccc}
\text{Model 1} & E^+ + S \xrightleftharpoons{\alpha K_s} E^+S & \\
& K_{e1} \big\Updownarrow \qquad \big\Updownarrow K_{es1} & \\
& H^+ \qquad\qquad H^+ & \\
& + \qquad\qquad + & \\
& E^\circ + S \xrightleftharpoons{K_s} E^\circ S \xrightarrow{k_p} E^\circ + P & \\
& K_{e2} \big\Updownarrow \qquad \big\Updownarrow K_{es2} & \\
& H^+ \qquad\qquad H^+ & \\
& + \qquad\qquad + & \\
& E^- + S \xrightleftharpoons{\beta K_s} E^-S &
\end{array}
$$

where αK_s, K_s, and βK_s are the dissociation constants of the enzyme–

substrate complexes (E^{n}S), k_p is the rate constant for product formation, the ionization states of E^n and E^{n}S are described by K_{e1} and K_{e2} and K_{es1} and K_{es2}, respectively, and n represents any possible ionization state of E (E$^+$, E$^\circ$, or E$^-$).

Similar to the derivation of other enzyme rate expressions, the ratio of reaction velocity (v) and total enzyme present (E$_t$) can be described by Eq. 1:

$$\frac{v}{E_t} = \frac{k_p E^\circ S}{E^\circ + E^+ + E^- + E^\circ S + E^+ S + E^- S} \tag{1}$$

Expressing the various equilibria in terms of a single species of enzyme (E$^\circ$) and the dissociation constant, K_s, yields Eq. 2:

$$\frac{v}{V_{max}} = \frac{S}{K_s\left(1 + \dfrac{H^+}{K_{e1}} + \dfrac{K_{e2}}{H^+}\right) + S\left(1 + \dfrac{H^+}{K_{es1}} + \dfrac{K_{es2}}{H^+}\right)} \tag{2}$$

This equation resembles the common Michaelis–Menten expression for reaction velocity (see Chapter 2). The terms in parentheses represent the Michaelis pH functions, f_E for K_s and f_{ES} for S, describing the ionization states of E and ES, respectively, at any pH. The Michaelis pH functions are always ≥ 1, and depend on the relative values of pH and the ionization constants for E and ES. From this relationship, the pH dependency of V_{max} and K_s also can be predicted using the Michaelis pH functions:

$$V_{max}^{H^+} = \frac{V_{max}}{\left(1 + \dfrac{H^+}{K_{es1}} + \dfrac{K_{es2}}{H^+}\right)} \tag{3}$$

$$K_s^{H^+} = \frac{K_s\left(1 + \dfrac{H^+}{K_{e1}} + \dfrac{K_{e2}}{H^+}\right)}{\left(1 + \dfrac{H^+}{K_{es1}} + \dfrac{K_{es2}}{H^+}\right)} \tag{4}$$

As can be seen in Eqs. 3 and 4, the values of V_{max} ($V_{max}^{H^+}$) and K_s ($K_s^{H^+}$) as functions of pH depend on the ionization states of 1/E^{n}S, and E^n/E^{n}S, respectively.

Referring to Eq. 3, when [H$^+$] is halfway between the two ionization steps (K_{es1} and K_{es2}) of the enzyme, $V_{max}^{H^+}$ will approach the intrinsic V_{max} (V_{max} at optimum pH) of the enzyme. (An analogous behavior of $K_s^{H^+}$ can be predicted, provided $K_{e1} \sim K_{es1}$ and $K_{e2} \sim K_{es2}$; see Eq. 4.) This effect can be appreciated most easily by referring to Fig. 1, in which $K_{es1} = 10^{-4}$ (pK = 4.0) and $K_{es2} = 10^{-8}$ (pK = 8.0) are assumed. The typical bell-shaped pH dependency of $V_{max}^{H^+}$ is observed with a pH optimum for activity at 6.0 (curve A). The pK values corresponding to the ionizable groups vital to enzyme activity (assuming Model 1) are identified as the pH values at which $V_{max}^{H^+} = 0.5 V_{max}$.

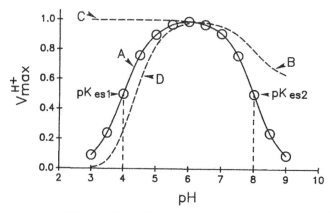

Figure I Effect of pH on $V_{max}^{H^+}$ of an enzyme characterized by Model 1 (see text for explanation).

Included in Fig. 1 are plots that represent some deviations from Model 1. Curve B in the alkaline pH region represents the predicted $V_{max}^{H^+}$ if the second ionization step (E°S → E⁻S + H⁺) yields an enzyme species capable of transforming substrate to product at 60% of the efficiency of the monoprotic form (E°S). Curve C, which remains at V_{max} in the acidic pH region, represents the predicted behavior if both the E⁺ and E° enzyme species can bind and transform substrate to product with equal efficiency. Curve D, which descends similarly to the original plot (curve A) in the acidic pH region represents the predicted behavior if two amino acid residues of a similar pK (4.0) are present and vital to enzyme catalysis.

Figure 1 also can be used to illustrate enzyme behavior when other models represent the effect of pH on enzyme activity. For instance, if the ordinate is $V_{max}^{H^+}/K_s^{H^+}$, then curve A represents the pH-dependent behavior of the free enzyme (E⁻, E°, and E⁺). This conclusion is implicit from the ratio of Eqs. 3 and 4, yielding the expression:

$$\frac{V_{max}^{H^+}}{K_s^{H^+}} = \frac{V_{max}}{K_s\left(1 + \dfrac{H^+}{K_{e1}} + \dfrac{K_{e2}}{H^+}\right)} \tag{5}$$

where the only Michaelis pH function is the one that represents the ionization state of the free enzyme (Eⁿ). Thus, if the plot (curve A, Fig. 1) of $V_{max}^{H^+}/K_s^{H^+}$ vs pH yielded a bell-shaped curve, then the pH dependency would represent a situation in which the active form of the free enzyme is E°. These types of plots identify which ionization states of Eⁿ or EⁿS are active, with activity depending on changes in protein conformation, substrate binding, or substrate transformation. To determine if the pH effect is on substrate binding, a similar plot using −log $K_s^{H^+}$ as the ordinate is useful. A bell-shaped curve, similar to curve A in Fig. 1, indicates that substrate binding depends on the ionization of two amino acids residues and that the corresponding decline in $V_{max}^{H^+}$ is caused by the inability of the free enzyme in states E^+ and E^- to bind substrate, rather than by a direct effect of pH on substrate transformation.

Tentative assignments of the pK values of 4 and 8 from Fig. 1 can be verified using the Dixon–Webb plot (Fig. 2). This plot is based on the logarithm of Eq. 3 to yield:

$$\log V_{max}^{H^+} = \log V_{max} - \log\left(1 + \frac{H^+}{K_{es1}} + \frac{K_{es2}}{H^+}\right) \tag{6}$$

The behavior of an enzyme represented by Model 1 is shown in Fig. 2 (curve A), where V_{max} is defined as 100. The pK values for the enzyme are identified by taking the tangents to the curve from the point at which $\log V_{max}^{H^+}$ is negligible (≤ 0), locating the horizontal intercept that represents V_{max}, and drawing a perpendicular line to the pH axis. A slope of 1 for this line indicates the presence of a single amino acid residue vital to enzyme activity, whereas a slope of 2 indicates the presence of 2 vital amino acid residues with similar pK values. Alternatively, if the slope of the tangent is 1, the pK values can be identified by the points on the curve that represent 0.3 units on the ordinate below the value corresponding to V_{max}. A similar analysis using the Dixon–Webb plot can be done for the free enzyme species (E^n), based on the logarithm of Eq. 5. For Model 1, the plot of $-\log K_s^{H^+}$ would yield a horizontal line on a Dixon–Webb plot, showing no pH dependency.

The assignments of pK values for enzyme activity provide a preliminary identification of the amino acid residues critical to enzyme activity. The characteristic pK values for amino acid residues in the range of physiological pH for most enzymes are provided in Table I. Many of these ionizable groups exhibit a range of pK values because of microenvironmental effects that influence the ionization potential of the amino acid side chains. Verification

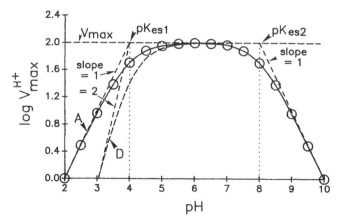

Figure 2 Dixon–Webb plot of an enzyme characterized by Model 1 ($V_{max} = 100$). Curves A and D represent the same conditions appearing in Fig. 1 (see text for explanation).

TABLE I
Ionization Constants and Enthalpies of Ionization of Amino Acid Side Chains in the Range of Physiological pH[a]

Functional group	Amino acid residue	Range of pKs	Range of enthalpy (kcal /mole)
Carboxyl	C terminus	3.0–3.2	±1.5
	Asp, Glu	3.0–4.7	
Imidazolium	His	5.6–7.0	6.9–7.5
Sulfhydryl	Cys	8.0–8.5	6.5–7.0
Ammonium	N terminus	7.6–8.4	10–13
	Lys	9.4–10.6	
Phenolic	Tyr	9.8–10.4	6

[a]Adapted from Whitaker, 1972, with permission.

of the role of these amino acids in activity can be achieved by evaluating the pH dependency of enzyme photoinactivation (Spikes, 1981), enzyme activity following specific chemical modifications (Means and Feeney, 1971), and temperature dependency of the pH dependency of enzyme activity. This last feature is based on the enthalpies of ionization of amino acid functional groups, provided in Table I. Studies on point mutations in the polypeptide chains of enzymes also can lead to an improved understanding of the requirement for specific amino acid residues in enzyme catalysis. The identification of the vital amino acid residues can be useful because modification of these residues, by chemical or biotechnological means, potentially yields an added dimension of control over enzyme-catalyzed reactions.

2. Effect of pH on Ligand Ionization

For some enzymes, the ionization of ligands (substrate, product, cofactor, or inhibitor) must be considered to explain the pH dependency of activity. A Michaelis pH function similar to the one developed for the ionization of the various species of enzymes (Eq. 2) can be derived for any ligand with one or more ionizable groups vital to its interaction with the enzyme. Inhibition of mushroom tyrosinase by hydrofluoric acid, hydrazoic acid, and 4-nitrocatechol strongly depends on pH (Robb et al., 1966). The dissociation constants for inhibition, K_i, increase as pH increases above the pK for all inhibitors. Therefore, the inhibitory action of these compounds is directly proportional to the concentration of inhibitor in the protonated state. Another possible example of the effect of ligand ionization on enzyme action can be drawn from the nature of inhibition of potato invertase by an endogenous proteinaceous inhibitor (Fig. 3). This inhibitor has been suggested to play a role in low-temperature sugar accumulation in cold-stored potatoes (Pressey, 1967). Plots of $V_{max}^{H^+}$ (E^{n}S species) and $V_{max}^{H^+}/K_s^{H^+}$ (E^n species) yield

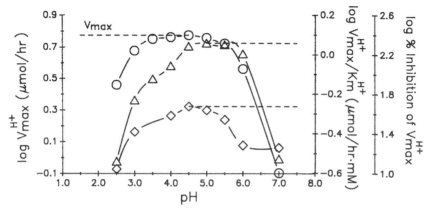

Figure 3 Effect of pH on activity and inhibition of potato invertase. Level of inhibitor is 1.8 units. Dashed lines represent maximal values for each curve. Log V_{max}, O; log $V_{max}^{H^+}/K_m$, $\triangle$; log % inhibition of $V_{max}^{H^+}$, $\diamond$. (Redrawn from Pressey, 1967, with permission of American Society of Plant Physiologists.)

two different curves. Tentative identification of the pK values for the ES and E species are about 2.5 and 6.2 and 3.0 and 6.5, respectively, indicating that the ionization states of either aspartate or glutamate and histidine residues are important to invertase activity. These data also demonstrate that the pK values of an enzyme are not always the same in the presence and absence of substrate; binding of substrate can modify the microenvironment of the ionizable groups. The pH dependency of enzyme–inhibitor interaction is characterized by pK values of 3.0 and 5.6. Whether the inhibitory action depends on the ionization state of the inhibitor, the enzyme, or both is not clear. Since this inhibitor is of the noncompetitive type (Pressey, 1967), the difference in pK values simply may reflect the ionization properties of the different binding sites for substrate and inhibitor. However, since the inhibitor is also a protein, its ionization state is likely to be related to its inhibitory action.

3. Effect of pH on Enzyme Stability

In some cases, the effect of pH on enzyme activity can be explained in terms of enzyme conformational stability. This effect is assessed most easily by preincubating an enzyme solution at a range of pH values, then assaying enzyme activity at the optimum pH. In the case of peroxidase extracted from red beet root, the loss of activity in the acidic pH region can be attributed largely to a lack of enzyme stability (Fig. 4). The pH dependency of beet peroxidase above the optimum pH appears to be caused by a direct effect of pH on enzyme activity, rather than on enzyme stability. The composition of the medium also can affect the pH stability of enzymes. For β-amylase, the presence of calcium stabilizes the enzyme at pH 5–10 (Whitaker, 1972).

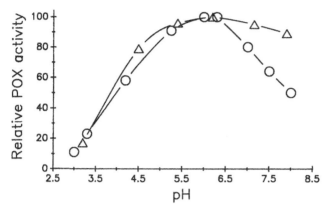

Figure 4 Effect of pH on stability (△) and activity (○) of peroxidase isolated from red beet roots. Incubation period was 5 min; activity was measured within 2 min at 25°C. Data from K. L. Parkin's laboratory.

4. Effect of pH on Enzyme Reaction Equilibria

An effect of pH on enzyme activity can be observed for many dehydrogenases that use NAD to catalyze the reversible oxidation of a substrate (AH_2):

$$AH_2 + NAD^+ \rightleftharpoons A + NADH + H^+$$

and

$$K_{eq} = \frac{[A]\,[NADH]\,[H^+]}{[AH_2]\,[NAD^+]}$$

If the K_{eq} of the reaction remains fairly constant over a broad range of pH, the equilibrium is shifted to the left and in the direction of reduction of substrate A as pH is reduced. Therefore, at fixed levels of initial substrate and cofactor concentrations, the extent of the reaction (the equilibrium state) at any pH can be predicted by simple calculation. In doing so, the effect of temperature on K_{eq} also must be considered (see Section IV,C).

III. Effect of the Aqueous Environment

A. General Considerations

Water is usually the continuous phase in most foods and food systems. It functions as a diffusion medium for solutes, controls the concentration of dissolved solutes, allows protein and enzyme function and serves as a cosubstrate in hydrolytic reactions. Dehydration and freezing are used to preserve foods by the removal or "immobilization" of water, thereby restricting microbial growth. Usually, enzyme activity is attenuated greatly by such processes. However, some enzymes remain active and often display altered patterns of activity in low moisture environments (Schwimmer, 1980).

B. Water Activity

1. Effect on Enzyme Stability

Generally, enzymes in solutions or in food systems are stable to dehydration, provided the temperature is controlled to prevent thermal inactivation (Hahn-Hagerdahl, 1986). Indeed, lyophilization is used routinely to prepare active enzyme powders from solutions. During the controlled drying of grains, many endogenous enzymes remain active (or dormant). Removal of water from biomaterials with organic solvents (acetone, alcohols, ether) also can lead to retention of enzyme activity, provided the temperature is maintained at $0°C$ or less. The composition of the medium in which the enzyme is suspended can influence retention of enzyme activity; methylamines, carbohydrates, and amino acids can stabilize proteins in dry or high osmotic environments (Yancey et al., 1982).

2. Effect on Enzyme Activity

a. Hydrolytic enzymes Hydrolytic reactions are reversible, although in foods in which water is $35–50 M$, hydrolysis is favored heavily. Several enzymes of importance in food processing—including carbohydrases, proteases, and lipases—have been shown to catalyze reverse reactions in water-restricted environments. The rate and extent of many enzyme reactions also depend on the availability of water.

In a model system with microcrystalline cellulose, invertase activity on sucrose is observed at an a_w as low as 0.57 (Silver and Karel, 1981). The initial rates of the reaction increase with increasing a_w from 0.57 to 0.90; however, the extent of the reaction is similar at a_w 0.75–0.90 (Fig. 5). An influence of a_w on the temperature dependency of invertase activity was noted also. The activation energy (E_a) increased from 4.8 kcal/mol at a_w 0.90 to 15.7 kcal/mol at a_w 0.58. This effect was attributed to a shift toward diffusion control of the reaction as a_w increases.

An application of the control of invertase activity by controlling humidity (a_w) in foods is illustrated in the study of the occurrence of the defect known as "sugar wall" in dates (Smolensky et al., 1975). The "sugar wall" defect arises from the crystallization of sucrose during the storage and handling of dates. Invertase, added as a spray or by vacuum infiltration, can delay or eliminate the occurrence of this defect. Hydrolysis by invertase of sucrose into its constituent sugars, glucose and fructose, interferes with the crystallization of sucrose. A water spray devoid of invertase also was found to delay the onset of the "sugar wall" defect. The water spray probably elevated the a_w in the date tissue and potentiated the action of endogenous invertase.

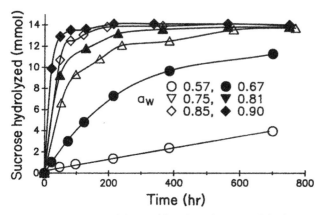

Figure 5 Invertase activity as a function of water activity in a model system containing 95 g Avicel and 5 g sucrose at pH 5.0 and 25°C. (Redrawn from Silver and Karel, 1981, with permission of Food & Nutrition Press.)

Glucoamylase, one of the enzymes used in the hydrolysis of starch, can catalyze synthetic reactions under the conditions used during industrial saccharification (30% starch, w/v; Adachi *et al.*, 1984). In the presence of 2.07 M glucose, glucoamylase forms a limited amount of maltose and appreciable quantities of isomaltose (Fig. 6). This behavior may account for the 6–8% oligosaccharide content that remains in corn syrup.

Proteases have been reported to synthesize peptide bonds in the presence of high concentrations (30–50% w/v) of protein hydrolysates (Arai *et al.*, 1975). This behavior is the basis of the plastein reaction, in which the rearrangement of protein structures is proposed to improve the functionality of relatively nonfunctional proteins, to reduce bitterness in protein hydroly-

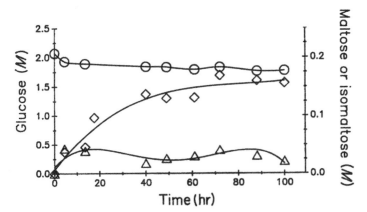

Figure 6 Time course of the conversion of glucose (○) to isomaltose (◇) and maltose (△) by glucoamylase (from *Aspergillus niger*) at pH 5.0 and 40°C. (Redrawn from Adachi *et al.*, 1984, with permission of John Wiley & Sons.)

sates, or to improve the nutritional quality of proteins through the incorporation of essential amino acids. The reversal of such proteolytic reactions may not be due to a true a_w effect, but may be driven more by the high concentration of reactants (Hahn-Hagerdahl, 1986).

The activity of lysozyme, a protease of commercial interest because of its ability to destroy gram-positive bacteria, also depends on a_w (Laretta-Garde *et al.*, 1988). When a_w is lowered by the addition of sucrose, glucose, or sorbitol, lysozyme activity displays a hyperbolic relationship with a_w (Fig. 7). After preincubation at reduced a_w (0.78–0.98) and immediate assay at a_w 0.99, activity is increased greatly. However, the enzyme appears to exhibit a "hydration memory" of the state existing during the preincubation period, as indicated by partial retention of the hyperbolic dependency on a_w of preincubation. The authors attributed this phenomenon to the retention of a microenvironment, near the protein active site, representative of the state of the enzyme in low a_w media. This hydration memory may be responsible for the hysteretic behavior of enzymes during reversible dehydration and rehydration.

Lipase action also depends on a_w. In a full-fat milk powder, endogenous lipolytic activity and activity from an added lipase isolated from *Pseudomonas fluorescens* are maximal at a_w 0.85 or greater; at a_w less than 0.6, lipolysis is impeded greatly (Fig. 8). A_w also can influence the kinetics of lipase action on triolein in a model system; V_{max} increases linearly with a_w, whereas K_m passes through a minimum at $a_w \sim 0.4$ (Drapron, 1972). A similar dependency on a_w for lipase action exists in grains. Lipase action in wheat bran held at 20°C at a_w 0.2 is only 15% of that observed at a_w 0.8 (Caillat and Drapron, 1974). The minimization of endogenous lipase activity in stored grains is important to

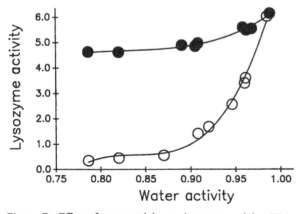

Figure 7 Effect of water activity on lysozyme activity. Water activity was controlled by the addition of 20–80% sucrose, sorbitol, or glucose. Activity (mg dried cells hydrolyzed per min per mg enzyme) was determined at pH 6.2 and 20°C at the indicated a_w values (O) or at a_w 0.99 after immediate transfer from the indicated a_w values (●). (Redrawn from Larreta-Garde *et al.*, 1988, with permission.)

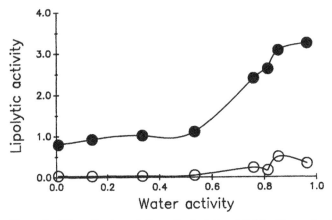

Figure 8 Effect of water activity on lipolysis in a full-fat milk powder. Lipase from *Pseudomonas fluorescens* was added at a level of 1 g/20 g milk powder. Incubation was at 20°C with lipolysis reported as milliequivalents –OH added per g milk powder. Exogenous lipase; ●; endogenous lipase, O. (Redrawn from Andersson, 1980, with permission of American Society of Microbiologists.)

preventing off-flavors in finished products (Schwimmer, 1981). The physical state of the acylglyceride substrates also regulates lipase action at a_w 0.65 (Acker and Wiese, 1972). Substrates in the fluid state (above their melting temperature) appear to be preferred over those in their solidified state (Fig. 9).

One commercial application of the control of reaction equilibria by a_w involves the action of lipase. Under conditions of restricted water content, lipase catalyzes the esterification of fatty acids to glycerol, monoacylglycerols, and diacylglycerols (Schuch and Mukherjee, 1989). Water-restricted environ-

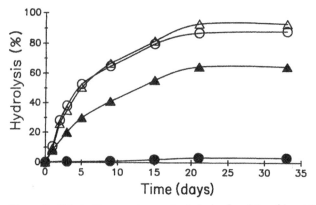

Figure 9 Effect of temperature on hydrolysis of acylglycerides at 0.65 water activity. Acylglycerides are triolein (△, 38°C; ▲, 25°C) and palmitoyloleoylstearylglycerol (O, 38°C; ●, 25°C). (Redrawn from Acker and Wiese, 1972, with permission.)

ments also can be used to catalyze the interesterification and transesterification of a mixture of acylglycerides. This process has been used to modify oils to improve their physical characteristics and to derive a cocoa butter substitute from a fraction of palm oil (Macrae, 1983). One method of refining food grade oils is treating them with lipase under restricted water conditions to esterify any free fatty acids present in the oil (Bhattacharyya *et al.*, 1989). Because of their specificity, enzymes often are preferred to nonenzymatic catalysts in obtaining directed modifications of lipids by transesterification.

b. Nonhydrolytic enzymes The activities of lipoxygenase and phenol oxidase in water-restricted environments have been the subject of much study. In food-related model systems, soy lipoxygenase activity is linearly dependent on increasing a_w until maximal activity is observed at a_w 0.55–0.65 (Brockmann and Acker, 1977a). At low a_w, capillary water was concluded to facilitate enzyme action on both free linoleate and linoleate esterified to phospholipids. Phenol oxidase activity exhibits a hyperbolic dependency on a_w; activity is observed initially at a_w as low as 0.5 (Tome *et al.*, 1978). A decrease in a_w from 1.0 to 0.85 also shifts the pH optimum of phenol oxidase almost 1 pH unit to the alkaline side. Progress curves of activity indicate that the extent of phenol oxidase action also is controlled by a_w (Fig. 10). Thus, a_w of the system appears to regulate activity in this case by controlling the diffusibility of reactants. Adjustment of the a_w from a lesser value to a greater one increases the extent of the reaction accordingly (Acker and Huber, 1969). The practical relevance of these phenomena is that, in the storage of intermediate moisture foods, fluctuation in a_w may potentiate undesirable chemical changes representative of storage at an intolerably high a_w.

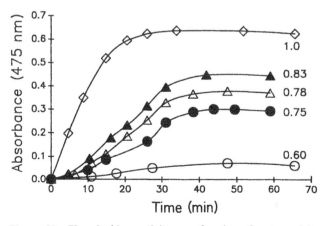

Figure 10 Phenoloxidase activity as a function of water activity measured as A_{475}. Water activity was adjusted with glycerol and enzyme activity was measured on tyrosine at pH 7.0 and 30°C. (Redrawn from Tome *et al.*, 1978, with permission.)

3. Alteration of Reaction Pathways

The nature of the products formed during enzyme reactions can be influenced by a_w. For soy lipoxygenase activity on free linoleate at a_w 0.65, the initial product formed is the linoleate hydroperoxide (Brockmann and Acker, 1977b). As the reaction progresses, the hydroperoxide appears to be converted, to a significant degree, into polar oligomeric secondary products. This behavior may be important in achieving the desirable effect on texture that lipoxygenase has in dough mixing and bread baking (Schwimmer, 1981), where a reduced a_w would prevail.

Altered reaction pathways at low a_w also have been noted for hydrolytic enzymes. At a_w 0.95, α-amylase (an endoamylase) forms a mixture of oligosaccharides (dextrins) from starch, whereas at a_w 0.75, glucose formation is highly favored (Drapron and Guilbot, 1962). The lowered a_w appears to restrict the access (diffusibility) of the enzyme to the internal sites of the starch polymer, reducing dextrin formation. In dry beans, the thermal potentiation of endogenous phytase activity converts phytate to inositol and inorganic phosphate, with little accumulation of the intermediate inositol phosphates (Chang et al., 1977). Again, the restriction of reactant diffusibility in the low a_w environment is likely to favor a more exhaustive hydrolysis of a single phytate molecule than a high a_w system.

C. Ionic Strength

Ionic strength of a food system can be increased through the addition of electrolytes or as a result of freezing or dehydration. The effects of ionic strength on enzyme activity can be caused by specific or nonspecific effects of electrolytes (Richardson and Hyslop, 1985). Ionic strength is well known to affect protein and, therefore, enzyme solubility by salting-in and salting-out phenomena. In most cases, enzyme solubility is a requisite for activity, since reactions require the diffusion of substrate to enzyme.

Specific effects of electrolytes depend on the relative concentrations of agents that stabilize, activate, inhibit, or inactivate a particular enzyme. Metal ions are particularly important, since many of them (e.g., magnesium, calcium, zinc) are cofactors or cosubstrates for enzyme reactions (see Chapter 2). Other metal ions such as the heavy metals (e.g., lead, mercury, cadmium) can inactivate enzymes by virtue of their ability to react with free sulfhydryl groups (cysteine residues).

Ion-specific effects may result from the presence of ions that are lyotrophs or chaotrophs (Hatefi and Hansen, 1969). Lyotrophs (principally cations) preserve water–solute interactions and water polarity and enhance the stability of soluble proteins. Chaotrophs (principally anions) disrupt water–solute interactions and reduce water polarity, often compromising the stability of soluble proteins. The relative chaotropic power of anions generally is related to charge density and follows the Hofmeister series (in decreasing order of strength):

$$SCN^- > I^- > ClO_4^- > NO_3^- > Br^- > Cl^- > CH_3COO^- > F^- > SO_4^-$$

Organisms that thrive in, or can adapt to, high osmotic environments have mechanisms to allow for normal enzyme functioning. Some enzymes have evolved to become halotolerant and are unimpaired by high osmotic environments. The evolution of osmolyte systems in organisms indigenous to high osmotic environments is a second mechanism of allowing normal enzyme functioning during osmotic stress (Yancey *et al.*, 1982). The mechanism(s) by which enzymes are stabilized by these osmolyte systems (amino acids, methylamines, and polyols) is unknown. However, the effect of polyhydroxy compounds apparently is caused by an increase in hydration of soluble proteins (Arakawa and Timasheff, 1982).

Exploiting enzyme adaptation to high osmotic conditions in food processing situations in which high concentrations of electrolytes exist may be possible. The addition of protein-stabilizing osmolytes is known to preserve enzyme function in systems that contain high levels of chaotropic ions or protein denaturants (urea) or have high osmolalities (Yancey *et al.*, 1982). Alternatively, enzymes adapted to high osmotic environments can be used in selected food processing situations. This adaptation is illustrated by the study of a neutral protease, with halophilic properties, isolated from *Bacillus subtilis* (El Mayda *et al.*, 1985). This enzyme is activated by the addition of NaCl and shows maximal activity on casein substrates at 4% NaCl (Fig. 11). At least 80% activity is retained on casein substrates at 10% NaCl. This behavior may prove valuable in cheesemaking, because high NaCl concentrations are known to inhibit chymosin proteolysis of β-casein.

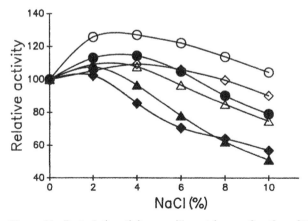

Figure 11   Proteolytic activity on milk proteins as a function of NaCl concentration. Substrates were β-(O), κ-($\diamond$), αs-($\triangle$), and whole casein ($\bullet$), whey protein ($\blacklozenge$), and β-lactoglobulin ($\blacktriangle$). Activity was measured by the production of TCA-soluble nitrogen at 35°C and pH 6.5 (caseins) or 8.5 (whey proteins). (Redrawn from El Mayda *et al.*, 1985, with permission of the Institute of Food Technologists.)

D. Freezing

1. General Considerations

Although the activities of many enzymes are attenuated by freezing, most enzymes survive freezing and some can be very active at high frozen storage temperatures (Fennema, 1975). The net effect of freezing on enzyme stability and activity is a function of the combined effects of "dehydration," concentration of solutes, and reduced temperature. The "dehydration" caused by freezing is not the same as that caused by dehydration of foods at temperatures above freezing (or in freeze-drying), since high a_w conditions prevail (> 0.82) in the former at or above recommended commercial frozen storage temperatures (Fennema, 1978).

The physical changes brought about by freezing include the concentration of solutes (including enzymes, substrates, inhibitors, and cofactors), an increase in viscosity of the unfrozen phase, and a disruption of (sub)cellular structures. The extent to which each of these changes takes place depends on the freezing temperature and freezing rate. Fast freezing tends to keep pools of unfrozen water segregated and minimizes the extent of cellular damage. The importance of cellular damage to the control of enzyme action lies in the decompartmentalization of previously separated enzymes and substrates. Although this effect may be of greatest importance during or after thawing, when the temperature becomes elevated, cellular systems (tissue foods) often exhibit an acceleration of certain, sometimes undesirable, enzymatic activities on freezing (Fennema, 1975).

2. Effect on Enzyme Stability

Much of our understanding of the behavior of enzymes in frozen foods comes from the study of model noncellular systems. For catalase, retention of activity after freezing is favored by high initial enzyme concentrations, quick thawing rates, and slow freezing rates (Fishbein and Winkert, 1977). Invertase also retains greater activity at lower freezing rates, although myosin B ATPase activity is more stable at faster freezing rates (Fennema, 1975). This difference may be caused by the macromolecular organization of the enzymes and by microenvironmental effects; invertase and catalase were studied as solutions whereas ATPase was associated with a complex muscle protein fraction. Retention of enzyme activity during freezing also depends on the composition of the medium. Retention of invertase (Tong and Pincock, 1969) and lactate dehydrogenase (Soliman and van den Berg, 1971) activities depend on pH. The concentration of solutes on freezing and their eutectic behavior can lead to changes in pH of the frozen system of as much as 1 pH unit in foods (van den Berg, 1968) and in simple solutions (van den Berg and Soliman, 1969). Polyhydroxy compounds are very effective in stabilizing phosphofructokinase to freeze-induced inactivation (Carpenter et al., 1986).

3. Effect on Enzyme Activity

The need to blanch vegetables prior to freezing is a constant reminder that food-deteriorating enzymes remain active in frozen environments. In some cases, freezing can accelerate enzyme action in simple solutions and in foods.

a. **Simple solutions** Freezing of dilute solutions of trypsin at $-23\,^\circ$C (Grant and Alburn, 1966) and invertase at $-5\,^\circ$C (Tong and Pincock, 1969) accelerates enzyme activity with respect to that occurring at nonfreezing or corresponding supercooled temperatures. This effect has been attributed to the several-fold increase in reactant concentrations brought about by freezing, and may offset any detrimental effects of freezing on enzyme stability.

Lipoxygenase and lipase exhibit activity at -12 to $-15\,^\circ$C in frozen solutions (Bengtsson and Bosund, 1966; Parducci and Fennema, 1978; Fennema and Sung, 1980). In both cases, elevation of the temperature of the frozen system results in an increased accumulation of reaction products (Fig. 12). These observations indicate that the extent of enzyme activity may be limited by the amount of unfrozen water available as a diffusion medium. Fast freezing only marginally reduced (by 10–20%) the extent of lipoxygenase activity in model systems at -5 to $-15\,^\circ$C, compared with slow freezing (Fennema and Sung, 1980).

b. **Tissue systems** Several examples of enzyme action that is accelerated in tissue systems at high freezing temperatures relative to low nonfreezing temperatures are presented in Table II. In muscle foods, textural and organoleptic qualities can be compromised during improper frozen storage. Significant extents of phospholipase action can be detected in fish stored at temperatures between -10 and $-14\,^\circ$C (Olley et al., 1969; Hanaoka and Toyomizu, 1979). This activity can produce free fatty acids and potentiate oxidation or possibly cause the insolubilization (denaturation) of muscle proteins (Anderson and Ravesi, 1970). Freezing of muscle of fishes of the Gadoid family is known to cause a rapid and unacceptable loss in textural quality if storage temperature is not maintained at $-18\,^\circ$C or below (Kelleher et al., 1982). This effect is proposed to be caused by the "activation" of the endogenous enzyme trimethylamine-N-oxide demethylase via decompartmentalization (Parkin and Hultin, 1982). Behnke et al. (1973) also observed an increase in the rate of ATP depletion and lactate accumulation in red muscle frozen at $-3\,^\circ$C, compared with unfrozen muscle. These investigators attempted to account for the relative contributions of temperature and solute concentration on reaction rates during the freezing of foods (Table III). The two forces controlling enzyme activity in frozen systems are temperature and solute concentration. For each of the four situations presented, the relative influence of these two forces and the overall effect on reaction rate can be predicted.

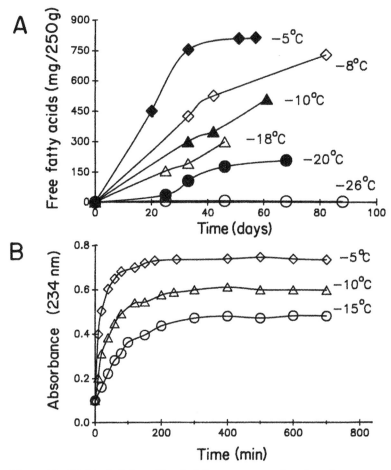

Figure 12 (A) Lipolysis in unblanched frozen peas as a function of temperature. (Redrawn from Bengtsson and Bosund, 1966, with permission of the Institute of Food Technologists.) (B) Lipid oxidation (conjugated diene formation) in frozen solutions of lipoxygenase and linolenic acid as a function of temperature. (Redrawn from Fennema and Sung, 1980, with permission.)

Enzyme reactions also can be accelerated on freezing in plant foods (Table II). However, because of the long shelf-life expectancy of frozen vegetables, the mere persistence of enzyme action at low temperatures can be devastating to quality. Lipase activity in unblanched frozen peas is significant at temperatures as low as $-20°C$ (Fig. 12a; Bengtsson and Bosund, 1966). After 7 months at $-15°C$, up to 32% of the total esterified fatty acid is hydrolyzed. Traditionally, peroxidase has been used as a blanching indicator. However, some research has indicated that lipid-degradative enzymes (palmitoyl-CoA hydrolase, Baardseth and Slinde, 1983; lipoxygenase, Williams *et al.*, 1986) may be more suitable as blanching indicators. To optimize

TABLE II
Examples of Potentiation of Enzyme Activity on Freezing of Food Tissue Systems[a]

Reaction catalyzed	Tissue	Temperature range of enzyme potentiation (°C)
Glycogen loss and lactate accumulation	Muscle	−2.5 to −6
Loss of high energy phosphates	Muscle	−2 to −8
Phospholipid hydrolysis	Muscle	−4 to −14
Peroxide decomposition	Vegetable	−0.8 to −5
L-Ascorbic acid oxidation	Vegetable and fruit	−2.5 to −10
Trimethylamine-N-oxide demethylation	Muscle	−5 to −18

[a]Compiled from Fennema (1975), Parkin and Hultin (1982), Kelleher et al. (1982), and Hanaoka and Toyomizu (1979).

blanching, the endogenous enzyme that limits the retention of quality must be identified first. Because of differences in enzyme composition and thermal resistance, the target enzyme is likely to vary between specific products.

IV. Effect of Temperature

A. General Considerations

All chemical reactions depend on temperature. The relationship between reaction rates and temperature is described by the Arrhenius equation:

$$k = Z \, e^{-E_a/2.3RT} \tag{7}$$

where k is the rate constant, T is the temperature (K), R is the gas constant,

TABLE III
Rates of Reactions as Influenced by Temperature and Concentration of Solutes during Freezing[a]

Situation	Change in rate of reaction caused by		Relative influence of the two effects	Total effect of freezing on reaction rate
	Lowering of temperature (effect T)	Concentration of solutes (effect S)		
1	Decrease	Decrease	Cooperative	Decrease
2	Decrease	Slight increase	T > S	Slight decrease
3	Decrease	Moderate increase	T = S	None
4	Decrease	Great increase	T < S	Increase

[a]Reproduced from Behnke et al. (1973), with permission.

and Z is the frequency of favorable collisions between reactants. For most nonenzymatic reactions, reaction rates continually increase with increasing temperature. For enzymatic reactions, a temperature optimum is observed. This optimum results from the opposing effects of increasing temperature on the reaction rate and on the rate of enzyme thermal inactivation (both as predicted by Eq. 7). Other effects of temperature on the rate of enzymatic reactions involve changes in environmental conditions (e.g., pH), the nature or availability of substrate(s), the ionization state of prototropic groups, kinetic constants of the reaction (e.g., K_m), the association of the subunits of oligomeric enzymes, and the thermal sensitivity of substrate(s). The effect of temperature on enzyme reactions is related intimately to the composition of the medium. Ionic strength, a_w, and pH can exert profound effects on the temperature response of enzyme activity. Therefore, although the general effects of temperature on enzyme reactions will be presented here, the reader should bear in mind that the effect of temperature on a specific enzyme depends on the environment. For example, an enhanced thermal stability and deviation from first-order kinetics destruction is observed for endogenous pectinesterase activity in orange juice as solids are increased from $10-30°$ to $35-50°$ Brix (Marshall *et al.*, 1985). In addition, alkaline phosphatase becomes less heat resistant when lactose is removed from milk by ultrafiltration (Mistry, 1989).

B. Effect on Enzyme Activity and Stability

1. Increasing Temperature

Although many enzymes are described as having an optimum temperature for activity, this optimum is defined within certain constraints. These constraints can be appreciated most easily by referring to Fig. 13, which

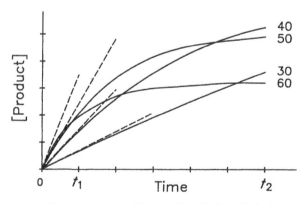

Figure 13 Time course of the reaction of a hypothetical enzyme as a function of temperature (°C). Initial velocities are represented as dashed lines.

shows the progress of a hypothetical enzyme reaction as a function of temperature. In this example, enzyme "reactivity" can be measured in at least three ways: initial velocity and the extent of reaction at times (t) 1 and 2. If one uses initial velocity as a criterion, the temperature optimum is identified as 60°C. However, if t_1 or t_2 is used, the temperature optima for the reaction are identified as 50°C and 40°C, respectively. The decline in reaction progress with time (dP/dt) at the higher temperatures can be explained by the progressive inactivation of enzyme during incubation, provided the substrate does not become rate limiting. Therefore, in a food processing setting, optimum temperature for a reaction often should be identified in the context of the extent of substrate transformation to product desired.

The opposing effects of increasing temperature on reaction rate acceleration and enzyme inactivation are illustrated in Fig. 14 for the caseinolytic activity of a protease isolated from melon fruit. The enzyme is stable (after a 10 min incubation) at temperatures up to 50°C, whereas the optimum temperature for activity (amount of product formed after 10 min) is defined as 70°C.

Often, the Arrhenius plot is used to analyze the temperature dependency of enzyme activation and inactivation. When the rate constants of an enzyme reaction are being compared, the logarithm of the Arrhenius equation (Eq. 7) is taken and integrated between the limits of k_1 and k_2 and a specific temperature range $[T\ (K)]$ to yield the following relationship:

$$\log \frac{k_2}{k_1} = \frac{E_a(T_2 - T_1)}{2.3RT_2T_1} \tag{8}$$

The slope of a plot of $\log k$ vs $1/T$ has a value of $\pm E_a/2.3R$. (This example is for illustrative purposes only and does not advocate the use of two points to determine a straight line for calculation of an E_a value for inactivation!) Using

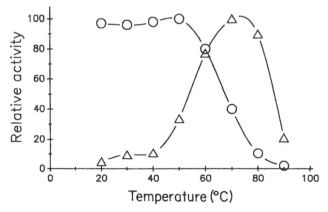

Figure 14 Caseinolytic activity ($\triangle$) and stability ($\bigcirc$) of a 54-kDa protease isolated from melon fruit as a function of temperature. (Redrawn from Yamagata *et al.*, 1989, with permission of The Agricultural Chemical Society of Japan.)

the data on enzyme activity in Fig. 14 and transforming to an Arrhenius plot yields Fig. 15. Taking the slopes of the plot on either side of the temperature "optimum" allows the calculation of $E_{a_{act}}$ and $E_{a_{inact}}$ as 13 and 37 kcal/mol, respectively. This pattern is generally representative of all enzymes; the temperature coefficient for thermal inactivation is greater than that for the acceleration of reaction rates. The range of values for $E_{a_{act}}$ and $E_{a_{inact}}$ for most enzymes is 6–15 and 50–150 kcal/mol, respectively. This feature accounts for the general observation that, as temperature increases, reaction rate increases exponentially to a point, after which it declines more rapidly (cf. Figs. 14 and 15). Although the Arrhenius plot defines the ordinate as log k, any reaction rate parameter that is directly proportional to k (e.g., V_{max}) can be used, as long as the relationship between the substituted parameter and k does not change over the temperature range evaluated.

A better alternative to using the Arrhenius plot of enzyme activity, described in Fig. 15, exists to evaluate the temperature effect on enzyme inactivation. The approach described in Fig. 15 does not distinguish between the effects of temperature on reaction activation and enzyme inactivation. A method that evaluates only the effect of temperature on enzyme deactivation involves preincubation of the enzyme solution at preselected temperatures and times, followed by assaying residual enzyme activity under a standard (and often optimum) condition under which no further enzyme deactivation is expected. When residual enzyme activity is compared with incubation time on a semi-log plot, often straightline relationships exist over a range of temperatures, indicating that enzyme thermal inactivation can be modeled as a first-order process. The slope of the plots at each temperature is proportional to the rate constant k for inactivation. The k values then can be

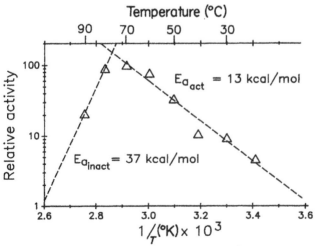

Figure 15 Arrhenius plot of caseinolytic activity of a melon protease as a function of temperature. Data are from the activity curve of Fig. 14.

presented on an Arrhenius plot and the E_a for enzyme inactivation determined.

Temperature sensitivity for enzyme inactivation is related directly to the water content of the system. Enzymes in dry or semi-moist food systems tend to be more heat stable. As water is removed near the monolayer of an enzyme, the enzyme can become remarkably stable at $100°C$, as illustrated for lipase (Zaks and Klibanov, 1984). Water is required to facilitate the unfolding of proteins during thermal denaturation. The enhanced thermal stability of enzymes in dry systems has potential food applications. This tendency is illustrated in the comparison of the temperature dependencies of hydrolytic and transesterification activities of lipase in aqueous and non-aqueous environments, respectively (Fig. 16). Transesterification reactions can be achieved at high temperatures in nonaqueous media, taking full advantage of the temperature coefficient of the reaction, whereas hydrolytic reactions possess a lower temperature optimum.

Although increased temperatures often are used to denature enzymes, several cases show it to be used to potentiate enzyme activity thermally and yield beneficial consequences. Probably the most common example used in the food industry is the preincubation of green beans at moderate temperatures prior to canning or blanching. This step potentiates endogenous pectin methylesterase activity and the attendant calcium reactivity with deesterified pectin polymers to induce a firming of the tissue and prevent sloughing in the final product (Steinbuch, 1976). The same effect has been reported in other vegetable tissues, including potatoes (Bartolome and Hoff, 1972) and carrots (Lee et al., 1979). Similarly, the control of exogenously added α-amylase

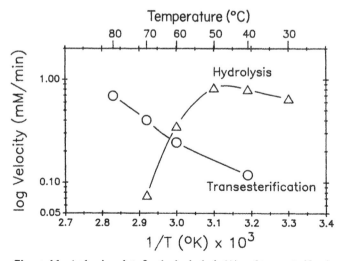

Figure 16 Arrhenius plots for the hydrolysis ($\triangle$) and transesterification ($\bigcirc$) reactions catalyzed by pancreatic lipase. Hydrolysis was measured on a 2% tributyrin suspension at pH 8.0 and transesterification was measured with myristic acid and 5% butter oil in hexane without the addition of water. Data from K. L. Parkin's laboratory.

activity at 70°C has been reported to improve the consistency of a sweet potato puree product (Szyperski *et al.*, 1986).

The thermal sensitivities of enzymes also have been exploited as indicators of adequate thermal processing of food materials. Peroxidase and alkaline phosphatase are indicator enzymes commonly used to insure the adequate blanching of vegetables and the pasteurization of dairy products, respectively. Proper and timely estimations of residual peroxidase and alkaline phosphatase activities are important in these situations, since both are known to renature or become reactivated after thermal processing (Schwimmer, 1981).

In general, enzymes with low molecular weights, a single polypeptide chain, and the most extensive disulfide cross-links (cystine residues) are more thermally stable than others. The mechanisms of thermal inactivation have begun to be investigated (Ahearn and Klibanov, 1985). Protein unfolding appears to be responsible for the initial loss of enzyme activity during exposure to high temperatures. However, irreversible inactivation appears to depend on the extent to which chemical reactions are allowed to proceed during extended high temperature exposure. Some of the chemical reactions involved in irreversible thermal inactivation are deamidation of asparagine residues, hydrolysis of Asp–X peptide bonds, and the destruction of cystine residues (β elimination). On cooling, the refolding of "thermally scrambled" enzymes to yield inactive structures can also contribute to irreversible inactivation. The relative contributions of these various mechanisms to thermal inactivation are pH dependent. This type of fundamental information is critical to the design of enzymes with enhanced thermal stability. The replacement of asparagine and aspartate residues by genetic engineering with residues that are less chemically reactive or have greater nonpolar character should enhance enzyme stability (Mozhaev and Martinek, 1984; Ahearn and Klibanov, 1985). Polyhydroxylation and glycosylation are proposed chemical modifications that can enhance enzyme thermal stability (O'Fagain *et al.*, 1988).

2. Decreasing Temperature

Low and nonfreezing temperatures also can have profound effects on enzyme activity. Generally, the prediction of reaction rates at low temperatures follows the relationship described by the Arrhenius equation (Eq. 7). However, several cases of deviation from this predicted behavior are known.

Phosphofructokinase (PFK) from potato tubers has been shown to undergo a reversible dissociation of subunits at 0–5°C (ap Rees *et al.*, 1981). The low temperature is likely to weaken the hydrophobic interactions involved in the assembly of the PFK subunits. The resultant impairment of enzyme activity and, thus, glycolysis may be partially responsible for the tendency of potato tubers to accumulate sugars during low-temperature storage. This low-temperature sweetening is well known to be associated with the propensity of cold-stored tuber stocks to darken extensively during thermal processing.

"Breaks" in the Arrhenius plots at low temperatures have been observed also. Such breaks imply a change in the physical state of an enzyme or its microenvironment, so the reaction becomes impaired (has a higher E_a) at low temperatures. The relevance of this observation to food science is in the proposed roles that membrane rigidification and the resultant impairment of membrane-associated enzyme activities (e.g., mitochondrial oxidase activity) have in the development of chilling injury in sensitive fruits and vegetables (Parkin *et al.*, 1989). Whether or not these "breaks" truly represent a "change in state" of a membrane enzyme remains to be determined. However, a possible explanation for such "breaks" at low temperatures may be the temperature dependence of K_m for some enzymes (Silvius *et al.*, 1978). K_m for ATP of a membrane-associated ATPase increases with increasing temperature. Thus, when reaction velocity at a fixed level of substrate is plotted on an Arrhenius plot, a "break" occurs at $20-25°C$, showing a higher E_a at the lower temperatures. In this case, the "break" is more likely to be a function of the dependency of reaction velocity on a change in $[S_0]/K_m$ than of a temperature-dependent change in the intrinsic activity of the ATPase. This example reemphasizes the need to use the appropriate values for the ordinate on an Arrhenius plot (see Section IV, B, 1).

C. Other Effects

As mentioned in Section II, B, 1, temperature influences the ionization potential of prototropic groups of the enzyme, which may influence activity. The ΔH of these ionizable groups is provided in Table I. As temperature increases, the pK of ionization of a prototropic group essential to enzyme activity decreases and affects the pH dependency of the reaction. Similarly, any enzyme reaction that exhibits an equilibrium condition between reactants and products will depend on the temperature dependency of the equilibrium constant (K_{eq}), which is predicted by the van't Hoff relationship:

$$K_{eq} = C \; e^{-\Delta H/2.3RT} \tag{9}$$

where C is an internal constant, ΔH is the heat of reaction, and T is temperature (K).

Temperature also can affect the disposition of the substrates for a reaction. In multienzyme systems, competing reactions may influence the overall pattern of reactivity as controlled by the relative temperature dependencies of the constituent enzymes. Substrates or cofactors that are temperature sensitive may also become rate limiting to reactions at high temperatures. Temperature also affects the solubility of substrates. Although substrate solubilities generally increase with increasing temperature, an exception to this trend is exhibited by gases. Therefore, the reactivity of oxygen-requiring

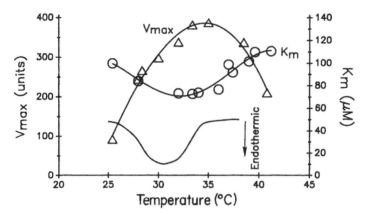

Figure 17 Effect of the physical state of a phospholipid–fatty acid ternary dispersion on the kinetics of phospholipase A₂ activity. Bottom curve represents the endothermic profile of the lipid dispersion. (Redrawn from Jain and Jahagirdar, 1985, with permission of Elsevier Science.)

enzymes may be attenuated at high temperatures not only by a direct effect on intrinsic enzyme activity or stability, but also by a decrease in the availability of substrate.

One last example of the effect of temperature on substrates can be found for enzymes acting at interfaces. Lipases are generally more active on acylglycerides in the fluid state (cf. Fig. 9). This feature is central to the concept of directed transesterification reactions in which reaction equilibria can be "pulled" toward the formation of acylglycerides that solidify below a specific reaction temperature, thereby effectively removing them from the reaction mixture (Macrae, 1983). Phospholipases exhibit a somewhat different dependency on the physical state of phospholipid micellar substrates. Reactivity appears to be most efficient when a mixture of liquid-phase and solid-phase phospholipids exists (Jain and Jahagirdar, 1985). This tendency is demonstrated in Fig. 17, which shows the dependency of K_m and V_{max} on the phase transition behavior of ternary dispersions of dipalmitoylphosphatidylcholine with a 0.18-mol fraction of 1-palmitoyllysophosphatidylcholine plus palmitic acid.

V. Effect of Other Environmental Conditions

Viscosity, sonic energy, pressure, and irradiation are other environmental parameters that can influence enzyme reactivity. Viscosity probably plays a role in the reactivity of enzymes in frozen foods since more than 90% of the "free" water is frozen at commercial storage temperatures. Although the direct evaluation of the effect of viscosity in frozen systems is obviously problematic, the observations of enzyme systems in nonfrozen media provide some insight. Invertase activity is known to deviate from the behavior predicted by the Michaelis–Menten relationship (substrate saturation and attainment of V_{max}) at sucrose levels above 6%. Bowski et al. (1971) concluded that this phenomenon could be explained by the increased viscosity of the system and by a direct inhibitory effect of sucrose on the enzyme.

Pressure also can have an influence on enzyme stability. Very high pressures (up to 6100 bar) often are required to inactivate enzymes (Schwimmer, 1981). This requirement is likely to preclude any widespread commercial application of pressure to control enzyme activity in foods (Richardson and Hyslop, 1985).

Sonic energy can be used to inactivate enzymes. Sonic energy-induced enzyme inactivation is most likely caused by cavitation and the resultant interfacial denaturation of proteins (Schwimmer, 1981). This effect has been shown for the protease papain (Ateqad and Iqbal, 1985). The extent of destruction of papain activity by sonic irradiation is inversely proportional to protein concentration and cannot be characterized by a first-order process.

Because of the resurgent interest in irradiation, some general comments on its effect on enzyme activity are in order. The damaging effect of ionizing radiation is directed randomly and is based on probability. Larger molecular weight structures are more likely to be damaged by ionizing radiation than smaller molecular weight structures. For enzymes, first-order rate constants for inactivation by ionizing radiation increase with increasing molecular weight (Miller and Robyt, 1986). Thus, the eradication of microorganisms in spices and the prevention of sprouting in vegetables can take place at doses of irradiation lower than those required to inactivate enzymes fully. DNA presents a larger "target" for γ-rays than most enzymes.

Several enzymes have been shown to be sensitive to γ-irradiation in model systems (Schwimmer, 1981). Dry enzymes are more stable than those in solution. For example, 7 Mrad are required to inactivate 63% peroxidase activity in a "dry" system, whereas only 0.24 Mrad are required to inactivate 85% peroxidase activity in a moist system. This result is caused by the added effect of the production of radiolytic products from water, resulting in the indirect destruction of enzymes by ionizing radiation. Enzyme damage is enhanced at high oxygen tensions also, because of the production of oxy radical species. The secondary radiolytic products formed from water and oxygen also contribute to the continuing loss of enzyme activity often observed during postirradiation incubation.

In foods, the effect of ionizing radiation is less predictable (Schwimmer, 1981). Although endogenous enzymes often are destroyed, their activity in foods depends on several other factors. The destruction of inhibitors, activators, substrates, or cofactors can affect residual enzyme activity. In addition, the radiation-induced damage to endogenous proteases may impair the natural turnover process for endogenous enzymes, allowing some to accumulate to levels higher than those observed in nonirradiated tissues. Finally, ionizing radiation is certain to result in some degree of cellular damage and the possible potentiation of enzyme activities on previously segregated substrates. One study illustrates the complexity of the effects that γ-irradiation has on food systems. Sweet potato roots irradiated at 0.1 – 0.2 Mrad showed a transient increase in β-amylase, sucrose synthase, sucrose-phosphate synthase, phosphoglucomutase, phosphoglucose isomerase, and phosphorylase activities, and an increase in sucrose content during the postirradiation period (Ajlouni and Hamdy, 1988). Higher levels of irradiation (0.3 – 0.5 Mrad) led to a decrease in these enzyme activities, although the sucrose content still increased during the postirradiation period.

VI. Summary

To exercise maximum control over enzyme-mediated food modification, a thorough understanding of the environmental parameters that can be manipulated and how they affect enzyme action is required. In tissue foods, the manipulation of the environment is limited to rudimentary parameters such as humidity (a_w), temperature, and the "immobilization" of water by freezing. Even these simple controls can have a great impact on maintaining the quality of foods. The proper conditioning of muscle into edible meat relies on the fine balance of control of temperature, rate of postmortem glycolysis, and progressive decline in pH. The proper control of temperature and humidity during the malting of barley in preparation for mashing results in the accumulation of endogenous amylolytic activities and the partial hydrolysis of the starch granule. The application of water (increasing a_w) to dates is a simple yet partially effective means of delaying the "sugar wall" defect (Section III, B, 2, a).

In nontissue food systems, added dimensions of reaction control are available. Exogenous enzymes of varying thermal stability can be chosen on the basis of the extent of activity that is desired, an example being the use of α-amylase from various sources in dough conditioning and bread baking. Water content can be changed to control or alter the progress of an enzyme reaction; acidification has long been used as a mean of attenuating enzyme action in food systems. However, with the ability to manipulate the environment comes the added complexity of the interaction between environmental factors. The effects of pH, temperature, a_w, and other environmental conditions on enzyme action are interdependent. Thus, the optimization of any enzyme-mediated food modification requires the integration of all the environmental effects on enzyme action for any process to be achieved consistently and uniformly.

Acknowledgments

This work was supported by the College of Agricultural and Life Sciences of the University of Wisconsin, Madison. The author is grateful for the assistance of B. Scullion in preparing this manuscript.

References

Acker, L., and Huber, L. (1969). Behavior of polyphenoloxidase in a water-deficient environment. *Lebensm. Wiss. Technol.* **3**, 82–85 *(in German).*

Acker, L. and Wiese, R. (1972). The behavior of lipase in systems of low water content. I. Influence of the physical state of the substrate on enzymatic hydrolysis. *Lebensm. Wiss. Technol.* **5**, 181–184, *(in German).*

Adachi, S., Ueda, Y., and Hashimoto, K. (1984). Kinetics of formation of maltose and isomaltose through condensation of glucose by glucoamylase. *Biotechnol. Bioeng.* **26**, 121–127.

Ahearn, T. J., and Klibanov, A. M. (1985). The mechanism of irreversible enzyme inactivation at 100°C. *Science* **228**, 1280–1284.

Ajlouni, S., and Hamdy, M. K. (1988). Effect of combined gamma-irradiation and storage on biochemical changes in sweet potato. *J. Food Sci.* **53**, 477–481.

Anderson, M. L., and Ravesi, E. M. (1970). On the nature of the association of protein in frozen-stored cod muscle. *J. Food Sci.* **35**, 551–558.

Andersson, R. E. (1980). Microbial lipolysis at low temperatures. *Appl. Environ. Microbiol.* **39**, 36–40.

ap Rees, T., Dixon, W. L., Pollock, C. J., and Franks, F. (1981). Low temperature sweetening of higher plants. *In* "Recent Advances in the Biochemistry of Fruits and Vegetables" (J. Friend and M. J. C. Rhodes, eds.), pp. 41–61. Academic Press, New York.

Arai, S., Yamashita, K. A., and Fujimaki, M. (1975). Plastein reaction and its application. *Cereal Foods World* **20**, 107–112.

Arakawa, T., and Timasheff, S. N. (1982). Stabilization of protein structure by sugars. *Biochemistry*, **21**, 6536–6544.

Ateqad, N., and Iqbal, J. (1985). Effect of ultrasound on papain. *Ind. J. Biochem. Biophys.* **22**, 190–192.

Baardseth, P., and Slinde, E. (1983). Peroxidase, catalase, and palmitoyl-CoA hydrolase activity in blanched carrot cubes after storage at −20°C. *J. Sci. Food Agric.* **34**, 1257–1262.

Bartolome, L. G., and Hoff, J. E. (1972). Firming of potatoes: Biochemical effects of preheating. *J. Agric. Food Chem.* **20**, 266–270.

Behnke, J. R., Fennema, O., and Cassens, R. G. (1973). Rates of postmortem metabolism in frozen animal tissues. *J. Agric. Food Chem.* **21**, 5–11.

Bengtsson, B., and Bosund, I. (1966). Lipid hydrolysis in unblanched frozen peas *(Pisum sativum). J. Food Sci.* **31**, 474–481.

Bhattacharyya, S., Bhattacharyya, D. K., Chakraborty, A. R., and Sengupta, R. (1989). Enzymatic deacidification of vegetable oils. *Fat Sci. Technol.* **91**, 27–30.

Bowski, L., Saini, R., Ryu, D. Y., and Veith, W. R. (1971). Kinetic modeling of the hydrolysis of sucrose by invertase. *Biotechnol. Bioeng.* **13**, 641–656.

Brockmann, R., and Acker, L. (1977a). The behavior of lipoxygenase in systems of low water content. I. Influence of water activity on the enzymatic oxidation of lipids. *Lebensm. Wiss. Technol.* **10**, 24–27 *(in German).*

Brockmann, R., and Acker, L. (1977b). The behavior of lipoxygenase in systems of low water content. II. Investigation of reaction products of enzymatic lipid oxidation. *Lebensm. Wiss Technol.* **10**, 332–336 *(in German).*

Caillat, J. M., and Drapron, R. (1974). The lipase of wheat. Characteristics of its action in aqueous and low-moisture media. *Ann. Technol. Agric.* **23**, 273–286 *(in French).*

Carpenter, J. F., Hand, S. C., Crowe, L. M., and Crowe, J. H. (1986). Cryoprotection of phosphofructokinase with organic solutes: Characterization of enhanced protection in the presence of divalent cations. *Arch. Biochem. Biophys.* **250**, 505–512.

Chang, R., Schwimmer, S., and Burr, H. K. (1977). Phytate: Removal from whole dry beans by enzymatic hydrolysis and diffusion. *J. Food Sci.* **42**, 1098–1101.

Drapron, R. (1972). Enzymatic reactions in systems of low water content. *Ann. Technol. Agric.* **21**, 487–499 *(in French).*

Drapron, R., and Guilbot, A. (1962). Contribution to the study of enzymatic reactions in poorly hydrated biological environments: The degradation of starch by amylases as a function of water activity and temperature. *Ann. Technol. Agric.* **11**, 275–371.

El Mayda, E., Paquet, D., and Ramet, J. P. (1985). Enzymatic proteolysis of milk proteins, in a salt environment, with a *Bacillus subtilis* neutral protease preparation. *J. Food Sci.* **50**, 1745–1746.

Fennema, O. (1975). Activity of enzymes in partially frozen aqueous systems. In "Water Relations of Foods" (R. B. Duckworth, ed.), pp. 397–413. Academic Press, New York.

Fennema, O. (1978). Water and protein hydration. *In* "Dry Biological Systems" (J. H. Crowe and J. S. Clegg, eds.), pp. 297–322. Academic Press, New York.

Fennema, O., and Sung, J. C. (1980). Lipoxygenase-catalyzed oxidation of linolenic acid at subfreezing temperatures. *Cryobiol.* **17**, 500–507.

Fishbein, W. N., and Winkert, J. W. (1977). Parameters of biological damage in simple solutions: Catalase. I. The characteristic pattern of intracellular freezing damage exhibited in a membraneless system. *Cryobiol.* **14**, 389–398.

Grant, N. H., and Alburn, H. E. (1966). Acceleration of enzyme reactions in ice. *Nature (London)* **212**, 194.

Hahn-Hagerdal, B. (1986). Water activity: A possible regulator in biotechnical processes. *Enz. Microb. Technol.* **8**, 322–327.

Hanaoka, K., and Toyomizu, M. (1979). Acceleration of phospholipid decomposition in fish muscle by freezing. *Bull. Jap. Soc. Sci. Fish.* **45**, 465–468 *(in Japanese)*.

Hatefi, Y., and Hansen, W. G. (1969). Solubilization of particulate proteins and nonelectrolytes by chaotropic agents. *Proc. Natl. Acad. Sci. U.S.A.* **62**, 1129–1136.

Jain, M. K., and Jahagirdar, D. V. (1985). Action of phospholipase A2 on bilayers. Effect of fatty acid and lysophospholipid additives on the kinetic parameters. *Biochim. Biophys. Acta* **814**, 313–318.

Kelleher, S. D., Buck, E. M., Hultin, H. O., Parkin, K. L., Licciardello, J. J., and Damon, R. A. (1982). Chemical and physical changes in red hake blocks during frozen storage. *J. Food Sci.* **47**, 65–70.

Larreta-Garde, V., Xu, Z. F., Lamy, L., Malthlouthi, M., and Thomas, D. (1988). Lysozyme kinetics in low water activity media. A possible hydration memory. *Biochem. Biophys. Res. Commun.* **155**, 816–822.

Lee, C. Y., Bourne, M. C., and Van Buren, J. P. (1979). Effect of blanching treatments on the firmness of carrots. *J. Food Sci.* **44**, 615–616.

Macrae, A. R. (1983). Lipase-catalyzed interesterification of oils and fats. *J. Amer. Oil Chem. Soc.* **60**, 291–294.

Marshall, M. R., Marcy, J. E., and Braddock, R. J. (1985). Effect of total solids level on heat inactivation of pectinesterase in orange juice. *J. Food Sci.* **50**, 220–222.

Means, G. E., and Feeney, R. E. (1971). "Chemical Modification of Proteins." Holden-Day, San Francisco.

Miller, A. W., and Robyt, J. F. (1986). Functional molecular size and structure of dextransucrase by radiation inactivation and gel electrophoresis. *Biochim. Biophys. Acta* **870**, 198–203.

Mistry, V. V. (1989). Thermal inactivation characteristics of alkaline phosphatase in ultrafiltered milk. *J. Dairy Sci.* **72**, 1112–1117.

Mozhaev, V. V., and Martinek, K. (1984). Structure–stability relationships in proteins: New approaches to stabilizing enzymes. *Enz. Microb. Technol.* **6**, 50–59.

O'Fagain, C., Sheehan, H., O'Kennedy, R., and Kilty, C. (1988). Maintenance of enzyme structure. Possible methods for enhancing stability. *Proc. Biochem.* **23**, 166–171.

Olley, J., Farmer, J., and Stephen, E. (1969). The rate of phospholipid hydrolysis in frozen fish. *J. Food Technol.* **4**, 27–37.

Parducci, L. G., and Fennema, O. (1978). Rate and extent of enzymatic lipolysis at subfreezing temperatures. *Cryobiol.* **15**, 199–204.

Parkin, K. L., and Hultin, H. O. (1982). Some factors influencing the production of dimethylamine and formaldehyde in minced and intact red hake muscle. *J. Food Proc. Preserv.* **6**, 173–197.

Parkin, K. L., Marangoni, A., Jackman, R. L., Yada, R. Y., and Stanley, D. W. (1989). Chilling injury. A review of possible mechanisms. *J. Food Biochem.* **13**, 127–153.

Pressey, R. (1967). Invertase inhibitor from potatoes: Purification, characterization, and reactivity with plant invertases. *Plant Physiol.* **42**, 1780–1786.

Richardson, T., and Hyslop, D. (1985). Enzymes. *In* "Food Chemistry" (O. Fennema, ed.), 2d Ed., pp. 371–476. Marcel Dekker, New York.

Robb, D. A., Swain, T., and Mapson, L. W. (1966). Substrates and inhibitors of the activated tyrosinase of broad bean *Vicia faba*. *Phytochemistry* **5**, 665–675.

Schwimmer, S. (1980). Influence of water activity on enzyme reactivity and stability. *Food Technol.* **34(5)**, 64–74, 82.

Schwimmer, S. (1981). "Source Book of Food Enzymology." AVI, Westport.

Schuch, R., and Mukherjee, D. (1989). Lipase-catalyzed reactions of fatty acids with glycerol and acylglycerols. *Appl. Microbiol. Biotechnol.* **30**, 332–336.

Silver, M., and Karel, M. (1981). The behavior of invertase in model systems at low moisture contents. *J. Food Biochem.* **5**, 283–311.

Silvius, J. R., Read, B. D., and McElhaney, R. N. (1978). Membrane enzymes: Artifacts in Arrhenius plots due to temperature dependence of substrate-binding affinity. *Science* **199**, 902–904.

Smolensky, D. C., Raymond, W. R., Hasegawa, S., and Maier, V. P. (1975). Enzymatic improvement of date quality. Use of invertase to improve texture and appearance of "sugar wall" dates. *J. Sci. Food Agric.* **26**, 1523–1528.

Soliman, F. S., and van den Berg, L. (1971). Factors affecting freezing damage of lactic dehydrogenase. *Cryobiol.* **8**, 73–78.

Spikes, J. D. (1981). The sensitized photooxidation of biomolecules. *In* "Oxygen and Oxy-Radicals in Chemistry and Biology" (M. A. J. Rogers and E. L. Powers, eds.), pp. 421–424. Academic Press, New York.

Steinbuch, E. (1976). Technical note: Improvement of texture of frozen vegetables by stepwise blanching treatments. *J. Food Technol.* **11**, 313–316.

Szyperski, R. J., Hamann, D. D., and Walter, W. M. (1986). Controlled alpha amylase process for improved sweet potato puree. *J. Food Sci.* **51**, 360–363, 377.

Tome, D., Nicolas, J., and Drapon, R. (1978). Influence of water activity on the reaction catalyzed by polyphenol oxidase (E.C. 1.14.18.1.) from mushrooms in organic liquid media. *Lebensm. Wiss. Technol.* **11**, 38–41.

Tong, M. M., and Pincock, R. E. (1969). Denaturation and reactivity of invertase in frozen solutions. *Biochemistry* **8**, 908–913.

van den Berg, L. (1968). Physiochemical changes in foods during freezing and subsequent storage. *Recent Adv. Food Sci.* **4**, 205–219.

van den Berg, L., and Soliman, F. S. (1969). Composition and pH changes during freezing of solutions containing calcium and magnesium phosphate. *Cryobiol.* **6**, 10–14.

Whitaker, J. R. (1972). "Principles of Enzymology for the Food Sciences." Marcel Dekker, New York.

Williams, D. C., Miang, H. L., Andi, O. C., Pangborn, R. M., and Whitaker, J. R. (1986). Blanching of vegetables for freezing—Which indicator to choose. *Food Technol.* **40(6)**, 130–140.

Yamagata, H., Ueno, S., and Iwasaki, T. (1989). Isolation and characterization of a possible native cucumisin from developing melon fruits and its limited autolysis to cucumisin. *Agric. Biol. Chem.* **53**, 1009–1017.

Yancey, P. H., Clark, M. E., Hand, S. E., Bowlus, R. D., and Somero, G. N. (1982). Living with water stress: Evolution of osmolyte systems. *Science* **217**, 1214–1222.

Zaks, A., and Klibanov, A. M. (1984). Enzymatic catalysis in organic media at 100°C. *Science* **224**, 1249–1251.

CHAPTER 4

Modern Methods of Enzyme Expression and Design

JO WEGSTEIN and HENRY HEINSOHN

I. Summary

Until the advent of genetic engineering, enzyme producers were limited in their ability to produce innovative products for the marketplace. They were constrained to isolate enzymes from organisms approved for the food industry. The desired characteristics could be enhanced only using classical mutagenesis techniques. When these methods failed, no alternatives were available. Commercialization depended on incremental yield improvements gained by continuous programs of strain development. The ability to use recombinant DNA (rDNA) techniques has removed many of these barriers.

Enzyme producers recognized early the potential for commercialization of new products using genetic manipulation. They worked with a wide variety of single-celled organisms that were simpler and, thus, easier to understand than the higher orders of plants and animals. The organisms already were well characterized for growth and expression rates. Short life cycles allowed rapid testing. These systems were ideal for genetic manipulation using rDNA techniques. Genetic engineering, combined with an understanding of biocatalysis to predict alterations for enzyme improvements, is revolutionizing the production and use of enzymes in the marketplace.

Offering a recombinantly produced product represents the culmination of a long and complex effort on the part of a multitude of disciplines: molecular and microbiology, X-ray crystallography, enzymology, protein and organic chemistry, biochemistry, fermentation and formulation engineering, assay chemistry, and technical service/applications, marketing, and sales (Edelson, 1989). Because of the variety of disciplines required, a "critical mass" is needed to innovate products successfully and bring them to market. The continued proliferation of novel enzyme products requires development of core technologies so complex and expensive that they can be justified only if they can be used in multiple applications. In addition, companies pursuing rDNA technology must consider regulatory issues, ownership protection, and consumer acceptance.

The purpose of this chapter is to review the current art and science of enzyme production in a simplified and understandable manner; to describe genetically engineered (GE) products available in the marketplace; and to present the power and implications of genetic engineering for new food-related applications.

II. Background

A. Historical Perspective

In the late 1800s, discoveries in biology, chemistry, and enzymology culminated in what is now recognized as the beginning of the biotechnology age. On one hand were researchers who concentrated on understanding the mechanics of life. On the other were chemists who sought to elucidate the structures of organic products. At that time, these disciplines were considered vastly different.

The first theories concerning life functions were drawn from observations of natural systems. Two of the earliest contributors were the German microscopists Matthias Schleiden and Theodor Schwann, who proposed that all plants and animals are composed of fundamental units called cells. In 1865, Gregor Mendel proposed the basic rules of heredity, that traits were controlled by elements we call genes. As early as 1908, the gene–enzyme relationship was postulated by A. E. Garrod, who suggested that genes work by controlling the synthesis of specific enzymes.

As analytical procedures improved, progress was made in defining the actual structure of proteins. Early in the 20th century, German chemist Emil Fischer established that proteins were polymers of nitrogen-containing organic molecules called amino acids. Proteins could contain up to 20 different amino acids in a highly varied and irregular order, but any given type of protein maintained a unique amino acid sequence. If the sequence determined the functionality of the protein, then a mechanism for choosing and ordering the subunits was needed in the cell.

Since all enzymes were shown to be proteins, the key question became how genes participated in the synthesis of enzymes. Although it was known that genes possessed a unique molecular constituent, deoxyribonucleic acid (DNA), there was no way to show that DNA carried genetic information. In 1953, James D. Watson and Francis Crick sorted out the molecular structure of DNA. More importantly, they recognized and proposed its function: the storage and replication of genetic data in the sequence of the base pairs. The tools were close at hand for manipulating genes in organisms, producing proteins of interest, and altering them for different characteristics. Methodologies were soon developed to implement the postulated ideas.

Our understanding of enzyme catalysis also advanced. In the early 1950s, Frederick Sanger established the first definitive order of amino acids in a protein by sequencing insulin. In 1967, D. C. Phillips and colleagues, knowing the primary structure and the three-dimensional conformation, proposed a precise chemical mechanism for the hydrolysis of a substrate by an enzyme (lysozyme). With succeeding work on other enzymes, this effort indicated that the chemical mechanisms of enzyme-catalyzed reactions could be understood.

B. Simplified Model

Crick, in 1958, hypothesized a central dogma: that chromosomal DNA functions as the template for synthesis of ribonucleic acid (RNA) molecules (transcription), which move to the cytoplasm where they determine the arrangement of amino acids in proteins (translation):

$$\text{Replication} \left(\text{DNA} \xrightarrow{\text{Transcription}} \text{RNA} \xrightarrow{\text{Translation}} \text{Protein} \right.$$

The arrows indicate the direction for transfer of genetic information. The arrow encircling DNA shows that it is the template for its own replication. The arrow between DNA and RNA shows that RNA molecules are made on (transcribed off) DNA templates. Correspondingly, all protein sequences are dictated by (translated on) RNA templates. Thus, the genetic code, the relationship between nucleotide sequence and amino acid sequence, has been deciphered and appears to be structurally and functionally identical in all living organisms: microbes, fungi, plants, and animals (Watson *et al.*, 1987).

The DNA molecule consists of two molecular chains coiled in a helix and held together by hydrogen bonds (a weak noncovalent chemical bond). The subunits of each chain are nucleotides, which contain phosphoric acid and the sugar deoxyribose coupled to one of four nitrogenous bases: adenine (A), guanine (G), cytosine (C), and thymine (T). In the double helical DNA molecule, A always pairs with T and G always pairs with C in the opposite chain. Watson and Crick first postulated the significance of this base pairing: the sequence of one strand exactly defines that of its partner and allows self-replication.

A gene is a DNA segment that carries information for synthesis of a specific RNA molecule and, therefore, usually of a specific protein (see

Fig. 1). The gene defines the amino acid sequence in the form of triplets of bases, called codons. The gene also contains various signal or control regions, called promotors and terminators, that are involved in the regulation and expression of the gene product (when and how often the RNA corresponding to the protein will be synthesized). Start and stop codons define the beginning and end of the amino acid sequence. The terminator region defines the end of the gene in the DNA molecule.

Protein production requires several steps. First, the double-stranded DNA unwinds and separates in the vicinity of the gene. Specific enzymes move along a single strand, "reading" the bases. Using base pairing, they construct another nucleic acid molecule, RNA, as a complementary copy of

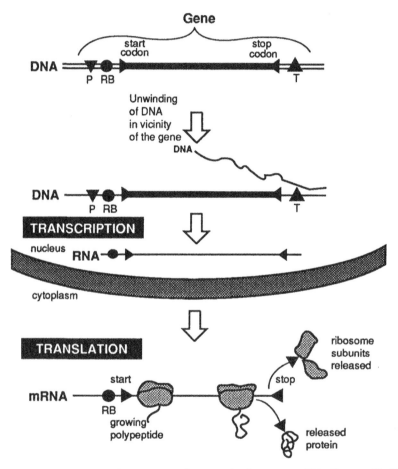

Figure 1 Expression of a gene to produce protein. P, promoter; RB, ribosome binding site; T, terminator.

the information in the original sequence (a process called transcription). RNA is chemically similar to DNA, but contains the sugar ribose rather than deoxyribose and the base uracil (U) rather than thymine. Once RNA containing genetic information from DNA is constructed, it moves from the nucleus to the ribosomes located in the cell cytoplasm and functions as a template for protein synthesis (translation). When the RNA reaches the cytoplasm, it is called messenger RNA (mRNA). The ribosomes attach themselves at the ribosome binding sequence and synthesize the protein from individual amino acids available in the cytoplasm.

C. Classical Mutagenesis

Since DNA contains all the information required for proper growth and maintenance of the cell, copying DNA must be essentially foolproof to insure correct reproduction and replication of the organism. Mutation is a change in DNA, a variation in the nucleotide sequence, that is passed to succeeding generations. This process happens continuously in nature at very low levels. In the laboratory, mutations can be induced by:

- errors in base pairing at transcription or translation
- interference with the enzymes that synthesize or repair DNA
- mutagens, chemicals that interfere with DNA function (an analog incorporated into DNA instead of a base)
- exposure to ultraviolet radiation that is absorbed by the DNA and damages it
- direct chemical attack on the DNA (Bailey and Ollis, 1986)

A program of directed mutation plays an important role in enzyme production improvement. Such a program is very effective and used extensively. Most production organisms were developed this way. Mutagenesis programs in the past have given a steady increase in productivity over time by selecting higher yielding mutants or preferential producers of a selected activity.

The mutagenesis process involves stressing an organism under near-fatal conditions and growing the survivors, which often show a relatively high rate of mutation. A successful program requires a predictive assay for identification and selection of promising mutants as well as rapid laboratory methods. Modern manufacturers often use automated methods with robots performing the tedious procedures.

The main disadvantage of classical mutagenesis is that it is random and time-consuming. Organisms can be manipulated and the selection process strictly adhered to, and still the objective is not achieved. Also, a characteristic cannot be manipulated if it is not expressed in the organism at all. Classical methods combined with genetic engineering offer a more complete approach to enzyme improvements.

III. Genetic Engineering

A. Recombinant DNA Techniques

In the late 1970s, a set of methodologies emerged that permitted isolation of discrete DNA segments that could then be inserted into living cells. The key to this method was the discovery of restriction and ligase enzymes to cut and join DNA. The ability to manipulate and recombine DNA allowed a more focused approach to mutagenesis, and created the potential for the manipulation of industrial microorganisms to synthesize valuable new enzymes or food ingredients.

1. Gene Cloning

Figure 2 summarizes the steps in cloning (producing an identical copy of) a gene in a new host organism. The remainder of this section provides a more detailed description of those steps.

Step I. Obtain an enzyme of interest with desired characteristics Often enzymes with characteristics needed for a specific application are produced at low levels, in obscure organisms unsuitable for modern fermentation processes, or in hosts unacceptable for food applications. Such enzymes are likely candidates for cloning into production organisms.

Step 2. Determine the amino acid sequence of the enzyme Automatic analyzers are now available that perform the chemical analysis steps once done manually. Phenylisothiocyanate is used to react with the N terminal of the amide bond. Trifluoroacetic acid is added to cleave one amino acid residue at a time, releasing a cyclic compound. This isothiocyanate derivative is analyzed, usually by high performance liquid chromatography (HPLC). The protein sequence then can be read by comparing the results against standards. Because of degradation of the protein in the system over time, less than 30 amino acids in a sequence can be analyzed in a single run. Overlapping sections of the enzyme protein must be analyzed so the sequence can be assembled. The process often takes weeks of effort; a degree of luck also helps accomplish the complete sequencing of an enzyme.

In the DNA molecule, three bases, called a codon, encode a particular amino acid. Using Table I, a DNA sequence can be suggested to match an enzyme's protein sequence. The start codon is always AUG, which also encodes methionine. (All polypeptides start with methionine.) Three separate codons (UAA, UAG, and UGA) serve as stop signals. Unfortunately, many amino acids are specified by more than one codon. This characteristic of the code is called degeneracy and makes tracing backward to find the correct DNA sequence of a gene difficult. The number of possible sequences can become astronomical.

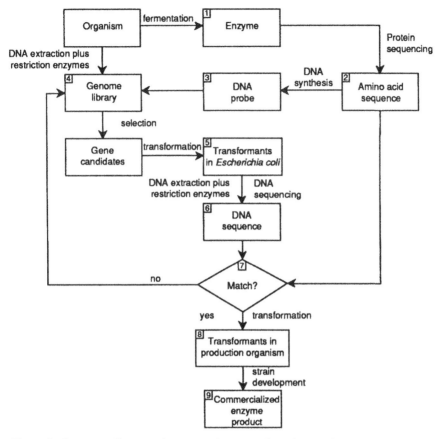

Figure 2 Summary of steps to clone a gene into a new host. See text for explanation.

Some mutations (in nature or induced in a laboratory) may result in codon changes so a different amino acid is inserted into a protein. One base change, from UUU to UUG, changes the resultant amino acid from phenylalanine to leucine. A change from UCU to UCG, however, still specifies serine. Once in a great while, mutagenesis may improve catalytic activity and benefit the organism. Other changes give rise to stop codons that terminate proteins prematurely and confer no benefit at all.

Step 3. Construct a DNA probe A probe is a single strand of DNA that encodes a section of the desired gene. A probe is made by synthesizing a short segment of DNA using radiolabeled nucleotides. Since the probe will be used to find a specific gene in a pool of DNA, a sequence is chosen to match a unique section of the gene of interest. Sequences containing codons for methionine or tryptophan are often used because only one codon specifies each of these amino acids. Sometimes sequences can be guessed because of a species tendency to use some codons preferentially.

TABLE I
The Genetic Code[a,b]

		Second Position			
		U	**C**	**A**	**G**	
First Position	**U**	UUU ⎤ Phe UUC ⎦ UUA ⎤ Leu UUG ⎦	UCU ⎤ UCC ⎥ Ser UCA ⎥ UCG ⎦	UAU ⎤ Tyr UAC ⎦ UAA Stop UAG Stop	UGU ⎤ Cys UGC ⎦ UGA Stop UGG Trp	U C A G
	C	CUU ⎤ CUC ⎥ Leu CUA ⎥ CUG ⎦	CCU ⎤ CCC ⎥ Pro CCA ⎥ CCG ⎦	CAU ⎤ His CAC ⎦ CAA ⎤ Gln CAG ⎦	CGU ⎤ CGC ⎥ Arg CGA ⎥ CGG ⎦	U C A G
	A	AUU ⎤ AUC ⎥ Ile AUA ⎦ AUG Met	ACU ⎤ ACC ⎥ Thr ACA ⎥ ACG ⎦	AAU ⎤ Asn AAC ⎦ AAA ⎤ Lys AAG ⎦	AGU ⎤ Ser AGC ⎦ AGA ⎤ Arg AGG ⎦	U C A G
	G	GUU ⎤ GUC ⎥ Val GUA ⎥ GUG ⎦	GCU ⎤ GCC ⎥ Ala GCA ⎥ GCG ⎦	GAU ⎤ Asp GAC ⎦ GAA ⎤ Glu GAG ⎦	GGU ⎤ GGC ⎥ Gly GGA ⎥ GGG ⎦	U C A G

(Right side label: **Third Position**)

[a] From J. D. Watson, *et al.* (1987). *Molecular Biology of the Gene,* 4th Ed., The Benjamin/Cummings Publishing, Menlo Park, California. Reprinted with permission.
[b] Abbreviations for amino acids: glycine, Gly; alanine, Ala; valine, Val; leucine, Leu; tryptophan, Trp; methionine, Met; phenylalanine, Phe; isoleucine, Ile; serine, Ser; threonine, Thr; tyrosine, Tyr; asparagine, Asn; glutamine, Gln; aspartic acid, Asp; glutamic acid, Glu; lysine, Lys; arginine, Arg; histidine, His; cysteine, Cys; proline, Pro.

Step 4. Screen a genome library A genome library is a pool of DNA segments that is likely to contain the gene of interest. Using standard procedures, DNA is extracted from the organism that makes the desired enzyme. The DNA is cut using restriction enzymes. Segments too small to contain a full gene are discarded. The mix is separated by size using gel electrophoresis. A quantity of the DNA probe is added and the material is warmed. Under controlled heating, the double-stranded DNA will "melt" and disassociate. After cooling, it will come back together; however, the single-stranded probes will bind to the sections they match. If bands can be detected autoradiographically, they contain sequences similar to those of the gene being sought. The band is cut out and the DNA it contains is replicated.

Probing becomes more difficult when dealing with DNA from eukaryotic cells because of introns (noncoding sections of DNA in a gene). The presence

TABLE II
Selection of Transformed Mutants

Marker	Description
Antibiotic resistance	Addition of a gene to produce a protein that hydrolyzes an antibiotic and allows the organism to grow; candidates are cultured in the presence of the antibiotic
Supplementation of an auxotroph	Addition of a gene to supply a needed growth factor (a vitamin or specific amino acid) that compensates for a deficiency in the host whose gene has been deleted; candidates are cultured in the absence of the growth factor
Hydrolysis of a specific sugar	Addition of a gene to hydrolyze a specific carbon source (e.g., lactose); candidates are cultured on the single carbon source

of introns creates more complexity when working with higher species; this is one reason bacteria are preferred.

Step 5. Transformation The process of genetic transfer of DNA is called transformation. DNA fragments are inserted into plasmids or phages, which are used because they are easy vectors with which to transfer foreign DNA into intact cells. The DNA of interest, inserted into the plasmid or phage, is mixed with organisms whose cell walls have been made receptive by chemical treatment. These organisms are usually special strains of *Escherichia coli* that have been characterized and modified carefully so that they can be transformed easily. The DNA enters some of the organisms and is replicated with the host DNA. The transformation process is random and is not successful for every organism in the experiment. Therefore, an important step is screening the population for the organism that has taken up the new DNA from among the many who have not.

Tranformed organisms are selected by adding markers to the gene before transformation. Markers are sections of DNA inserted with the desired gene to make selection and detection of successfully transformed organisms a straightforward process. Often the marker confers an ability to survive under specific growth conditions and must be used with specially modified hosts that lack the specific trait. The markers will be removed before transformation of the gene into a production organism. Table II shows some common markers. The transformed organisms, which contain the DNA of interest, are grown to provide a source of recombinant DNA for further testing and manipulation.

Step 6. Sequencing the chosen DNA The ability to sequence DNA was one of the more important breakthroughs that made genetic engineering practical. Two methods for accurately establishing the order of the four bases A, G,

C, and T along a given DNA segment have been developed. DNA strands are separated (since only one strand needs to be sequenced) and labeled at one end by enzymatically removing the phosphorus atom and replacing it with a radioisotope of phosphorus. The solution of single-stranded DNA is then divided into four pools.

In Maxam and Gilbert's method, each pool is treated with chemical reagents, for example, dimethysulfate or hydrazine, using conditions that selectively destroy one or two of the four nucleotides in each pool. The conditions are adjusted so the reaction does not proceed to completion; at most one nucleotide per fragment is destroyed. These reactions result in four pools of radioactively tagged strands of all possible lengths, ending in the nucleotide destroyed. In practice, nucleotides destroyed are G, G + A, T + C, and C, each in a separate pool.

Sanger's method makes use of the enzyme DNA polymerase to produce fragments ending in specific bases. The pieces of single-stranded DNA serve as templates for synthesis of complementary strands of DNA. The nucleotides necessary to form the complementary strand are added to each of the four pools along with the polymerase enzyme. In addition, a small amount of a nucleotide analog, a radioactive 2′,3′-dideoxynucleotide, containing one of the four base types is added to each pool. The concentration is controlled so only one base will be incorporated per strand. The polymerase builds the complementary strand along the DNA fragments until one of the base analogs is inserted in the chain. Further strand elongation is blocked, since the analog interferes with placement of the next nucleotide. The result is four pools of radioactive complementary strands of DNA of all possible lengths, ending in a specific base.

Gel electroporesis is used in both methods to separate each pool by length, with a size resolution of a single base. The gels are slabs of poly acrylamide with sample applied at one end. An electric potential causes the fragments to move through the gel in a track with the smallest fragment moving most quickly. The gel is then used to expose X-ray photographic film. The resulting autoradiograph shows four tracks of bands. Since the gel has a resolution of one base, and all possible fragment lengths exist in the four pools, the DNA sequence can be read directly from the gel (see Fig. 3). Because of the ease of the enzymatic method and the ready availability of dideoxynucleotide reagents, the Sanger method is becoming the method of choice for sequencing DNA.

Step 7. Verify the DNA has the right sequence to make the enzyme The DNA sequence is compared with the known protein sequence of the desired enzyme. If the sequence does not match, a new genome library must be screened. The DNA fragment is examined to determine if the entire gene is present. Start and stop codons allow recognition of whether the gene is intact. With some additional effort, the gene can be isolated using restriction enzymes to cut and trim it until only the coding sequence remains.

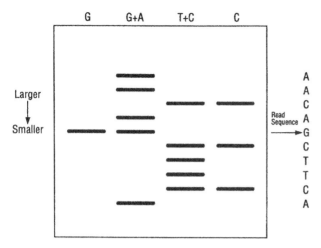

Figure 3 Reading the sequence of a DNA fragment. See text for explanation.

Step 8. Transform the gene into an appropriate host A toxin-free organism with a demonstrated high production capacity for foreign gene products is chosen. The highly produced native genes are deleted, while the transcriptional, translational, and secretory control elements are retained. The gene of interest is transferred into the chosen production organism. Transformed organisms are selected (see Step 5) and cultured. The expressed enzyme is purified and analyzed using general protein chemistry methods to insure that it is identical to the native enzyme.

Step 9. Improve the yield to levels acceptable for commercialization Living cells possess intricate control systems to insure efficient use of material and energy resources. Often, cell regulatory elements limit the amount of enzme produced, making commercialization economically unfeasible. However, changes in the control systems can be used to boost yield. This area is of great importance to enzyme producers and will be discussed in greater detail in a later section.

Instability of recombinant production strains is always a possibility. Monitoring of stock cultures is necessary to avoid costly production failures.

2. Complementary DNA

Another option for obtaining DNA suitable for enzyme production is the construction of complementary or copy DNA (cDNA). Messenger RNA is a complementary copy of a DNA sequence. When an organism is actively making a protein, reasonable amounts of mRNA are available in the cytoplasm and can be recovered. The mRNA can be used as a template to make a single strand of DNA using reverse transcriptase enzymes. Polymerase enzymes add the complementary strand, yielding a DNA fragment that encodes the desired gene. This DNA is not complete because it does not have introns or transcription signals (promotors), but often cDNA is the only alternative available for generating a gene.

3. Site-Directed Mutagenesis

The commercial advantages of transferring a gene from one organism to a more productive one are obvious. The first products made available to consumers are of this type. The techniques can be used to manipulate a segment of DNA of any length (even a codon, to change one amino acid in a protein sequence), although the benefit of small changes was not immediately obvious. Bott *et al.* (1988) discuss the results of substituting a single amino acid for the glycine at the 166 position (in a peptide chain of 275 amino acids) in subtilisin from *Bacillus amyloliquefaciens*. The 166 position is not part of the catalytic site, but is implicated in substrate binding. Figure 4 shows the activity of the protease mutants containing various amino acids as well as that of the protease containing glycine. All 19 mutants retained measurable enzyme activity, but differed in activity from the naturally occurring enzyme. These results suggest the potential for the deliberate design of enzymes for specific applications.

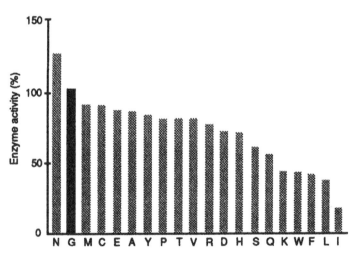

Figure 4 Substitution of various amino acids in subtilisin. The glycine at the 166 position in subtilism from *Bacillus amyloliquefaciens* was substituted with various amino acids using recombinant DNA techniques. The resulting protease mutants were assayed for their ability to hydrolyze casein at specific conditions. The activity of the native enzyme (glycine-166) is set to 100%. Abbreviations: G, glycine; A, alanine; V, valine; L, leucine; W, tryptophan; M, methionine; F, Phenylalanine; I, isoleucine; S, serine; T, threonine; Y, tyrosine; N, asparagine; Q, glutamine; D, aspartic acid; E, glutamic acid; K, lysine; R, arginine; H, histidine; C, cysteine; P, proline. (Reprinted with permission from Bott *et al.*, 1988.)

B. Enzyme Design

Organisms evolve functionally optimal enzyme systems for their environment because doing so confers a survival advantage. However, the optimal enzyme system for chemical or food processes may require characteristics not generally found in nature. The greatest opportunity for improvement, therefore, lies in engineering those properties that have not been selected naturally.

1. Structure–Function Relationships

Subtilisins are among the most highly studied and well-understood enzymes. They are secreted in large amounts by a variety of *Bacillus* species. The protein sequences from several commercially important species and their three-dimensional structures have been determined. The enzyme mechanism is known, and the catalytic and substrate binding residues have been identified. Extensive kinetic and chemical modification and crystallographic studies have been performed. Subtilism is an ideal model for the application of protein engineering to better understand the theoretical basis for catalytic systems.

X-ray crystallography provides a valuable tool for improving our understanding of structure–function relationships. The crystal structure can be incorporated into a three-dimensional modeling system that enables enzymologists to study the active site, the binding regions, how substrates fit, the possible conformations of the enzyme, and stearic hindrances of suggested changes. The active sites of similar enzymes from different organisms can be overlapped to see structural similarities that may correspond to similarities in function. By understanding the relationship between catalytic properties and structure, improvements in function eventually may be predicted for suggested changes in the structure (Wells *et al.*, 1987).

Over 400 site-specific structural changes have been made in subtilism to verify computer-modeled predictions of improvements to enzyme properties. Some results were predictable; others still need explanation. Almost every property of the enzyme has been altered by protein engineering: catalytic site, substrate specificity, pH/rate profile, and stability to oxidative, thermal, and alkaline inactivation. Examples from this work are presented in Figs. 5, 6, and 7 (Wells and Estell, 1988).

2. Random Mutagenesis

Enzymes are large and complicated catalysts. The three-dimensional structures of fewer than 100 (of ~10,000) have been determined to date. Painstaking study, prediction, and expression as described for subtilism is not always feasible. Researchers with limited resources perform random mutagenesis on a selected gene followed by expression and testing. When little is know about the enzyme structure, numerous mutagenized enzymes are generated rapidly and screened for indication of change in the characteristics.

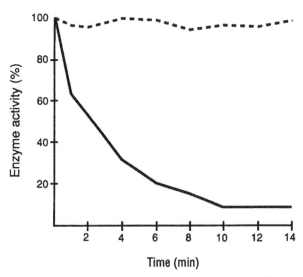

Figure 5 Engineering resistance to oxidizing agents. Effect of 1 *M* hydrogen peroxide (bleach) on the activity of the parent subtilisin, methionine-222 (——) and engineered mutant alanine-222 (– –). The mutant has dramatically improved resistance to the oxidizing agent. (Reprinted with permission from Estell *et al.*, 1985.)

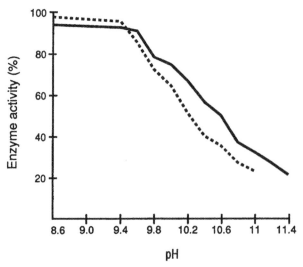

Figure 6 Engineering tolerance to alkaline conditions. Asparagine-166 subtilisin (——) works more efficiently at higher pH than the parent enzyme, glycine-166 (– –). (Reprinted with permission of Genencor, 1988.)

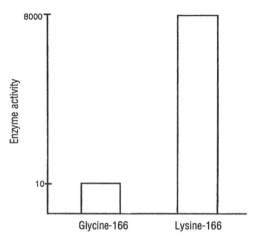

Figure 7   Engineering higher efficiency, lysine-166 subtilisin hydrolyzes a specific substrate 800 times faster than the parent enzyme (glycine-166). (Reprinted with permission of Genencor, 1988.)

The interesting mutants are sequenced to suggest the region in the enzyme that influences the characteristic. Although still a random process, this method is an improvement on classical mutagenesis, since it focuses on the gene of interest. Because this approach is more precise, fewer resources and less time are required to yield an acceptable result.

Successful enzyme engineering depends on the ability to screen for mutants. Development of a screen for rapid and accurate selection is central to the process of mutagenesis. Temperature stable mutants are among the first mutants commercialized from wild-type enzymes, because testing the enzymes at higher temperatures was an easy screen. Temperature stability demonstrated in the laboratory also scaled to the plant level. Some properties pose more difficulty in screening. One example is alteration of pH optima, because enzyme activities depend on the assay conditions (buffers, their ionic strength, and specific substrate conditions). Assays may not represent the variety of end-user processing conditions and substrates. Therefore, screening for commercializable candidates requires extensive testing, which is time and resource consuming.

Table III shows some common methods for detection and selection of mutants that allow the researcher to find the transformed organism. If selection methods can be employed during the growth phase, the number of assays for detection of the desired enzyme can be reduced significantly.

C. Metabolic Pathway Modification

Pathway engineering is the process of modifying metabolic pathways in organisms by changing or adding enzymes to an existing pathway or by creating

TABLE III
Some Methods of Detection and Selection of Mutants

Characteristic	Selection and detection
Higher enzyme yield	Organisms are plated on agar with dye-linked substrate; higher producers give larger clearing zones; enzymes are tested for desired activity
Higher temperature stability	Enzymes are assayed for activity at desired temperature
Different pH optima	Enzymes are assayed for activity at the desired pH using a variety of buffers, substrates, concentrations, etc.
Ability to use a cheaper carbon source	Organisms are cultured only on the new carbon source; enzymes are assayed for desired activity
Resistance or tolerance to inhibitory chemicals (e.g., sulfite tolerance)	Organisms are cultured in the presence of the chemicals; enzymes are assayed for desired activity

new ones. Food scientists can manipulate intermediate steps in the production of certain foods to alter final product characteristics. Many examples of pathway engineering encompassing numerous industries are recognized (see Table IV).

A specific example is the metabolic pathway engineering of *Lactobacillus bulgaricus* as a means of preventing postfermentation acidification of yogurt. Like many perishable foods, yogurt develops off-flavors during refrigerated storage. In this case, the phenomenon responsible is the continued production of lactic acid by the organism *L. bulgaricus* used in the production of yogurt. Specifically, lactose in the yogurt is converted to glucose and galactose by the enzyme β-galactosidase. Glucose, used by the organism to generate energy, results in the production of lactic acid, the metabolic end product, which is secreted into the yogurt and changes the flavor.

TABLE IV
Examples of Potentials for Pathway Engineering[a]

Introduction of nitrogen-fixing capabilities in certain plants to reduce/eliminate requirements for nitrogen-based fertilizers

Design of metabolic systems capable of degrading certain hazardous waste compounds

Control of methane-producing metabolism in bacteria to make fuel from organic material/waste

Introduction of gene encoding the enzyme necessary for ascorbic acid precursor synthesis in microorganisms

Manipulation of pathways for the creation of phage-resistant microbial strains used by the food industry

Alteration of sugar utilization pathways in microbes to allow fermentation of novel products

[a] Reprinted by permission from Yoast *et al.*, 1989.

If lactic acid production were limited, yogurt would be stable under refrigeration. Yoast *et al.* (1989) describe the *in vitro* random mutagenesis of the *L. bulgaricus* β-galactosidase gene using chemical mutagens. After the plasmid was cloned into an *Escherichia coli* expression system, enzymes were produced that had randomly changed amino acid sequences. They were screened to select for those that showed normal activity at yogurt production temperature (37°C), but little or no activity at refrigeration temperatures (<10°C).

A number of mutant cold-sensitive enzymes were identified to replace the wild-type *L. bulgaricus* β-galactosidase. The genetically engineered host organism will exhibit little or no ability to acidify yogurt during refrigerated storage, resulting in longer shelf life, because no off-flavor develops.

D. Engineering for Production Improvement

Current enzyme production encompasses fermentation, recovery, purification, and formulation technologies. An in-depth understanding of the fermentation parameters for each organism is required to maximize growth and enzyme production. Submerged-culture batch fermentation predominates, but fed-batch and continuous fermentation are taking hold rapidly. A different set of conditions generally is required for efficient recovery and purification of each enzyme. As a result, recovery processes frequently are designed as dedicated trains, reducing plant utilization and increasing capital investment.

Use of recombinant techniques has begun to influence how enzymes are manufactured. Innovations in rDNA techniques and protein engineering allow producers to design a strategy for future production strains: the development of well-defined and well-understood bacterial and fungal hosts for overproduction of foreign gene products. The ability to express new proteins in "workhorse" organisms using common induction systems allows use of fermentation processes similar to existing schemes and rapid scale-up of new products. Such consistencies will expedite transition from the laboratory to the marketplace.

Knowledge of repression and induction systems permits generation of targeted alterations within the organism to develop more desirable fermentation processes. Organisms can be modified to contain expression systems that are triggered by temperature shifts, certain growth rates, low oxygen concentrations, or inexpensive inducing agents, as well as being insensitive to feedback inhibition from a build-up of media components over long fermentation cycles. Protein will be expressed during earlier growth phases (necessary for continuous fermentation processes in fungi). Use of rDNA techniques allows dramatic changes in fermentation process design and economics.

Several other molecular biology options exist to boost productivity. Yields can be increased by inserting multiple genes copies with a combination of promotors to achieve optimal secretion. Nonactive sections of an enzyme can be deleted to reduce the biological burden to produce it in the cell.

Sporulation controls can be modified to make an organism produce longer than it would in its natural cycle.

Synthesis of extra protein represents a waste of valuable energy and intermediary metabolites for the host. In hyperproducing organisms, the new gene product may constitute a large portion of the total cellular protein. Since this product serves no useful purpose for the organism, diversion of so much biosynthetic activity seriously limits the capacity of the cell for growth and even survival. New fermentation schemes and higher yields make organisms more sensitive to processing excursions. These organisms will demand a higher level of process control than previously required.

Alterations can be made in the protein products themselves to improve production. Yields of a protease that autolyzes during fermentation and recovery can be improved by shifting the pH optimum of the enzyme away from the pH used during production. Changes in enzyme stability to temperature or certain compounds (oxidants) could allow wider selection of fermenter growth conditions. Amino acid substitutions that alter the surface charge may allow more efficient extraction, for example, by enhancing ultrafiltration membrane efficiency. Production strategies that incorporate improved enzyme characteristics for end-user requirements and increased yield for improved economics, and that better utilize recovery and purification equipment, will be a major new focus for rDNA technology application. (Arbige and Pitcher, 1989).

IV. Commercialization of Genetically Engineered Products

In the biotechnology age, the ability to offer a new product to the marketplace successfully will require the combination of enzyme design and product information capabilities of enzyme producers, and application information, requirements, and acceptance of enzyme users and consumers. This section cites examples of products available in the marketplace and, for one specific case, describes some of the complexities in development and production.

A. Example

Genetically engineered (GE) products are now available in the marketplace and will continue to proliferate. Table V shows a listing of GE enzymes in commerce. This list excludes the large number of restriction enzymes and other small-scale enzymes currently in use. Products with applications for GRAS (generally recognized as safe) status with the United States Food and Drug Administration (FDA) have been noted.

An example of the application of genetic engineering to solve an enzyme supply problem is illustrated by developments in the production of milk-

TABLE V
Genetically Engineered Enzymes in Commerce[a,b]

Enzyme	Source	Host	Use	Commercial producer
Subtilisin	*Bacillus* sp.	*Bacillus subtilis*	Detergents	Genencor
Subtilisin	*Bacillus alcalophilus*	*Bacillus alcalophilus*	Detergents	Gist–Brocades
Hydrolase	*Bacillus* sp.	*Bacillus* sp.		Gist–Brocades
Lipase	*Humicola*	*Aspergillus* sp.	Detergents	Novo
α-Amylase[b]	*Bacillus megaterium*	*Bacillus subtilis*	Starch modification	Enzyme Biosystems
α-Amylase[b]	*Bacillus stearothermophilus*	*Bacillus subtilis*	Starch modification	Enzyme Biosystems
Chymosin[b]	Calf stomach	*Escherichia coli*	Cheese production	Pfizer
Chymosin[b]	Calf stomach	*Kluyveromyces marxianus* var. *lactis*	Cheese production	Gist–Brocades
Chymosin[b]	Calf stomach	*Aspergillus niger* var. *awamori*	Cheese production	Genencor

[a] Adapted with permission from Arbige and Pitcher, 1989.
[b] Products with applications for GRAS status.

clotting enzymes for the manufacture of cheese (Hayenga *et al.,* 1989). Historically, the prevalent milk coagulant has been calf rennet, derived from the fourth stomach of suckling calves. The principle active proteases in calf rennet are chymosin (EC 3.4.23.4) and pepsin (EC 3.4.23.1). These two enzymes are closely related—both initiate the clotting process in milk by cleavage of casein. Chymosin is produced predominantly only in the first weeks after birth; thereafter, it is replaced increasingly by pepsin. Of the two enzymes, chymosin is preferred by cheesemakers because of its limited proteolysis, leading to minimal bitterness in aged cheese.

High-quality calf rennet is in short supply because of a decline in the market demand for veal and an increase in cheese production. These conditions have led to unstable rennet prices. Alternative sources of milk-clotting enzymes have been offered commercially to meet market demands, including microbial coagulants derived from *Mucor miehei, Mucor pusillus,* and *Endothia parasitica.* However, unweaned calf rennet, specifically chymosin with very low levels of pepsin, is considered the preferred coagulant by the cheese industry worldwide.

Three manufacturers have announced commercial availability of recombinantly produced calf chymosin. Cloning and producing GE chymosin by controlled fermentation can provide a consistent supply of enzyme with properties analogous to those of the native chymosin. The proteins are identical, despite the fact that the enzyme is produced in three different microorganisms: the bacterium *Escherichia coli,* the yeast *Kluyveromyces marxianus* var. *lactis,* and the fungus *Aspergillus niger* var. *awamori.* Table VI summarizes the advantages and disadvantages of each expression system.

TABLE VI
Heterologous Expression Systems for Calf Chymosin[a]

Advantages	Disadvantages
Bacteria, especially *Escherichia coli*	
Well characterized production strain	Chymosin is deposited as insoluble refractile bodies and must be solubilized and renatured to be active
	Products from *E. coli* must undergo extensive purification prior to use in food
Yeast, such as *Saccharomyces cerevisiae*	
Chymosin is secreted as an active enzyme	Current published production levels are uneconomical
Safe production organism	
Filamentous fungi, such as *Aspergillus nieger* var. *awamori*	
Chymosin is secreted as an active enzyme	Low frequency transformation system
Safe production organism	

[a] Reprinted by permission from Yoast *et al.,* 1989.

Companies producing the enzyme commercially by genetic engineering submitted data to the FDA that demonstrated that GE chymosin is chemically and kinetically identical to the native chymosin and that all safety and good manufacturing requirements are met. An additional advantage is the absence of the contaminating pepsin. The food industry and consumers are the beneficiaries of a supply of consistently high quality product at a stable price.

The rest of this section describes the heterologous (cross-species) expression in a food-grade filamentous fungus of the gene encoding chymosin from the fourth stomach of a calf, and the subsequent processing of the gene product to a commercially useful product (Genencor, 1989).

Fungi produce prodigious amounts of protein under proper fermentation conditions. For example, glucoamylase in *A. niger* is induced by the presence of maltose or isomaltose and can be secreted in excess of 50 g/liter in batch fermentations regulated by starch induction. Until recently, however, very little was known about how to transform fungi and use them as hosts. New expression systems had to be identified and developed. Since the bovine promotion system was not needed, the gene could be constructed from RNA (cDNA). Chymosin was secreted using expression units that contain prochymosin (the inactive chymosin precursor) cDNA coupled to the transcriptional, translational, and secretory control elements of the *A. niger* glucoamylase gene (Cullen *et al.*, 1987). Protein expression was regulated by starch induction and production processing was similar to that of glucoamylase.

1. Transformation

Expression and secretion of chymosin involved these steps.

a. Development of a host strain with markers for transformation experiments *Aspergillus niger* var. *awamori* was chosen as a host because it is toxin free and a hyperproducer of glucoamylase. The organism was subjected to uv mutagenesis, which produced a mutation in the *pryG* gene, resulting in a requirement for uridine. Chemical mutagenesis of the same precursor produced a strain with arginine auxotrophy (inability to synthesize a needed growth factor) caused by a mutation in the *argB* gene. A double auxotrophic strain was selected from the progeny of a parasexual cross of the two single auxotrophic strains.

b. Deletion of the host's protease, aspergillopepsin A, which causes degradation of chymosin or produces bitter peptides in cheese To generate strains that were deficient in the production of aspergillopepsin A, a gene replacement strategy similar to that reported by Miller *et al.* (1985) was employed. Briefly, the aspergillopepsin DNA coding sequence was replaced by the *Aspergillus nidulans argB* gene and used for transformation of the double auxotrophic strain. Of the transformants selected, four lacked the aspergillopepsin gene. Absence of aspergillopepsin activity was verified by several tests.

c. Construction of a plasmid vector containing a selectable marker and an expression unit for prochymosin A vector called pGAMpR was constructed (Fig. 8) in which prochymosin cDNA sequences encoding the prochymosin B isozyme were fused to the last codon of the *A. niger* var. *awamori* glucoamylase gene. The selectable marker used in the vector was the *Neurospora crassa pyr*4 gene, which corrects for uridine auxotrophy when its gene product (the enzyme orotidine-5'-monophosphate decarboxylase) is expressed. The gene product of this construction is a fusion protein consisting of the *awamori* glucoamylase enzyme covalently coupled to prochymosin.

d. Transformation of the host strain with the expression vector and subsequent molecular analysis of the transformed strains Aspergillopepsin-deficient strains were transformed and a number of chymosin-producing strains identified and selected. The highest producing strains were analyzed and shown to contain the glucoamylase prochymosin expression unit integrated into the genome, possibly in multiple copies.

e. Mutagenesis and screening of improved strains The strains were subjected to further rounds of classical mutagenesis, screening, and selection for overproduction of chymosin. In addition, all possible production strains were tested carefully to insure that they produced no mycotoxins or antibiotics.

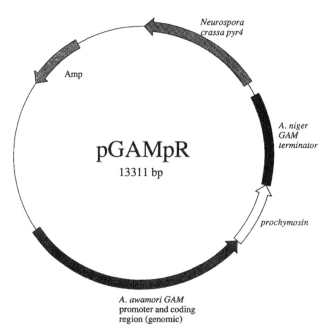

Figure 8 Map of the glucoamylase-prochymosin fusion vector. (Reprinted with permission from Genencor, 1989.)

2. Characterization

The enzyme product of the transformed strain is chymosin B, an isozyme containing aspartic acid at position 290 instead of glycine as found in chymosin A. Chymosin B has a lower specific activity than chymosin A but is reported to be more stable. Both isozymes are found in natural products.

Figure 9 illustrates the production of mature chymosin by the recombinant organism. Only mature chymosin and mature glucoamylase are observed in the fermentation medium. This suggests that the fusion protein undergoes an *in vivo* processing step, resulting in the cleavage of glucoamylase from prochymosin B and the subsequent self-activation of prochymosin to chymosin. The cleavage of the two enzymes can be effected either by prochymosin itself or by other endogenous *awamori* proteases.

Chymosin B was purified using standard biochemical methods. The enzyme was compared exhaustively to animal-derived chymosin using physical, immunological, and functional tests. All results confirmed there was no significant difference between native and recombinant chymosin expressed in *A. niger* var. *awamori*.

3. Fermentation and Recovery

Production of chymosin is accomplished by submerged fermentation. On completion, an *in situ* cell inactivation protocol is used in which the cells are killed before the fermenter broth leaves the contained fermenter. The pH of the broth is lowered by acid addition while maintaining normal agitation and aeration. After 1 h, viable cell counts are reduced to acceptable levels.

First, the broth is filtered to remove cells and other debris. Further processing is required because the harvested broth contains residual nutrients as well as endogenous enzymes produced by the organism. Two processing schemes were developed to recover high quality chymosin from the broth. The first consists of addition of sodium chloride followed by separation on a hydrophobic chromatography column. Chymosin binds to the resin while impurities pass through the column. The resin is washed and the chymosin eluted with a low ionic strength buffer. The product is then formulated, sterile filtered, and packaged.

The second process involves chemical extraction using a two-phase liquid-extraction system. Selected conditions allow efficient extraction of the chymosin into one phase, leaving the impurities in the other. The phases are separated by centrifugration. The chymosin-containing phase is passed over an ion exchange resin to bind the chymosin, which allows complete removal of the extraction phase chemicals. The resin is washed and the chymosin eluted. The product is then formulated, sterile filtered, and packaged. The finished product meets an extensive list of specifications chosen to insure that a consistent, safe, high-quality product is produced.

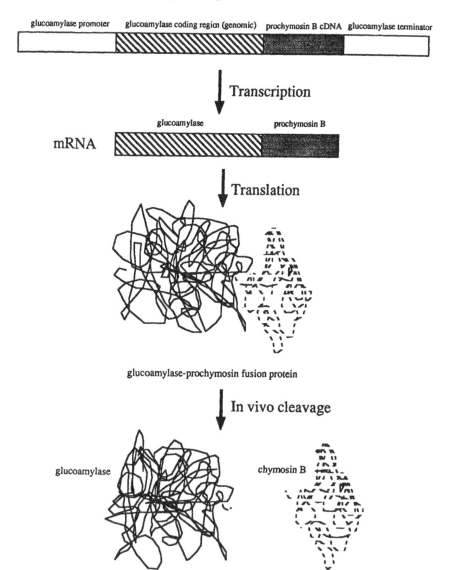

Figure 9   Schematic model for the production of chymosin from the glucoamylase-prochymosin expression unit. (Reprinted with permission from Genencor, 1989.)

B. Regulatory Considerations

The anticipated appearance in the marketplace of the first products made using rDNA techniques raised questions concerning whether or not new regulations were required for their sale. If new regulations were needed, the substantial time required might delay benefits that could be realized by the food industry. Also, the risk of investment for those companies actively pursuing the technology might be increased.

Food and food ingredients are regulated in the United States by the FDA under the Federal Food, Drug, and Cosmetic Act (1982). After careful deliberation by the FDA and other regulatory bodies, it was decided that current regulations were sufficiently comprehensive to apply to products involving rDNA technology (Office of Science and Technology Policy, 1986). In short, any product satisfying GRAS criteria could be eligible for GRAS status. Teague *et al.* (1990) give an overview of the interacting elements of the first GRAS petition to illustrate the process of safety evaluation by the industry and the FDA. Summaries of subsequent petitions are included also to elaborate on the differences in the rDNA technology and the safety evaluations applied. Teague and colleagues reported:

> In retrospect, the application of rDNA technology should not have generated undue concern about establishing the safety of food products. . . . [DNA], a chemical with no inherent activity of its own, is placed . . . where host organism enzymes will use it to synthesize proteins. . . . The source of the genetic information, be it natural or synthetic, prokaryote or eukaryote, closely related or distantly related to the chosen host, is not important because of the universality of the code contained in the [DNA] and the mechanisms for translating that code. . . . Safety evaluations have clearly indicated that enzyme producers can judiciously use rDNA technology to generate GRAS products. No host organism has acquired unexpected hazardous properties due to the expression of foreign DNA. The FDA's statement that no new regulations were required has been validated by their acceptance for filing of these various petitions. Since the regulations focus on the product safety and not on the method of production, the regulations and procedures proved durable to the introduction of this technology. (Teague *et al.*, 1990)

C. Ownership Protection

The use of biotechnology can be a justified expense only if the developer can prevent the invention from being used by the competition or, at least, cause the competition to pay the same costs to compete. For enzymes — as is the case to date for pharmaceuticals — the battle for market supremacy will be fought in the patent courts, not in the marketplace. Those companies with the best technology and the appropriate patent portfolio will be the leaders of tomorrow.

Industry has two alternatives for protecting valuable "intellectual property": trade secrets and patents. A "trade secret" is information, which, if kept secret, provides a competitive advantage to its owner. The advantage exists only as long as the owner actually keeps it secret and no other individual independently discovers and uses it. The decision thus depends on the ease with which the secret can be kept, meaning not only an assessment of a competitor's ability to invent the product on its own or to discover it through "reverse engineering," but also of the ability of employees and the company to maintain secrecy. Theoretically, secrecy can be maintained indefinitely if enough care is taken. Patents, on the other hand, are exclusive rights granted to an inventor to exclude others from making, using, or selling an invention and are enforceable against all other individuals, including independent discoverers, but are limited in the time of their enforceability (17 years in the United States). Industrial enzyme inventions, subject to either trade secrets or patents, can be of two types: compositions of matter (an enzyme, intermediate, or the like) and processes for making and using the composition of matter.

Trade secrets become the primary method of protection for enzyme production when the likelihood of another individual discovering the method or of the inventor policing others is low, or when the ability to engineer around the process cheaply and efficiently is high. Trade secrets are also sometimes useful for the protection of individual microbial strains that produce high amounts of protein product or have some unique characteristic, because of the need to deposit such strains publicly for patent purposes, which enables competitors to obtain and improve on these strains to avoid patent infringement. Trade secrets are also important when processes are fairly obvious but optimum production conditions are not. Since the patent must contain the best mode known of practicing the invention, the inventor may feel that a patent is not enough benefit to warrant disclosure.

Patents are useful for protecting novel enzymes and their intermediates as compositions of matter, as well as protecting certain processes of making and using the enzymes. Composition-of-matter patents have always been viewed as a major goal in research and development. They cover a product (regardless of the process used to make the product), are easy to police, and are very difficult (if not impossible) to engineer around without producing an entirely new compound.

Although process patents have been seen as relatively weak and often worthless, new patent legislation in the United States strengthens courts to enforce process patents. Process patents are useful as a roadblock when engineering around the process may be difficult, when the process is costly or time consuming, or when a core technology can be protected by the process patent. A key example is the basic patent that covers recombinant technology issued to Stanford University (U.S. Patent No. 4,237,224). The patent applies to all companies that want to produce products recombinantly in the United States. Stanford University licenses the company for an initial fee and receives royalties that currently amount to tens of thousands of dollars yearly, an amount that is expected to grow tremendously as more and more biotechnology products are introduced. All patent holders may not be so generous. There is no longer a requirement in the United States that a patent holder license a patent: the holder may exclude others from ever practicing the invention.

Until recently, enzymes were not subject to much active patenting. Most enzymes and their uses were known and process patents were viewed as weak. These factors combined to minimize the amount of research and development done on such systems. The advent of biotechnology changed the face of enzyme biochemistry. Now it is possible to produce mutant enzymes with new DNA sequences and processes for production that meet the United States Patent and Trademark Office requirements for novelty, nonobviousness, and utility.

The first patent issued for a novel mutant enzyme produced by genetic changes at certain positions in the DNA was U.S. Patent No. 4,760,025 issued to Genencor. The patent claims novel subtilisin enzymes with changes from a naturally occurring to a nonnaturally occurring amino acid residue at specific positions in the enzyme. These changes result in new enzymes with properties different from those of the original enzyme. Figure 5 shows data on the oxidative stability of one of those novel subtilisin enzymes.

Process patents also will become extremely important. Although a recombinant process may not be the only method by which to obtain a commercially desirable enzyme, it may be, by far, the cheapest method of producing a product, making it the only "commercially viable" method. The new aspects of process patent law also will discourage infringing uses of process patent technology, both in the United States and abroad.

D. Consumer Attitudes

Success of food industry ventures depends on the ability to interpret consumer attitudes accurately and produce products that meet their needs and desires. Some individuals suggest that processing in the most historically accurate and, therefore, "natural" way is what customers want. Laws such as Germany's Reinheitsgebot, which legislates that only pure water, hops, malt, and yeast can be used to make beer, is a case in point. In addition, Standards of Identity describe in detail what a product must contain and, sometimes, how it can be made to meet labeling requirements. Although these well-intentioned regulations are meant to protect consumers from unscrupulous manufacturers, they tie products to historical definitions of processing conditions and prevent improvements based on advancing technology.

Innovation provides consumers with specialty products that have functionalities that mean differentiation and high margins, for example, low-calorie, low-fat, high-fiber, low-alcohol, and improved flavor. Enzymes— especially the new enzymes made available by rDNA technology—can generate such innovation. A degree of comfort with and acceptance of the science by the food industry is needed. That comfort and acceptance also must be translated to the marketplace by educating general consumers about the benefits of biotechnology. Public concerns must be recognized and answered effectively. Harlander (1989) cites the food irradiation issue and warns of the dangers if public concerns over food biotechnology are ignored or trivialized.

V. Implications for the Future

Like many breakthrough technologies, the basic discoveries of the biotechnology age did not, at the time, seem to be of great industrial importance. In 1953, DNA structure and its implications were proposed. In the 1960s, the basic mechanics of translation, transcription, and gene expression were elucidated. The 1970s brought development of the techniques to manipulate, splice, and sequence DNA. However, these abilities have not led automatically to a large impact on the food processing sector. Only in the late 1980s were new products cautiously introduced into the food industry to test the regulatory climate and the acceptance by consumers, and to establish the benefits of the technology.

Less frequently discussed are the developments in technology necessary to exploit the new science, or the developments in consumer attitudes to accept the benefits of the science readily. The mechanics of life are fascinating, but do not necessarily allow production of a cheaper fuel, plastic, or foodstuff. Enzymatic processes proceed under milder conditions, give purer products, and impact the environment to a lesser extent, yet still must compete against established chemical methods (Paul, 1981).

A debt is owed to individuals who pioneered the discoveries leading to the commercial use of genetic engineering. Although the methods described in this chapter seem relatively simple, they have been developed painstakingly over a number of years. Once defined, however, they are straightforward and easily accomplished, requiring only care and attention to achieve results. Indeed, high school biology classes perform cloning experiments as part of their laboratory requirements. However, the directed use of these techniques to introduce commercially viable products has been a long struggle; practice is a combination of extensive training and understanding, careful persistent work, a large amount of both skill and art, and some luck as well.

The successful biotechnology company must be proficient in all phases of enzyme engineering and design, process development, and day-to-day production, marketing, distribution, and administration. The technology required to exploit microbes also includes development of more efficient and creatively engineered fermentation facilities and product separation processes. These areas all must be handled in-house, adding costs to already strained research and development budgets. The industry is still young and cannot support engineering, design, and construction firms similar to those available to well-established industries (for example, the petrochemical or food processing industries). For these reasons, the application of biotechnology has not yet reached its full potential.

Returns from research and development programs cannot be realized quickly. Therefore, target markets must be large to justify the expense involved. Non-food applications will be favored to avoid the risk associated with regulatory and consumer acceptance hurdles. High value-added industries will be the primary markets for enzyme manufacturers, for example, the petrochemical, energy, or environmental industries, in which historically favored processing requirements do not exist. Already the food industry has been neglected by biotechnical companies that have directed their resources to human health-care applications, in which the high value pays handsome returns and the life-saving benefits reduce the reluctance of the general consumer. The food industry may be among the last industries to embrace the benefits offered by rDNA technology (Paul, 1981).

The use of enzymes for alternative processes or new products has been discussed at great length. However, enzymes currently have limited application in the food industry. They are characterized by relatively undifferentiated nonproprietary products, forced to compete on a price basis. The majority of useful enzymes are hydrolytic in nature and are used as processing aids. Enzyme-modified products must compete with commodity processing of foodstuffs. Limited proprietary protection has reduced willingness to invest in research and development. Enzyme production and use often requires extensive capital expenditure.

The advent of rDNA technology changes this outlook. Genes can be transferred into high-yielding organisms for more economical applications. Enzyme engineering can alter substrate specificities and pH or temperature optima for improved efficiency. Enzymes for synthesis reactions are on the brink of commercialization.

However, biotechnology is a highly skilled and specialized science. The food industry cannot rely on microbiologists to propose new applications. Insight is needed to identify opportunities in which novel enzymes might excel. Such insight requires knowledge of the benefits and willingness by food scientists to work jointly with enzyme suppliers to pioneer new applications. The opportunities are unlimited.

Acknowledgments

We express our appreciation to Jim Passé, General Patent Counsel, for his section on ownership protection. Also, thanks to Raj Lad, Mike Knauf, Margie Mullins, Randy Berka, Tom Graycar, Karen Chu, and Alice Caddow for their suggestions and comments.

Supplemental Reading

Bailey, J. E., and Ollis, D. E. (1986). "Biochemical Engineering Fundamentals, "2d Ed. McGraw-Hill, New York.
  Engineering principles applied to biological science.
Maniatis, T., Fritsch, E. F., and Sambrook, J. (1982). "Molecular Cloning, A Laboratory Manual." Cold Spring Harbor Laboratory Press, Cold Spring Harbor, New York.
  Recombinant DNA protocols and procedures for those who wish to read about the details.
Watson, J. D. (1968). "The Double Helix." Atheneum, New York.
  Story of the events leading to the discovery of the structure and function of DNA. Inside glimpse of the human dynamics involved in modern research. Also available in video cassette.
Watson, J. D., Hopkins, N. H., Roberts, J. W., Steitz, J. A., and Weiner, A. M. (1987). "Molecular Biology of the Gene," 4th ed. Benjamin/Cummings Publishing, Menlo Park, California.
  Contains a highly readable historical review of discoveries and inventions in biological science. Detailed discussion of molecular and cell biology.

References

Arbige, M. V., and Pitcher, W. H. (1989). Industrial enzymology: A look towards the future. *Trends Biotech.* **17** (12), 330–335.
Bailey, J. E., and Ollis, D. E. (1986). "Biochemical Engineering Fundamentals," 2d Ed. McGraw-Hill, New York.
Bott, R., Ultsch, M., Wells, J., Powers, D., Burnier, J., Adams, J., Power, S., Miller, J., Graycar, T., and Estell, D. (1988). "Three Dimensional Structures of Enzymes Engineered to Have Altered Specificities." Genencor Technical Literature, South San Francisco.
Crick, F. H. C. (1958). On protein synthesis. *Symp. Soc. Exp. Biol.* **12**, 548–555.
Cullen, D., Gray, G. L., Wilson, L. J., Hayenga, K. J., Lamsa, M. H., Rey, M. W., Norton, S., and Berka, R. M. (1987). Controlled expression and secretion of bovine chymosin in *Aspergillus nidulans. Bio/Technology* **5**, 369–376.
Edelson, S. S. (1989). "Commercialization of Engineered Industrial Enzymes." Presentation at *Tenth Engineering Foundation Conference on Enzyme Engineering*, Kashikojima, Japan, Sept. 25, 1989
Estell, D. A., Graycar, T. P., and Wells, J. A. (1985). Engineering an enzyme by site-directed mutagenesis to be resistant to chemical oxidation. *J. Biol. Chem.* **260**, 6518–6521.
Federal Food, Drug, and Cosmetic Act (1982). 21 U.S. Code §§301–392.
Genencor (1988). Private communication. South San Francisco, CA.
Genencor (1989). Filing of a petition for affirmation of GRAS status (Petition No. GRASP 9G0352). *Fed. Reg.* **54**, 40910.
Harlander, S. (1989). Food biotechnology: Yesterday, today and tomorrow. *Food Technol.* **43**(9), 196–206.
Hayenga, K., Heinsohn, H., Arnold, R., and Murphy, M. (1989). "Calf Chymosin Production from *Aspergillus niger* var. *awamori*." Genencor Technical Literature, South San Francisco.
Miller, B. L., Miller, K. Y., and Timberlake, W. E. (1985). Direct and indirect gene replacements in *Aspergillus nidulans. Mol. Cell. Biol.* **5**(7), 1714–1721.
Office of Science and Technology Policy (1986). Coordinated framework for regulation of biotechnology. *Fed. Reg.* **51**, 23302–23350.
Paul, J. K. (1981). "Genetic Engineering Applications for Industry." Noyes Data Corporation, Park Ridge, New Jersey.
Teague, W. M., Metz, R. J., and Zeman, N. W. (1990). Safety evaluation of genetically engi-

neered enzymes for food use. *In* "Biotechnology and Food Safety" (D. D. Bills and Shain-dav Kung, eds.). Butterworth-Heinmann, Boston, Massachusetts.

Watson, J. D., Hopkins, N. H., Roberts, J. W., Steitz, J. A., and Weiner, A. M. (1987). "Molecular Biology of the Gene," 4th Ed. Benjamin/Cummings Publishing, Menlo Park, California.

Wells, J. A., and Estell, D. A. (1988). Subtilism—An enzyme designed to be engineered. *Trends Biochem. Sci.* **13**, (8), 291.

Wells, J. A., Powers, D. B., Bott, R. R., Katz, B. A., Ultsch, M. H., Kossiakoff, A. A., Power, S. D., Adams, R. M., Heyneker, H. H., Cunningham, B. C., Miller, J. V., Graycar, T. P., and Estell, D. A. (1987). Protein engineering of subtilisin. *In* "Protein Engineering Symposium Syllabus" (Dale L. Oxender and C. Fred Fox, eds). Alan R. Liss, Inc. New York.

Yoast, S., Mainzer, S. E., Palombella, A., Adams, R. M., Poolman, B., and Schmidt, B. F. (1989). "Metabolic Pathway Engineering as a Means of Preventing Post-Fermentation Acidification in Yogurt." Genencor Technical Literature, San Francisco.

CHAPTER 5

Immobilized Enzymes

PATRICK ADLERCREUTZ

I. Introduction

Enzymes are biological catalysts and, like other catalysts, are not consumed in the reaction they catalyze. However, in many applications, soluble enzymes are used in a manner similar to any other chemical and no efforts are made to recover them after the reaction. Obviously, process economics can be improved by reuse of the enzymes, provided enzyme recovery can be achieved in a convenient way.

Immobilized enzymes have been defined by Kennedy and Cabral (1987) as "enzymes that are physically confined or localized in a certain defined region of space with retention of their catalytic activities, and that can be used repeatedly and continuously." Some authors use a more narrow definition, including only enzymes that are linked to solid materials. In this chapter, the broader definition will be used.

The first publication on enzyme immobilization appeared as early as 1916; it was reported that invertase had been adsorbed on to aluminum hydroxide with retention of its catalytic activity (Nelson and Griffin, 1916). Since then, especially during the last decades, a vast number of methods for immobilization of enzymes have been presented. The first publicly announced commercial process using an immobilized enzyme was started in Japan in 1969 for the production of L-amino acids by amino acylase (Chibata *et al.*, 1976). Several industrial processes using immobilized enzymes or cells are now in operation. This chapter concentrates on the use of enzymes, but applications of nongrowing cells will be included also, since these in most respects can be regarded as crude enzyme preparations. In fact, some of the commercially available enzymes are cell homogenates or cells that have been processed to make them practical to use. The literature on immobilized biocatalysts is somewhat confusing since authors describe their preparations in widely differing ways. In an effort to unify these descriptions, guidelines for the characterization of immobilized biocatalysts have been published (European Federation of Biotechnology, 1983).

This chapter includes a short overview of enzyme immobilization methods, a discussion of how immobilization influences the performance of the enzyme, and a brief presentation of some applications in the food industry. Finally, some factors concerning the use of free or immobilized enzymes are discussed.

II. Immobilization Methods

Enzyme immobilization can be accomplished by chemical or physical means. Using the former, the enzymes are linked by chemical bonds to an insoluble support material. This modification simplifies the handling of the enzyme; the immobilized preparation can be filtered off easily after the reaction or it can be used in a packed bed reactor. Sometimes a covalent linkage is not necessary; physical adsorption to a support or entrapment in a polymer gel can be sufficient. If the entire reaction mixture is liquid, the enzyme can be retained in the reactor by a membrane that is permeable to substrates and products but not to enzymes. Another possibility is the use of a two-phase system in which the enzyme is confined to one phase only. A schematic classification of different types of immobilization methods is given in Table I. A detailed description of the methods used for immobilizing enzymes is far beyond the scope of this chapter. Instead, some representative examples will be presented with emphasis on methods used in the food industry. Several

TABLE I
Classification of Immobilization Methods

Chemical methods
  Covalent binding (Section II, A)
  Cross-linking (Section II,B)
Physical methods
  Adsorption (Section II,C)
  Physical deposition for use in organic media (Section II,D)
  Entrapment
    Entrapment in polymer gels (Section II,E,1–2)
    Entrapment in microcapsules (Section II,E,3)
  Membranes (Section II,F)
  Two-phase systems (Section II,G)
    Organic/aqueous
    Aqueous/aqueous

more extensive reviews have been published (Mosbach, 1976, 1987; Kennedy and Cabral, 1987).

A. Covalent Immobilization on a Solid Support

Many methods for the covalent coupling of enzymes to solid materials have been developed. Enzymes are proteins, and the amino acid side chains contain several functional groups suitable for coupling reactions. In most methods, amino groups of lysine residues are the principal reactive groups. However, in many cases, coupling occurs also with sulfhydryl groups of cysteines, phenolic hydroxyl groups of tyrosines, carboxyl groups of aspartic and glutamic acids, and, occasionally, other groups (Srere and Uyeda, 1976). Further, the terminal amino and carboxyl groups are reactive. Some enzymes are glycoproteins; for these, the carbohydrate moieties can be used for coupling.

Many solid materials can be used as supports in enzyme immobilization. The main prerequisites are that the material has suitable properties for use in the reactor of the process and is stable under operation conditions, both chemically and mechanically. Further, the support should contain functional groups suitable for binding the enzyme. Normally, a small particle size is advantageous from the perspective of avoiding mass transfer limitations (see Section III,B), but in packed bed reactors small particles give large pressure drops, so a compromise is sometimes required. Often the use of porous particles is beneficial since these have large specific surface areas and can bind large amounts of enzyme, resulting in highly active preparations.

Several polysaccharides have been used for enzyme immobilization, for example, agarose, dextran, starch, and cellulose; all contain many hydroxyl groups that can be used for coupling. Synthetic polymers such as polyacrylates are used also; for these molecules, the coupling chemistry can vary widely because different functional groups can be introduced into the polymers. Several inorganic supports, for example, different kinds of glass, kieselgur, and bentonite, have been used for enzyme immobilization; the mechanical properties of these materials are often good.

Normally, enzyme immobilization is carried out in two steps. In the first step, the support is treated with a reagent that activates some of its functional groups; in the second step, the activated support is mixed with the enzyme and coupling is achieved. Some of the reagents used for activation of supports containing hydroxyl or amino groups are shown in Fig. 1. Cyanogen bromide was used frequently in the past but has a high toxicity; also leakage of enzyme often occurs from the immobilized preparations because the bonds are not completely stable.

Glutaraldehyde is a reagent often used for the immobilization of enzymes in the food industry. Amino groups of the support material and the enzyme can form aldimines (Schiff bases) with aldehyde groups of glutaraldehyde

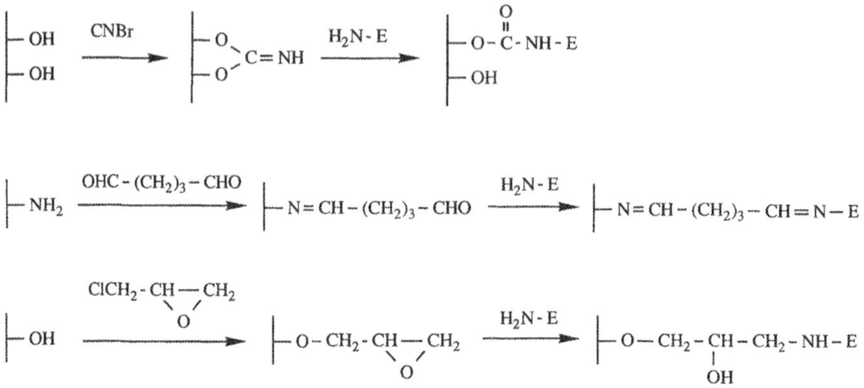

Figure 1 Examples of methods for covalent attachment of enzymes on solid supports. E-NH$_2$ represents the enzyme, which is immobilized via amino groups.

(Fig. 1). However, the reactions are somewhat more complex. Glutaraldehyde is known to polymerize spontaneously and these polymers probably play an important role in the coupling reaction.

When enzymes are immobilized covalently on inorganic supports, such as glass or alumina, silane coupling techniques are used often (Weetall, 1976). In these methods, the support is treated first with a silane that introduces suitable functional groups; often amino groups are introduced using 3-aminopropyltriethoxysilane (Fig. 2). Then different coupling methods can be used to bind the enzyme; in the case of amino groups, glutaraldehyde coupling can be employed.

Other coupling reagents used for enzyme immobilization include epoxides, cyanuric chloride, divinyl sulfone, and many others. Several more extensive reviews on the subject have been published (Kennedy and Cabral, 1987; Rosevear *et al.*, 1987).

B. Cross-linking

By cross-linking the enzyme molecules, large aggregates are formed that eventually become so large that they are no longer soluble. In this way, the enzyme functions both as the catalyst and as the support material. Sometimes quite crude enzyme preparations are cross-linked, in which case other pro-

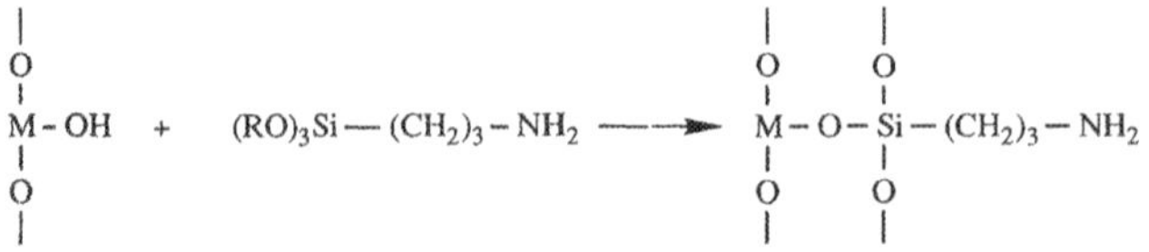

Figure 2 Silanization of inorganic supports for subsequent immobilization of enzymes.

teins and cell components constitute the support. Glutaraldehyde is the most commonly used cross-linking reagent. Often cross-linking is used in combination with other immobilization methods, for example, to stabilize preparations of enzymes immobilized by physical adsorption.

C. Adsorption

Adsorption is perhaps the simplest of the methods for enzyme immobilization. The support is mixed with a solution of the enzyme. After some time, the support with adsorbed enzyme is filtered off. The attractions between the support and the enzyme can be van der Waals forces, ionic interactions, or specific interactions. Commonly used adsorbents are alumina, activated carbon, diatomaceous earth, and glass. Several ion exchange materials have been used to immobilize enzymes by ionic interactions. Carbohydrate-based materials (for example, cellulose and Sephadex) and other materials (for example, polystyrene) with different ionic groups have been used.

The main advantage of the adsorption method for immobilization is its simplicity. A drawback is that leakage can occur, especially if the conditions (pH, ionic strength, etc.) in the reaction mixture are drastically different from those used during adsorption.

D. Enzymes in Organic Media

Enzymes traditionally have been used in aqueous media. However, research has shown that enzymes can work very well as catalysts in organic media with only minute amounts of water present (Laane et al., 1987). A main advantage of this technique compared with the use of aqueous reaction media is the increased efficiency in the conversion of hydrophobic substances due to increased substrate solubility. Further, hydrolytic reactions can be reversed, for example, to achieve synthesis of peptides from amino acids. A practical way of preparing the enzyme for use in organic media is to immobilize it on a solid support. The enzyme is adsorbed on the support or simply deposited by drying a mixture of the support and an aqueous solution of the enzyme. These preparations can be used as catalysts in organic solvents; small amounts of water (often less than 1% of the total reactor volume) are needed to activate the enzyme. Since enzymes are insoluble in almost all organic solvents, the risk of enzyme leakage is normally negligible. The supports used in organic media are Celite, glass, and many others.

Another technique for using enzymes in organic media is taking the solid enzyme powder and adding it directly to the solvent. However, the catalytic activity of the enzyme often is increased by immobilization on a suitable support.

E. Entrapment

1. Entrapment in Polymer Gels

By polymerizing a suitable monomer in the presence of an enzyme, the enzyme can be entrapped in the three-dimensional network of the polymer. The polymer most commonly used for this purpose is polyacrylamide. The enzyme is dissolved in an aqueous buffer containing acrylamide and a bisacrylate to act as a cross-linker. The polymerization is initiated by generating radicals in the solution. The polymerization involves two risks for the enzyme: first, the monomers and the radicals formed during the reaction can react with the enzyme and inactivate it, and second, the heat generated during polymerization can inactivate the enzyme. The simplest way to carry out the polymerization is to cast the gel as a block that is then disintegrated to a suitable particle size. However, in this procedure heat evolution can be a serious problem. An alternative is to carry out the polymerization in an emulsion of the aqueous solution in an organic solvent. The heat generated is, to a large extent, taken up by the solvent. The size of the aqueous droplets determines the size of the gel particles formed.

Another way to decrease heat evolution during enzyme entrapment is to use oligomers instead of monomers as reactants. One method is to use photocrosslinkable prepolymers such as polyethyleneglycol dimethacrylate (Fukui *et al.*, 1976). Polymerization is initiated by ultraviolet light; mechanically stable and chemically inert polymers are formed under mild conditions. Suitable functional groups can be introduced into the prepolymers. A method based on the same principle uses urethane prepolymers that polymerize in water with evolution of carbon dioxide, giving a polyurethane foam with entrapped enzyme (Fukushima *et al.*, 1978). The properties of the polymer can be varied by choosing different urethane prepolymers (Omata *et al.*, 1979).

2. Entrapment of Whole Cells in Polysaccharide Gels

Some polysaccharide gels have proved to be quite useful for entrapment of biocatalysts, especially of whole cells. The most widely used polysaccharide is alginate, which is produced by certain algae. Alginate is a linear copolymer of D-mannuronic and L-guluronic acid. The immobilization procedure starts with dissolution of a soluble alginate, usually the sodium or potassium salt, followed by mixing with the biocatalyst (Kierstan and Bucke, 1977; Bucke, 1987). The mixture is dropped into a solution containing Ca^{2+}, Ba^{2+}, or other ions that cross-link the polysaccharide chains, thereby forming a gel. The conditions are very mild since only cross-linking and no polymerization occurs during the immobilization process. Another gel of this type is carrageenan. In this case, the gel is formed by contact with potassium or ammonium ions and/or lowering of the temperature (Chibata *et al.*, 1987).

3. Entrapment in Microcapsules

Enzymes can be immobilized in microcapsules formed of polymer membranes permeable to substrates and products but not to enzymes. In one method, an emulsion is prepared in which small droplets of water containing the enzyme and a water-soluble monomer are formed in a bulk organic phase containing the other monomer. Polymerization occurs at the interface, resulting in entrapment of the enzyme. The size of the microcapsules can be varied using different conditions during emulsification. Polymers prepared by this method include polyamides, polyurethanes, and polyesters. Most work on these techniques was done to develop medical applications (Chang, 1977), not large scale industrial bioconversions.

F. Immobilization by Membranes

Membranes can be used to retain enzymes in the reactor while allowing products to be removed. Several types of membrane devices have been used for enzymatic reactions, for example, flat membranes and hollow fiber devices. Many different membrane materials can be used so the properties can be optimized with respect to permeability to molecules of different size, charge, and so forth. An advantage of this method is that it is quite mild.

Membrane reactors have been used for several coenzyme-dependent reactions. In this application, the coenzyme NADH was coupled to polyethylene glycol to increase its size, so it would be retained with the enzyme by the membrane (Kula and Wandrey, 1987). Reactions carried out using this technique include reductions of ketoacids to L-amino acids.

G. Immobilization in Two-Phase Systems

By carrying out enzymatic conversions in two-phase systems, the enzyme can be retained in one phase and the product can be removed from the reactor with the other phase. Both organic/aqueous and aqueous/aqueous two-phase systems have been used. Organic/aqueous two-phase systems have proved to be quite useful for conversion of hydrophobic substances such as steroids (Carrea, 1984). The enzyme partitions to the aqueous phase and the substrates and products partition mainly to the organic phase. The solvent should be chosen by taking into consideration its capacity to dissolve substrates and products and its influence on the enzyme. Enzyme denaturation can occur at the interphase, especially if the phases are mixed vigorously to achieve effective mass transfer. Otherwise the use of two-phase systems is a mild immobilization method.

If aqueous solutions of two incompatible polymers are mixed, a two-phase system is formed (Albertsson, 1971). These aqueous two-phase systems can be used for enzymatic conversions (Mattiasson, 1983). The system is often designed so the enzyme partitions predominantly to one of the phases; smaller molecules normally are distributed more evenly. The interfacial tension in these systems is quite low, meaning small amounts of energy are needed to create an emulsion. Further, aqueous two-phase systems are quite mild and the extent of enzyme inactivation is normally small.

III. Effects of Immobilization

A. Inactivation during Immobilization

Immobilization methods in which chemical bonds are formed involve the risk that essential functional groups in the active site of the enzyme will take part in the reactions, which normally results in loss of catalytic activity. To reduce this effect, one can carry out the immobilization in the presence of a suitable substrate or substrate analog that binds the active site, thereby protecting it while coupling is achieved with other parts of the enzyme molecule. After immobilization, the protecting substance can be washed away. Inactivation during immobilization can occur also because of high temperatures or other unsuitable conditions during the procedure.

B. Mass Transfer Effects

The reaction rate of a chemical process is determined by mass transfer effects and kinetic effects. Under some conditions, mass transfer limits the overall reaction rate whereas under other conditions the kinetics is rate limiting. The situation is principally the same for traditional chemical processes and biotechnological processes.

When soluble enzymes are used as catalysts, mass transfer limitations normally can be neglected, except when the substrate is present mainly in another phase, that is, as a gas or a solid. However, when enzymes bound to solid supports are used, mass transfer limitations often occur; the local substrate concentration around the enzyme is normally different from the bulk substrate concentration. In the simplest case, the enzyme is immobilized on a flat surface. Near the surface is a stagnant liquid film through which substrates and products are transported by diffusion. The concentration difference across the stagnant film is the driving force for the diffusion process. Depending on the relative rates of diffusion and reaction, the concentration difference between the bulk phase and the microenvironment of the enzyme can be large or small; the faster the reaction, the larger the concentration difference.

The situation becomes more complex if the enzyme is present inside the solid material. One of the simplest examples is of the enzyme evenly distributed in spherical particles (Fig. 3). As in the previous example, a linear concentration gradient for the substrate is formed outside the particles and the substrate is transported from the bulk liquid to the particle surface by diffusion. Within the particles, the substrate is still transported by diffusion. (This process is called internal mass transfer to distinguish it from mass transfer outside the particles, which is called external mass transfer.) The diffusion coefficient inside the particles can be significantly different from that in free solution. Normally diffusion is slower inside the particles, but in highly hydrated gels such as alginate it can be almost equal to that in water.

The substrate concentration around the enzyme molecule will depend on its position in the particle. The reaction rate obtained in a small section of the particle can be expressed as follows, provided the free enzyme obeys Michaelis–Menten kinetics:

$$v = k[\mathrm{E}]\frac{c}{c + K_{\mathrm{m}}} \tag{1}$$

where k and K_{m} are kinetic constants for the enzyme, E is the enzyme concentration, and c is the local substrate concentration. The observed reaction rate (v_{obs}) with the immobilized enzyme will be an average of the local reaction rates in different parts of the particles. In the absence of mass transfer limitations, the substrate concentration would have been equal to the bulk substrate concentration throughout the particles; the reaction rate for this hypothetical situation is called v'. Because the substrate concentration in the particles is lower than the bulk substrate concentration, v_{obs} will be lower

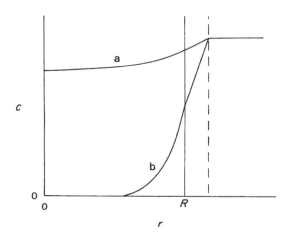

Figure 3 Schematic presentation of substrate concentration (c) profiles in spherical particles containing evenly distributed enzyme. The distance from the center of the particle is denoted r. Curve **a** represents an immobilized preparation with low catalytic activity and curve **b** a more active preparation. The dashed line represents the end of the stagnant liquid film around the particles; R is the particle radius. The substrate partition coefficient between the bulk phase and the particles is 1.

than v'. The degree of mass transfer limitations often is expressed as the overall effectiveness factor, η, which is defined as:

$$\eta = \frac{v_{obs}}{v'} \tag{2}$$

The effectiveness factor depends on different parameters. Some of these factors are characteristics of the enzyme chosen, whereas others depend on how immobilization is achieved and on the conditions during the reaction. The influence of some parameters in a typical example is shown in Fig. 4. As expected, η decreases with increasing particle size because of the longer diffusion distances in larger particles. Further, η decreases with increasing enzyme concentration, which means that each enzyme molecule is more effectively used when low enzyme loadings are used.

Mass transfer in immobilized enzyme preparations has been subject to extensive theoretical work, so models of varying complexity have been developed. Assuming that the enzyme is distributed evenly in spherical particles of equal size and that the partition coefficient for the substrate between the liquid and the particle is 1, the following equation is applicable:

$$D\left(\frac{d^2c}{dr^2} + \frac{2dc}{rdr}\right) = k[E]\,\frac{c}{c + K_m} \tag{3}$$

where D is the diffusion coefficient of the substrate in the particles and r is the distance from the center of the particle. The boundary conditions that apply are:

$$r = 0,\ \frac{dc}{dr} = 0,\ r = R,\ c = c_R$$

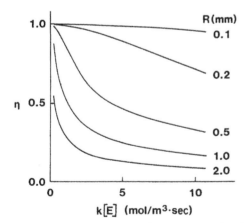

Figure 4 The effectiveness factor (η) as a function of the enzyme loading (expressed as the maximal reaction rate in the particles, $k[E]$) for spherical particles with different radii (R) and uniformly distributed enzyme. The diffusivity of the substrate in the particles is assumed to be 7.2×10^{-10} m²/sec and its concentration in the bulk phase 40 mM. (Data from Adlercreutz, 1986.)

where R is the particle radius and c_R is the substrate concentration at the particle surface. This equation cannot be solved analytically, but numerical methods can be employed (Fink *et al.*, 1973). From the calculations, substrate concentration profiles in the catalyst particles as well as the overall reaction rate can be deduced. The concentration profiles of some substrates, for example oxygen, have been measured directly using microelectrodes (Kasche and Kuhlmann, 1980). These measurements show that the models normally used to describe the mass transfer and the enzymatic reaction in immobilized enzyme preparations are adequate.

C. pH Shift

The support material can influence the microenvironment of an enzyme and thereby influence its catalytic activity. The pH optimum of enzymes was noticed to shift when the enzyme was immobilized on charged support materials (Goldstein *et al.*, 1964). This result was interpreted as attraction of hydrogen ions to a negatively charged support, leading to a more acidic microenvironment and a shift in pH optimum towards higher pH values. The opposite situation occurred with positively charged supports, which generated a more basic microenvironment.

For enzymatic reactions in organic media, the protonation state of the enzyme is important. To obtain maximal activity, the enzyme should be in contact with an aqueous solution with suitable pH before use in an organic medium (Zaks and Klibanov, 1985). Once the enzyme has been transferred to the organic medium, its protonation state is not changed unless it comes into contact with acidic or basic substances.

D. Substrate and Product Partitioning

The concentration of substrates and products in the microenvironment of the immobilized enzyme can be quite different from the bulk concentrations, not only because of mass transfer limitations but also because of partitioning of the substances between the bulk liquid and the immobilized preparation. In some steroid conversions, the properties of the gel used for entrapment of the biocatalyst influence the reaction rate to a large extent (Omata *et al.*, 1979). The increased rate of cholesterol oxidation with increasing gel hydrophobicity was interpreted to be caused by partitioning of this hydrophobic substrate to the hydrophobic gels. With less hydrophobic substrates, the influence of gel hydrophobicity on the reaction rate was less pronounced.

Partitioning effects are important when enzymes are used in organic media. In many cases, the system can be regarded as a two-phase system with an aqueous phase containing the enzyme surrounded by a bulk organic liquid. The aqueous phase can be extremely small; sometimes less than a monolayer of water is present around the enzyme. Normally, organic media are used for the conversion of hydrophobic substances; in these cases the substrates and often also the products partition mainly to the organic phase. Thus, the local substrate concentration in the microenvironment of the enzyme is lower than the bulk concentration. If the local substrate concentration approaches the K_m value of the enzyme or is even lower, the reaction rate is reduced, but if the K_m value is low enough, no negative effects occur. Instead, substrate and product partitioning to the solvent can increase the reaction rate when the enzyme is subject to substrate or product inhibition. Finally, the use of organic media offers a great advantage because the organic phase contains a large amount of dissolved substrate that is transferred rapidly to the enzyme in the aqueous phase.

In addition to the effects on the reaction rate, partitioning effects also can influence the equilibrium position of the reaction, both in aqueous and in organic media. Normally, it is advantageous to design the system so the substrates are enriched in the vicinity of the enzyme and the products are extracted away from the enzyme.

E. Stabilization by Immobilization

Numerous publications in the literature report on stabilization of enzymes by immobilization. Some of these reports are misleading because the effects described are caused by mass transfer limitations. If the reaction rate in the immobilized preparation is governed mainly by mass transfer limitations, inactivation of the enzyme will not have as large an effect on the observed reaction rate as in the case with free enzyme; this effect often has been interpreted as stabilization due to immobilization. However, immobilization can lead to true stabilization of enzymes (Klibanov, 1979). By introducing several covalent bonds or noncovalent interactions between the enzyme and the support, the enzyme can be stabilized by decreased flexibility, which slows down denaturation reactions.

Several specific stabilization effects have been observed in the immobilization of certain enzymes. For example, dissociation of oligomeric enzymes can be prevented by immobilization and proteolytic enzymes are stabilized by attachment to supports because their autolysis is prevented. Enzymes inside solid particles are protected from microbial degradation and proteolytic breakdown. The support can be designed to contain certain functional groups or additives, giving protection from certain inactivating reagents. For example, metallic oxides or catalase have been added to decompose hydrogen peroxide.

Water has been observed to take part in the most important mechanisms for irreversible enzyme inactivation (Ahern and Klibanov, 1985). Accordingly, very good thermostability has been observed for enzymes in organic media; the stability decreases strongly with increasing water content (Zaks and Klibanov, 1984).

F. Catalytic Properties of Immobilized Enzymes in Organic Media

Apart from the effects discussed already, which are important in both aqueous and organic media, immobilization of enzymes for use in organic media involves some additional effects. The water content during enzymatic conversions in organic media is normally low, but it plays an important role (Reslow *et al.*, 1987, 1988). A small amount of water is needed by the enzyme to make it flexible and, thereby, activate it. Some of the water in the system is dissolved in the solvent and some is bound to the enzyme and to other solid materials present (support material, other proteins in the enzyme preparation, etc.). Accordingly, the amount of water needed strongly depends on the solvent used. Often an amount of water close to water saturation is optimal for enzymatic activity in water-immiscible solvents. When hydrolytic side reactions should be suppressed, lower amounts of water can be useful. The properties of the support material greatly influence the catalytic activity of the immobilized enzyme. A good correlation was found between the water-attracting capacity ("aquaphilicity") of the support and the reaction rate obtained with enzymes deposited on that support; low aquaphilicity resulted in high catalytic activity (Reslow *et al.*, 1988).

IV. Applications

Some of the most important applications of immobilized enzymes in the food industry are presented briefly in this section, with emphasis on the immobilization methods used. A more extensive list of processes using immobilized enzymes in different industries has been published(Gestrelius and Mosbach, 1987). For most of the applications, different companies have developed and patented strategies for enzyme immobilization. Often, a combination of immobilization methods is used to achieve a catalyst with suitable properties.

The largest industrial process using immobilized enzymes is the production of high fructose corn syrup (HFCS) catalyzed by glucose isomerase. A solution containing 40–45% (w/w) glucose is processed to HFCS, in which about 42% of the glucose is converted to fructose. Several different organisms have been used as enzyme sources and several immobilization methods have been used (Jensen and Rugh, 1987), among which are:

1. Cross-linking of cell homogenate with glutaraldehyde followed by extrusion
2. Flocculation of whole cells followed by glutaraldehyde cross-linking
3. Adsorption on an ion exchanger followed by embedding in polystyrene

Advantages achieved using immobilized enzymes in packed bed reactors include decreased by-product formation because of short residence time in the reactor and increased productivity. Immobilized enzyme preparations are exhibiting a trend toward containing enzyme of higher purity and higher catalytic activity.

Lactase (β-galactosidase) is used industrially to hydrolyze lactose in whey and similar substrates (Baret, 1987). Immobilization methods were developed to reduce the enzyme cost of the process. One process uses enzyme immobilized on a controlled-pore SiO_2 ceramic using silylation with 3-aminopropyltriethoxysilane, followed by glutaraldehyde coupling.

Isomaltulose is a disaccharide that can be prepared enzymatically from sucrose. In one industrial process, whole cells are immobilized in calcium alginate gel (Cheetham *et al.*, 1985; Cheetham, 1987). This immobilization method provides highly active preparations, and the operational stability increased 350 times compared with free cells. Another immobilization method involves flocculation, extrusion into ropes followed by drying, and cross-linking with glutaraldehyde (Cheetham, 1987).

Lipases have been used to modify the fatty acid composition of triglycerides (Macrae, 1985). The reactions catalyzed by lipases and other hydrolytic enzymes can give different products depending on the reaction conditions. In aqueous media, hydrolysis is the dominating reaction: triglycerides are hydrolyzed to glycerol and fatty acids. However, in organic media, the amount of water available is limited so the reactions can be reversed to synthesize triglycerides. Further, because of the mechanism of the reaction, via an acyl-enzyme intermediate, several exchange reactions can take place, for example, transesterification reactions. A new fatty acid can be introduced into triglycerides; it is possible to exchange fatty acids between different triglyceride fractions. Transesterification reactions have been exploited on a pilot plant scale for the production of cocoa butter substitutes. In this application, lipases specific for the 1- and 3-positions of triglycerides are used to incorporate stearic acid into a suitable oil, often a palm oil fraction. The lipases are immobilized by adsorption or similar methods on solid materials such as Celite or ion exchange resins.

Aspartame (α-L-Asp–L-Phe–OMe) is a sweetener that is about 200 times sweeter than sucrose. It can be produced from its amino acid building blocks, either chemically (which yields a fairly large amount of the β-isomer with a bitter taste) or enzymatically using thermolysin (Oyama *et al.*, 1987). Immobilization of the enzyme has been accomplished by adsorption on hydrophobic adsorbents or ion exchangers or by covalent linkage. The reaction is carried out in an organic/aqueous two-phase system designed so the aqueous phase contains high concentrations of the reactants and the product is effectively extracted into the organic phase (ethyl acetate), giving a favorable equilibrium position.

L-Aspartic acid has been produced enzymatically in Japan since 1973 using aspartase (L-aspartate ammonia lyase), which catalyzes the conversion of ammonium fumarate to L-aspartic acid (Sato *et al.*, 1975; Fusee, 1987). The biocatalyst used contains whole cells of *Escherichia coli* entrapped in polyacrylamide, carrageenan, or polyurethane.

V. Free or Immobilized Enzyme?

Some advantages and drawbacks of immobilization of enzymes are listed in Table II. The cost of the enzyme is a key parameter. If the enzyme is very cheap, it is difficult to find an immobilization procedure that can improve the economics of the process, but for high priced enzymes immobilization is an attractive alternative. Another key question is whether the enzyme can be tolerated in the final product. If this is not the case, immobilization is a good idea since the enzyme must be removed from the product anyway. Recovery is then advantageous so the enzyme can be reused. Mass transfer limitations and enzyme inactivation during immobilization can decrease the reaction rate obtained with immobilized biocatalysts. However, often the conditions can be chosen so these effects largely can be overcome.

As mentioned earlier, immobilization can increase the stability of the enzyme. The immobilized enzyme must have good operational stability; if not, one must soon replace it although it is immobilized. Then free enzyme probably will be more economical. The operational stability of the enzyme is influenced by the purity of the substrate solution, which sometimes contains inactivating agents. As immobilized enzyme preparations come into contact with larger amounts of substrate, they are more sensitive to this inactivation effect than their free counterparts. The presence of solid material in the substrate makes it difficult to use enzymes immobilized on solid supports.

Immobilization of the enzyme makes it easier to construct continuous processes. The residence time in the reactor can be shorter in a continuous process than in a batchwise process, which can be an advantage if noncatalyzed reactions interfere with the enzymatic reactions. Using a very high enzyme concentration and a short residence time, the interference of spontaneous reactions can be minimized (Wehtje *et al.*, 1988).

Many factors must be considered when deciding whether to use free or immobilized enzyme for a process. The examples in this book show that, for most applications in the food industry, free enzymes are used but, for some

TABLE II
Advantages and Drawbacks of Enzyme Immobilization

Advantages
  Lower enzyme cost
  No enzyme contaminates the product
  Enzyme stabilization (sometimes)

Drawbacks
  Cost of immobilization
  Enzyme inactivation (sometimes)
  Mass transfer limitations

---

applications, immobilization has proven to be advantageous. Increasing numbers of industrial processes using immobilized enzymes are in operation.

Acknowledgment

The author thanks Bärbel Hahn-Hägerdal for giving him access to her extensive literature in the field.

References

Adlercreutz, P. (1986). Oxygen supply to immobilized cells. 5. Theoretical calculations and experimental data for the oxidation of glycerol by immobilized *Gluconobacter oxydans* cells with oxygen or *p*-benzoquinone as electron acceptor. *Biotechnol. Bioeng.* **28**, 223–232.

Ahern, T. J., and Klibanov, A. M. (1985). The mechanism of irreversible enzyme inactivation of 100°C. *Science* **228**, 1280–1284.

Albertsson, P.-Å. (1971). "Partition of Cell Particles and Macromolecules," 2d Ed. Almqvist and Wiksell, Stockholm.

Baret, J. L. (1987). Large-scale production and application of immobilized lactase. *Meth. Enzymol.* **136**, 411–423.

Bucke, C. (1987). Cell immobilization in calcium alginate. *Meth. Enzymol.* **135**, 175–189.

Carrea, G. (1984). Biocatalysis in water–organic solvent two-phase systems. *Trends Biotechnol.* **2**, 102–106.

Chang, T. M. S. (1977). "Biomedical Applications of Immobilized Enzymes and Proteins." Plenum, New York.

Cheetham, P. S. J. (1987). Production of isomaltulose using immobilized microbial cells. *Meth. Enzymol.* **136**, 432–454.

Cheetham, P. S. J., Garrett, C., and Clark, J. (1985). Isomaltulose production using immobilized cells. *Biotechnol. Bioeng.* **27**, 471–481.

Chibata, I., Tosa, T., Sato, T., and Mori, T. (1976). Production of L-amino acids by aminoacylase adsorbed on DEAE–Sephadex. *Meth. Enzymol.* **44**, 746–759.

Chibata, I., Tosa, T., Sato, T., and Takata, I. (1987). Immobilization of cells in carrageenan. *Meth. Enzymol.* **135**, 189–198.

European Federation of Biotechnology (1983). Guidelines for the characterization of immobilized biocatalysts. *Enz. Microb. Technol.* **5**, 304–307.

Fink, D. J., Na, T. Y., and Schultz, J. S. (1973). Effectiveness factor calculations for immobilized enzyme catalysts. *Biotechnol. Bioeng.* **15**, 879–888.

Fukui, S., Tanaka, A., Iida, T., and Hasegawa, E. (1976). Application of photo-crosslinkable resin to immobilization of an enzyme. *FEBS Lett.* **66**, 179–182.

Fukushima, S., Nagai, T., Fujita, K., Tanaka, A., and Fukui, S. (1978). Hydrophilic urethane prepolymers: Convenient materials for enzyme entrapment. *Biotechnol. Bioeng.* **20**, 1465–1469.

Fusee, M. C. (1987). Industrial production of L-aspartic acid using polyurethane-immobilized cells containing aspartase. *Meth. Enzymol.* **136**, 463–471.

Gestrelius, S., and Mosbach, K. (1987). Overview. *Meth. Enzymol.* **136**, 353–356.

Goldstein, L., Levin, Y., and Katchalski, E. (1964). A water-insoluble polyanionic derivative of trypsin. II. Effect of the polyelectrolyte carrier on the kinetic behavior of the bound trypsin. *Biochemistry* **3**, 1913–1919.

Jensen, V. J., and Rugh, S. (1987). Industrial-scale production and application of immobilized glucose isomerase. *Meth. Enzymol.* **136**, 356–370.

Kasche, V., and Kuhlmann, G. (1980). Direct measurement of the thickness of the unstirred diffusion layer outside immobilized biocatalysts. *Enz. Microb. Technol.* **2**, 309–312.

Kennedy, J. F., and Cabral, J. M. S. (1987). Enzyme Immobilization *In* "Biotechnology Vol. 72" (H. J. Rehm and G. Reed, eds.). pp. 347–404. VCH Verlagsgesellschaft, Weinheim, Germany.

Kierstan, M., and Bucke, C. (1977). The immobilization of microbial cells, subcellular organelles, and enzymes in calcium alginate gels. *Biotechnol. Bioeng.* **19,** 387–397.

Klibanov, A. M. (1979). Enzyme stabilization by immobilization. *Anal. Biochem.* **93,** 1–25.

Kula, M.-R., and Wandrey, C. (1987). Continuous enzymatic transformation in an enzyme–membrane reactor with simultaneous NADH regeneration. *Meth. Enzymol.* **136,** 9–21.

Laane, C., Tramper, J., and Lilly, M. D. (eds.) (1987). "Biocatalysis in Organic Media." Elsevier, Amsterdam.

Macrae, A. R. (1985). Interesterification of fats and oils. *In* "Biocatalysts in Organic Synthesis" (J. Tramper, H. C. van der Plas, and P. Linko, eds.), pp. 195–208. Elsevier, Amsterdam.

Mattiasson, B. (1983). Applications of aqueous two-phase systems in biotechnology. *Trends Biotechnol.* **1,** 16–20.

Mosbach, K. (ed.) (1976). "Methods in Enzymology," Vol. 44. Academic Press, New York.

Mosbach, K. (ed.) (1987). "Methods in Enzymology," Vol. 135. Academic Press, New York.

Nelson, J. M., and Griffin, E. G. (1916). Adsorption of invertase. *J. Am. Chem. Soc.* **38,** 1109–1115.

Omata, T., Iida, T., Tanaka, A., and Fukui, S. (1979). Transformation of steroids by gel-entrapped *Nocardia rhodocrous* cells in organic solvent. *Eur. J. Appl. Microbiol. Biotechnol.* **8,** 143–155.

Oyama, K., Irino, S., and Hagi, N. (1987). Production of aspartame by immobilized thermoase. *Meth. Enzymol.* **136,** 503–516.

Reslow, M., Adlercreutz, P., and Mattiasson, B. (1987). Organic solvents for bioorganic synthesis. 1. Optimization of parameters for a chymotrypsin-catalyzed process. *Appl. Microbiol. Biotechnol.* **26,** 1–8.

Reslow, M., Adlercreutz, P., and Mattiasson, B. (1988). On the importance of the support material for bioorganic synthesis. Influence of water partition between solvent, enzyme, and solid support in water-poor reaction media. *Eur. J. Biochem.* **172,** 573–578.

Rosevear, A., Kennedy, J. F., and Cabral, J. M. S. (1987). "Immobilized Enzymes and Cells." Adam Hilger, Bristol, England.

Sato, T., Mori, T., Tosa, T., Chibata, I., Furui, M., Yamashita, K., and Sumi, A. (1975). Engineering analysis of continuous production of L-aspartic acid by immobilized *Escherichia coli* cells in fixed beds. *Biotechnol. Bioeng.* **17,** 1797–1804.

Srere, P. A., and Uyeda, K. (1976). Functional groups on enzymes suitable for binding to matrices. *Meth. Enzymol.* **44,** 11–19.

Weetall, H. H. (1976). Covalent coupling methods for inorganic support materials. *Meth. Enzymol.* **44,** 134–148.

Wehtje, E., Adlercreutz, P., and Mattiasson, B. (1988). Activity and operational stability of immobilized mandelonitrile lyase in methanol/water mixtures. *Appl. Microbiol. Biotechnol.* **29,** 419–425.

Zaks, A., and Klibanov, A. M. (1984). Enzymatic catalysis in organic media at 100°C. *Science* **224,** 1249–1251.

Zaks, A., and Klibanov, A. M. (1985). Enzyme-catalyzed processes in organic solvents. *Proc. Natl. Acad. Sci. U.S.A.* **82,** 3192–3196.

CHAPTER 6

Carbohydrases

RAMUNAS BIGELIS

I. Introduction

Carbohydrases play an important role as processing aids in the food industry. This chapter examines their role in the production of food ingredients, the enhancement of product quality, and the improvement of the efficiency of food processing steps. The chapter considers the classical uses of carbohydrases and also some emerging applications that may be of importance in the future.

II. Food-Processing Carbohydrases

A. Amylases

1. Starch Processing

The food industry uses enzymes as processing aids to convert starch-bearing raw materials to starches, starch derivatives, and starch saccharification products (MacAllister *et al.*, 1975; Radley, 1976; MacAllister, 1979; Norman, 1979,1981; Bucke, 1981,1983a,b; Luesner, 1983; Ostergaard, 1983; Reichelt, 1983; Coker and Venkatasubramanian, 1985; Kempf, 1985; Van Beynum and Roels, 1985; Newsome, 1986; Pedersen and Norman, 1987; Peppler and Reed, 1987; Luallen, 1988; Spradlin, 1989; Fogarty and Kelly, 1990; Gerhartz, 1990). Table I shows some examples of hydrolyzed starch products that have applications in food processing (Corn Refiners Association (CFA), 1979a,b; Lineback and Inglett, 1982; Van Beynum and

TABLE I
Properties and Industrial Applications of Hydrolyzed Starch Products[a]

Type of syrup	DE	Composition (%)	Properties	Application
Low DE maltodextrins	15–30	1–20 D-glucose, 4–13 maltose, 6–22 maltotriose, 50–80 higher oligomers	Low osmolarity	Clinical feed formulations; raw materials for enzymic saccharification; thickeners, fillers, stabilizers, glues, pastes
Maltose syrups	40–45	16–20 D-glucose, 41–44 maltose, 36–43 higher oligomers	High viscosity, reduced crystallization, moderately sweet	Confectionary; soft drinks; brewing and fermentation; jams, jellies, and conserves, ice cream; sauces
High maltose syrups	48–55	2–9 D-glucose, 48–55 maltose, 15–16 maltotriose	Increased maltose content	Hard confectionary; brewing and fermentation
High DE syrups	56–68	25–35 D-glucose, 40–48 maltose	Increased moisture holding, increased sweetness, reduced content of higher sugars, reduced viscosity, higher fermentability	Confectionary; soft drinks; brewing and fermentation; jams and conserves; sauces
Glucose syrups	96–98	95–98 D-glucose, 1–2 maltose, 0.5–2 isomaltose	Commercial liquid 'dextrose'	Soft drinks; caramel; baking; brewing and fermentation; raw material
Fructose syrups	98	48 D-glucose, 52 D-fructose	Alternative industrial sweeteners to sucrose	Soft drinks; conserves; sauces; yogurts; canned fruits

[a] Reprinted with permission from Kennedy et al. (1988).

Roels, 1985; Kennedy *et al.*, 1988; Koch and Roper, 1988; Hacking, 1991; Pomerantz, 1991).

The common sources for raw starch are maize, wheat, and potatoes, although barley, cassava, and other plant crops may be important in the future. The dominant position of corn as a carbohydrate source in the United States suggests that it will continue to be an important raw material for the food industry. The following discussion on starch processing therefore will focus primarily on the enzymatic processing of corn starch to industrial sweeteners and other food ingredients.

a. Corn starch Corn starch is composed of about 25% amylose and 75% amylopectin. Amylose is a linear homopolymer of glucose bound by α-1,4 glycosidic linkages. Amylopectin is a branched glucose polymer that bears α-1,6 branchpoints (Whistler *et al.*, 1984; Guilbot and Mercier, 1985; Koch and Roper, 1988; Pomerantz, 1991). Amylases hydrolyze starch and play an important role in biotechnological approaches to starch saccharification (Kulp, 1975; Radley, 1976; Fogarty and Kelly, 1979,1980; Norman, 1979; Fogarty, 1983; Van Beynum and Roels, 1985; Bigelis and Lasure, 1987; Linko, 1987; Shetty and Allen, 1988; Spradlin, 1989; Vihinen and Mäntsälä, 1989). Although various enzymatic processes may be used for starch saccharification, bacterial α-amylases usually are employed as endoenzymes to hydrolyze internal α-1,4 glycosidic linkages and glucoamylases from fungal sources are used as exoenzymes to release terminal glucose monomers. A debranching enzyme, such as bacterial pullulanase, may be added during starch saccharification to cleave α-1,6 branchpoints, although glucoamylase itself possesses some debranching activity. As will be discussed in a later section, glucose isomerases may be used in conjunction with amylases to convert glucose to fructose in the manufacture of high fructose corn syrup. The flow chart in Fig. 1 illustrates the application of these enzymes for the production of commercial carbohydrate products.

Industrial processes for starch hydrolysis to glucose rely on inorganic acids or enzyme catalysis. The use of enzymes is preferred currently and offers a number of advantages associated with improved yields and favorable economics. Enzymatic hydrolysis allows greater control over amylolysis, the specificity of the reaction, and the stability of the generated products. The milder reaction conditions involve lower temperatures and near-neutral pH, thus reducing unwanted side reactions. Fewer off-flavor and off-color compounds are produced, especially 5-hydroxy-2-methylfurfuraldehyde, anhydroglucose compounds, and undesirable salts. Enzymatic methods are favored because they also lower energy requirements and eliminate neutralization steps.

b. Gelatinization and liquefaction of starch Current starch processing methods require the gelatinization of corn starch after wet milling. Such a pretreatment step is essential before an extended period of enzymatic hydrolysis. The starch, which is present as $5-25\ \mu$m particles, is exposed to

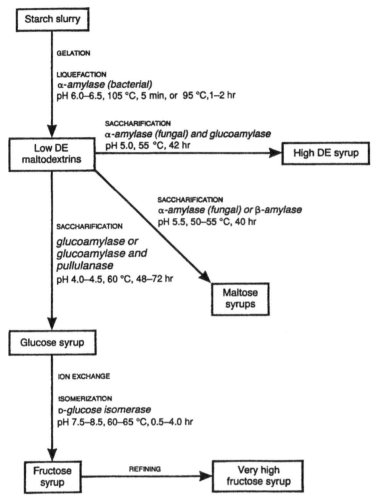

Figure 1   Starch processing using enzymes. (Reprinted with permission from Kennedy *et al.*, 1988.)

temperatures above 60°C that swell and disrupt the granules. The addition of heat-stable α-amylase from strains of *Bacillus subtilis* at this stage causes random hydrolysis of glycosidic bonds and significantly thins the starch slurry. *Bacillus* amylases tolerate temperatures of up to 70°C and can withstand the harsh conditions associated with gelatinization. The end product of the thinning step is a starch slurry of manageable viscosity that can be processed further by enzymatic means.

After gelation of the starch, a process leading to its liquefaction begins. The thinned starch slurry composed of 30–40% solids is treated with a thermostable *Bacillus licheniformis* α-amylase able to function at pH 6.5. The starch is contained in a steam jet cooker operated in a continuous mode for

5–10 min at 103–107°C. Afterward, a 1–2-hr treatment at a lower temperature of about 95°C permits hydrolysis of the thinned starch to a syrup of 0.5–1.5 DE (dextrose equivalents). Digestion is allowed to continue until a DE of 10–15 is attained. The liquefaction process at both higher and lower temperatures depends on efficient action of the heat-stable bacterial amylase. The net yield of dextrose after complete enzymatic digestion of the corn starch with bacterial amylase is 95–97% if hydrolysis is coupled with a fungal amylase treatment, such as that described in Section I,A,3.

Starch derivatives obtained via the liquefaction process have commercial value as food ingredients (Table I). The products have a DE value of about 10 and are sold as maltodextrins, which are carbohydrate preparations composed of 3% maltose, 4% maltotriose, and 93% tetraoses or larger polysaccharides. Maltodextrins are valuable to the food industry since they serve as thickening agents and additives able to promote drying of hygroscopic food components.

c. Saccharification of starch The saccharification of starch, that is, the near-total hydrolysis of starch to glucose, is accomplished with a fungal glucoamylase. The enzyme is effective since it has both exoamylase and debranching activity and is important for the production of a corn syrup of high DE or of crystalline dextrose. A debranching enzyme such as pullulanase may be used also. Fungal glucoamylase, which is obtained by fermentation of *Aspergillus niger, A. awamori, A. oryzae,* or *Rhizopus oryzae,* can hydrolyze starch to glucose almost completely after a treatment time of 3–4 days at pH 4 and 60°C. The activity and dose of the enzyme preparation will determine the appropriate incubation period. The low pH optimum of fungal amylolytic enzymes permits the convenient use of acid conditions for the saccharification. Such conditions reduce unwanted isomerization reactions to fructose and other sugars that may reduce the glucose yield. Moreover, acid conditions restrict the growth of contaminating microorganisms in the saccharification reactors.

Transglucosylation reactions can lead to undesirable side products during saccharification. Under conditions of high glucose concentration, side reactions can give rise to maltose, isomaltose, and panose. Industrial enzyme preparations of glucoamylase with low levels of transglucosylase activity are preferred and are manufactured with special production organisms that synthesize very little of this activity. After a maximum hydrolysis of starch has been reached, the starch hydrolysate is treated with heat to inactivate the glucoamylase activity and the accompanying transglucosylase that later may generate unwanted disaccharides (McCleary *et al.,* 1989).

Pullulanases, which can serve as debranching enzymes during starch processing, are obtained from bacterial sources, typically from *Bacillus* species. Pullulanases are useful in starch processing since they can improve glucose yields and decrease reaction times. Further, they can be used to produce maltose syrups that have food applications. The *Bacillus* enzyme is thermostable and acid tolerant and thus well suited for reactor conditions during saccharification (Takasaki *et al.*, 1981; Slominska and Maczynski, 1985; Boyce, 1986; Sheppard, 1986).

2. Baking

The supplementation of doughs with α-amylase affects the functional properties of the doughs and may determine characteristics that are critical for automated manufacturing processes. Largely for this reason, α-amylase, sometimes in conjunction with protease, is commonly used by the baking industry (Amos, 1955; Waldt, 1965,1969; Pyler, 1969; Guilbot, 1972; Barrett, 1975; Monnier and Godon, 1975; Gams, 1976; Marston and Wannan, 1976; Schwimmer, 1981; Reichelt, 1983; Drapron and Godon, 1987; ter Haseborg, 1988; Stauffer, 1990; Vandam and Hille, 1992).

Malted wheat, barley, bacteria, and fungi are typical sources of α-amylase for baking purposes. Fungal α-amylase is added to bread doughs in the form of diluted powders, prepacked doses, or water dispersible tablets. The enzyme may be added to flours at the bakery or, more rarely, at the mill itself. Malted wheat and barley also can serve as sources of amylolytic activity when flours from these grains are blended with the final product at the mill. The properties of bacterial α-amylase permit its application to the production of coffee cake, fruit cake, brownies, cookies, snacks, and crackers. Fungal α-amylase, usually from *A. oryzae, A. niger, A. awamori*, or species of *Rhizopus*, is used to supplement the amylolytic activity in flour. Enzymes from these sources can raise the levels of fermentable monosaccharides and disaccharides of dough from a native level of 0.5% to concentrations that promote yeast growth. The sustained release of glucose and maltose by added fungal and endogenous enzymes provides the nutrients essential for yeast metabolism and gas production during panary fermentation. The *A. oryzae* α-amylase is sometimes favored for baking applications over the bacterial enzyme obtained from *Bacillus* species since the fungal enzyme is heat labile at $60-70\,^{\circ}C$ and does not survive the baking process. Its thermolability prevents enzymatic action on the gelatinized starch in the finished loaf which would cause a soft or sticky crumb. Bacterial α-amylase is also used with good results, but its dose must be measured carefully to avoid a bread with a gummy mouthfeel. Amylase supplementation is also beneficial and sometimes essential, since white bread flours contain $6.7-10.5\%$ damaged starch. The added enzyme degrades damaged, ruptured starch granules that usually are present in bread flour more efficiently than does wheat β-amylase.

Amylase supplementation can improve other characteristics of bread quality, in addition to improving the quality of rolls, buns, and crackers, when used during manufacturing processes for these baked goods (Selman and Sumner, 1947; Johnson and Miller, 1949; Conn *et al.*, 1950; Berger and Granvoinnet, 1974; Barrett, 1975; Kaur and Bains, 1978; Brabender and Seitz, 1979; Maninder and Jorgensen, 1983; Reichelt, 1983; Boyce, 1986; Cauvain and Chamberlain, 1988; ter Haseborg, 1988; Spradlin, 1989; Hebeda *et al.*, 1990,1991). In bread baking, treatment with fungal or bacterial amylase lowers the viscosity of bread dough, thereby improving the ease of manipulation by manual workers or machines. Measured doses of enzyme also lower the compressibility of the loaf, producing a softer bread (Fig. 2). Further, such processing increases the bread volume (Fig. 3) by reducing the viscosity of the gelling starch and allowing greater expansion during baking before protein denaturation and enzyme inactivation fix the volume of the loaf. Favorable effects on taste, crust properties, and toasting qualities are observed. The storage characteristics of breads are changed also, yielding a product with a softer, more compressible crumb that firms more slowly and keeps longer, as determined by taste panels (Table II). Amylolytic activity also may elevate the sugar concentration in bread and yield a preferred sweeter product with sensory advantages.

3. Brewing

The brewing of beer depends on yeast fermentation and may use microbial enzyme supplements at several stages of the brewing process (Broderick, 1977; Westermann and Huige, 1979; Briggs *et al.*, 1981; Gillis-van Maele, 1982; Godfrey, 1983a; Priest and Campbell, 1987). Crude enzyme preparations of β-glucanase, cellulase, α-amylase, glucoamylase, and protease obtained by fungal or bacterial fermentation are employed most commonly (Fig. 4). The application of one class of enzymes, amylases, to the early stages and the late stages of the brewing process is discussed in detail in this section.

The early stage of brewing involves wort production, a process that generates the constituents needed for the subsequent yeast fermentation. Enzymes often are supplemented at this stage, especially during the mashing

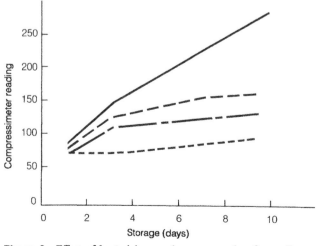

Figure 2 Effect of bacterial α-amylase on crumb softness. Dosage: 0 SKB unit/100 g flour (——); 1 SKB unit/100 g flour (———); 3 SKB units/100 g flour (—-—); 6 SKB units/100 g flour (---). (Reprinted with permission from Boyce, 1986.)

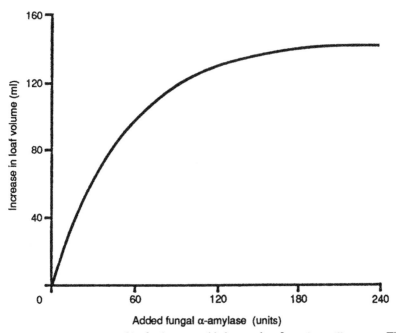

Figure 3 The response of loaf volume to added α-amylase from *Aspergillus oryzae*. The average response curve is fitted to 174 data points obtained with 58 commercially milled white flours. (Reprinted with permission from Cauvain and Chamberlain, 1988.)

step. Enzymes also are used during special brewing procedures that involve barley brewing, cereal cooking, or the production of low-carbohydrate beers (Saletan, 1968; Nielsen, 1971; Allen and Spradlin, 1973; Knoepfel and Pfenninger, 1974; Bass and Cayle, 1975; Woodward, 1978; Briggs *et al.*, 1981; Denault *et al.*, 1981; Marshall *et al.*, 1982; Godfrey, 1983a; Hough, 1985; Slaughter, 1985).

The mashing step during the early phases of brewing involves the libera-

TABLE II
Taste Panel Evaluation of White Pan Breads[a]

Bacterial α-amylase dose (SKB units/100 g flour)	Expert panel rating on day 9[b]	
	Chewability	Overall
0 (control)	Too firm	Stale bread
1	Good	Best bread
3	Slightly gummy	Borderline
6	Very gummy	Unacceptable

[a] Reprinted with permission from Boyce (1986).
[b] Breads were stored for 9 days in plastic bags at 75° F.

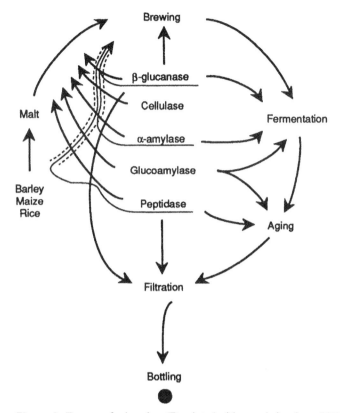

Figure 4 Enzymes for brewing. (Reprinted with permission from Gillis-Van Maele, 1982.)

tion of fermentable sugar from starch. Added enzymes contribute to the action of endogenous barley β-amylase and aid in the starch digestion process. Such added enzymes are especially important when nonmalted cereal grains such as corn and rice, termed adjuncts, are used. Since these adjunct grains are deficient in carbohydrases, fungal α-amylase and glucoamylase can increase starch digestion, reduce the proportion of unmalted grain, and insure a consistent quality of the mash. Amylase solubilizes barley amylose and amylopectin, exposing these substrates to further degradation by barley β-amylase. As a result, the levels of maltose and small dextrins are raised, eventually yielding the wort ingredients that promote yeast fermentation. Amylase preparations with low transglucosidase activity are favored since trace levels of this enzyme generate isomaltose and panose, both of which are nonfermentable by yeast (McCleary *et al.*, 1989). The source of amylase activity for brewing applications is generally enzyme from *Aspergillus* species such as *A. niger* or *A. oryzae*. Protease from these sources may be added in concert with amylase to solubilize protein and release amino acids essential for yeast proliferation.

Amylases from fungal sources also may be used in barley brewing, a brewing method that relies on unmalted barley and a mixture of enzymes in the mash tun (Fig. 4). The barley brewer simulates a malt wort and avoids the mashing step by using exogenous enzymes to generate a yeast fermentation medium. The enzymes α-amylase, β-glucanase, and proteinase are selected for their favorable thermostability characteristics. Generally, enzyme additions equal 0.05–0.10% of the grist volume. The fermentors usually are cooled after addition of enzyme since β-glucanase and proteinase act in the temperature range of 45–50°C. After an incubation period, the temperature is raised to a range that is optimal for starch hydrolysis, most commonly 65–68°C. The careful management of temperature conditions is critical since excessive heating will reduce the levels of fermentable sugars and thus damage the wort. In barley brewing, the enzymes are inactivated during the boiling step to avoid hydrolytic activity in the fermenting wort. The use of a barley and enzyme combination rather than malt yields economic benefits for the brewer and allows greater control over enzymatic digestion and mashing conditions.

Enzymes are indispensable to the production of low-carbohydrate beers, brews that are increasingly popular and known as "light" beers. Worts for such beers normally contain 4% (w/v) of their carbohydrates as unfermentable dextrins. Yeast can ferment maltose well and maltotriose only poorly, but is incapable of metabolizing the larger dextrin oligosaccharides. Thus, dextrins are carried over to the beer and elevate its caloric content. The digestion of wort dextrins with fungal glucoamylase added directly to the fermentor eliminates the unfermentable carbohydrate fraction almost completely and yields a reduced-calorie beer. The alcohol content of the beer remains the same if the gravity of the starting material is adjusted. Hydrolysis of the α-1,4 linkages and α-1,6 branchpoints yields glucose, which also must be metabolized if a *sweet* reduced-calorie beer is to be avoided. An alternative approach to treatment with fungal glucoamylase is the addition of a mixture of fungal α-amylase and bacterial pullulanase to the fermentor (Willox *et al.*, 1977; Sheppard, 1986). Pullulanase can hydrolyze α-1,6 linkages efficiently so these two enzymes added in combination can hydrolyze wort dextrins to fermentable sugars effectively.

Carbohydrases are used in cereal cooking in the United States. The enzymes are added during brewing processes that employ unmalted cereal adjuncts such as rice, maize, or sorghum in the mash. Adjunct cereal starch is gelatinized incompletely under mashing conditions and requires an additional cereal cooking step to increase the exposure of adjunct starch to malt enzymes. Amylase treatment promotes gelatinization and liquefaction of the starch, allowing shorter cooking times. Treatment of the mash with thermostable bacterial α-amylase also permits a temperature reduction in the cereal cooker, leading to energy savings and more uniformity of cooker operations. Additional benefits of enzyme treatment include flavor and color characteristics of the final product. Both may be improved by reducing the severity of heat treatments that would release undesirable compounds.

Once the wort has been produced, commercial enzymes also may be used during the late stages of brewing, that is, during maturation, processing, and finishing (Fig. 4). Thus, added amylases can play a role in chillproofing, the production of sweet beers, and natural conditioning (Nielsen, 1971; Bass and Cayle, 1975; Briggs *et al.*, 1981; Marshall *et al.*, 1982; Godfrey, 1983a; Slaughter, 1985; Beckerich and Denault, 1987).

Chillproofing, a common practice in brewing, involves the addition of papain or protease from strains of *Aspergillus* and is used to reduce a colloidal haze. The haze is composed of polypeptides and a polyphenolic complex. This colloidal complex, which is intensified by chilling, is undesirable and can be a serious problem since most beers are served cold. The proteolytic preparation is added after fermentation and clarification, but before pasteurization. If the temperature is maintained below 60°C, some of the enzyme may continue to remove traces of the colloidal haze after pasteurization. A fungal α-amylase sometimes is blended with chillproofing proteases. This enzyme combination then can solubilize the remaining starch and the peptides that may cause turbidity in the final product.

Natural conditioning is a process that requires a second fermentation and is fueled by residual dextrins that increase the carbonation of draught beer. Glucoamylase or α-amylase can be added with a yeast inoculum to initiate the natural conditioning process.

The production of sweet beers also relies on an amylase supplement. Glucoamylase, usually from *Aspergillus* species, is added directly to the fermentor. The subsequent hydrolysis of residual dextrins releases glucose and generates the desired sweetness without a need for additive sweeteners.

4. High Fructose Corn Syrup Production

High fructose corn syrup (HFCS) is widely used as a caloric sweetener in foods and beverages. The worldwide market has increased about 20% every year since 1980 and now accounts for over 30% of the total nutritive sweetener consumption in the United States (Anonymous, 1988a; Angold *et al.*, 1989). When appropriate for use in product formulations, HFCS has several advantages. It is cheaper and less caloric than sucrose, but retains the same level of sweetness as common sugar, since it is 1.2–1.8 times sweeter than sucrose on a dry weight basis (Table III). The replacement of invert sugar by 42% (fructose w/w) HFCS in various food products and by 55% (fructose w/w) HFCS in soft drinks is expected to continue and probably expand, as it has in the last two decades, as long as its advantages remain and new products continue to emerge. The availability of HFCS containing 90% fructose (Anonymous, 1988a) and a convenient crystalline form (Anonymous, 1987; Bos, 1990) suggests new applications for this sweetener, as does the development of new technology for the conversion of HFCS to a crystalline form by sonic drying (Swientek, 1986; Davenport, 1988).

Immobilized glucose isomerases applied to the production of HFCS are prepared from various bacterial sources and used in packed bed reactors to

TABLE III
Relative Sweetness of Sweetening Agents Compared with Sucrose

Sweetening agent	Relative sweetness
Lactose	0.27
Galactose	0.35
D-Glucose	0.5–0.6
Maltose	0.60
Invert sugar	0.8–0.9
Sucrose	1.0
D-Fructose	1.2–1.8
Sodium cyclamate	30
Saccharin	200–700
Aspartame	200
Aryl ureas and trisubstituted guanidines	up to 200,000

convert glucose syrups to fructose. In the United States, corn is the preferred raw material for the production of such glucose syrups by the hydrolysis of starch with microbial amylases (Casey, 1977; Antrim *et al.*, 1979; MacAllister, 1979,1980; Pancoast and Junk, 1980; Bucke, 1981,1983a,b; Luesner, 1983; Carasik and Carroll, 1983; Landis and Beery, 1984; Barker and Petsch, 1985; Coker and Venkatasubramanian, 1985; Verhoff *et al.*, 1985; Long, 1986; Linko, 1987; Swaisgood and Horton, 1989; Pomerantz, 1991).

New alternative low-caloric sweeteners could join saccharine, xylitol, aspartame, and acesulfame K as substitutes for cane-, beet-, and corn-based caloric sweeteners in the United States and abroad. Sweeteners that fall into two related classes of aryl ureas and trisubstituted guanidines have been unveiled already (Anonymous, 1990). These molecules are up to 200,000 times sweeter than sucrose (Table III). The development of new products using such compounds could have significant impact on consumer preferences and future markets for sweetening agents, especially ones derived from traditional plant sources (Anonymous, 1986; Newsome, 1986; Kulp *et al.*, 1991; Marie and Piggott, 1991).

5. Distilled Alcoholic Beverage Production

Many alcoholic beverages are produced from plant starches or sugar-rich raw materials, for example, whiskey, bourbon whiskey, vodka, and brandies. The large-scale manufacture of these and many other alcoholic beverages often relies on the use of carbohydrases as aids for processing of starting materials. Malt, barley, corn, milo, and rye are common starch-bearing substrates for fermentations in the United States, whereas barley, malt, maize, potatoes, and rye are preferred in Europe. Potatoes, rye, and wheat serve as sources in the C.I.S. Rye and sweet potatoes are commonly used substrates in the Orient and cassava is used in tropical countries. The starch in these raw materials first must be digested to fermentable sugars that can fuel the ethanolic fermentation. Enzymes of vegetable origin, such as malt, or enzymes obtained by microbial surface culture, such as koji, have been used in traditional methods and are still employed today. Most modern processes, however, rely primarily on microbial enzymes for the starch conversion steps essential for alcohol production (Brandt, 1979; Poulson, 1983).

The starch conversion processes for distilled alcohol production are often similar to those used in beer brewing. Malt or one of various microbial enzymes is added in the mashing step. Local regulations may specify conditions related to the permissible amounts and proportions of the enzyme supplements. Glucoamylases from industrial strains of *Aspergillus* spp., typically *A. niger* or *A. awamori*, or *Rhizopus* spp., are used to reduce malt requirements to levels as low as 2% of the grain by weight. The raw plant material is cooked first, yielding a gelatinized starch susceptible to enzymatic degradation. The cooked mash is cooled to 20–25°C and saccharified with amylases, directly in fermentors when possible. Whereas malt enzymes are labile, the added fungal enzymes are stable under distillery conditions and do not lose activity after 96 hr in fermentors, even at pH 3.5 or lower. Treatment with microbial enzymes may have other advantages over processing exclusively with malt. Fungal glucoamylase saccharifies the starch more rapidly and completely than does malt. It generates less maltose, isomaltose, and oligosaccharides, the latter being virtually nonfermentable by yeast, and thus elevates the yield of fermentable sugars. Accompanying proteases in the enzyme preparation degrade grain protein and raise the levels of available nitrogen needed for yeast propagation. Glucoamylase treatment also increases the rate of fermentation, as well as the number of proof gallons per bushel of grain. Thus, the use of glucoamylase and α-amylase in conjunction with malt can convert an inexpensive grain mash efficiently into an excellent medium for the growth of distiller's yeast and the production of alcohol (Brandt, 1975; Maisch *et al.*, 1979; Poulson, 1983; Berry, 1984; Sobolov *et al.*, 1985).

6. Vinegar Fermentation

The production of vinegar depends on microbial fermentation as well as carbohydrase supplements to produce food-grade acetic acid. A yeast fermentation is used first to generate the alcohol that serves as substrate for the subsequent fermentation by special strains of *Acetobacter*. Large-scale vinegar fermentations today rely on the processing of vegetable biomass or fruit waste with added industrial enzymes to generate the substrate for the first stage of vinegar production. The suspension of fruit or vegetable pulp is treated with fungal or bacterial α-amylase and glucoamylase to digest the starch to glucose necessary for yeast growth. Acid-tolerant pectinases along with accompanying carbohydrases in enzyme mixtures obtained from *Aspergillus* strains complement the action of amylases and aid in the disruption of structural tissues in the plant cell wall. In Japan, koji made by the surface culture of *A. oryzae* is used as the source of amylolytic enzymes for rice vinegar production, whereas *Mucor* and *Rhizopus* strains are common sources of saccharifying enzymes in China (Adams, 1985; Godfrey, 1985).

B. Pectic Enzymes

Pectic enzymes, often collectively termed "pectinases," are a mixture of enzymes that act on pectic substances, plant polysaccharides that maintain the integrity of the cell wall or middle lamella. Pectic substances are acidic heteropolysaccharides of about 30,000 to 300,000 molecular weight and consist mainly of pectin, a polymer of D-galacturonic acid. At least 75% of the monomers of D-galacturonic acid are esterified with methanol (methoxylated) or with rhamnogalacturonans, galacturonans, galactans, arabinogalactans, and arabinans (Fogarty and Ward, 1974; MacMillan and Sheiman, 1974; Whitaker, 1984).

The presence of pectic substances during fruit and juice processing may lead to serious technical problems. Fungal pectic enzyme mixtures are used in commercial applications to remove pectic substances, typically serving as processing aids in fruit juice extraction and clarification. Several key enzymes present in such industrial enzyme preparations, mainly pectinase, pectase, and pectin lyase, act on specific plant cell wall polysaccharides present in the food material. Pectinases, also termed polygalacturonases, degrade polygalacturonic acid and other polymers composed of D-galacturonic acid to soluble oligosaccharides. Pectinases are endoenzymes that are able to split interior linkages and prefer low-methoxyl pectin or completely deesterified pectin, termed pectate. Pectases, also called pectin methylesterases or pectinesterases, remove methanol from esterified carboxyl groups and convert pectins to low-methoxyl pectins and eventually to completely deesterified pectate. Pectin lyases are endoenzymes that randomly break glycosidic linkages next to a methyl ester group, and prefer high-methoxyl pectins. Commercial pectic enzyme preparations can contain pectinases, pectases, and pectin lyases, as well as other polysaccharidases that may act on pectic substances. For example, a common commercial source of pectic enzymes such as *A. niger* may contain activities many of which can play a role in degrading plant cell wall substances (Table IV). These accompanying enzymatic activities also can have a beneficial effect on the final characteristics of the food material after enzyme treatment (Pilnik and Rombouts, 1981; Whitaker, 1984; Linhardt *et al.*, 1986; Voragen and Pilnik, 1989; Ward, 1989; Walter, 1991).

Pectic enzymes usually are produced on an industrial scale with improved strains of filamentous fungi. *Aspergillus* species are especially common sources, although *Coniothyrium diplodiella*, *Sclerotinia libertiana*, and species of *Botrytis*, *Penicillium*, and *Rhizopus* also are used for enzyme production. The fungal strains are grown by submerged fermentation in deep-tank fermentors or by semisolid substrate fermentation in rotary drums or trays. Submerged fermentation is more convenient and is favored by many industrial produc-

TABLE IV
Enzymes Produced by *Aspergillus niger* that Hydrolyze Polymers[a]

Substrate	Enzymes
Arabinans	α-L-arabinofuranosidases
Cellulose	C_1-, C_X-type cellulases
Dextran	Dextranase
DNA, RNA	Deoxyribonuclease, ribonuclease
β-Glucans	β-glucanase
Hemicellulose	Hemicellulases
Inulin	Inulinase
Mannans	β-mannanase
Pectic substances	Pectin methylesterase, pectate lyase, polygalacturonase
Proteins	Proteases
Starch	α-amylase, glucoamylase
Xylans	Xylanase

[a] Reprinted with permission of Butterworth–Heinemann from Whitaker (1984), *Enzyme and Microbial Technology* **6**, 341–349.

ers. Generally, an acid medium of about pH 3.5 is supplemented with pectin and a cheap carbon source containing sucrose, lactose, glucose, or carbohydrate mixtures. The added pectin, which serves as an inducer of pectic enzymes, is typically waste material from fruit or vegetable processing, for example, apple pomace, citrus peel, or beet pulp. The deep-tank fermentations take 3–6 days. Pectic enzyme production usually reaches a maximum when the carbohydrate is exhausted. Cell-free fermentation fluid from the submerged culture, which bears extracellular pectic enzymes, usually is sold as is or concentrated. In contrast, the culture mass obtained after semisolid fermentation is extracted to produce an enzyme solution. Pectic enzyme mixtures from submerged or semisolid substrate culture can be precipitated with inorganic salts or organic solvents and dried to a powder, milled, and finally mixed with stabilizer or inert ingredients (Nyiri, 1969; Rombouts and Pilnik, 1980; Fogarty and Kelly, 1983).

1. Fruit and Juice Processing

Fruit and juice processors have relied on fungal pectic enzymes to facilitate processing of raw material and to improve the properties of the final product. The production of a variety of sparkling clear juices and cloud juices depends on such an enzyme treatment (Fogarty and Ward, 1974; Neubeck, 1975; Rombouts and Pilnik, 1980; Bauman, 1981; Pilnik and Rombouts, 1981; Pilnik, 1982; Whitaker, 1984; Voragen and Pilnik, 1989). An example of an industrial process for making fruit juice with pectic enzymes is shown in Fig. 5.

Juices extracted from fruits can contain solids composed mainly of pectic substances. This material may constitute 5–10% of the fresh weight of tree-

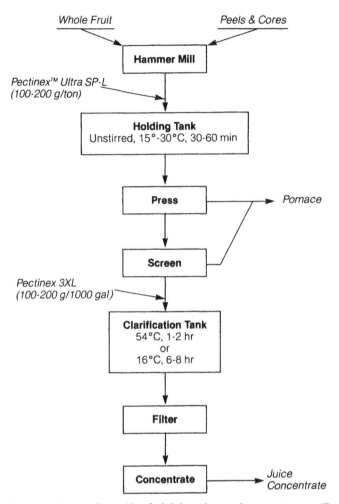

Figure 5 Process for making fruit juice using pectinase treatments. (Reprinted with permission from Boyce, 1986.)

borne fruit and may affect juice appearance and texture. The pectic substances may interfere with further processing or give rise to undesirable
characteristics in the finished food product. In the United States and other
countries, pectic enzyme mixtures from *A. niger* or other fungi commonly are
used in apple and grape juice processing to reduce the cloudiness formed by
colloidal suspension of pulp, to increase the filterability of the juice, and to
prevent gelling of pectin in concentrated juice products. When added to
juices, pectic enzymes partially hydrolyze the soluble pectin to smaller particles, which can then flocculate and be removed easily. A nonenzymatic process that involves neutralization of electrostatic charges on the partially
degraded particles also plays a role in pectin removal. Depectinization of the

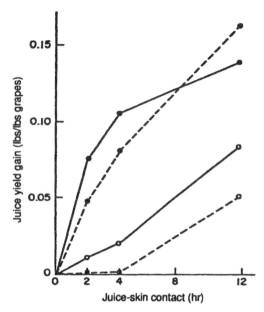

Figure 6 Effect of added pectic enzymes on juice yield. O, no added enzymes; ●, added enzymes; ——, Chenin blanc grapes; –––, Muscat of Alexandria grapes. (Reprinted with permission from Ough and Crowell, 1979.)

pulp not only improves the clarity and viscosity of sparkling juices such as grape and apple juices, but also increases yield by 10–20% (Fig. 6). Enzyme treatment can increase the efficiency of pressing and juice extraction and even enhance flavor and color release, especially during continuous pressing operations (Neubeck, 1975; Ough and Crowell, 1979; Bauman, 1981; Pilnik, 1982; Whitaker, 1984; Downing, 1988; Voragen and Pilnik, 1989).

Many noncitrus tree fruits, berry fruits, and tropical fruits are manufactured with the aid of pectinases. These fruits include black currants, cherries, raspberries, strawberries, and bananas. The type, variety, and maturity of the fruit can determine the content of pectic substances, the ratio of soluble to insoluble material, and, thus, the processing variables important for juice extraction. These factors also influence the nature of the enzyme treatment and the stage in the manufacturing process at which it is important. The concentration, temperature, and reaction time must be adjusted in response to these variables (Neubeck, 1975).

Pectinase treatment can benefit citrus fruit processing in several ways. Such treatment can improve the stability of the cloud, a desirable feature in many citrus juices (Pilnik, 1982; Whitaker, 1984). Pectinase injected into citrus fruits dissolves the albedo and loosens peels, according to studies by the U.S. Department of Agriculture. The peels then can be removed more easily by machines, yielding high quality sectioned fruit that can be canned or eaten as is (Anonymous, 1988b). In addition, citrus oils such as lemon oil can be extracted with the help of pectic enzymes. Pectic enzymes are useful in these applications since they destroy the emulsifying properties of pectin, which interferes with the collection of oil from citrus peel extracts (Scott, 1978).

2. Wine Making

Pectic enzymes from fungi, typically A. *niger, Penicillium notatum,* or *Botrytis cinerea,* are useful in wine making. Although juices of berries, peaches, apples, pears, and other fruits are employed in wine making (Cruess and Besone, 1941; Martiniere *et al.,* 1973; Neubeck, 1975; Ferenczi and Asvany, 1977; Robertson, 1977; Ough and Crowell, 1979; Haupt, 1981; Pilnik, 1982; Felix and Villettaz, 1983; Fogarty and Kelly, 1983; Villettaz, 1984), grape wines are produced in the greatest volume and will be considered here.

Pectic enzymes are used to reduce haze or gelling of grape juice at various stages of the wine making process. Enzymes may be added at any one of three stages: at the first stage, while the grapes are being crushed; at the second stage, which involves the must (free-run juice) before its fermentation or after; or at the last stage, once the fermentation is complete, when the wine is ready for transfer or bottling. The addition of pectic enzyme at the first stage is preferred since it increases the volume of the free-run juice and reduces pressing time. It also aids in juice filtration and must clarification. The increase in juice yield per ton of grapes can be economically significant: a 10% increase can generate 16 additional gallons of juice per ton of grapes. Further, pectic enzyme treatment promotes extraction of pigments when the grapes are heat-extracted or fermented on the skin. The color of the juice is generally superior and the color of red wines is usually more intense (Martiniere *et al.,* 1973; Ough and Berg, 1974; Ough *et al.,* 1975; Ferenczi and Asvany, 1977; Ough and Crowell, 1979). Pectic enzyme preparations from A. *niger* and A. *oryzae* even may contain an anthocyanase that hydrolyzes colored anthocyanins to colorless derivatives, thereby lightening the shade of red wines (Huang, 1955).

Treatment of the juice with pectic enzymes at the second stage before or during fermentation settles out many suspended particles and often some undesirable microorganisms. A firmer yeast sediment and clearer wine results. Addition of pectic enzymes to the fermented wine at the last stage increases the filtration rate and clarity, but the level of enzyme supplement must be adjusted owing to the inhibitory effect of alcohol on pectinases. Supplementation at this point promotes flocculation and precipitation of pectin particles, floating microoganisms, and protein. Elimination of protein improves the stability of the wine and reduces the reliance on bentonite for juice clarification. The use of pectic enzyme during all three stages of wine making promotes a faster aging of the wine. However, the effect of enzyme treatment on the flavor and bouquet is more difficult to evaluate. Some sensory tests indicate that wine quality is enhanced whereas others reveal little improvement (Neubeck, 1975; Ferenczi and Asvany, 1977; Ough and Crowell, 1979).

3. Coffee and Tea Fermentation

Pectinases play a role in coffee and tea fermentation. The processing of coffee depends on fermentation with pectinolytic microorganisms to remove the mucilage coat from the coffee beans. Fungal pectic enzymes are sometimes added to the fermentation to remove the pulpy layer of the bean, three-fourths of which consists of pectic substances. Cellulase and hemicellulase present in the enzyme preparation aids the digestion of the mucilage. A diluted commercial enzyme preparation is sprayed onto the cherries at a dose of 2–10 g per ton at 15–20°C. The fermentation stage of coffee processing is accelerated by the enzymatic treatment and is reduced from 40–80 hr to about 20 hr. The decanted liquid from the fermentation can be reused for several additional enzyme treatments. However, since such large-scale treatments with commercial pectinases can be costly and uneconomical, inoculated waste mucilage often is fermented in tanks and used as a source of microbial pectic enzymes. The fermentation liquid is washed, filtered, and then sprayed onto the cherries. About 10 kg crude enzyme is sufficient to treat 1000 kg ripe cherries. The subsequent coffee fermentation is reduced to about 12 hr after application of crude enzyme solution (Amorim and Amorim, 1977; Arunga, 1982; Castelein and Verachtert, 1983; Jones and Jones, 1984; Carr, 1985; Godfrey, 1985).

Fungal pectinase is used in the manufacture of tea. Pectinase treatment accelerates tea fermentation, although the enzyme dose must be adjusted carefully to avoid damage to the tea leaf. The addition of pectinase also improves the foam-forming property of instant tea powders by destroying tea pectins (Sanderson and Coggon, 1977; Sanderson, 1983; Carr, 1985; Willson and Clifford, 1992).

C. Lactases

Lactose is the main sugar in milk and whey, constituting about 5% (w/v). The enzyme lactase acts as a β-galactosidase and hydrolyzes this sugar to glucose and galactose, two monosaccharides that are sweeter (Table III), more digestible, and more soluble than lactose (Shukla, 1975; Klostermeyer et al., 1978; Severinsen et al., 1979; Barker and Shirley, 1980; Coughlan and Charles, 1980; Fox, 1980; Miller and Brand, 1980; Ahn and Kim, 1981; Linko and Linko, 1983; Gekas and Lopez-Leiva, 1985; Mahoney, 1985; Gerhartz, 1990).

Commercial lactases have a number of uses in the food industry. These applications often are related to the development of new milk-derived products. Lactases currently are used to transform cheese whey, formerly a discarded waste product of cheese manufacture, into a sweet syrup used in food processing and as a component of fermentation media, as well as to produce low-lactose milk and other dairy products. Lactose-derived syrups can serve as replacements for corn syrup or sucrose in bakery, dairy, soft drink, dessert, or confectionary products since they have favorable properties, among them good humectancy and the ability to promote Maillard browning. When added to dairy foods such as yogurt, sour cream, and buttermilk, lactases improve the taste without significantly increasing caloric content. Lactases also are used to reduce lactose crystallization in dairy products. Lactase treatment of three-fold concentrated milk before freezing reduces thickening and extends its storage life. Enzyme treatment also improves the flavor of the reconstituted product. Addition of lactase to milk destined for yogurt manufacture accelerates acid development by the starter cultures and increases the sweetness, viscosity, and shelf life of the fermented product. In a similar way, lactase treatment hastens the acidification process during cheese manufacture. Enzyme treatment promotes a firmer, more elastic cottage cheese curd, in addition to reducing set time. In the case of ripened cheeses, lactase supplementation significantly shortens the ripening process and reduces costs, while also enhancing flavor development. Other enzymes such as proteases that accompany lactase also may contribute to the ripening process (Woychik and Holsinger, 1977; Barker and Shirley, 1980; Coughlan and Charles, 1980; Moore, 1980; Nijpels, 1981; Böing, 1982; Burgess and Shaw, 1983; Crueger and Crueger, 1984; Moulin and Galzy, 1984; Mahoney, 1985; Swaisgood and Horton, 1989).

Commercial lactase is produced from *A. niger* or *A. oryzae* as well as from the yeasts *Kluyveromyces marxianus* var. *lactis*, *K. fragilis*, and *Candida pseudotropicalis*. The enzymes from these organisms have been well characterized. The yeast enzyme functions best at neutral pH and is used to treat milk or sweet whey, whereas *A. niger* lactase is more thermostable than the yeast enzyme and has a broader but lower pH optimum of 4–5. This enzyme is used primarily to process acid wheys, but still can be used at pH 6.5. Such fungal lactases in immobilized form are used for industrial hydrolysis of lactose in milk or whey. The systems are designed with the quality of the input material in mind, it is sensitive to the presence of milk proteins and impurities. Pretreatment, quite often pasteurization, ultrafiltration, and demineralization, are necessary (Coughlan and Charles, 1980; Greenberg and Mahoney, 1981; Richmond *et al.*, 1981; Gekas and Lopez-Leiva, 1985; Mahoney, 1985; Van Griethuysen-Dilber *et al.*, 1988; Axelsson and Zacchi, 1990).

Processes with the yeast or *A. niger* lactase are in commercial operation. Immobilized lactase from *K. marxianus* var. *lactis* in triacetate fibers can produce up to 10 tons of low-lactose milk per day (Moulin and Galzy, 1984). Immobilized mycelia of *A. niger* with high lactase activity are able to convert cheese whey into a sweet syrup that can serve as an ingredient in baked goods, ice cream, yogurt, canned fruit, candies, and other foods (Finnocchario *et al.*, 1980). The Corning process (Fig. 7), which uses lactase immobilized on porous glass beads, yields approximately 1.7 tons of hydrolyzed syrup from approximately 30,000 liters of whey. More than 80% of the input lactose is

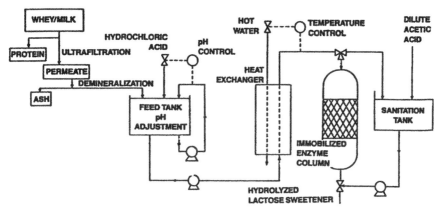

Figure 7 Commercial process with immobilized lactase that hydrolyzes lactose using whey, whey permeate, or milk permeate as starting material. (Reprinted with permission from Moore, 1980), *Food Product Development,* now *Prepared Foods,* Delta Food Group, Delta Communications, Inc., Chicago, Illinois.

hydrolyzed to a mixture of glucose and galactose, which has about 70% of the sweetness associated with sucrose (Table V). Each unit for lactose hydrolysis operates continuously at pH 5.0 and 50°C, and is able to process 360–500 liters of lactose per hr (Moore, 1980; Mahoney, 1985; Swaisgood and Horton, 1989).

New methods in packaging and aseptic processing may lead to novel applications for lactase in milk products in the future. Research on "process-in-package" approaches has considered the incorporation of lactase into

TABLE V
Proximate Composition and Properties of Hydrolyzed Lactose Syrup[a]

Total solids	61%
Total ash	0.8%
Protein (N × 6.38)	0.55%
Lactose	12%
Glucose	22 g/liter
Galactose	24 g/liter
Total plate count/ml	—
Yeast, m/liter	—
Sweetness (sucrose = 100)	70
Viscosity at 20°C	380 cps
Solubility at 25°C	60%

[a] Reprinted with permission from Moore (1980), *Food Product Development,* now *Prepared Foods,* Delta Food Group, Delta Communications, Inc., Chicago, Illinois.

special milk cartons that could serve as enzyme reactors. The milk cartons would convert lactose into glucose and galactose during transit or storage. It has been demonstrated that lactase can be retained in the container material and thus would not be consumed along with the dairy products (Rice, 1988). Such built-in enzyme components of food packaging could yield a lactose-free milk product and thereby make nutritional dairy products available to many lactose-intolerant people, especially lactase-deficient individuals. Lactase supplements taken orally or added to milk products already serve as a digestive aid for lactose intolerance, a phenomenon characteristic of a large percentage of the world's population (Houts, 1988).

In another approach, an enzyme-dosing system designed for individual packages of aseptically processed foods, especially milk products, is expected in the near future. The technology allows the introduction of measured doses of a liquid enzyme such as lactase into a sterilized milk line and the direct piping into an aseptic filling machine. Hydrolysis of lactose in the milk product would occur within the package in several days. The postprocess supplementation of lactase to the packaged product averts possible heat inactivation of the enzyme; it also avoids the caramelization of sugars released by the enzymatic process. These new approaches in lactase treatment of dairy products would not alter the nutritional quality of the milk, but would permit the conversion of lactose to more digestible sugars in addition to increasing the sweetness of the final product (Duxbury, 1989).

D. Invertases

Yeast invertase hydrolyzes the terminal nonreducing portion of β-fructofuranosides and can be used to digest disaccharides, trisaccharides, and fructans. The primary use of invertase is the conversion of sucrose to fructose and glucose, both termed invert sugars (Kulp, 1975; Wiseman, 1981; Pedersen and Norman, 1987). Yeast invertases are used to manufacture confections, syrups, desserts, and artificial honey. In the confectionary industry, they are important for the production of fondants, chocolate coatings, and chocolate-coated candies with a soft center. In the last application, a solid sucrose core containing a dose of invertase is coated; liquefaction is caused by the action of the enzyme during storage (Ingleton, 1963; Szekely, 1970; Wiseman, 1981).

Commercial invertase preparations from *Saccharomyces cerevisiae* or *S. carlsbergensis* are obtained from yeast cultivated on molasses. The cells are treated with organic solvents to remove the enzyme from the periplasmic space (Böing, 1982), although autolysis may prove to be a superior method of enzyme extraction in the future. Subsequent purification of the enzyme preparation is essential to eliminate objectionable flavors originating from the yeast culture. The final product has a broad pH optimum and is relatively heat stable, but can be stabilized further with cross-linking agents. (Wiseman, 1981; Crueger and Crueger, 1984; Monsan and Combes, 1984).

In addition to its main confectionary applications, yeast invertase is used to make syrups of invert sugars, melibiose from raffinose, gentiobiose from gentianose, and D-fructose from inulin. The enzyme also is used in the recovery of scrap sucrose products and for the hydrolysis of sucrose in molasses-based microbial fermentations (Barker and Shirley, 1980; Woodward and Wiseman, 1982; Wiseman, 1983; Buchholz and Rabet, 1987). Immobilized fungal invertases have been investigated as a means for converting sucrose to sugar mixtures (Reilly, 1980; Wiseman, 1981; Crueger and Crueger, 1984; Monsan and Combes, 1984; Pedersen and Norman, 1987). The resulting syrups containing glucose, fructose, and sucrose have advantages: they do not crystallize as readily when present as mixtures and they are free of the colored by-products generated by acid hydrolysis of sucrose. Conversions of 50–72% of the substrate can be achieved in reactors containing invertase immobilized in ceramic membranes or onto agricultural by-products (Monsan and Combes, 1984; Nakajima *et al.*, 1988). Although high yields are possible, the application of invertase has been limited because of the success of processes that use cheaper raw materials and the amylase/glucose isomerase process for starch conversion to invert sugar.

E. α-Galactosidases

The enzyme α-galactosidase from *Mortierella vinaceae* var. *raffinose utilizer*, also termed raffinase or melibiase, is used to hydrolyze raffinose to sucrose and galactose during beet sugar refining (Suzuki *et al.*, 1969; Kobayashi and Suzuki, 1972; Kaneko *et al.*, 1990). Raffinose interferes with the crystallization of sucrose obtained from molasses and α-galactosidase reduces the levels of this trisaccharide. Concentrations of 0.05–1.5% raffinose are detrimental to the refining process since they significantly reduce yields by promoting the formation of fine needles of sucrose rather than the desired large crystals (Reilly, 1980; Lindley, 1982; Mattes and Beaucamp, 1982; Crueger and Crueger, 1984). In sucrose processing, α-galactosidase is used in the form of mycelial pellets obtained from a strain of *M. vinaceae* that lacks invertase activity. Suspensions of the pellets with bound enzyme are maintained in horizontal enzyme reactors and molasses containing raffinose is directed through the chambered troughs (Shimizu and Kaga, 1972). Such a continuous process can convert about 65% of the raffinose in molasses to sucrose, thus making it available for crystallization and recovery. A sugar processing plant that refines 600 tons of sucrose per day from 3000 tons of sugar beets reportedly can convert up to 3.25 tons of raffinose to sucrose (Scott, 1975; Obara *et al.*, 1977; Linden, 1982; Blanch, 1984).

The enzyme α-galactosidase also may be used in the processing of legume products, especially soy products, which are rich in galacto-oligosaccharides such as stachyose, verbascose, and raffinose (Liener, 1977). These compounds cause flatulence and gastric distress when ingested and can be degraded by supplemented α-galactosidase. The enzyme can be obtained from the food-approved fungi *A. oryzae* (Annunziato *et al.*, 1986) and *A. niger* (Agnantiari *et al.*, 1991) or can act on food substances during fermentation mediated by species of *Rhizopus* (Liener, 1977).

F. Cellulases

Cellulases are carbohydrases that cleave the β-1,4 linkages of cellulose or its chemically modified forms, in addition to degrading cellodextrin or cello-biose. Typically, they are multienzyme complexes bearing endo-1,4-β-glucanase, cellobiohydrolase, and β-glucosidase activity (Enari, 1983a; Wood, 1985; Kubicek *et al.*, 1990; Goyal *et al.*, 1991; Teeri *et al.*, 1992). Fungal cellulases are used alone or in conjunction with pectinases, β-glucanases, and starch-degrading enzymes in brewing, cereal processing, fruit and juice pro-cessing, food fermentations, wine production, and alcohol fermentation. Cellulases also have been used to improve the palatability of low-quality vegetables, increase the flavor of mushrooms, promote the extraction of natural products, and alter the texture of foods. Research efforts are being devoted to the conversion of food processing wastes to food ingredients, single-cell protein, or substrates for microbes that convert biomass to fuels (Ghose and Pathak, 1973; Emert *et al.*, 1974; Fox, 1974; Halliwell, 1979; Goksøyr and Ericksen, 1980; Enari, 1983b; Coughlan, 1985,1992; Mandels, 1985; Montenecourt and Eveleigh, 1985; Ward, 1985; Wood, 1985; Beguin and Gilkes, 1987; Beguin *et al.*, 1988; Wood and Kellogg, 1988; Aubert *et al.*, 1988; Sukan, 1988; Scott, 1989; Marek *et al.*, 1990; Pokorny *et al.*, 1990; Walker and Wilson, 1990).

Generally, *A. niger* is the source of cellulase destined for food use and *Trichoderma viride* is used for nonfood applications, although both enzymes can fulfill many tasks. *T. viride* is grown by submerged fermentation in the presence of inducers, typically cellulose, whereas large-scale cultivation of *A. niger* usually involves surface fermentation. The enzymes are concentrated and partially purified before being sold as solutions or vacuum-dried powders (Scott, 1978; Böing, 1982; Vanbelle *et al.*, 1982; Enari, 1983a; Frost and Moss, 1987; Gerhartz, 1990; Pokorny *et al.*, 1990; Esterbauer *et al.*, 1991).

G. Hemicellulases

Hemicelluloses are alkali-soluble polysaccharides exclusive of cellulosic or pectic substances found in plant cell walls. Hemicellulases obtained from various microorganisms can break down these polysaccharides. Fungi are the most common industrial sources for hemicellulases such as glucanases, xylan-ases, galactanases, mannanases, galactomannanases, and pentosanases. These enzymes are often by-products of commercial processes for cellulase or pectinase production and some are valuable aids in food processing (Dekker,

1979, 1985; Fincher and Stone, 1981; Meier and Reed, 1981; Woodward, 1984; Biely, 1985; Aubert *et al.*, 1988; Wong *et al.*, 1988; Wood and Kellogg, 1988; Ward and Moo-Young, 1989; Araujo and Ward, 1990).

Microbial hemicellulases are employed to reduce the levels of barley β-glucans, compounds that can be a nuisance during beer brewing. Barley β-glucans are unbranched polymers of β-linked D-glucosyl residues and often are termed barley gums, mixed linkage β-glucans, or $(1-3)$ $(1-4)$-β-D-glucans (Bamforth, 1982; McCleary, 1986). These carbohydrates are major constituents of barley endosperm cell walls, representing about 75% of their total carbohydrate content. The glucans may be degraded completely to glucose by barley β-glucanases during the malting of barley grain. However, if they are broken down only partially, a high molecular weight viscous material consisting primarily of glucans is released into solution. This incompletely digested cell wall material may impede starch digestion. Supplemented enzyme may be essential to degrade the residual β-glucans, especially if the malting process is accelerated or if unmodified barley adjuncts are used. *Bacillus subtilis* or another bacterial β-glucanase may be added to the wort to aid the digestion of cell wall material from barley endosperm. This supplement also will promote exposure of starch granules to amylases. *A. niger*, *Trichoderma reesei*, *T. viride*, and *Penicillium emersonii* are other microorganisms that can serve as sources of β-glucanase for brewing applications. Glucanases from these fungi also are used to reduce wort viscosity by digesting viscous residual barley β-glucans that may clog pumps and filters in breweries. Preparations of *T. viride* xylanase can perform a similar task by degrading xylans that cause viscosity problems during transfer operations (Scott, 1972; Bathgate and Dalgliesh, 1975; Stentebjerg-Olesen, 1980; McCleary, 1986; Anonymous, 1989; Todo *et al.*, 1989).

As during the early stages of brewing that lead to wort production, viscosity problems can arise during the late stages that involve beer transfer and filtration operations (Takayanagi *et al.*, 1969; Enkenlund, 1972; Bass and Cayle, 1975; Leedham *et al.*, 1975; Bournes *et al.*, 1976; Narziss, 1981; Bamforth, 1982; McCleary, 1986; Canales *et al.*, 1988). The addition of β-glucanase to beer increases filtration throughput, reduces the need for filter aids, and improves its clarity and stability (Fig. 8). In addition, enzyme treatment shortens lautering times and increases brewhouse yield. Processing with enzymes is especially important when the brewer uses unmalted barley containing elevated levels of particulate matter composed of nonstarch polysaccharides, primarily hemicelluloses. The source of the β-glucanase for this brewing application is typically *B. subtilis*, *A. niger*, or *T. reesei*.

Hemicellulases also are used to improve the properties of doughs used in the production of baked goods (Kulp, 1968). Fungal pentosanase has been used to hydrolyze wheat hemicellulose that could lead to a coarser bread crumb. Treatment of bread dough with purified fungal xylanase significantly decreases dough strength, yielding loaves with an open crumb structure. Bread made with wheat and guar flour treated with *T. viride* xylanase reportedly has a 12% greater loaf volume and height than control loaves. The

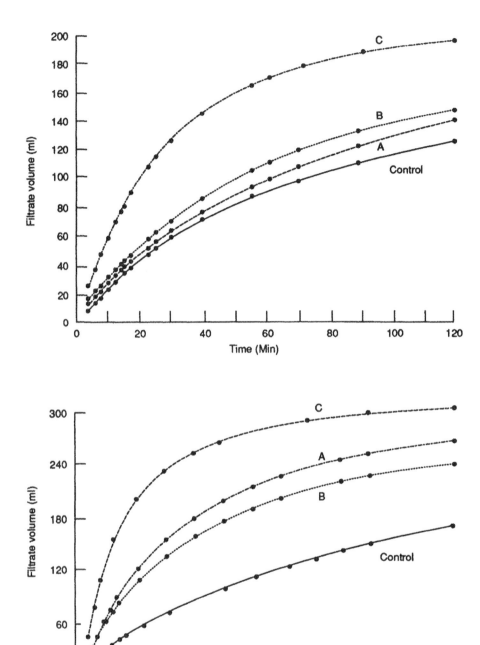

Figure 8 Wort filtrate volume with respect to time for a 50% malt/15% barley/35% rice laboratory mash supplemented with β-glucanase from *Bacillus subtilis* (– · –), *Aspergillus niger* (——), and *Trichoderma reesei* (– – –). The control mash (——) was untreated. (Reprinted with permission from Canales *et al.*, 1988.)

crumb texture is finer and the crumb color is superior. *T. viride* pentosanase and xylanase preparations are also useful in degrading wheat gums and fibers during the processing of wheat starch and gluten (McCleary *et al.*, 1986; McCleary, 1987; Wong *et al.*, 1988; Scott, 1989).

Hemicellulases have been used to process several other plant-derived materials. The enzyme β-glucanase has been applied to grape wine clarification and filtration (Villettaz *et al.*, 1984). In yet another application, fungal β-glucanase is used during the manufacture of instant coffee to degrade coffee mucilage and prevent gelation of liquid coffee concentrate (Amorim and Amorim, 1977; Arunga, 1982; Jones and Jones, 1984). Mannanases and galactanases have been used to hydrolyze coffee bean mannans and galactans during coffee production (Hashimoto and Fukumoto, 1969; Godfrey, 1983b).

H. Dextranases

Dextran is a polysaccharide synthesized by *Leuconostoc mesenteroides* or *L. dextranicum*. Both microorganisms can convert sucrose to fructose and then dextran via the enzyme dextransucrase. Dextran is a glucan of 1.5×10^4 to 2×10^7 molecular weight or greater. It contains primarily α-1,6 linkages and branches consisting of α-1,3 and α-1,4 linkages, although the degree of branching is variable. During the manufacture of sugar, species of *Leuconostoc* can contaminate sugar cane juice or juice from temperature-abused beets, resulting in the appearance of dextran. The polysaccharide interferes with beet sugar refining and reduces the economic efficiency of the mill. The increased viscosity of the contaminated juice is associated with slower heating, a reduction in the rate of sucrose crystallization, increased turbidity, slower filtration, and the presence of elongated sugar crystals. The application of fungal dextranase during the sugar refining process greatly reduces these difficulties. Treatment of the juice with a dextranase preparation from *Penicillium funiculosum* or *P. lilacinum* for 20 min at 40°C and pH 5.4 degrades 68% of the dextran, mainly to isomaltose and isomaltotriose, sugars that do not interfere with sucrose crystallization. The specific viscosity of the juice is reduced significantly by the enzyme treatment and the processing rate is increased (Keniry *et al.*, 1967; Foster, 1969; Imrie and Tilbury, 1972; Tilbury, 1972; Kosaric *et al.*, 1973; Scott, 1975; Abram and Ramage, 1979; Godfrey, 1983c; Barfoed and Møllgaard, 1987).

References

Abram, J. C., and Ramage, J. S. (1979). Sugar refining: Present technology and future developments. *In* "Sugar Science and Technology" (G. G. Birch and K. J. Parker, eds.), pp. 49–95. Elsevier Applied Science, London.

Adams, M. R. (1985). Vinegar. *In* "Microbiology of Fermented Foods" (B. J. B. Wood, ed.), Vol. 1, pp. 1–47. Elsevier Applied Science, London.

Agnantiari, G., Christakopoulos, P., Kekos, D., and Macris, B. J. (1991). A purified α-galactosidase from *Aspergillus niger* with enhanced kinetic characteristics. *Acta Biotechnol.* **11**, 479–484.

Ahn, J. K., and Kim, H. U. (1981). A review on beta-galactosidases used in milk and milk products. *Kor. J. Dairy Sci.* **3**, 47–74.

Allen, W. G., and Spradlin, J. E. (1973). Amylases and their properties. *Brewers Dig.* **48**, 48–50, 52–53, 65.

Amorim, H. V., and Amorim, V. L. (1977). Coffee enzymes and coffee quality. *In* "Enzymes in Food and Beverage" (R. L. Ory and A. J. St. Angelo, ed.), pp. 27–56. American Chemical Society, Washington, D.C.

Amos, A. J. (1955). The use of enzymes in the baking industry. *J. Sci. Food Agric.* **6**, 489–495.

Angold, R., Beech, G., and Taggart, J. (1989). "Food Biotechnology." Cambridge University Press, Cambridge.

Annunziato, M. E., Mahoney, R. R., and Mudgett, R. E. (1986). Production of α-galactosidase from *Aspergillus oryzae* grown in solid state culture. *J. Food Sci.* **51**, 1370–1371.

Anonymous (1986). Sweeteners. *Food Technol.* **40**(1), 112–130.

Anonymous (1987). Crystalline fructose: A breakthrough in corn sweetener process technology. *Food Technol.* **41**(1), 66–67, 72.

Anonymous (1988a). High fructose corn syrup comes of age. *Dairy Field* **March**, 43, 45.

Anonymous (1988b). USDA develops vacuum infusion process for citrus. *Food Bus.* **Jan.**, 16–17.

Anonymous (1989). "Product Information: Bio-xylanase." Biocon, Lexington, Kentucky.

Anonymous (1990). World's sweetest chemicals? *Chem. Ind.* **March**, 153.

Antrim, R. L., Colilla, W., and Schnyder, B. J. (1979). Glucose isomerase production of high-fructose syrups. *In* "Applied Biochemistry and Bioengineering. Enzyme Technology" (L. M. Winegard, Jr., E. Katchalski-Katzir, and L. Goldstein, eds.), pp. 97–155. Academic Press, New York.

Araujo, A., and Ward, O. P. (1990). Extracellular mannanases and galactanases from selected fungi. *J. Ind. Microbiol.* **6**, 171–178.

Arunga, R. C. (1982). Coffee. *In* "Economic Microbiology. Fermented Foods" (A. H. Rose, ed.), pp. 259–292. Academic Press, London.

Aubert, J.-P., Beguin, P., and Millet, J. (1988). Biochemistry and Genetics of Cellulose Degradation." Academic Press, San Diego.

Axelsson, A., and Zacchi, G. (1990). Economic evaluation of the hydrolysis of lactose using immobilized beta-galactosidase. *Appl. Biochem. Biotechnol.* **24**, 679–694.

Bamforth, C. W. (1982). Barley β-glucans. Their role in malting and brewing. *Brewers Dig.* **57**, 22–27.

Barfoed, S., and Møllgaard, A. (1987). Dextranase solved dextran problems in DDS' beet sugar factory. *Zuckerind.* **112**, 391–395.

Barker, S. A., and Petch, G. S. (1985). Enzymatic processes for high-fructose corn syrup. *In* "Enzymes and Immobilized Cells in Biotechnology" (A. I. Laskin, ed.), pp. 93–107. Benjamin/Cummings, Menlo Park, California.

Barker, S. A., and Shirley, J. A. (1980). Glucose oxidase, glucose dehydrogenase, glucose isomerase, β-galactosidase, and invertase. *In* "Economic Microbiology. Microbial Enzymes and Bioconversions" (A. H. Rose, ed.), pp. 171–226. Academic Press, London.

Barrett, F. F. (1975). Enzyme uses in the milling and baking industries. *In* "Enzymes in Food Processing" (G. Reed, ed.), 2d Ed., pp. 301–330. Academic Press, New York.

Bass, E. J., and Cayle, T. (1975). Beer. *In* "Enzymes in Food Processing" (G. Reed, ed.), 2d Ed., pp. 455–471. Academic Press, New York.

Bathgate, G. N., and Dalgliesh, C. E. (1975). The diversity of barley and malt β-glucans. *Proc. Am. Soc. Brew. Chem.* **33**, 32–36.

Bauman, J. W. (1981). Application of enzymes in fruit juice technology. *In* "Enzymes and Food Processing" (G. G. Birch, N. Blakebrough, and K. J. Parker, eds.), pp. 129–147. Applied Science, London.

Beckerich, R. P., and Denault, L. J. (1987). Enzymes in the preparation of beer and fuel alcohol. *In* "Enzymes and Their Role in Cereal Technology" (J. E. Kruger, D. Lineback, and C. E. Stauffer, eds.), pp. 335–355. American Association of Cereal Chemists, St. Paul, Minnesota.

Beguin, P., and Gilkes, N. R. (1987). Cloning of cellulase genes. *CRC Crit. Rev. Biotechnol.* **6,** 129–162.

Beguin, P., Grepinet, O., Millet, J., and Aubert, J. (1988). Recent aspects in the biochemistry and genetics of cellulose degradation. *8th Int. Biotechnol. Symp.* **2,** 1015–1029.

Berger, M., and Granvoinnet, P. (1974). Comparison de l'action des alpha-amylases bacterienne, fongique, et de malt d'orge en panification. *Ann. Technol. Agric.* **23,** 161–174.

Berry, D. R. (1984). The physiology and microbiology of scotch whiskey production. *Prog. Ind. Microbiol* **19,** 199–243.

Biely, P. (1985). Microbial xylanolytic systems. *Trends Biotechnol.* **3,** 286–290.

Bigelis, R., and Lasure, L. L. (1987). Fungal enzymes and primary metabolites used in food processing. *In* "Food and Beverage Mycology" (L. R. Beuchat, ed.), pp. 473–516. Van Nostrand Reinhold, New York.

Blanch, H. W. (1984). Immobilized microbial cells. *Ann. Rep. Ferm. Process.* **7,** 81–105.

Böing, J. T. P. (1982). Enzyme production. *In* "Prescott and Dunn's Industrial Microbiology" (G. Reed, ed.), pp. 634–708. AVI Publishing, Westport, Connecticut.

Bos, C. (1990). The production of crystalline fructose. *Zuckerind.* **115,** 771–780.

Bournes, D. T., Jones, M., and Pierce, J. S. (1976). Beta-glucan and beta-glucanases in malting and brewing. *Master Brew. Assoc. Am. Tech. Quart.* **13,** 3–7.

Boyce, C. O. L. (1986). "Novo's Handbook of Practical Biotechnology." Bagsvaerd, Denmark.

Brabender, M., and Seitz, W. (1979). Effect of fungal amylases and malt flour on baking properties of flour. *Muehle Mischfuttertechn.* **116,** 219–220.

Brandt, D. A. (1975). Distilled alcoholic beverages. *In* "Enzymes in Food Processing" (G. Reed, ed.), 2d Ed., pp. 443–453. Academic Press, New York.

Briggs, D. E., Hough, J. S., Stevens, R., and Young, T. W. (1981). Adjuncts, sugars, wort-syrups, and industrial enzymes. *In* "Malting and Brewing Science," 2d Ed., Vol. 1, pp. 222–253. Chapman and Hall, New York.

Broderick, H. M. (1977). "The Practical Brewer — A Manual for the Brewing Industry." Master Brewers Association of the Americas, Madison, Wisconsin.

Buchholz, K., and Rabet, D. (1987). The effect of cell and microbial invertases on juice extraction from sugar beet. *Zuckerind.* **112,** 792–795.

Bucke, C. (1981). Enzymes in fructose manufacture. *In* "Enzymes and Food Processing" (G. G. Birch, N. Blakebrough, and K. J. Parker, eds.), pp. 51–72. Applied Science, London.

Bucke, C. (1983a). Glucose-transforming enzymes. *In* "Microbial Enzymes and Biotechnology" (W. M. Fogarty, ed.), pp. 93–129. Applied Science, London.

Bucke, C. (1983b). Carbohydrate transformations by immobilized cells. *Biochem. Soc. Symp.* **48,** 25–38.

Burgess, K., and Shaw, M. (1983). Dairy. *In* "Industrial Enzymology. The Application of Enzymes in Industry" (T. Godfrey and J. Reichelt, eds.), pp. 260–283. Nature Press, New York.

Canales, A. M., Garza, R., Sierra, J. A., and Arnold, R. (1988). The application of a beta-glucanase with additional side activities in brewing. *Master Brew. Assoc. Am. Tech. Quart.* **25,** 27–31.

Carasik, W., and Carroll, J. O. (1983). Development of immobilized enzymes for production of high-fructose corn syrup. *Food Technol.* **37(10),** 85–91.

Carr, J. G. (1985). Tea, coffee, and cocoa. *In* "Microbiology of Fermented Foods" (B. J. B. Wood, ed.), Vol. 2, pp. 133–154. Elsevier Applied Science, London.

Casey, J. P. (1977). High fructose corn syrup. *Die Stärke* **29,** 196–204.

Castelein, J., and Verachtert, H. (1983). Coffee fermentation. *In* "Biotechnology. A Comprehensive Treatise" (H. J. Rehm and G. Reed, eds.), Vol. 5, pp. 587–598. Verlag-Chemie, Deerfield Beach, Florida.

Cauvain, S. P., and Chamberlain, N. (1988). The bread improving effect of a fungal α-amylase. *J. Cereal Sci.* **8,** 239–248.

Coker, L. E., and Venkatasubramanian, K. (1985). Starch conversion processes. *In* "Comprehensive Biotechnology. The Principles, Applications, and Regulations of Biotechnology in Industry, Agriculture, and Medicine" (H. W. Blanch, S. Drew, and D. I. C. Wang, eds.), Vol. 3, pp. 777–787. Pergamon Press, Oxford.

Conn, J. F., Johnson, J. A., and Miller, B. F. (1950). An investigation of commercial fungal and bacterial alpha-amylase preparations in baking. *Cereal Chem.* **27,** 191–205.

Corn Refiners Association (1979a). "Corn Starch." Corn Refiners Association, Washington, D.C.

Corn Refiners Association (1979b). "Nutritive Sweeteners from Corn." Corn Refiners Association, Washington, D.C.

Coughlan, M. P. (1985). The properties of fungal and bacterial cellulases with comment on their production and application. *Biotechnol. Genet. Eng. Rev.* **3,** 39–111.

Coughlan, M. P. (1992). Enzymic hydrolysis of cellulose—An overview. *Bioresource Technol.* **39,** 107–115.

Coughlan, R. W., and Charles, M. (1980). Applications of lactase and immobilized lactase. *In* "Immobilized Enzymes for Food Processing" (W. H. Pitcher, ed.), pp. 153–173. CRC Press, Boca Raton, Florida.

Crueger, A., and Crueger, W. (1984). "Biotechnology. A Textbook of Industrial Microbiology." Sinauer Associates, Sunderland, Massachusetts.

Cruess, W. V., and Besone, J. (1941). Observations on the use of pectic enzymes in wine making. *Fruit Prod. J.* **20,** 365–367.

Davenport, R. (1988). Sonic drying offers new product potential. *Food Business* Jan., 20.

Dekker, R. F. H. (1979). The hemicellulase group of enzymes. *In* "Polysaccharides in Food" (J. M. V. Blanshard and J. R. Mitchell, eds.). pp. 93–108. Butterworth, London.

Dekker, R. F. H. (1985). Biodegradation of hemicelluloses. *In* "Biosynthesis and Biodegradation of Wood Components" (T. Higuchi, ed.), pp. 505–533. Academic Press, Orlando, Florida.

Dekker, R. F. H., and Richards, G. N. (1976). Hemicellulases: Their occurrence, purification, properties, and mode of action. *Adv. Carbohydr. Chem.* **32,** 277–352.

Denault, L. J., Glenister, P. R., and Chau, S. (1981). Enzymology of the mashing step during beer production. *J. Am. Soc. Brew. Chem.* **39,** 46–52.

Downing, D. L. (1988). "Processed Apple Products." Van Nostrand Reinhold/AVI, New York.

Drapron, R., and Godon, B. (1987). Role of enzymes in baking. *In* "Enzymes and Their Role in Cereal Technology" (J. E. Kruger, D. Lineback, and C. E. Stauffer, eds.), pp. 280–324. American Association of Cereal Chemists, St. Paul, Minnesota.

Duxbury, D. D. (1989). Liquid enzyme provides economical milk sugar sweetness replacement. *Food Process.* **50(2),** 77–78.

Emert, G. H., Gum, E. K., Jr., Lang, J. A., Liu, T. H., and Brown, R. D., Jr. (1974). Cellulases. *In* "Food Related Enzymes" (J. R. Whitaker, ed.), pp. 79–100. American Chemical Society, Washington, D.C.

Enari, T. M. (1983). Cellulases. *In* "Food Related Enzymes" (J. R. Whitaker, ed.), pp. 183–224. American Chemical Society, Washington, D.C.

Enari, T. M. (1983). Microbial cellulases. *In* "Microbial Enzymes and Biotechnology" (W. M. Fogarty, ed.), pp. 183–223. Applied Science Publishers, London.

Enkenlund, J. (1972). Externally added beta-glucanase. *Proc. Biochem.* **7(8),** 27–29.

Esterbauer, W., Steiner, H., Labudova, I., Hermann, A., and Hayn, M. (1991). Production of *Trichoderma* cellulase in laboratory and pilot scale. *Bioresource Technol.* **36,** 51–66.

Felix, R., and Villettaz, J.-C. (1983). Wine. *In* "Industrial Enzymology. The Application of Enzymes in Industry" (T. Godfrey and J. Reichelt, eds.), pp. 410–421. MacMillan, New York.

Ferenczi, S., and Asvany, S. (1977). Quelques aspects de l'intervention des enzymes en oenologie. Cas des enzymes pectinolytique et de l'invertase. *Bull. Off. Int. Vigne Vin* **50(551),** 43–49.

Fincher, G. B., and Stone, B. A. (1981). Metabolism of non-cellulosic polysaccharides. *In*

"Encyclopedia of Plant Physiology. New Series" (W. Tanner and F. A. Loewus, eds.), Vol. 13, pp. 69–132. Springer-Verlag, Berlin.

Finnocchario, T., Olson, N. F., and Richardson, T. (1980). Use of immobilized lactase in milk systems. *Adv. Biochem. Eng.* **15**, 71–88.

Fogarty, W. M., and Kelly, C. T. (1979). Starch-degrading enzymes of microbial origin, Part 1. *Prog. Ind. Microbiol.* **15**, 89–150.

Fogarty, W. M., and Kelly, C. T. (1980). Amylases, amyloglucosidases, and related glucanases. *In* "Economic Microbiology. Microbial Enzymes and Bioconversions" (A. H. Rose, ed.), Vol. 5, pp. 115–170. Academic Press, London.

Fogarty, W. M., and Kelly, C. T. (1983). Pectic enzymes. *In* "Microbial Enzymes and Biotechnology" (W. M. Fogarty, ed.), pp. 131–182. Applied Science Publishers, London.

Fogarty, W. M., and Kelly, C. T. (1990). "Microbial Enzymes and Biotechnology," 2d. Ed., Elsevier, Amsterdam.

Fogarty, W. M., and Ward, O. P. (1974). Pectinases and pectic polysaccharides. *Prog. Ind. Microbiol.* **13**, 59–113.

Foster, D. H. (1969). Deterioration of chopped cane. *Proc. Queensl. Soc. Sugar Cane Technol.* **36**, 21–28.

Fox, P. F. (1974). Enzymes in food processing. *In* "Industrial Aspects of Biochemistry" (B. Spencer, ed.), pp. 213–239. North Holland, Amsterdam.

Fox, P. F. (1980). Enzymes other than rennets in dairy technology. *J. Soc. Dairy Technol.* **33**, 118–128.

Frost, G. M., and Moss, D. A. (1987). Production of enzymes by fermentation. *In* "Biotechnology. Enzyme Technology" (J. F. Kennedy, ed.), Vol. 7a, pp. 65–211. VCH Verlagsgesellschaft, Weinheim, Germany.

Gams, T. C. (1976). Der Einsatz von mikrobiellen Enzymen in der Bäckerei. *Getr. Mehl Brot* **30**, 113–116.

Gekas, V., and Lopez-Leiva, M. (1985). Hydrolysis of lactose: A literature review. *Proc. Biochem.* **20(2)**, 2–12.

Gerhartz, W. (1990). "Enzymes in Industry." VCH Verlagsgesellschaft, Weinheim, Germany.

Ghose, T. K., and Pathak, A. N. (1973). Cellulases—2: Applications. *Proc. Biochem.* **8(5)**, 20–21, 24.

Gillis-van Maele, A. (1982). Actual achievements and future prospects concerning the use of enzymes in the brewery. *Cerevisia* **7**, 31–41.

Godfrey, A. (1985). Production of industrial enzymes and some applications in fermented foods. *In* "Microbiology of Fermented Foods" (B. J. B. Wood, ed.)., Vol. 1, pp. 345–371. Elsevier Applied Science, London.

Godfrey, T. (1983a). Brewing. *In* "Industrial Enzymology. The Application of Enzymes in Industry" (T. Godfrey and J. Reichelt, eds.), pp. 221–259. Nature Press, New York.

Godfrey, T. (1983b). Edible oils. *In* "Industrial Enzymology. The Application of Enzymes in Industry" (T. Godfrey and J. Reichelt, eds.), pp. 424–427. Nature Press, New York.

Godfrey, T. (1983c). Dextranases and sugar processing. *In* "Industrial Enzymology. The Application of Enzymes in Industry" (T. Godfrey and J. Reichelt, eds.), pp. 422–423. Nature Press, New York.

Goksøyr, J., and Eriksen, J. (1980). Cellulases. *In* "Economic Microbiology. Microbial Enzymes and Bioconversions" (A. H. Rose, ed.), Vol. 5, pp. 283–330. Academic Press, London.

Goyal, A., Ghosh, B., and Eveleigh, D. (1991). Characteristics of fungal cellulases. *Bioresource Technol.* **36**, 37–50.

Greenberg, N. A., and Mahoney, R. R. (1981). Immobilization of lactase (β-galactosidase) for use in dairy processing: A review. *Proc. Biochem.* **16(2)**, 2–8.

Guilbot, A. (1972). Use of enzymes in the bakery industry. *Ann. Technol. Agric.* **21**, 237–252.

Guilbot, A., and Mercier, C. (1985). Starch. *In* "Polysaccharides" (G. O. Aspinall, ed.), pp. 209–282. Academic Press, Orlando, Florida.

Hacking, A. J. (1991). Biocatalysis in the production of carbohydrates for food uses. *In* "Biocatalysis for Industry" (J. S. Dordick, ed.), pp. 63–82. Plenum, New York.

Halliwell, G. (1979). Microbial β-glucanases. *Prog. Ind. Microbiol.* **15**, 3–61.

Hashimoto, Y., and Fukumoto, J. (1969). Studies on the enzyme treatment of coffee beans. Purification of mannanase of *Rhizopus niveus* and its action on coffee mannan. *Nippon Nogei Kagaku Kaishi* **43**, 317–322.

Haupt, W. (1981). Pectolytic enzymes in wine production. *Weinwiss.* **117**, 1014–1017.

Hebeda, R. E., Bowles, L. K., and Teague, W. M. (1990). Developments in enzymes for retarding staling of baked goods. *Cereal Foods World* **35**, 453–457.

Hebeda, R. E., Bowles, L. K., and Teague, W. M. (1991). Use of intermediate stability enzymes for retarding staling in baked goods. *Cereal Foods World* **36**, 619–625.

Hough, J. S. (1985). "The Biotechnology of Malting and Brewing." Cambridge University Press, Cambridge.

Houts, S. S. (1988). Lactose intolerance. *Food Technol.* **42(3)**, 110–113.

Huang, H. T. (1955). Decolorization of anthocyanins by fungal enzymes. *J. Agric. Food Chem.* **3**, 141–146.

Imrie, F. K. E., and Tilbury, R. H. (1972). Polysaccharides in sugar cane and its products. *Sugar Technol. Rev.* **1**, 291–361.

Ingleton, J. F. (1963). The use of invertase in the confectionery industry. *Confect. Prod.* **29**, 773–774, 776–777, 790.

Johnson, J. A., and Miller, B. S. (1949). Studies on the role of alpha-amylase and proteinase on breadmaking. *Cereal Chem.* **26**, 371–383.

Jones, K. L., and Jones, S. E. (1984). Fermentations involved in the production of cocoa, coffee, and tea. *Prog. Ind. Microbiol.* **19**, 411–456.

Kaneko, R., Kusakabe, I., Sakai, Y., and Murakami, K. (1990). Substrate specificity of α-galactosidase from *Mortierella vinacea*. *Agric. Biol. Chem.* **54**, 237–238.

Kaur, M., and Bains, G. S. (1978). Amylase supplementation of Indian wheat flours for improving bread potential. *Indian Miller*, **8**, 38–41.

Kempf, W. (1985). New possible outlets for starch and starch products in chemical and technical industries. *Food Technol. (Australia)* **37**, 241–245.

Keniry, J. S., Lee, J. B., and Davis, C. W. (1967). Deterioration of mechanically harvested chopped-up cane. Part I. Dextran—A promising quantitative indicator of the processing quality of chopped-up cane. *Int. Sugar J.* **69**, 330–333.

Kennedy, J. F., Cabalda, V. M., and White, C. A. (1988). Enzymic starch utilization in genetic engineering. *Trends Biotechnol.* **6**, 184–189.

Klostermeyer, H., Werlitz, E., Juegens, R. H., Reimerdes, E. H., and Thomasow, J. (1978). Lactasebehandlung von Magermilch zur Herstellung lactosereduzierten Magermilch-pulvers. *Kieler Milch Forsch.* **30**, 295–340.

Knoepfel, H. P., and Pfenninger, H. B. (1974). Enzymatic treatment of unmalted grain. *Schweiz. Brau. Rundsch.* **85**, 213–220.

Kobayashi, H., and Suzuki, H. (1972). Studies on the decomposition of raffinose by α-galactosidase of mold. *J. Ferm. Technol.* **50**, 625–632.

Koch, H., and Roper, H. (1988). New industrial products from starch. *Starch/Stärke* **40**, 121–131.

Kosaric, N., Yu, K., and Zajic, J. E. (1973). Dextranase production from *Penicillium funiculosum*. *Biotechnol. Bioeng.* **15**, 729–741.

Kubicek, C. P., Eveleigh, D. E., Esterbauer, H., Steiner, W., and Kubicek-Pranz, E. M. (1990). "*Trichoderma reesei* Cellulases. Biochemistry, Genetics, Physiology, and Application." Royal Society of Chemistry, Cambridge.

Kulp, K. (1968). Enzymolysis of pentosans of wheat flour. *Cereal Chem.* **45**, 339–350.

Kulp, K. (1975). Carbohydrases. *In* "Enzymes in Food Processing" (G. Reed, ed.), pp. 53–122. Academic Press, New York.

Kulp, K., Lorenz, K., and Stone, M. (1991). Functionality of carbohydrate ingredients in bakery products. *Food Technol.* **45(3)**, 136, 138–140, 142.

Landis, B. H., and Beery, K. E. (1984). High fructose corn syrup. *Dev. Soft Drink Technol.* **3**, 85–120.

Leedham, P. A., and Savage, D. J., and Crabb, D., and Morgan, G. T. (1975). Materials and methods of wort production that influence beer filtration. *Proc. Eur. Brew. Conv. Congress (Nice)* 201–216.

Liener, I. E. (1977). Removal of naturally occurring toxicants through enzymatic processing. *In* "Food Proteins. Improvement through chemical and enzymatic modification" (R. E. Feeney and J. R. Whitaker, eds.), pp. 283–299.

Linden, J. C. (1982). Immobilized alpha-D-galactosidase (EC 3.2.1.22) in the sugar beet industry. *Enz. Microb. Technol.* **4,** 130–136.

Lindley, M. G. (1982). Cellobiase, melibiase, and other disaccharidases. *Dev. Food Carbohydr.* **3,** 141–165.

Lineback, D. R., and Inglett, G. E. (1982). "Food Carbohydrates." AVI Publishing, Westport, Connecticut.

Linhardt, R. J., Galliher, P. M., and Cooney, C. L. (1986). Polysaccharide lyases. *Appl. Biochem. Biotechnol.* **12,** 135–176.

Linko, P. (1987). Enzymes in the industrial utilization of cereals. *In* "Enzymes and Their Role Cereal Technology" (J. E. Kruger, D. Lineback, and C. E. Stauffer, eds.), pp. 280–324. American Association of Cereal Chemists, St. Paul, Minnesota.

Linko, P., and Linko, Y.-Y. (1983). Applications of immobilized microbial cells. *In* "Immobilized Microbial Cells" (I. Chibata and L. B. Wingard, Jr., eds.), Vol. 4, pp. 53–151. Academic Press, New York.

Long, J. E. (1986). High fructose corn syrup. *Cereal Foods World* **31,** 863–865.

Luallen, T. E. (1988). Structure, characteristics, and uses of some typical carbohydrate food ingredients. *Cereal Foods World* **33(11),** 924–927.

Luesner, S. J. (1983). Microbial enzymes for industrial sweetener production. *Dev. Ind. Microbiol.* **24,** 79–96.

MacAllister, R. V. (1979). Nutritive sweeteners made from starch. *Adv. Carbohydr. Chem.* **36,** 15–56.

MacAllister, R. V. (1980). Manufacture of high fructose corn syrup using immobilized glucose isomerase. *In* "Immobilized Enzymes for Food Processing" (W. H. Pitcher, Jr., ed.), pp. 81–111. CRC Press, Boca Raton, Florida.

MacAllister, R. V., Wardrip, E. K., and Schnyder, B. J. (1975). Modified starches, corn syrups containing glucose and maltose, corn syrups containing glucose and fructose, and crystalline dextrose. *In* "Enzymes and Food Processing" (G. Reed, ed.), 2d Ed., pp. 331–359. Academic Press, New York.

McCleary, B. V. (1986). Problems caused by barley beta-glucans in the brewing industry. *Chem. Aust.,* **53,** 306–308.

McCleary, B. V. (1987). Enzymatic modification of polysaccharides in brewing, baking and syrup manufacture. *Food Hydrocolloids* **1,** 445–448.

McCleary, B. V., Gibson, T. S., Allen, H., and Gams, T. C. (1986). Enzymic hydrolysis and industrial importance of barley β-glucans and wheat flour pentosans. *Starch/Stärke* **38,** 433–437.

McCleary, B. V., Gibson, T. S., Sheehan, H., Casey, A., Horgan, L., and O'Flaherty, J. (1989). Purification, properties, and industrial significance of transglucosidase from *Aspergillus niger. Carbohydr. Res.* **185,** 147–162.

MacMillan, J. D., and Sheiman, M. I. (1974). Pectic enzymes. *In* "Food Related Enzymes" (J. R. Whitaker, ed.), pp. 101–130. American Chemical Society, Washington, D.C.

Mahoney, R. R. (1985). Modification of lactose and lactose-containing dairy products with beta-galactosidase. *Dev. Dairy Chem.* **3,** 69–109.

Maisch, W. F., Sobolov, M., and Petricola, A. J. (1979). Distilled beverages. *In* "Microbial Technology. Fermentation Technology" (H. J. Peppler and D. Perlman, eds.), 2d Ed., Vol. 2, pp. 79–94. Academic Press, New York.

Mandels, M. (1985). Applications of cellulases. *Biochem. Soc. Trans.* **13,** 414–416.

Maninder, K., and Jorgensen, O. B. (1983). Interrelations of starch and fungal alpha-amylase in bread making. *Starch/Stärke* **35,** 419–426.

Marek, E., Schalinatus, E., Weigelt, E., Mieth, G., Kerns, G., and Kude, J. (1990). On the application of enzymes in the production of vegetable oil. *Prog. Biotechnol.* **6,** 471–474.

Marie, S., and Piggott, J. R. (1991). "Handbook of Sweeteners." Blackie and Son, Glasgow.

Marshall, J. J., Allen, W. G., Denault, L. J., Glenister, P. R., and Power, J. (1982). Enzymes in brewing. *Brewers Dig.* **57,** 14–18.

Marston, P. E., and Wannan, T. L. (1976). Bread baking. The transformation of dough to bread. *Bakers Dig.* **50**(4), 24–28, 49.

Martiniere, P., Sapis, J. C., Guimberteau, G., and Ribereau-Gayon, J. (1973). Utilisation d'une preparation enzymatique pectinolytique en vinification. *C. R. Seances Acad. Agr. Fr.* **59,** 267–273.

Mattes, R., and Beaucamp, K. (1982). Verfahren zur Herstellung eines Mikroorganismus, welcher α-Galaktosidase, aber keine Invertase bildet, so erhaltener Mikroorganismus und seine Verwendung *Dtsch. Offenlegungsschr.* DE 3122216 Al.

Meier, W., and Reid, J. S. G. (1981). Reserve polysaccharides other than starch in higher plants. *In* "Encyclopedia of Plant Physiology, New Series" (A. Pirson and M. H. Zimmermann, eds.), Vol. 13A, pp. 418–471. Springer-Verlag, Berlin.

Miller, J. J., and Brand, J. C. (1980). Enzymic lactose hydrolysis. *Food Technol. (Australia)* **32,** 144–146.

Monnier, B., and Godon, B. (1975). Use of proteases in baking industry. *Ind. Aliment. Agric.* **92,** 521–529.

Monsan, P., and Combes, D. (1984). Application of immobilized invertase to hydrolysis of concentrated sucrose solutions. *Biotechnol. Bioeng.* **26,** 347–351.

Montenecourt, B. S., and Eveleigh, D. E. (1985). Fungal carbohydrases: Amylases and cellulases. *In* "Gene Manipulations in Fungi" (J. W. Bennet and L. L. Lasure, eds.), pp. 491–512. Academic Press, New York.

Moore, K. (1980). Immobilized enzyme technology commercially hydrolyzes lactose. *Food Prod. Devel.* **14**(1), 50–51.

Moulin, G., and Galzy, P. (1984). Whey, a potential substrate for biotechnology. *In* "Biotechnology and Genetic Engineering Reviews" (G. E. Russell, ed.), Vol. 1, pp. 347–374. Intercept, Newcastle, England.

Nakajima, M., Jimbo, N., Nishizawa, K., Nabetani, H., and Watanabe, A. (1988). Conversion of sucrose by immobilized invertase in an asymmetric membrane reactor. *Proc. Biochem.* **23**(2), 32–35.

Narziss, L. (1981). Beta-glucan and beta-glucanases. *Proc. Eur. Brew. Conv. Barley Malt Symp. (Helsinki)* 99–117.

Neubeck, C. E. (1975). Fruit, fruit products. *In* "Enzymes in Food Processing" (G. Reed, ed.), 2d Ed., pp. 397–442. Academic Press, New York.

Newsome, R. L. (1986). Sweeteners: Nutritive and non-nutritive. *Food Technol.* **40**(8), 195–206.

Nielsen, E. B. (1971). Brewing with barley and enzymes—A review. *Eur. Brew. Conv. Proc. Congr. 13th Estoril.* 149–170.

Nijpels, H. H. (1981). Lactases and their application. *In* "Enzymes and Food Processing" (G. G. Birch, N. Blakebrough, and K. J. Parker, eds.), pp. 89–104. Applied Science, London.

Norman, B. (1979). The application of polysaccharide degrading enzymes in the starch industry. *In* "Microbial Polysaccharides and Polysaccharases" (R. C. W. Berkeley, G. W. Gooday, and D. C. Ellwood, eds.), pp. 339–376. Academic Press, London.

Norman, B. E. (1981). New developments in starch syrup technology. *In* "Enzymes and Food Processing" (G. G. Birch, N. Blakebrough, and K. J. Parker, eds.), pp. 15–50. Applied Science, London.

Nyiri, L. (1969). Manufacture of pectinases. *Proc. Biochem.* **4**(8), 27–30.

Obara, J., Hashimoto, S., and Suzuki, H. (1977). Enzyme applications in the sucrose industries. *Sugar Technol. Rev.* **4,** 209–258.

Ostergaard, J. (1983). Enzymes in the carbohydrate industry. *In* "Utilisation des Enzymes en Technologie Alimentaire" (P. Dupuy, ed.), pp. 57–79. Technique et Documentation Lavoisier, Paris.

Ough, C. S., and Berg, H. W. (1974). The effect of two commercial pectic enzymes on grape musts and wines. *Am. J. Enol. Vitic.* **25,** 208–211.

Ough, C. S., and Crowell, E. A. (1979). Pectic-enzyme treatment of white grapes: Temperature, variety and skin-contact time factors. *Am. J. Enol. Vitic.* **30,** 22–27.

Ough, C. S., Noble, A. C., and Temple, D. (1975). Pectic enzyme effects on red grapes. *Am. J. Enol. Vitic.* **26,** 195–200.

Pancoast, H. M., and Junk, J. R. (1980). "Carbohydrate Sweetners in Foods and Nutrition." Academic Press, London.

Pedersen, S., and Norman, B. E. (1987). Enzymatic modification of food carbohydrates. *In* "Chemical Espects of Food Enzymes" (A. T. Andrews, ed.), pp. 156–187. Royal Society Chemistry, London.

Peppler, H. J., and Reed, G. (1987). Enzymes in food and feed processing. *In* "Biotechnology. Enzyme Technology" (J. F. Kennedy, ed.), Vol. 7a, pp. 547–603. VCH Verlagsgesellschaft, Weinheim, Germany.

Pilnik, W. (1982). Enzymes in the beverage industry (fruit juices, nectar, wine, spirits, beer). *In* "Utilization des Enzymes en Technologie Alimentaire" (P. Dupuy, ed.), pp. 425–450. Technique et Documentation Lavoisier, Paris.

Pilnik, W., and Rombouts, F. M. (1981). Pectic enzymes. *In* "Enzymes and Food Processing" (G. G. Birch, N. Blakebrough, and K. J. Parker., eds.), pp. 105–128. Applied Science, London.

Pokorny, M., Zupancic, S., Steiner, T., and Kreiner, W. (1990). Production and down-stream processing of cellulases on a pilot scale. *In* "*Trichoderma reesei* Cellulases. Biochemistry, Genetics, Physiology and Application" (C. P. Kubicek, D. E. Eveleigh, N. Esterbauer, W. Steiner, and E. M. Kubicek-Pranz, eds.), pp. 168–184. Royal Society of Chemistry, Oxford.

Pomerantz, Y. (1991). "Functional Properties of Food Components." Academic Press, San Diego.

Poulson, P. B. (1983). Alcohol-potable. *In* "Industrial Enzymology. The Application of Enzymes in Industry" (T. Godfrey and J. Reichelt, eds.), pp. 170–178. Nature Press, New York.

Priest, F. G., and Campbell, I. (1987). "Brewing Microbiology." Elsevier, Amsterdam.

Pyler, E. J. (1969). Enzymes in baking—Theory and practice. *Bakers Dig.* **43**(1), 36–40, 42–44, 46–47.

Radley, J. A. (1976). "Starch Production Technology." Applied Science, London.

Reichelt, J. R. (1983). Baking. *In* "Industrial Enzymology. The Application of Enzymes in Industry" (T. Godfrey and J. Reichelt, eds.), pp. 210–220. Nature Press, New York.

Reilly, P. J. (1980). Potential and uses of immobilized carbohydrases. *In* "Immobilized Enzyme for Food Processing" (W. H. Pitcher, Jr., ed.), pp. 113–151. CRC Press, Boca Raton, Florida.

Rice, J. (1988). Enzymology and food packaging. *Food Process.* **49**(12), 188–189.

Richmond, M. L., Gray, J. I., and Stine, C. M. (1981). β-Galactosidase: A review of recent research related to technological application, nutritional concerns, and immobilization. *J. Dairy Sci.* **64,** 1759–1771.

Robertson, G. L. (1977). Pectic enzymes and winemaking. *Food Technol. (New Zealand)* **12,** 32, 34–35.

Rombouts, F. M., and Pilnik, W. (1980). Pectic enzymes. *In* "Economic Microbiology, Microbial Enzymes and Bioconversions" (A. H. Rose, ed.), Vol. 5., pp. 227–283. Academic Press, London.

Saletan, L. T. (1968). Carbohydrases of interest in brewing, with particular reference to amyloglucosidase. *Wallerstein Lab. Commun.* **31**(104), 33–44.

Sanderson, G. W. (1983). Tea manufacture. *In* "Biotechnology. A Comprehensive Treatise" (H.-J. Rehm and G. Reed, eds.), Vol. 5, pp. 577–586. Verlag-Chemie, Deerfield Beach, Florida.

Sanderson, G. W., and Coggon, P. (1977). The use of enzymes in the manufacture of black tea and instant tea. *In* "Enzymes in Food and Beverage Processing" (R. L. Ory and A. J. St. Angelo, eds.), pp. 12–26. American Chemical Society, Washington, D.C.

Schwimmer, S. (1981). "Source Book of Food Enzymology." AVI Publishing, Westport, Connecticut.

Scott, D. (1975). Miscellaneous applications of enzymes. *In* "Enzymes in Food Processing" (G. Reed, ed.), 2d Ed., pp. 493–517. Academic Press, New York.

Scott, D. (1978). Enzymes, industrial. *In* "Kirk-Othmer Encyclopedia of Chemical Technology" (M. Grayson and D. Ekroth, eds.), pp. 173–224. John Wiley and Sons, New York.

Scott, D. (1989). Specialty enzymes and products for the food industry. *In* "Biocatalysis in Agricultural Biotechnology" (J. R. Whitaker and P. E. Sonnet, eds.), pp. 176–192. American Chemical Society, Washington, D.C.

Scott, R. W. (1972). The viscosity of worts in relation to their content of β-glucan. *J. Inst. Brewing* **78,** 179–186.

Selman, R. W., and Sumner, R. J. (1947). The use of the amylograph for flour malt control. *Cereal Chem.* **24,** 291–299.

Severinsen, S. G., Jebsen, S.-G., Gicquiaux, Y., and Regimer, J. (1979). Use of lactases. *Alimenta* **72,** 60, 63, 66–69.

Sheppard, G. (1986). The production and uses of microbial enzymes in food processing. *Prog. Ind. Microbiol.* **23,** 237–283.

Shetty, J. K., and Allen, W. G. (1988). An acid-stable thermostable alpha-amylase for starch liquefaction. *Cereal Foods World* **33,** 929–934.

Shimizu, J., and Kaga, T. (1972). Apparatus for continuous hydrolysis of raffinose. *U.S. Patent* No. 3,664,927.

Shukla, T. P. (1975). Beta-galactosidase technology. A solution to the lactose problem. *CRC Crit. Rev. Biotechnol.* **5,** 325–356.

Slaughter, J. C. (1985). Enzymes in the brewing industry. *In* "Alcoholic Beverages" (G. G. Birch and M. G. Lindley, ed.), pp. 15–27. Elsevier, Amsterdam.

Slominska, L., and Maczynski, M. (1985). Studies on the application of pullulanase in starch saccharification process. *Die Stärke* **37,** 386–390.

Sobolov, M., Booth, D. M., and Aldi, R. G. (1985). Whiskey. *In* "Comprehensive Biotechnology. The Principles, Applications, and Regulations of Biotechnology in Industry, Agriculture, and Medicine" (H. W. Blanch, S. Drew, and D. I. C. Wang, eds.), Vol. 3, pp. 383–393. Pergamon Press, Oxford.

Spradlin, J. E. (1989). Tailoring enzyme systems for food processing. *In* "Biocatalysis in Agricultural Biotechnology" (J. R. Whitaker and P. E. Sonnet, eds.), pp. 24–43. American Chemical Society, Washington, D.C.

Stauffer, C. E. (1990). "Functional Additives for Bakery Foods." Van Nostrand Reinhold/AVI, New York.

Stentebjerg-Olesen, B. (1980). Application of beta-glucanases in the brewing industry. *Proc. 16th Conv. Inst. Brew. (Aust. Section) Sydney* 127–134.

Sukan, S. S. (1988). Challenges in bioconversion of cellulosic and partially soluble plant materials in submerged culture. *Dev. Food Microbiol.* **3,** 109–140.

Suzuki, H., Ozawa, Y., Oota, H., and Yoshida, H. (1969). Studies on the decomposition of raffinose by α-galactosidase of mold. *Agric. Biol. Chem.* **33,** 507–513.

Swaisgood, H. E., and Horton, H. R. (1989). Immobilized enzymes as processing aids and analytical tools. *In* "Biocatalysis in Agricultural Biotechnology" (J. R. Whitaker and P. E. Sonnet, eds.), pp. 242–261. American Chemical Society, Washington, D.C.

Swientek, R. J. (1986). Sonic technology applied to food drying. *Food Process.* **47**(7), 62–63.

Szekely, P. (1970). Use of invertase enzyme in confectionery. *Edesipar* **21,** 80–83.

Takasaki, Y., and Yamanobe, T. (1981). Production of maltose by pullulanases and β-amylase. *In* "Enzymes in Food Processing" (G. G. Birch, N. Blakebrough, and K. J. Parker, eds.), pp. 73–88. Applied Science, London.

Teeri, T. T., Penttilä, M., Keränen, S., Nevalainen, H., and Knowles, J. K. C. (1992). Structure, function, and genetics of cellulases. *In* "Biotechnology of Filamentous Fungi" (D. B. Finklestein and C. Ball, eds.), pp. 417–445. Butterworth-Heineman, Stoneham, Massachusetts.

ter Haseborg, E. (1988). Enzymatic treatment of flour. *Alimenta* **27,** 2–10.

Tilbury, R. H. (1972). Sucrose from sugar cane—Addition of dextranase during processing to remove dextran. *Brit. Patent No.* 1290694.

Todo, V., Carbonell, J. V., and Sendra, J. M. (1989). Kinetics of β-glucan degradation in wort by exogenous β-glucanases treatment. *J. Inst. Brew.* **95**, 419–422.

Vanbelle, M., Meurens, M., and Crichton, R. R. (1982). Enzymes in foods and feeds. *Rev. Ferm. Ind. Aliment.* **37**, 124–135.

Van Beynum, G. M. A., and Roels, J. A. (1985). "Starch Conversion Technology." Marcel Dekker, New York.

Vandam, H. W., and Hille, J. D. R. (1992). Yeast and enzymes in breadmaking. *Cereal Foods World* **37**, 245–252.

Van Griethuysen-Dilber, E., Flaschel, R., and Renken, A. (1988). Process development of the hydrolysis of lactose in whey by immobilized lactase of *Aspergillus oryzae*. *Proc. Biochem.* **23**(2), 55–59.

Verhoff, F. H., Boguslawski, G., Lantero, O. J., Schlager, S. T., and Jao, Y. C. (1985). Glucose isomerase. *In* "Comprehensive Biotechnology" (C. L. Cooney and A. E. Humphrey, eds.), Vol. 3, pp. 837–859. Pergamon Press, Oxford.

Vihinen, M., and Mäntsälä, P. (1989). Microbial and amylolytic enzymes. *Crit. Rev. Biochem. Mol. Biol.* **24**, 329–418.

Villettaz, J. C. (1984). Les enzymes en oenologie. *Bull. Off. Int. Vigne Vin* **57**(635), 19–29.

Villettaz, J. C., Steiner, D., and Trogus, W. (1984). The use of beta glucanase as an enzyme in wine clarification and filtration. *Am. J. Enol. Vitic.* **35**, 253–256.

Voragen, A. G. J., and Pilnik, W. (1989). Pectin-degrading enzymes in fruit and vegetable processing. *In* "Biocatalysis in Agricultural Biotechnology" (J. R. Whitaker, and P. E. Sonnet, eds.), pp. 93–115. American Chemical Society, Washington, D.C.

Waldt, L. M. (1965). Fungal enzymes: Their role in continuous process bread. *Cereal Sci. Today* **10**, 447–450.

Waldt, L. M. (1969). Enzymes in baking. *Wallerstein Lab. Commun.* **32**(107), 39–48.

Walker, L. P., and Wilson, D. B. (1991). Enzymatic hydrolysis of cellulose—an overview. *Bioresource Technol.* **36**, 3–14.

Walter, R. H. (1991). "The Chemistry and Technology of Pectin." Academic Press, San Diego.

Ward, O. P. (1985). Hydrolytic enzymes. *In* "Comprehensive Biotechnology. The Principles, Applications, and Regulations of Biotechnology in Industry, Agriculture, and Medicine" (H. W. Blanch, S. Drew, and D. I. C. Wang, eds.), Vol. 3., pp. 819–835. Pergamon Press, Oxford.

Ward, O. P. (1989). Enzymatic degradation of cell wall and related plant polysaccharides. *CRC Crit. Rev. Biotechnol.* **8**, 237–274.

Ward, O. P., and Moo-Young, M. (1989). Enzymatic degradation of cell wall and related plant polysaccharides. *Crit. Rev. Biotechnol.* **8**, 237–274.

Westermann, D. H., and Huige, N. J. (1979). Beer brewing. *In* "Microbial Technology. Fermentation Technology" (H. J. Peppler and D. Perlman, eds.), Vol. 2, 2d Ed., pp. 1–37. Academic Press, New York.

Whistler, R. L., BeMiller, J. W., and Paschall, E. F. (1984). "Starch. Chemistry and Technology," 2d Ed. Academic Press, New York.

Whitaker, J. R. (1984). Pectic substances, pectic enzymes, and haze formation in fruit juices. *Enz. Microb. Technol.* **6**, 341–349.

Willox, I. C., Rader, S. R., Riolo, J. M., and Stern, W. (1977). The addition of starch debranching enzyme to mashing and fermentation and their influence on attenuation. *Master Brew. Assoc. Am. Tech. Quart.* **14**, 105–110.

Willson, K. C., and Clifford, M. N. (1992). "Tea. Consumption to Cultivation." Chapman and Hall, London.

Wiseman, A. (1981). New and modified invertases—And their applications. *In* "Topics in Enzyme and Fermentation Biotechnology" (A. Wiseman, ed.), Vol. 3, pp. 265–288. Ellis Horwood, Chichester.

Wong, K. K. Y., Tan, L. U. L., and Saddler, J. N. (1988). Multiplicity of β-1,4-xylanase in microorganisms: Functions and applications. *Microbiol. Rev.* **52**, 305–317.

Wood, T. M. (1985). Properties of cellulolytic enzyme systems. *Biochem. Soc. Trans.* **13**, 407–410.

Wood, W. A., and Kellogg, S. T. (1988). "Cellulose and Hemicellulose. Methods in Enzymology. Biomass," Vol. 160. Academic Press, San Diego.

Woodward, J. D. (1978). Enzymes in practical brewing. *Brewers Dig.* **53,** 38, 40, 42–44.

Woodward, J. (1984). Xylanases: Functions, properties and applications. *In* "Topics in Enzyme and Fermentation Biotechnology" (A. Wiseman, ed.), Vol. 8, pp. 9–30. Ellis Horwood, Chichester.

Woodward, J., and Wiseman, A. (1982). Invertase. *Dev. Food Carbohydr.* **3,** 1–21.

Woychik, J. H., and Holsinger, V. H. (1977). Use of lactose in the manufacture of dairy products. *In* "Enzymes in Food and Beverage Processing" (R. L. Ory and A. J. St. Angelo, eds.), pp. 67–79. American Chemical Society, Washington, D.C.

Zobel, H. F. (1988). Molecules to granules: A comprehensive starch review. *Starch/Stärke* **40,** 44–50.

CHAPTER 7

Proteases

JENS ADLER-NISSEN

I. Introduction

The preparation of a large variety of highly esteemed food products, for example cheese, beer, and shoyu (soy sauce), involves an enzymatic degradation of the proteins in the food as a key reaction. The manufacture of these foods is based on traditions of good craftsmanship made sophisticated through many years of scientific studies of the principles behind these processes. The scientific interest in proteases and their actions on various food proteins not only led to better cheese, beer, and shoyu, but also spurred the development of a number of new applications for proteases in the production of foods and food ingredients. This development also has been made possible by the increased availability of commercial proteases with suitable properties for use in the food industry.

The proteases catalyzing the degradation of food proteins originate from three different sources:

1. They may already be present in the food material.
2. They may be excreted by microorganisms growing in the food.
3. They may be added as isolated enzyme preparations.

Examples of processes in the first category are aging of meat, autolysis of fish to make Southeast Asian fish sauce, or autolysis of yeast to produce the British food condiment Marmite®. Cheese manufacture is a combination of categories 3 (a preparation of rennet is added) and 2 (proteases are excreted by the lactic acid bacteria after they have finished growing and undergo lysis). In the production of protein hydrolysates for use in protein-enriched beverages for clinical nutrition, the protease is added in a high concentration as a standardized preparation (category 3). However, in this case, the reaction is completed in a few hours, in contrast to the other processes mentioned, which may take weeks or even months.

The scope of this chapter is to treat the general characteristics of proteases for food use, whereas the specific applications of, for example, rennet in cheese manufacture or papain in meat tenderization are dealt with in the respective chapters. These general aspects concern not only the enzymes themselves but also the reactions they catalyze, that is, the hydrolysis of proteins. Also, the substrate and general concepts in protein hydrolysis processes will be treated in this chapter. Included in these general topics is a discussion of the important issue of peptide bitterness.

In the sections describing the individual enzymes, care has been taken to include those proteases that are used in food while omitting reference to the proteases occurring in animals outside the digestive tract (e.g., the groups of cathepsins and the coagulation activation factors), since the activity of these enzymes is not of direct concern in food (with the possible exception of aging of meat). With respect to the plant and microbial enzymes, the criterion for inclusion also is based on their use in food processing.

The field of food protein hydrolysis and the use of proteases in food processing has received considerable attention in the last 20 years: the literature references are numerous. Protein hydrolysis to produce new nutritional and functional food ingredients resulted in the publication of *Enzymic Hydrolysis of Food Protein* (Alder-Nissen, 1986a). This chapter includes a concise presentation of some of the concepts and results of that work, but also contains much more information on proteases than is found in the book. With respect to information on the individual proteases, the major source for any work in enzymology is the standard handbook series, *Methods in Enzymology*, of which Volume 19, *Proteolytic Enzymes*, (Perlmann and Lorand) appeared in 1970. All proteases that are regularly used today in food processing are described in individual chapters in this volume. Even for the microbial proteases, for which development should be expected to have occurred most rapidly, the more recent, useful review by Ward (1983) supplements but does not extend substantially the range of proteases relevant to food use. The newer biochemical knowledge of the proteases is covered well in the volume *Hydrolytic Enzymes* (Neuberger and Brocklehurst, 1987). This text contains extensive chapters on each of the major classes of proteases, with particular emphasis on their structure and catalytic mechanism. A most valuable source of information on proteases (and other enzymes) used in industrial practice is *Industrial Enzymology* (Godfrey and Reichelt, 1983). This book contains, among other things, a comparative characterization of important proteases with respect to pH and temperature activity profiles, a summary that is hard to find elsewhere (Godfrey, 1983b). Sigmund Schwimmer's *Source Book of Food Enzymology* (1981) also must be mentioned in this context because it contains a wealth of information, particularly on the biochemical events that take place in foods when enzymes are acting on the various food constituents. Finally, the textbook *Proteins, Structures, and Molecular Properties* by Creighton

(1984) is recommended as a coherent and in-depth treatment of the protein and enzyme chemistry that underlies this chapter.

Obviously, this chapter draws heavily on these books. Individual references to general information extracted from them will, therefore, not be given, except in cases of direct quotation.

II. Substrates

A. Protein Chemistry in Brief

From a chemical perspective, proteins are linear polymers of amino acids. The general formula for amino acids is $R-CHNH_2-COOH$, that is, an amino group and a carboxyl group are both linked to the same carbon atom, the α-carbon atom. R denotes a side chain which, in the case of the simplest amino acid glycine, is simply a hydrogen atom. The amino acids, of which approximately 20 occur widely in nature, are thus different with respect to the side chain only, which in turn determines the particular chemical properties of the individual amino acid.

The α-carbon atom is asymmetric (except in the case of glycine). The amino acids, therefore, exist as two optically active isomers, the L and the D form. The natural amino acids derived from proteins are always L isomers.

In proteins, the amino acids are linked together by a covalent bond from the nitrogen atom of the amino group to the carbon atom of the preceding carboxyl group forming the so-called peptide bond. The formation of this bond involves the loss of one molecule of water; the OH comes from the carboxyl group and the H from the amino group. Usually, fifty to several hundred amino acids are linked together in one unbranched chain. The general structure of such a polypeptide is shown in Fig. 1. The backbone of the polypeptide chain is the repetitive sequence $N-C-C-$ with one of the carbon atoms carrying a double-bonded oxygen and the other carrying the side chain R. By convention, the amino acid residues are numbered consecutively from the amino end of the polypeptide chain.

Since the side chains do not appear to take part in peptide bond formation, the chemical properties of the particular side chains are to a large extent

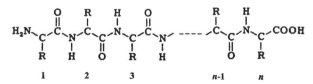

Figure I Polypeptide chain.

preserved in the protein. The sequence of different side chains, called the primary structure of the protein, is encoded in the genes of the organism synthesizing the particular protein. The differences in the relative occurrence and distribution of the different R groups explain to a great extent the enormous differences in the physical, chemical, and biological properties of proteins, including the large range of catalytic power and high specificity of the many thousand known enzymes.

In addition to the polypeptide chain, many proteins contain nonpeptide matter to exert their biological function. The porphyrin–metal complexes in hemoglobin and chlorophyll are outstanding examples, but many enzymes, including an entire class of proteases, contain an inorganic ion as a bound complex in their active site. Proteins from higher organisms (eukaryotes) often have carbohydrates linked covalently to the polypeptide chain; these sugar chains apparently function, among other things, as recognition signals in the interaction of cells (Paulson, 1989).

Two of the most important chemical properties of the side chains are charge and polarity. Some side chains have a carboxyl group at the free end and, thus, are weak acids, whereas others have a nitrogen atom that can take up a proton and are, therefore, basic. With respect to polarity, some side chains are aliphatic or aromatic hydrocarbons and therefore poorly soluble in water but readily soluble in organic solvents, that is they are hydrophobic. A hydrophobicity value, Δf_t, can be assigned to each of the side chains, where Δf_t is a measure of the difference in free energy associated with the transfer of the side chain from a nonpolar to an aqueous environment (Tanford, 1962; Nozaki and Tanford, 1971).

Hydrophobicity is an important concept in understanding the forces that hold the polypeptide chain in a particular spatial structure (the tertiary structure) that is necessary for the protein to express a biological function. In brief, the hydrophobic forces can be explained by the fact that water molecules have a strong attraction to each other but only a weak attraction to the hydrophobic side chains. When such a side chain is transferred from the interior of a protein molecule to the aqueous environment, the transfer involves the breaking of strong attractive forces between water molecules or between water molecules and polar or charged protein groups (Hvidt, 1983). This effect creates a situation that is thermodynamically unfavorable. However, if two side chains are able to aggregate, their overall interface with water decreases so, consequently, aggregation (hydrophobic interaction) is favored. Thus no bonds form between the hydrophobic side chains, unlike the formation of secondary bonds such as hydrogen bonds between hydrophilic side chains that also contribute to the stabilization of protein structures. The hydrophobic forces in proteins should be perceived as a manifestation of a net gain in thermodynamic free energy from the hydrophobic interaction of many of the hydrophobic side chains. For an average globular protein molecule, this gain is a major factor in the stabilization of its native conformation.

In relation to food applications of proteases, hydrophobicity is important to our understanding of the technical properties of protein hydrolysates, with respect to both physical performance and taste. This subject will be discussed in more detail later.

Table I presents the formulas and Table II the most important physicochemical properties of the common amino acids. One amino acid, cysteine,

TABLE I
Structure of the Common Amino Acids in Proteins

Type of side chain	Name	Structural formula
Carboxylic acid	Aspartic acid	$HOOC-CH_2-CNH_2-COOH$
	Glutamic acid	$HOOC-CH_2-CH_2-CNH_2-COOH$
Amide	Asparagine	$H_2N(CO)-CH_2-CNH_2-COOH$
	Glutamine	$H_2N(CO)-CH_2-CH_2-CNH_2-COOH$
Large hydrophobic side chain	Valine[a]	$(CH_3)_2-CH-CNH_2-COOH$
	Leucine[a]	$(CH_3)_2-CH-CH_2-CNH_2-COOH$
	Isoleucine[a]	$CH_3-CH_2-CH(CH_3)-CNH_2-COOH$
	Phenylalanine[a]	
	Tyrosine	
	Tryptophan[a]	
	Proline	
	Hydroxyproline[b]	
S-Containing side chain	Cysteine	$HSCH_2-CNH_2-COOH$
	Methionine[a]	$CH_3-S-CH_2-CH_2-CNH_2-COOH$
Basic side chain	Lysine[a]	$H_2N-CH_2-CH_2-CH_2-CH_2-CNH_2-COOH$
	Arginine	$H_2N-(CNH)-NH-CH_2-CH_2-CH_2-CNH_2-COOH$
	Histidine	
Small neutral side chain	Glycine	$H-CNH_2-COOH$
	Alanine	CH_3-CNH_2-COOH
	Serine	$HOCH_2-CNH_2-COOH$
	Threonine[a]	$CH_3-CHOH-CNH_2-COOH$

[a] Nutritionally essential to humans.
[b] Occurs in gelatin, which is produced from collagen.

TABLE II
Properties of the Amino Acids

Type of side chain	Name	Abbreviations	Mole weight (MW)	Hydrophobicity $(\Delta f_t)^a$	Dissociation constant of side chain (pK)^b
Carboxylic acid	Aspartic acid	Asp, D	133.1	0.0	4.5
	Glutamic acid	Glu, E	147.1	0.0	4.6
Amide	Asparagine	Asn, N	132.1	0.0	
	Glutamine	Gln, Q	146.1	0.0	
Large hydrophobic side chain	Valine	Val, V	117.2	1.5	
	Leucine	Leu, L	131.2	1.8	
	Isoleucine	Ile, I	131.2	2.95	
	Phenylalanine	Phe, F	165.2	2.5	
	Tyrosine	Tyr, Y	181.2	2.3	9.7
	Tryptophan	Trp, W	204.2	3.4	
	Proline	Pro, P	115.1	2.6	
	Hydroxyproline	Hyp	133.1	1.8	
S-Containing side chain	Cysteine	Cys, C	121.2^c	1.0	9.1–9.5
	Methionine	Met, M	149.2	1.3	
Basic side chain	Lysine	Lys, K	146.2	1.5	10.4
	Arginine	Arg, R	174.2	0.75	12
	Histidine	His, H	155.2	0.5	6.2
Small neutral side chain	Glycine	Gly, G	75.1	0.0	
	Alanine	Ala, A	89.1	0.5	
	Serine	Ser, S	105.1	−0.3	
	Threonine	Thr, T	119.1	0.4	

^{a} kcal/mole (Adler-Nissen, 1986b).
^{b} pK (25°C) for α-amino groups in free amino acids is 6.8–7.9 (in polypeptides, 7.5–7.8). pK for α-carboxylic groups is 3.5–4.3 (in polypeptides, 3.1–3.6).
^{c} MW of cystine is 2×120.1.

has the unique property of linking two polypeptide chains together through a covalent bond. The cysteine side chain is $—CH_2—SH$. Since the sulfhydryl group is oxidized easily to form a disulfide with another sulfhydryl group, disulfide cross-links are found in many proteins (Fig. 2). These cross-links stabilize the tertiary structure of the protein against forces that tend to cause an unfolding or denaturation of the protein.

    Denaturation of proteins is, ideally, a reversible process in which the polypeptide chain can unfold to a random coil and refold (renature) to the proper tertiary structure. Such reversible denaturation is observed, for example, when a solution of ribonuclease is heated in dilute solution at pH 2.1 to above 40°C, which leads to complete unfolding and loss of enzyme activity. After cooling, the activity and the native state are restored quantitatively (Ginsburg and Carroll, 1965). However, in most cases, the initial unfolding is followed by further reactions that impair proper refolding. Examples of such

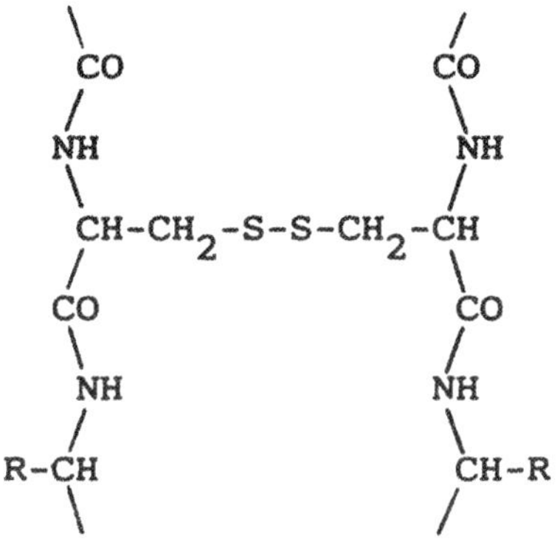

Figure 2 Disulfide cross-link in proteins.

reactions are splitting and incorrect rejoining of disulfide bonds, splitting of the peptide chain by hydrolysis,[1] or aggregation of two or more polypeptides that occurs when the hydrophobic regions of the peptide bonds join through hydrophobic interaction (Volkin and Klibanov, 1989).

Denaturation of the substrate also is a very important factor in the kinetic behavior of proteases, as will be discussed in Section III,C,3. Denaturation of the proteolytic enzyme itself is the primary cause of loss, voluntarily or involuntarily, of protease activity. Usually, the denaturation of proteases leads to an irreversible destruction of the enzyme through autolysis (Volkin and Klibanov, 1989; Ottesen and Svendsen, 1970). The kinetics and thermodynamics of the denaturation and autodigestion reaction of a protease (a subtilisin) are treated in detail elsewhere (Adler-Nissen, 1986a).

B. Food Proteins as Substrates

The preceding description of proteins and their chemistry is oriented to proteins as biochemicals. However, proteins are also an indispensable part of our diet, serving as a source of amino acids and amino nitrogen. At least eight of the amino acids (valine, leucine, isoleucine, phenylalanine, tryptophan, threonine, methionine and lysine) are well known to be essential to humans. The subject of nutrition shall not be pursued further. Note, however, that the average food product often contains a very large number of different proteins, including all the enzymes in the tissue of plants or animals. Therefore, it is impractical to consider the individual molecular species in the food. In food science and nutrition, the definition of "protein" is based on analytical procedures for constituents of protein, most prominently Kjeldahl nitrogen.

---

[1]Peptide bonds of aspartic acid are particularly susceptible to hydrolysis under weakly acidic conditions, whereas most other peptide bonds require either strong acid or the presence of a protease to be broken.

The protein content of food is calculated from the content of nitrogen, as assayed by the Kjeldahl method, multiplied by an appropriate factor, f_N. The standard conversion factor is 6.25, which should be used unless another factor is specified. The absolute size of the factor for various proteins has been much debated, but in practice this problem is rarely critical and can be disregarded.

Analysis of the amino acid composition of proteins by automatic ion exchange chromatography has become routine. From a nutritional perspective, it is evidently important to know the amino acid composition of food proteins, but this composition is also necessary information for the calculation of two important entities in hydrolysis experiments, namely, $H\Phi$ and h_{tot}.

$H\Phi$ is the average molecular hydrophobicity (Bigelow and Channon, 1976) and is calculated as:

$$H\Phi = \Sigma \, (x \, \Delta f_t)_i \tag{1}$$

where x is the mole fraction and Δf_t is the hydrophobicity value of the individual amino acids (cf. Table II). The units of $H\Phi$ are kJ/mol, but figures in the literature usually are given as cal/mol. $H\Phi$ is important for evaluating, among other things, the risk of bitter peptide formation during hydrolysis.

h_{tot} is the number of peptide bonds per gram protein and has units of meq/g N · f_N. This number is calculated from the amino acid composition as the sum of the concentrations (in mmol/g N · f_N) of each amino acid. These concentrations are obtained immediately from the amino acid analysis, in which the concentrations usually are given as g per 100 g protein (16 g N using the standard factor 6.25). h_{tot} is used to calculate the degree of hydrolysis (DH), as discussed in more detail later.

Table III gives f_N, $H\Phi$, and h_{tot} for a number of common food proteins. If no data on a particular protein are available, f_N is considered to be 6.25 and h_{tot} 8.0 meq/g N · 6.25.

TABLE III
Key Data for Common Food Proteins[a]

Protein material	Kjeldahl conversion factor (f_N)	h_{tot} meq/g (N · f_N)	$H\Phi$ (kcal/mol)
Casein	6.38	8.2	1.14
Whey protein concentrate	6.38	8.8	1.03
Meat	6.25	7.7	0.94
Hemoglobin	6.25	8.3	0.95
Gelatin	5.55	11.1	0.89
Fish protein concentrate	6.25	8.6	0.89
Soya proteins	6.25	7.8	0.90
Wheat gluten	5.7	8.3	0.97
Maize protein isolate	6.25	9.2	1.06

[a] Reprinted with permission from Adler-Nissen (1986a).

III. Enzymes

A. Classification of Proteases

Proteases are classified according to their source (animal, plant, microbial), their catalytic action (endopeptidase or exopeptidase), and the nature of the catalytic site. In the EC system for enzyme nomenclature, all proteases (or peptide hydrolyases) are in subclass 3.4, which is further divided into 3.4.11–19, the exopeptidases, and 3.4.21–24, the endopeptidases or proteinases. Endopeptidases are the proteases most commonly used in food processing, but in some cases their action is supplemented with exopeptidases.

Endopeptidases cleave the polypeptide chain at particularly susceptible peptide bonds distributed along the chain, whereas exopeptidases hydrolyze one amino acid (or dipeptide, in the case of 3.4.14 and 15) at a time from either the N terminus (aminopeptidases) or the C terminus (carboxypeptidases).

1. Endopeptidases

The four major classes of endopeptidases are serine proteases (EC 3.4.21), cysteine proteases (EC 3.4.22), aspartic proteases (EC 3.4.23), and metalloproteases (EC 3.4.24).

As the names imply, serine, cysteine, and aspartic proteases have serine, cysteine, and aspartic acid side chains, respectively, as an essential part of the catalytic site. Modification or blocking of this side chain usually leads to complete inactivation of the enzyme, and is a standard way of determining the nature of an unknown protease produced by a new strain of microorganism.

The serine proteases have maximum activity at alkaline pH; the closely related cysteine proteases usually show maximum activity at more neutral pH values. The aspartic proteases generally have maximum catalytic activity at acid pH. Among our digestive enzymes, the aspartic protease pepsin is secreted in the stomach and the serine proteases trypsin and chymotrypsin are excreted in the duodenum, in accordance with pH values of the digestive tract (acid in the stomach and alkaline in the gut).

The metalloproteases contain an essential metal atom, usually Zn, and have optimum activity near neutral pH. Ca^{2+} stabilizes these enzymes and strong chelating agents, such as EDTA, inhibit them. Such enzymes are common in microorganisms.

2. Exopeptidases

The aminopeptidases (EC 3.4.11) are ubiquitous, but less readily available as commercial products, since many of them are intracellular or membrane bound. The commercial enzyme preparation Pronase®, isolated from *Streptomyces griseus,* contains endopeptidase plus aminopeptidase and carboxypeptidase activities. Pronase often is used in the laboratory to achieve thorough hydrolysis of proteins, but the preparation is too costly for food use. Animal tissue homogenates also have been applied on a semi-industrial scale as sources of aminopeptidase activity (Clegg *et al.,* 1974; Clegg, 1978).

Carboxypeptidases are subdivided into serine carboxypeptidases (EC 3.4.16), metallocarboxypeptidases (EC 3.4.17), and cysteine carboxypeptidases (EC 3.4.18) according to the nature of the catalytic site. Many commercial proteases, in particular from fungi, contain appreciable amounts of carboxypeptidase activity. In the digestive tract, the metallocarboxypeptidases A and B are excreted in conjunction with the major serine digestive proteases (trypsin, chymotrypsin, and elastase). Therefore, they occur in the impure, commercially available preparation largely composed of trypsin that is called pancreatin.

For the sake of completeness, the exopeptidase subgroup also includes the dipeptide hydrolases (EC 3.4.13) that are specific for dipeptide substrates and, consequently, cannot be classified as amino- or carboxypeptidases. Dipeptide hydrolases are, however, not significant for food use.

From a food-processing perspective, the exopeptidases are mainly attractive as a means (among several others) of debittering protein hydrolysates. The most systematic studies have been carried out on hydrolysates of casein, which is particularly prone to yielding bitter peptides. Aminopeptidases (Clegg, 1978; Minagawa *et al.,* 1989) and carboxypeptidases (Umetsu *et al.,* 1983) can be applied successfully. In this context, note that the peptidase activity of the lysed lactic acid bacteria in freshly prepared cheese curd is essential for proper development of the flavor during maturation of cheese (Thomas and Pritchard, 1987).

B. Peptide Bond Cleavage and Protease Specificity

The question of specificity was raised already in the previous section, since proteases were divided into two groups according to their catalytic specificity (exo or endo). Further attempts at systematization do not, however, lead to quite as unambiguous a classification. In fact, the peptide hydrolases are notorious for the complexity of their substrate specificity.

The specificity of endoproteases usually is considered in terms of the two amino acids forming the peptide bond that is susceptible to hydrolysis by the enzyme in question. Depending on the catalytic mechanism, the specificity is associated with the side chain on the carboxyl side only, with the side chain on the amino side only, or with both side chains in combination. The simplest example of side-chain specificity is the serine protease trypsin, which readily cleaves all peptide bonds with lysine or arginine on the carboxyl side. The specificity of trypsin and other trypsin-like serine proteases is easy to understand in light of the mechanism of the hydrolytic reaction, which will be addressed in the following section.

1. Mechanism of Peptide Bond Cleavage Exemplified by the Action of Trypsin

The cleavage of peptide bonds by trypsin or chymotrypsin is known in great detail. The action of these two enzymes provides excellent examples for illustrating the chemistry and physical chemistry of the proteolytic reaction. Despite the lack of originality in choosing this example, we shall proceed with a presentation of the details of tryptic hydrolysis.

1. The enzyme binds to the polypeptide chain of the substrate through secondary binding forces (electrostatic attraction, hydrogen bonding, hydrophobic interaction). Of particular significance for this binding is that trypsin has a negatively charged pocket near its active site. This pocket can attract and accommodate the basic end group of the lysine or arginine side chain. The polypeptide chain is now fixed through the strong attraction of the lysine or arginine side chain, so the peptide bond on the carboxyl side comes close to the active site. In kinetic terms, the enzyme–substrate complex, ES, is the result of this step.

2. Through a series of electron shifts, a covalent bond is formed between the active site Ser-195 oxygen atom and the carbon atom of the carbonyl group. Simultaneously, the original covalent bond between the same carbon atom and the nitrogen atom is weakened and broken by the withdrawal of the electrons. A so-called tetrahedral intermediate involving Ser-195 and His-57 results from these electron shifts. The intermediate is fixed through hydrogen bonding to backbone amine groups of the enzyme, notably to an oxyanion binding site, the so-called oxyanion hole (Polgár, 1987). A buried aspartyl group (Asp-102) aids in stabilizing the tetrahedral intermediate (Warshel *et al.*, 1989). The completion of the electron shifts liberates the polypeptide chain on the amino side of the peptide bond, taking with it the now free hydrogen atom from the serine hydroxyl group. In kinetic terms, the enzyme–product complex, EP, is the result of this step.

3. The acyl-enzyme complex is subjected to a nucleophilic attack by a water molecule. A hydrogen atom is donated to the serine oxygen, while the hydroxyl group of the water binds to the carbonyl group, thereby regenerating the carboxyl group of the bound lysine or arginine.

4. The carboxyl group dissociates immediately (pH is neutral or slightly alkaline) and the associated free energy gain in forming the carboxylate anion more than compensates for the binding forces between the enzyme and the lysine or arginine side chain. The polypeptide chain with the now C-terminal lysine or arginine is free. In kinetic terms, Steps 3 and 4 are the reactions leading to a free product, P, and a free enzyme, E.

These reaction steps are all reversible and well described in thermodynamic terms. The mechanism is general, even in its details, for all serine proteases;

only the shape, size, and charge of the substrate binding pocket varies (Polgár, 1987).

Note that the net difference in free energy from Step 1 to Step 3 is positive. The large negative free energy term associated with Step 4 is largely responsible for driving the hydrolysis reaction under normal conditions (Fruton, 1982). Thus, if the dissociation of the carboxyl group can be suppressed (for example, by addition of dioxane or other solvents that shift the pK value of the carboxyl group to higher values), synthesis of a peptide bond can be achieved readily (Fruton, 1982).

In this connection, in reaction Step 3, the amino group of another peptide also may act as nucleophile. This action results in the reformation of a peptide bond linking the acyl-enzyme complex with this other peptide, a so-called transpeptidation reaction. In aqueous solutions of protein hydrolysates, competition exists between H_2O and H_2N-; although hydrolysis is favored by the high concentration of H_2O, the nucleophilic attack by H_2N- often proceeds many times faster (Fastrez and Fersht, 1973). Thus, transpeptidation is of considerable importance in practice. For example, human insulin can be synthesized by tryptic transpeptidation (Markussen, 1987). Transpeptidation also plays a key role in the plastein reaction, which is discussed in Section IV,D.

2. Specificity and Mechanism of other Peptidases

From the discussion of the tryptic hydrolysis reaction, it is evident why trypsin is specific for peptide bonds with lysine and arginine on the carbonyl side. Only such peptides are able to form a sufficiently stable ES complex. However, for most other endopeptidases, the case is less simple.[2] Considering Alcalase® (subtilisin Carlsberg) as an example, it is reported that it "has a broad specificity, hydrolyzing most peptide bond types and some ester bonds, preferentially those containing aromatic amino acid residues" (Ward, 1983). This description is not very specific. Subtilisin Carlsberg is also a serine protease, so the hydrolytic reaction obeys the same mechanism as trypsin, but subtilisin Carlsberg has a rather open hydrophobic pocket with no particular affinity for lysine or arginine. The preference to aromatic and other hydrophobic amino acids can be explained by the presence of this hydrophobic binding area on the subtilisin Carlsberg molecule.

The issue of peptide bond specificity is moderated further by the fact that secondary binding sites farther from the peptide bond to be cleaved also exert influence on the binding of the polypeptide chain (Svendsen, 1976; Polgár, 1987). This interaction complicates the interpretation of hydrolysis

[2] Trypsin has, in fact, a unique position among the commonly applied endoproteases because of its well-defined specificity. However, a serine protease has been isolated on a large scale as a by-product in the commercial preparation of subtilisin Carlsberg (Alcalase). This enzyme is essentially specific for cleavage at the carboxyl side of glutamate and aspartate, with strong preference for glutamic acid (Svendsen and Breddam, 1992).

experiments carried out on one particular substrate, for example, the often quoted comparative studies of different proteases and their degradation pattern of the oxidized B-chain of insulin (see Morihara, 1974).

The catalytic mechanisms of the three other classes of endopeptidases have not been studied as extensively as that of the serine proteases. The reaction catalyzed by the cysteine proteases seems to follow the mechanism of serine proteases closely, differing mainly because the acyl–enzyme complex is formed through the —SH group instead of the —OH group (Polgár and Halász, 1982). However, the possible role of the oxyanion binding site still is not clarified in the case of the cysteine proteases (Brocklehurst et al., 1987).

The aspartate proteases have a common structure consisting of two homologous halves with a long hydrophobic cleft between the two domains (Foltman, 1981). Each of the domains harbors one of the active site aspartic acid residues (32 and 215, respectively). The substrate polypeptide chain binds to the cleft stretching on both sides of the peptide bond to be cleaved. Again, the reaction consists of the four steps described earlier, in which Asp-32 acts as a base to attack the acyl carbon of the substrate (Fruton, 1987). However, both aspartate residues may form covalent bonds with the substrate (Foltman, 1981), which may explain why the aspartic proteases also can transfer the amino portion of the substrate. Peptide bond specificity is, therefore, not associated only with the carbonyl side of the substrate but also with the amino side, unlike in serine and cysteine proteases. Another difference is, of course, that at the very low pH value of peptic hydrolysis, the liberated amino group and not the carboxyl group becomes charged.

Most current knowledge of the metalloproteases is concluded from studies on bovine carboxypeptidase A, which has been investigated in much more detail than the microbial metalloendopeptidases. The active site Zn^{2+} is likely to play a role in fixing the peptide bond during cleavage by a metal-complex coordination bond to the carbonyl oxygen, analogous to the role of the oxyanion binding site. The nucleophilic attack apparently is carried out by a glutamic acid residue, but the mechanism is not well elucidated (Auld and Vallee, 1987). An alternative mechanism has been proposed also (Mock and Zhang, 1991).

As inferred from the previous text, it is generally assumed that the hydrolytic mechanism of exopeptidases is similar to that of the endopeptidases, depending on whether they belong to the serine class or the metalloenzyme class.

The mechanistic discussion of the concept of peptide bond specificity is complicated further by the fact that considerable differences are evident in the rates at which different peptide bonds are cleaved by the same enzyme. An outstanding example is the aspartate protease chymosin which, in the milk-clotting process, rapidly hydrolyzes the bond Phe-105–Met-106 in κ-casein, thus apparently displaying a high specificity for this bond. However, on other substrates chymosin exhibits general proteolytic capability, acting as any other ordinary aspartate protease (Visser, 1981). In fact, apart from simple examples such as trypsin and the other trypsin-like serine proteases, the concept of peptide bond specificity for endopeptidases is not very useful. Indeed, many endopeptidases are not specific for the peptide bond alone but are also very efficient esterases hydrolyzing a broad range of ester bonds.

C. Peptide Bond Cleavage and Kinetics

In the preceding section, the mechanism of the reaction of the serine proteases was described in detail and the mechanism of the other endoproteases was discussed briefly. If we neglect the reverse reactions leading to peptide synthesis, the kinetic scheme describing the action of endopeptidases can be simplified as:

$$E + S \underset{k_{-1}}{\overset{k_{+1}}{\rightleftharpoons}} ES \xrightarrow{k_{+2}} EP + H-P' \xrightarrow[+H_2O]{k_{+3}} E + P-OH + H-P' \qquad (2)$$

where S is the substrate and P and P' are the resulting peptides.

For serine proteases, it is generally accepted that the acylation step, Step 2, is rate determining (Svendsen, 1976; Polgár, 1987). Thus, the overall reaction rate constant, k_{cat}, is determined by k_{+2} and K_m is approximately equal to the true dissociation constant, k_{-1}/k_{+1}. In other words, classical Michaelis–Menten kinetics are obeyed, simplifying the kinetic considerations of the protein hydrolysis reaction considerably.[3]

1. Degree of Hydrolysis

The net result of the hydrolytic reaction is that, for each peptide bond cleaved, one free amino group and one free carboxyl group are formed. In aqueous solutions, these groups will be more or less ionized, depending on pH. The pK value (at 25°C) of the carboxyl group is 3.1–3.6 and that of the amino group is 7.5–7.8 (cf. Table II; Section II,A). Thus, hydrolysis of proteins is accompanied by a release or uptake of H^+, as already noted by Sörensen (1908). If the hydrolysis mixture is titrated with acid or base during the reaction so the pH is kept constant, the consumption of titrant can be related quantitatively to the number of peptide bonds cleaved. This measure is the basis for the pH-stat technique developed at the Carlsberg Laboratory (Jacobsen et al., 1957) which has been adapted successfully to food protein hydrolysis processes. A detailed discussion of the use of the pH-stat technique for food protein hydrolysis experiments is given elsewhere (Adler-Nissen, 1986a).

The number of peptide bonds cleaved also can be determined by other means. The use of a freezing point osmometer is a convenient way of monitoring the process when the pH-stat principle cannot be used (Adler-Nissen,

---

[3] For the hydrolysis of amino acid esters, k_{+2} is not a negligible term in K_m. The reported K_m values for such esters therefore are not applicable to the hydrolysis of peptide bonds (Svendsen, 1976). This illustrates one of the difficulties in determining the kinetics of protein hydrolysis from the hydrolysis kinetics of synthetic substrates.

1984). Samples also may be drawn and the concentration of α-amino groups assayed by a colorimetric reaction (Adler-Nissen, 1979).

From the number of peptide bonds cleaved (h, meq per g protein), the degree of hydrolysis may be calculated:

$$DH = h/h_{tot} \times 100\% \tag{3}$$

where h_{tot} is the total number of peptide bonds in the protein, as discussed previously (Section II,B).

2. Protein Hydrolysis Relative to Enzyme Kinetics

In the hydrolysis of a food protein, the reaction rate obtained from the total number of peptide bonds cleaved per unit of time is, of course, the sum of a multitude of reactions involving different polypeptides. Each of these reactions might be described by the kinetic variables in the Michaelis–Menten equation, that is, K_m and k_{cat} (k_{+2}), so the overall reaction might be described as a competition between all these reactions. Such an approach is discouraged, however, because of the large number of competing molecular species in the reaction. In practice, the total reaction curve, or hydrolysis curve, for a given enzyme–substrate system must be obtained experimentally. Then, kinetic considerations usually are applied in two cases, either for a qualitative description of the shape of the curve and the composition of the resulting hydrolysate or for a quantitative assessment of the effect of changing the hydrolysis conditions, for example, temperature or enzyme concentration.

The hydrolysis curve depends on both the enzyme and the substrates, as illustrated by the hydrolysis curves in Figs. 3 and 4. The quantitative relationship between DH and time, which the hydrolysis curves represent in graphical

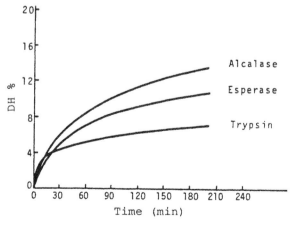

Figure 3 Hydrolysis curves for casein with different serine proteases. For casein, S = 8%, E/S = 12 AU/kg, pH = 8, T = 50°C. (Reprinted with permission from Adler-Nissen, 1986a.)

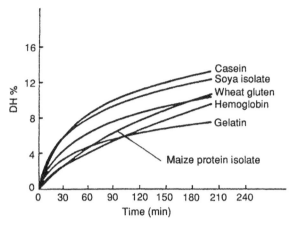

Figure 4 Hydrolysis curves of Alcalase with different substrates. For Alcalase, S = 8%, E/S = 12 AU/kg, pH = 8, T = 50°C. (Reprinted with permission from Adler-Nissen, 1986a.)

form, cannot be predicted theoretically, but must be determined by experiment. On the other hand, the qualitative similarity between the curves, notably the smooth downward curve, can be predicted from kinetic considerations.

For simple enzyme–substrate systems under analytical conditions, the shape of the reaction curve often follows the integrated Michaelis–Menten equation. Under industrial conditions, when the substrate concentration is usually very high, this approach does not apply to simple enzyme–substrate systems and certainly not to the hydrolysis of proteins. In fact, it has been shown that, during the typical conditions of food protein hydrolysis processes, substrate saturation prevails throughout the entire course of reaction (Adler-Nissen, 1986a). The kinetic basis for this effect is that the peptides resulting from the initial attack of the protease on the protein molecules also act as substrate for further degradation to smaller peptides. This process results in substrate competition between the original protein, S, and the peptides, P, as shown in the simplified kinetic scheme in Eq. 4.

$$S \xrightarrow{v_1} P \xrightarrow{v_2} R$$

$$v_1 = \frac{V_s}{1 + \dfrac{K_s}{S}\left(1 + \dfrac{P}{K_P}\right)} \qquad v_2 = \frac{V_P}{1 + \dfrac{K_P}{P}\left(1 + \dfrac{S}{K_S}\right)} \tag{4}$$

For this model, V_P has been shown to be considerably smaller than V_S, and $K_P < K_S$. This situation will result in a rapid build-up of P, the intermediary product. Because K_P is smaller than K_S, substrate saturation with respect to P will soon develop. Consequently, v_2 will increase until it approaches V_P. On the other hand, v_1 decreases for two reasons: the decrease in S and, more importantly, the quickly attained large value of P/K_P. The overall reaction rate, $v = v_1 + v_2$, therefore will decrease to a rate not much higher than v_2. This model complies qualitatively with endoprotease–protein systems investigated to date. For example, the model explains why unconverted protein is still generally observed, even after quite extensive hydrolysis.

3. Protein Hydrolysis Relative to Protein Conformation

Many proteins are known to be more easily hydrolyzed if they are denatured by, for example, heat treatment. This effect is understood immediately when the mechanism of denaturation is considered. Denaturation implies an unfolding of the polypeptide chain(s) from the particular tertiary structure of the native, biologically active protein to a less well defined, open state, the so-called random coil. In the native state, many of the peptide bonds are concealed in the interior of the tertiary structure; these peptide bonds now become exposed to the solvent. Many more peptide bonds of the substrate therefore are available for attack by a protease, which qualitatively explains the increased rate at which denatured proteins are hydrolyzed.

The relatively high resistance to proteolysis of many globular proteins led Linderstrøm-Lang (1952) to propose a scheme by which an initial reversible denaturation step is essential for hydrolysis to occur. The reversible denaturation is understood today as a temporarily distorted form of the native tertiary structure that, in aqueous solution, is twisting rapidly and fluctuating around the thermodynamic equilibrium. At any time, only a few of the protein molecules are in this denatured state, in which they can be attacked by the enzyme according to Eq. 5.

$$\text{Native} \underset{v_{-0}}{\overset{v_{+0}}{\rightleftarrows}} \text{Denatured} \xrightarrow[v_{\mathrm{I}}]{\text{enzyme}} \text{Intermediate} \xrightarrow[v_{\mathrm{II}}]{\text{enzyme}} \text{End product} \quad (5)$$

The equilibrium constant for reversible denaturation is:

$$K_0 = k_{-1}/k_{+1}$$

The reaction rate of denaturation is:

$$v_0 = v_{+0} - v_{-0}$$

As soon as the first peptide bond is cleaved, the protein molecule (the intermediate) becomes unstable and unfolds so many more peptide bonds become exposed. Therefore, the intermediate is degraded rapidly to a large number of smaller peptides. In many cases, the first reversible reaction is rate determining for the overall reaction, which means that the intermediate is degraded as soon as it is formed. The reaction mixture consists of native protein and end product (peptides) only; in Linderstrøm-Lang's own terms, the protein molecules are hydrolyzed to peptides completely, one by one.

The difference in susceptibility to hydrolysis of the native and denatured state is often considerable. For some globular proteins, it is several orders of magnitude. For example, the tryptic hydrolysis of a particular peptide bond in unfolded (denatured) ribonuclease occurs at least 1700 times faster than when the ribonuclease is still folded in the native configuration (Pace and Barret, 1984).

In some cases, the intermediate is stable. For example, the digestive proteases trypsin and pepsin are formed from their precursors, the zymogens, by an autocatalytic splitting of one or more peptide bonds close to the N terminus. The rest of the zymogen then undergoes the slight change in conformation that is necessary to exert its proteolytic activity.

These considerations of protein structure and reaction rate are important in protein hydrolysis processes. An industrial example of a virtually pure one-by-one reaction is the process for enzymatic decolorization of slaughterhouse blood. In this process, the red cell fraction is separated from the valuable plasma and hemolyzed by dilution with water. This aqueous solution of native hemoglobin is hydrolyzed at alkaline pH by Alcalase to a high DH value (DH 18–20%); Adler-Nissen, 1986a). Gel chromatography of the hydrolysate at intermediate DH (DH 10%) shows that only high molecular weight and very low molecular weight peptides are present, indicating a one-by-one reaction pattern (cf. Fig. 5). At DH 20%, the high molecular weight peak has disappeared completely. This particular degradation pattern is the cause of the successful application of ultrafiltration to separation of blood hydrolysates (Olsen, 1983).

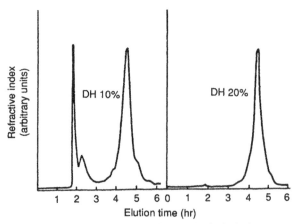

Figure 5 Gel chromatography of hemoglobin hydrolysate. Gel used is BioRad P-10 at pH 3.5. (Reprinted with permission from Adler-Nissen, 1986a.)

The hydrolysis of soy protein with Alcalase is also an illustrative example. If undenatured soy protein is used as substrate, the gel chromatogram on Fig. 6 indicates that the hydrolysate largely consists of high molecular weight and low molecular weight proteinaceous material, although the case is not as clear cut as the hydrolysis of hemoglobin. However, when the substrate is denatured by a heat treatment, the degradation pattern is changed and the hydrolysate now contains appreciable amounts of peptides of medium chain length.

These observations are in agreement with Linderstrøm-Lang's one-by-one reaction scheme discussed earlier. When the protein is in the native state, the first reaction step is rate limiting. Conventional reaction kinetics for sequential reactions predicts that the reaction mixture contains both unconverted substrate and end products, but negligible amounts of intermediate products. However, denaturation unfolds the protein molecule, many more peptide bonds are now available for enzymatic attack, and the initial reaction step may no longer be rate limiting. Therefore, a significant concentration of intermediate products builds up during the process.

These changes in molecular composition of protein hydrolysates with the state of denaturation of the substrate are important for the food use of protein hydrolysates. For example, the emulsifying properties of the hydrolysates may be expected to be superior if the hydrolysate contains appreciable amounts of medium- and long-chain soluble peptides (Adler-Nissen, 1988b; cf. Adler-Nissen, 1986a). This expectation suggests using a partially denatured substrate in the process.

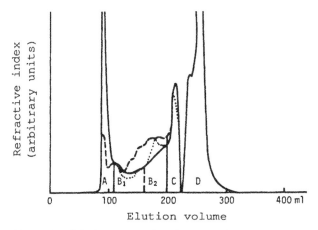

Figure 6 Gel chromatography of soy protein hydrolysates (DH, 6%), soluble fraction, at pH 5. Gel used is BioRad P-10 at pH 7.6. Undenatured (——) acid denatured (····), or heat denatured (– – –) samples. A: proteins and large peptides; B: medium-sized peptides; C: small peptides; D: dipeptides, amino acids, salt, sugars. (Reprinted with permission from Adler-Nissen, 1986a.)

D. Activity and Stability Profiles of Proteases

Commercial suppliers of industrial enzymes usually give information on four properties:

- pH activity
- pH stability
- temperature activity
- temperature stability

Typically, this information is presented as curves which, in the case of the activity curves, show the relative activity at different pH values or temperatures under modified assay conditions. The stability curves show the residual relative activity after exposure to the given values of pH or temperature for a defined period.

The meaning of these activity and stability curves and their utility for the application of proteases will be discussed briefly in the following sections.

1. pH Activity and Stability

As mentioned previously (Section III,A), the different proteases have different pH activity optima. On this basis, it is usual to classify proteases as alkaline, neutral, or acid. This classification and the pH optima themselves should, however, not be accepted dogmatically. It is advisable not to delimit the experimental work by insisting on using only optimal pH conditions. As an illustrative example, the metalloprotease Neutrase® is denoted as a neutral protease and the serine protease Alcalase is considered alkaline, in accordance with the pH- values for maximum activity. Nevertheless, when the pH–activity curves of these two enzymes are compared, Alcalase appears to exert significant hydrolytic activity even below pH 5. At the same pH values, the relative activity of Neutrase has decreased to virtually no activity.

The example in Fig. 7 demonstrates that the activity of these enzymes (and most other proteases) is not limited to a narrow range around the pH optimum. Proteases often are applied with success at conditions that nominally differ 1–2 pH units from the declared optimum.

Another point to consider is that the pH–activity curves show the relative initial activity under assay conditions, so they cannot be extrapolated quantitatively to performance under practical conditions. In particular, the use of substrate other than the denatured hemoglobin or casein prescribed in the assay influences the pH–activity profile considerably. For example, when native hemoglobin is used as a substrate for Alcalase, the relative activity of Alcalase almost doubles if the pH is changed from pH 8.0 to pH 8.5, a change that hardly affects the relative activity when casein or soy isolates are used as substrates (Adler-Nissen, 1986a).

Another example of the influence of the substrate on the pH–activity profile is papain, for which the optimum varies between pH 5 (for gelatin)

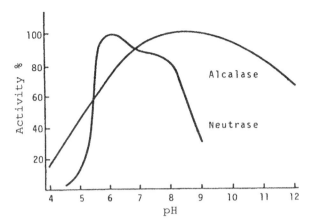

Figure 7 pH activity curves for Alcalase and Neutrase acting on denatured hemoglobin. (Reprinted with permission from Adler-Nissen, 1986a.)

and pH 7 (for egg albumin and casein) (Yamamoto, 1975). In fact, the only really reliable way of assessing the performance of the proteases under application conditions is to carry out experiments at varying pH values.

The pH – stability curves are obviously useful for choosing suitable application conditions. In some cases, a protease is also stable at pH values at which there is no activity. A well-known example of such behavior is trypsin, which has maximum stability at pH 3. In other cases, exposure to extreme pH values will cause the enzyme molecule to unfold, which increases its susceptibility to proteolytic degradation. Therefore, the protease is inactivated rapidly through an autolytic reaction, a phenomenon observed with subtilisin Carlsberg (Alcalase) (Ottesen and Svendsen, 1970) at acid pH, for example.

The possibility of inactivating proteases at extreme pH values is, in some cases, of great advantage since the change in pH can be carried out rapidly, even in large reactor volumes.

2. Temperature Activity and Stability

Proteases, like other enzymes, will exhibit increasing specific activity with increasing temperature, that is, there will be an increase in the rate at which the substrate is hydrolyzed. The reaction rate depends on both the velocity constant, k_{cat} (k_{+2}),[4] and on K_m. k_{cat} follows the Arrhenius equation, whereas K_m changes with temperature (it usually increases) according to the Gibbs – Helmholtz equation. At high substrate concentrations, at which the effect of changes in K_m with temperature can be neglected, the temperature activity curve should reflect approximately the activation energy of the hydrolysis reaction, in accordance with the Arrhenius equation.

[4] In the case of insoluble substrates, such as denatured hemoglobin, the reaction rate is influenced also by the adsorption characteristics of the enzyme (cf. Section IV,B).

At high temperatures, the thermal inactivation of the enzyme will become pronounced. The decrease in activity because of denaturation soon will outweigh the effect of the increased specific activity of the enzyme itself. The further increase in temperature therefore will lead to a decrease in the activity as measured under assay conditions. This effect leads to an apparent optimal temperature for the enzyme — a concept that has led to much confusion. In contrast to the pH – activity profile in which a true optimal value of pH can be defined, the optimal temperature is a function of the reaction time used in the assay. The longer the reaction time, the lower the optimal temperature. Figure 8 illustrates this phenomenon for Alcalase and soy protein. The curves are based on an analytical expression for the temperature activity curve derived from the thermodynamics of enzyme catalysis and enzyme inactivation.

The temperature stability curves simply show the relative decrease in activity with time at various temperatures. The stability curves usually are determined in a buffer solution at a pH value close to the value for optimal stability. In the presence of substrate, as in food protein hydrolysis processes, the stability is increased considerably, so the stability curves can only be indicative. A change to pH values other than the optimum means that the stability will be lowered, which sometimes can be useful in practice to achieve inactivation through a combined change in pH and temperature.

E. Protease Preparations in Practice

Commercial proteases are supplied as liquid preparations that contain the active enzyme protein in a dilute solution or as solid preparations with inert fillers. Several good reasons exist for not using pure crystalline proteases in

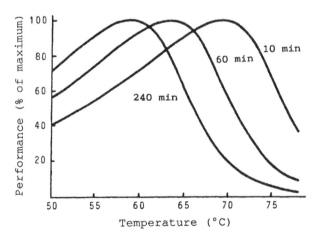

Figure 8 Temperature activity curves for different reaction times of soy isolate (S = 8%) and Alcalase, pH 8.0. (Reprinted with permission from Adler-Nissen, 1986a.)

industrial practice. First, they are usually not very stable during storage. Consequently, the specific activity of the enzyme preparation would vary in an unpredictable manner unless stabilizers were added. The use of pure crystalline proteases on a larger scale also would present an unnecessary health hazard due to the risk of inhalation of dust from the handling of the enzyme. These two issues will be discussed in the following sections.

1. Activity Units and Declared Activity

The activity of proteases should, in principle, be given in international units (IU), that is, as microequivalent peptide bonds cleaved per minute. Then the activity units would have a meaningful relationship to DH (cf. Section III,C,1). However, for historical reasons, a bewildering number of arbitrary units based on particular assays are used for proteases. Most of these assays use denatured hemoglobin or casein as substrate. The activity unit is calculated from the amount of proteinaceous matter digested (that is, became soluble in TCA) after a given time. For example, the activity of neutral or alkaline proteases commonly is determined by an assay originally devised by Anson (1939). In this method, denatured hemoglobin is digested at pH 7.5 and 25°C for 10 min. The amount of TCA-soluble product is determined with Folin and Ciocalteau phenol reagent. One Anson Unit (AU) is the amount of enzyme that digests the hemoglobin to the extent that the amount of TCA-soluble product liberated per minute gives the same color as 1 mmol tyrosine.

Using the Anson Units as an example, the specific activity of a protease preparation is given in Anson Units per gram preparation. Food-grade Alcalase (subtilisin Carlsberg), for example, has a declared activity of 2.4 AU/g; for comparison, the crystalline enzyme has a specific activity above 25 AU/g (Adler-Nissen, 1986a). The declared activity is the nominal activity as guaranteed by the supplier; the actual activity, when measured on a particular sample, often will be slightly higher. This discrepancy exists because producers of industrial enzymes standardize the different batches so the actual activity is about 10% higher than the declared activity, to compensate for analytical uncertainty and for the slight loss of activity during longer storage.

In industrial applications of proteases (and other enzymes), the amount to use is always calculated on the basis of the declared activity. The amount is specified in activity units per kg substrate, for example, AU/kg. Since enzymes are sold on the basis of activity, this notation is generally much more informative than specifying the amount as a weight percentage of enzyme preparation or by other measures not related to the activity.

2. Handling and Safety Measures

Protease preparations, like other enzyme preparations, must be stored in a cool dry place or under refrigeration, in accordance with the instructions by the supplier. In industrial practice, the preparations are ready for use and should not be diluted or mixed with other reagents before use. In laboratory experiments, however, dilution may sometimes be necessary, because the small volumes may make accurate measurement of the amount of enzyme difficult. If dilution is needed, the dilute solution should be prepared immediately before use, because the stabilizing effect of the additives in the preparation may no longer be reliable. Experience bears out that dilute protease solutions may lose much activity, even after only a few days in a refrigerator, because of susceptibility of the proteases to inactivation by autolysis or, in the case of the cysteine proteases, inactivation by oxidation.

The handling instructions from the supplier also should be followed with respect to safety measures. With modern liquid or dry preparations for industrial use, the health risk from involuntary inhalation of enzyme dust has been eliminated (Farrow and Reichelt, 1983). However, this risk has not been eliminated for the purified enzymes supplied for laboratory use. Traditional sources of proteases (for example, crude papaya latex) also present a risk because of dust.

Proteases, like other proteins, are potential allergens. Allergenicity may be invoked after prolonged and uncritical exposure, particularly inhalation. Dust or aerosol formation therefore must be avoided. All spilled enzyme must be removed by washing with water. The proteases also present a hazard through their ability to digest living tissue; again, inhalation is the primary risk. However, these known hazards of proteases can be easily controlled; provided the enzymes are subjected to proper handling, proteases are safe and unproblematic reagents for use in food processing.

IV. General Issues in Protein Hydrolysis Processes

Some general aspects of protein hydrolysis processes are useful to consider in any systematic application experiments with proteases. The scope of this chapter permits only a short introduction to some of the issues with which the researcher usually is confronted when initiating work on protein hydrolysis. According to my experience in the field, the issues treated in these four subsections are the most pertinent.

A. Protein Hydrolysis Indices

Experimental work with protein hydrolysis is eased considerably through the systematic use of characteristic indices for describing the hydrolysis reaction. Given the substrate and the enzyme, there are four indices that define the initial conditions (S, E/S, pH, and T) and four indices that describe the reaction and the composition of the hydrolysate (DH, psi, PCL, and TCA index). The definition and significance of these eight indices will be summarized here.

S is the substrate concentration in the reaction mixture calculated on a weight basis. A molar concentration of the substrate is meaningless in the case of food proteins, and the concentration usually is based on a Kjeldahl analysis (cf. Section II,B). E/S is the enzyme–substrate ratio and is given in activity units per kg substrate. Knowing S, E/S, and the mass of the reaction mixture, the total number of activity units are calculated immediately. From the number of activity units and the declared specific activity of the enzyme preparation, the mass of enzyme preparation to be added can be calculated. pH and T are the initial values of pH and temperature, respectively.

DH was defined previously (Section III,C,1) as the percentage of peptide bonds cleaved. In the course of the hydrolysis reaction, DH increases until it reaches a preselected value, at which time the reaction is terminated by inactivation of the enzyme. DH is the controlling parameter of the hydrolysis reaction. The advantage of using DH, rather than time, t, as the controlling parameter is that, as a general rule, the properties of the hydrolysate depend on DH only and are independent of changes in S, E/S, and T (only pH exerts an independent, although minor, influence on the properties of the hydrolysate). This theorem of DH allows the experimenter to reduce considerably the number of experiments in any systematic work. Instead of working with five independent variables, as is done conventionally (S, E/S, pH, T, and t), these variables can be condensed to two—DH and pH—of which DH is by far the most significant. Having chosen the substrate and enzyme, DH is the principal variable to be used in the optimization of protein hydrolysis processes.

psi is the protein solubility index, defined as the mass of soluble protein relative to the total mass of protein in the reaction mixture. The solubility index is defined and measured under the practical conditions used for separating the soluble from the insoluble phase in a hydrolysis process (usually, pH is adjusted to near the isoelectric point of the substrate). psi indicates the maximum yield of soluble protein, independent on the efficiency of the separation (centrifuging) process. In the case of hydrolysis processes, in which the total reaction mixture is recovered without any subsequent separation, psi is an indication of how large a fraction of the substrate protein has been affected by proteolysis, which again influences the functional properties of the hydrolysate.

PCL is the average peptide chain length of the soluble phase. PCL is given as the average number of amino acid residues per peptide molecule in the soluble fraction. PCL can be calculated from an assay of the concentration of α-amino groups in the soluble phase (Adler-Nissen, 1979). Because the insoluble phase consists mainly of unconverted protein, that is, protein that has not undergone proteolysis, it can be shown that DH, psi, and PCL are approximately interrelated:

$$PCL = psi \times 100/DH\% \qquad (6)$$

PCL is an indicator of the functional properties of the peptide fraction. Thus, a peptide fraction with a low PCL value (2–3) usually will be just soluble, but if PCL is higher, the soluble phase also may exhibit surface active properties, making it an interesting emulsifier or foaming agent (Adler-Nissen, 1988b).

TCA index is the percentage of TCA-soluble peptides in the isoelectric soluble fraction. Those peptides that are soluble at the isoelectric pH but insoluble in TCA will be long peptides that usually exhibit interesting functional properties. The TCA index thus supplements the PCL value as an evaluation of the functional potential of the hydrolysate.

In Section III,C,2, the shape of the hydrolysis curve was explained by substrate competition between initially formed peptides and the still-unconverted substrate protein. The logical consequence of this competition is that, concomitant with the rise in DH, psi will rise but PCL will decrease during the hydrolysis reaction; this effect is observed in practice (Adler-Nissen, 1986a). The decrease in PCL also leads to a rise in TCA index, as the longest peptides are hydrolyzed gradually. Figure 9 shows, in principle, the time course of these four indices.

B. Application Screening of Proteases

An often encountered problem is making a systematic investigation of a number of different proteases to select the most suitable one for a particular purpose. As explained earlier, with increasing DH psi will increase but PCL

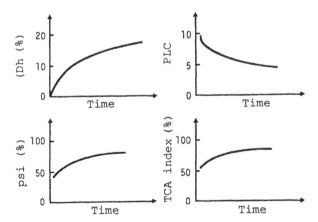

Figure 9 Time course of key indices in enzymatic protein hydrolysis. PCL and TCA index are given for soluble peptides. (Reprinted with permission from Adler-Nissen, 1986a.)

will decrease. Since, in many cases, both a high psi value and a high PCL value are desirable, a compromise must be sought.

Each protease will have its particular relationship between DH, PCL, and psi. A comparison of a number of different hydrolysates made with different proteases but hydrolyzed to the same DH requires a plot of psi versus PCL. This plot is exemplified in Fig. 10; the figure demonstrates that Alcalase is superior to the other enzymes tested with respect to producing protein hydrolysates with a high PCL value and in high yields.

The psi–PCL plot does not address the question of optimal concentration of enzyme because the same DH value and the same properties can be obtained using a low enzyme concentration and a long reaction time or a higher enzyme concentration for a shorter time. As a general rule, hydrolysis time is inversely proportional to E/S, other parameters being kept constant. However, for insoluble substrates for which adsorption and diffusion phenomena play a role, it has been shown that t is inversely proportional to $(E/S)^n$, where $n < 1$ (Adler-Nissen, 1986a).

A comparison of different proteases with respect to dosage raises the question of a meaningful basis for comparing the enzymes. Comparison on a simple weight basis does not take into account differences in specific activity. On the other hand, comparison on an activity unit basis also may be misleading, since the technical performance of different proteases is not a simple function of the specific activity (cf. Section III,D,1). Comparison on a cost basis, that is, dollars per ton of substrate converted product, seems immediately appealing, provided the enzymes produce a hydrolysate with the desired properties. However, that issue is trivial and needs no further comment.

In conclusion, intelligent screening of proteases does not follow a simple

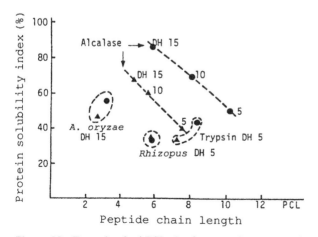

Figure 10 Example of psi-PCL plot for two substrates, casein (●) and soya isolate (▲). (Reprinted with permission from Adler-Nissen, 1986a.)

procedure, but requires a sound technical and economical evaluation of usually conflicting considerations.

C. Peptide Bitterness

The problem of peptide bitterness has been a dominating theme in the literature on protein hydrolysis during the 1970s and 1980s. The bitter taste has long been known as a fault during cheese maturation, but soon was perceived to be a major problem in the food use of protein hydrolysates. The bitter taste originates from the very products of proteolysis, the peptides. The problem of bitterness is, thus, inherent in any application of proteases for food.

The causes of the bitter taste have been studied and debated. There seems to be general consensus that the bitterness is caused by the presence of peptides of medium to short chain length, with an exceptionally high content of hydrophobic amino acid side chains. From reports on the taste and amino acid composition of peptides, Ney (1971) formulated the so-called Q-rule. This rule states that a peptide will be bitter if the average hydrophobicity, Q, is above $+1400$ cal/mol and nonbitter if the average hydrophobicity is below $+1300$ cal/mol.[5]

The crucial role of hydrophobicity for the bitter taste is further substantiated by theoretical considerations on taste receptor chemistry (Belitz *et al.*, 1979), as well as by quantitative taste studies. For bitter peptides, it is generally observed that the higher the hydrophobicity of a particular peptide, the more intense is its bitter taste. From data published by Wiezer and Belitz (1976), it appears that this relationship can be expressed as a linear (although crude) correlation between the logarithm of the threshold value for bitterness and the hydrophobicity (Adler-Nissen, 1986b,1988a).

Although the Q-rule (and similar correlations based on average hydrophobicity calculations) has been assessed and verified for individual small peptides, it cannot be extended to include mixtures of peptides, that is, protein hydrolysates, simply because the hydrolysate consists of peptides that show a broad distribution of average hydrophobicity. The majority of peptide material is, in fact, nonbitter according to the Q-rule; bitterness is caused solely by the presence of a small fraction of soluble highly hydrophobic peptides (Adler-Nissen, 1986b).

The bitterness determinant is verified, for example, by experiments in which protein hydrolysates are extracted with 2-butanol. This extraction results in an effective debittering, although only a small fraction of the peptide material is removed (Lalasidis, 1978). A debittering is observed also

[5] The hydrophobicity values for the individual amino acid side chains used in calculating Q are those given by Tanford (1962) in the original paper on hydrophobicity. If the revised values are used (Nozaki and Tanford, 1971; Bigelow, 1976), the numerical values will, in general, be smaller. Thus, for food proteins, $H\Phi$ is generally 10% lower than Q (Adler-Nissen, 1986b).

after isoelectric precipitation (Adler-Nissen, 1986b), a phenomenon utilized in the industrial production of isoelectric soluble soy protein hydrolysates (ISSPH). These studies demonstrate that the bitter peptides usually constitute only a minor weight fraction of the hydrolysate, but they occupy the extreme end of the (theoretical) hydrophobicity distribution function of all the peptides in the hydrolysate. The concentration of these peptides cannot, of course, be estimated from the average value of the hydrophobicity; thus, the Q-rule can not be applied to predict the risk of obtaining a bitter protein hydrolysate.

These considerations form the basis for establishing a quantitative correlation between the bitter taste of a protein hydrolysate and its chemical properties, primarily the amino acid composition and molar concentration of the 2-butanol-extractable peptides in protein hydrolysates (Adler-Nissen, 1988a). As shown in Fig. 11, the predicted bitterness correlates well with the observed bitterness except at very high or very low DH values, at which the bitterness is lowered by a complicated chain-length effect involving not only the size of the peptides but also the position (terminal or nonterminal) of the hydrophobic amino acid side chains.

All in all, the bitterness of protein hydrolysates is a complex phenomenon. Bitterness can be concluded to be influenced by at least five variables (Adler-Nissen, 1986a):

1. hydrophobicity of the substrate, since a high value of $H\Phi$ generally results in a relatively larger amount of highly hydrophobic peptides
2. DH, which influences both the concentration of soluble hydrophobic peptides and their chain length
3. enzyme, which in combination with DH influences the peptide hydro-

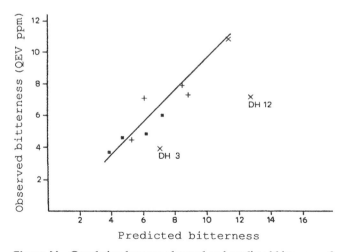

Figure 11 Correlation between observed and predicted bitterness of soluble protein hydrolysates of casein–alcalase (X), casein–trypsin (+), and soy isolate–alcalase (■). Slope = 1.02; $r = 0.94$. (Reprinted with permission from Adler-Nissen, 1988a.)

phobicity distribution in an unpredictable manner; further, the specificity of the protease also determines, to some extent, the position (terminal or nonterminal) of the hydrophobic side chains; there are indications that the terminal position gives rise to less bitterness; these specific effects of the enzyme are not negligible in practice, but must be investigated experimentally

4. any separation step included in the hydrolysis process; for example, by centrifuging at isoelectric pH, bitter peptides are removed with unconverted protein in the ISSPH process, as mentioned earlier

5. masking effects from components other than peptides; for example, organic acids, in particular citric and malic acids, exert a significant masking effect in soy protein hydrolysates (Adler-Nissen and Olsen, 1982); this masking effect can be used in the final formulation of a food or beverage product containing protein hydrolysates.

D. Plastein Formation

The plastein reaction is a peculiar phenomenon in the field of food protein hydrolysis. Plasteins are gel-like substances that can be produced from proteins by enzymatic hydrolysis, followed by concentration of the peptide solution and a renewed incubation with a protease at neutral pH (Fujimaki, 1978). The plastein gel usually is recovered by precipitation with ethanol or acetone.

The plastein reaction has received much attention because it can be used for debittering protein hydrolysates, an effect discovered by Fujimaki and co-workers (1970). The mechanism of the plastein reaction first was believed by the Japanese authors to be resynthesis of protein. Subsequent studies have disproved this hypothesis by both theoretical and empirical arguments. Resynthesis is thermodynamically unfavorable under plastein reaction conditions (You-shang, 1983). Chemical analyses of the plastein gels do not indicate the existence of high molecular weight material (Monti and Jost, 1979). The most likely mechanism is transpeptidation leading to the formation of new hydrophobic peptides that are barely soluble and therefore condense into a gel (Gololobov, 1981).

The Japanese group has applied the transpeptidation reaction successfully with papain to incorporate long-chain alkyl esters into food proteins. They were able to make ingredients with exceptionally good emulsifying properties (Arai et al., 1986).

V. Individual Proteases

Table IV gives a survey of the names and properties of common protease preparations for food use. Note that many of these preparations are actually

TABLE IV
Proteases for Food Protein Hydrolysis[a,b]

Type of protease	Source	Common names, tradenames[c]	Typical pH range[d]	Preferential specificity[e]
Serine protease	Ox, pig	**Trypsin**	pH 7–9	Lys–, Arg–COOH
		Chymotrypsin	pH 8–9	Phe–, Tyr–, Trp–COOH
	Bacillus licheniformis	**Subtilisin Carlsberg, Alcalase, Maxatase, Optimase**	pH 6–10	Broad specificity, mainly hydrophobic–COOH
	Bacillus amyloliquefaciens (Bacillus subtilis)	Subtilisin Novo, subtilisin BPN'	pH 6–10	
	Bacillus sp. alkalophilic	Subtilisin Esperase, subtilisin Savinase	pH 7–12	
Cysteine protease	Papaya latex	**Papain**	pH 5–8	Broad specificity, mainly hydrophobic–X–COOH
	Pineapple stem	**Bromelain**	pH 5–8	
	Fig latex	**Ficin**	pH 5–8	
Aspartic protease	Ox, pig	**Pepsin, pepsin A**	pH 1–4	Mainly hydrophobic–COOH and –NH₂
	Calf	**Chymosin, rennin**	pH 4–6	Rennet specificity on casein; mainly hydrophobic–COOH and –NH₂ on protein substrates in general
	Mucor miehei	**Rennilase**, Fromase, Marzyme, Morcurd	pH 4–6	
	Mucor pusillus	**Emporase**, Meito rennet, Noury lab	pH 4–6	
	Endothia parasitica	Surecurd, Suparen	pH 4–6	Also Glx–COOH
	Aspergillus niger (Aspergillus saitoi)	Aspergillopeptidase A (pure aspartic protease), Sumyzyme AP, Proctase, Molsin, Pamprosin (mixed with carboxypeptidase)	pH 2–5	Pure aspartic protease: like pepsin; mixed preparations: broad specificity
	Rhizopus sp.	Sumyzyme RP, Newlase	pH 3–6	Like pepsin
Metalloprotease	*Bacillus thermoproteolyticus*	**Thermolysin**, Termoase	pH 7–9	Ile–, Leu–, Val–, Phe–NH₂
	B. amyloliquefaciens (B. subtilis)	**Neutrase**	pH 6–8	Leu–, Phe–NH₂, and other

(continues)

TABLE IV (*continued*)

Type of protease	Source	Common names, tradenames[c]	Typical pH range[d]	Preferential specificity[e]
Mixture of trypsin, chymotrypsin, elastase, and carboxypeptidase A or B	Ox, pig	Pancreatin	pH 7–9	Very broad specificity
Mixture of papain, chymopapain, and lysozyme	Papaya fruit	Papain, crude	pH 5–9	Broad specificity
Mixture of *B. amyloliquefaciens* serine and metalloprotease	*B. amyloliquefaciens*	Biophrase, Nagase, Rapidermase, Rhozyme P 53, MKC protease	pH 6–9	Broad specificity
Mixture of aspartic protease, metalloprotease, serine protease, and carboxypeptidase	*Aspergillus oryzae*	***A. oryzae* protease**, Takadiastase, Sumyzyme LP, Veron P, Panazyme, Prozyme, Biozyme A, Sanzyme	pH 4–8	Very broad specificity
Mixture of alkaline and neutral protease plus aminopeptidase and carboxypeptidase	*Streptomyces griseus*	Pronase	pH 7–9	Very broad specificity
Aspartic protease, with some carboxypeptidase	*Penicillium duponti*	*P. duponti* protease	pH 2–5	Broad specifity

[a] The data have been compiled from the following sources: Aunstrup, 1980; Brocklehurst *et al.*, 1987; Dambmann and Aunstrup, 1981; Dixon and Webb, 1979; Emi *et al.*, 1976; Fruton, 1987; Godfrey, 1983b; Gould, 1975; Perlmann and Lorand, 1970; Morihara, 1974; Polgár, 1987.
[b] This table includes all common industrial proteases for food use as well as several nonindustrial proteases that are generally used. Enzymes printed in boldface are discussed further in the text.
[c] Not all tradenames are included.
[d] Indicated application conditions.
[e] Terminal amino acid after cleavage.

mixtures of different types of proteases; these mixed preparations are listed as such. The subsequent sections contain a brief discussion of the properties and uses of the most important proteases in the food industry (printed in boldface in the table). Table IV does not, of course, include all known proteases, and the selection is discretionary. Information on less well-known proteases of microbial origin may be found in the survey by Ward (1983). More detailed information on the individual enzymes in Table IV is given by Godfrey (1983b), whose comparative overview of industrial enzymes is extremely useful.

A. Serine Proteases

1. Trypsin

Trypsin is formed in the digestive tract by an autocatalytic reaction in which trypsinogen is hydrolyzed to trypsin by the action of trypsin itself. Trypsin is isolated in conjunction with insulin by extraction from the pancreatic glands of ox or pig. The commercial availability of trypsin has, therefore, been intimately bound to the market for insulin for many years.[6] The molecular mass of trypsin is ~ 24 kDa and the pI ~ 10.1, depending on the species from which the enzyme is derived.

As discussed in detail previously (Section III,B,1), trypsin specifically hydrolyzes peptide bonds with lysine and arginine on the carbonyl side. Like the other serine proteases, trypsin has maximum catalytic activity at alkaline pH, but its stability is limited by autolysis. Ca^{2+} enhances the stability against autolysis, but this tendency is much more pronounced with bovine than with porcine trypsin. Note, however, that when trypsin is used in protein hydrolysis processes the autolysis reaction is suppressed effectively by the large amounts of an alternative substrate, the food protein. Thus, the susceptibility to autolysis is of concern only when handling the enzyme preparation before it is added to the reaction mixture.

At acid pH, at which the catalytic activity is reduced to zero, trypsin is remarkably stable; it can be stored in the cold in dilute solutions at pH 3 for weeks without loss of activity. Trypsin is rather heat stable and typically can be applied at temperatures of $50-60°C$.

Trypsin has been used particularly for producing protein hydrolysates from whey protein (Jost and Monti, 1977) and appears to be superior to other proteases for that purpose (Monti and Jost, 1978). The fairly long peptides that result from tryptic hydrolysis because of the defined specificity of the enzyme give tryptic hydrolysates of casein and whey protein excellent emulsifying properties (Chobert et al., 1988).

---

[6] With the increasing production of insulin from genetically engineered microorganisms, this situation will change in the future.

A number of *Streptomyces* species produce proteases that are trypsin-like with respect to their catalytic properties (Ward, 1983). These enzymes also are inhibited by soybean trypsin inhibitors (cf. Section V,E). Because of the established commercial availability of the mammalian trypsins, no industrial application of their microbial counterparts is known.

2. Chymotrypsin

The term chymotrypsin includes a number of isozymes formed from their corresponding zymogens when they are excreted in the pancreatic juice in addition to trypsin. Although there are slight differences in the specificity of the different isozymes, these differences are usually of no important in the food use of chymotrypsin preparations. The molecular mass of chymotrypsin is ~ 25 kDa, depending on the species from which the enzyme is derived.

Purified chymotrypsin has been used experimentally to produce protein hydrolysates, for example, in making an infant formula with reduced allergenicity (Asselin *et al.*, 1989), but appears not to have any commercial food use. In its specificity and pH–activity profile, chymotrypsin resembles the subtilisins rather than trypsin. Supplies are limited, and chymotrypsin is much more expensive to produce, so there is little incentive to develop a commercial use for chymotrypsin.

3. Subtilisins

The subtilisins constitute a family of closely related bacterial serine proteases produced by *Bacillus* species. Alcalase (subtilisin Carlsberg) is produced by *Bacillus licheniformis* and is the protease that is produced in the largest amounts, thanks to its prominent use in detergents since the 1960s (Dambmann and Aunstrup, 1981). Alcalase also happens to be one of the best enzymes for producing soluble protein hydrolysates from a number of protein sources, for example, soy protein (Adler-Nissen, 1986a), fish protein (Poulsen and Adler-Nissen, 1989; Rebeca *et al.*, 1991), or the red cell fraction of slaughterhouse blood (Olsen, 1983). Other functional properties of soy proteins, such as the emulsifying properties (Adler-Nissen *et al.*, 1983; Adler-Nissen, 1988b; Mietsch *et al.*, 1989) or the foaming properties (Olsen and Adler-Nissen, 1981; Adler-Nissen *et al.*, 1983), also can be improved by a limited hydrolysis with this enzyme. The molecular mass of subtilisin Carlsberg is 27 kDa and the pI is ~ 9.4.

Alcalase has a broad specificity and cleaves many types of peptide bonds, preferentially those with a hydrophobic side chain on the carbonyl side. Its specificity is discussed in more detail by Svendsen (1976). It has a broad pH–activity profile with a maximum at pH 8–9 for typical food protein substrates. It is quite thermostable; the typical temperature of application is 55–60°C, but the enzyme may be applied at temperatures up to 70°C.

At pH values below 5, Alcalase loses its high thermostability; near pH 4 the enzyme can be inactivated readily, even at moderate temperatures. The mechanism of inactivation is a pH-induced unfolding of the native enzyme molecule followed by rapid autodigestion (Ottesen and Svendsen, 1970). This inactivation mechanism is widely applied in industrial practice, since it allows an almost instantaneous arresting of the proteolytic reaction in a batch hydrolysis process.

Microorganisms that are closely related to *B. licheniformis* produce enzymes that belong to the subtilisin group but have higher activity and stability at extremely alkaline pH values. These enzymes (subtilisin Esperase, subtilisin Savinase) were developed originally for use in detergents, but are now also becoming available as food-grade enzymes.

B. Cysteine Proteases

1. Papain

Papain is produced from papaya latex and is commercially available as a powder or a liquid preparation. Depending on the source, it contains more or less of the other principal papaya latex enzymes, chymopapain and lysozyme. The molecular mass of papain is ~21 kDa and the pI is ~8.8.

Papain has a broad specificity and a broad pH–activity profile from pH 5 to 9, depending on the substrate. Papain is quite stable to heat but is inactivated by oxidizing agents and by exposure to air.

Papain is used industrially in many protein hydrolysis processes. With subtilisin Carlsberg (Alcalase), it is the most effective general purpose protease for producing soluble food protein hydrolysates. Papain also is used in a process developed for solubilizing fish and fish offal for animal feed (Vega and Brennan, 1988). As mentioned in Section IV,D, papain is the enzyme of choice in the plastein process (Arai and Fujimaki, 1991). In addition to this application, papain is traditionally used in chillproofing of beer (Moll, 1987) and meat tenderizing (Schwimmer, 1981).

2. Bromelain and Ficin

Bromelain is isolated commercially from the crushed stems of pineapple plants. Ficin is obtained from dried fig latex. The molecular mass of bromelain is ~33 kDa and that of ficin, 26 kDa. Both enzymes resemble papain with respect to specificity and pH–activity profile, but they are slightly less thermostable.

Bromelain and ficin are used industrially for the same purposes as papain. However, quantitatively the market for these two plant proteases seems to be much smaller than the market for papain.

C. Aspartic Proteases

1. Pepsin

Like trypsin, pepsin is formed by an autocatalytic reaction from pepsinogen, which is found in the stomach mucosa of animals. Pepsin is produced commercially from this source and is used to a limited extent to produce protein hydrolysates. Quantitatively, the most important food use of pepsin is in a mixture with chymosin in rennet preparations. The molecular mass of porcine pepsin is ~35 kDa and the pI is between 2 and 3, but the data vary because of microheterogeneities in the preparations of pepsin (Fruton, 1987). This issue is discussed in more detail by Foltmann (1981).

The pH–activity profile of pepsin is remarkable because of the very low pH optimum at pH 2 for most substrates. Like trypsin, pepsin in pure solution is degraded slowly by autolysis at the pH values at which it is most active. Pepsin is relatively heat stable and can be used at temperatures up to 60°C. Its low pH–activity profile is an advantage in process hygiene.

2. Chymosin

Chymosin is activated by limited proteolysis of the zymogen prochymosin, and is the dominant gastric protease in young animals. The primary structures of chymosin and pepsin exhibit about 50% homology and their tertiary structures are quite similar (Foltmann, 1981). The molecular mass is about the same as that of pepsin (35 kDa), but the pI is higher (Fruton, 1987).

The primary use of chymosin is, of course, in the clotting of milk for cheese making, for which purpose it displays a remarkable specificity (cf. Section III,B). Apart from that use, chymosin has a weak proteolytic effect on proteins in general (see subsequent text).

3. Microbial Rennets

Microbial rennets are produced commercially from a number of organisms, primarily *Mucor* species *(M. miehei, M. pusillus)*. Their function as rennets shall not be pursued here, but the possible function of these proteases as "ordinary" proteases for the modification of food proteins should be recognized. The molecular masses are 38 kDa for the *M. miehei* protease and 30 kDa for the *M. pusillus* protease. Both enzymes have a pH optimum around pH 4 and are relatively heat labile, although they are more stable than chymosin. This stability is sometimes a disadvantage, however, but the heat stability of the *M. miehei* rennet can be reduced by chemical modification so it resembles animal rennet in that respect as well (Branner-Jørgensen *et al.*, 1981).

The relatively high costs of the microbial rennets compared with most other proteases may, however, limit their possible use in practice outside cheese manufacture.

4. Aspergillus *Proteases*

Aspergillus oryzae and other *Aspergillus* species produce a number of different proteases. Commercial preparations often contain at least three enzymes: a serine protease, a metalloprotease, and an aspartic protease. This mixture gives an unusually broad pH–activity range for practical applications.

The pH optima of the three enzymes are pH 4.5 for the aspartic protease, pH 6.8 for the metalloprotease, and pH 7.5–9.0 for the serine protease.

The most widespread industrial use of these proteases is in the form of koji, a surface-grown culture of *A. oryzae* on wheat. Koji is used as the enzyme source in the brewing of soy sauce and in the mashing of rice in sake production. As commercial enzyme preparations, the proteases are used mainly in the baking industry (Reichelt, 1983).

D. Metalloproteases

1. Thermolysin

Thermolysin, which is produced by *Bacillus stearothermophilus*, is the most heat stable of the commercially available proteases (Ward, 1983). In the presence of Ca^{2+} it is only slowly inactivated, even at temperatures near 80°C. Its pH optimum is slightly alkaline (pH 7–9). Like other metalloproteases, it is inhibited by chelating agents such as EDTA, citrate, and phosphate, which is of concern in food uses. With respect to peptide bond specificity, thermolysin preferentially hydrolyzes peptide bonds with a hydrophobic side chain on the amino side. The molecular mass of thermolysin is 34 kDa.

2. Bacillus amyloliquefaciens *Neutral Protease*

The metalloprotease produced by *Bacillus amyloliquefaciens* is commercially available under the trade name Neutrase. In its specificity, Neutrase resembles thermolysin, except that its maximum relative activity occurs at slightly lower pH values (pH 6–8). Neutrase has much lower temperature stability, so the practical operating temperature should not be much above 50°C. The relatively low heat stability is an advantage in some processes, however, since it allows a relatively easy arresting of the proteolytic reaction at the desired DH value. The molecular mass of Neutrase is 37 kDa.

Neutrase finds primary use in the brewing industry, in which it is used during mashing to increase the amount of soluble nitrogen. In this application, it is an advantage that the enzyme is not inhibited by the naturally occurring serine protease inhibitors of barley (Godfrey, 1983a). Neutrase also is applied for the hydrolysis of meat and gelatin. In the baking industry, it is used for modification of cracker and biscuit doughs (Reichelt, 1983). Neutrase also is suggested for improving the yield of protein in soy milk processing (Eriksen, 1983).

E. Naturally Occurring Protease Inhibitors

Many proteins are natural inhibitors of proteases. For example, trypsin inhibitors are found in animal and plant tissues. Also, some microorganisms produce protease inhibitors, for example, the subtilisin inhibitor from *Streptomyces albogriseolus* that strongly binds the proteases of the subtilisin family (Laskowski and Kato, 1980).

Comprehensive reviews of the proteins with protease inhibitor activity are given by Laskowski and Kato (1980) and by Birk (1987). The inhibitors of serine proteases are ubiquitous. Approximately 10 families of these proteins can be recognized on the basis of primary sequence homology and the position of the disulfide bridges. Since the inhibitors are common constituents of cereals and legumes, their presence is of concern, both from a nutritional and a technological perspective.

The most well-known protease inhibitors are probably the trypsin inhibitors of soybeans. The antitryptic activity of raw soybeans has been recognized for a long time. In nearly all soybean foods (and feed materials), a partial destruction of the inhibitor activity is an inherent part of the manufacturing process. The inhibitor activity is caused by two different proteins with different heat inactivation characteristics: the Kunitz soybean trypsin inhibitor (KSBI) and the Bowman-Birk inhibitor (BBI). Both inhibitors are well described and their adverse physiological significance and general inactivation characteristics have been discussed extensively (Liener and Kakade, 1980; Liener, 1986; Rackis *et al.*, 1986). In commercial soy protein isolate, the trypsin inhibitor activity is not destroyed completely (because that would also destroy the functional food properties of the protein). This constraint hinders the use of trypsin for the hydrolysis of soy isolate in practice (Adler-Nissen, 1986a).

VI. Summary of the Food Uses of Proteases

Broadly speaking, the uses of industrial proteases in foods fall in two different categories: (1) processing aids in the manufacture of traditional food products, and (2) production of new food protein ingredients with particular technological, so-called food functional, properties.

A. Proteases as Processing Aids

The use of rennet (both animal and microbial) in the manufacture of cheese stands out as, economically, the most important use of proteases in the food industry. This process also happens to be the oldest example of industrial food enzymology. Christian Hansen's marketing in 1874 of standardized calf rennets was a pioneering event. For the first time, an enzyme preparation with predictable performance was available commercially. Today, the company bearing his name is still one of the leading producers of commercial rennets.

Another example of a processing aid is the use of proteases in the baking industry for the modification of hard wheat flours. The mixed protease preparations from *Aspergillus oryzae* are widely used, but bacterial proteases also are marketed for this application (Reichelt, 1983).

Papain has been used since 1911 in the beer industry for the prevention of haze formation during cold storage of the beer, the so-called chillproofing of beer. The effect of the enzyme is believed to be the proteolytic degradation of ill-defined complexes between proteins and tannins (Moll, 1987). Neutral bacterial protease (cf. Section V,D,2) can be used during the mashing process to increase the nitrogen content of the wort when brewing with a mixture of malt and raw grain.

The deliberate action of proteases on muscle tissue to make it softer has been practiced for centuries in the form of cooking meat wrapped in papaya leaves. Papain is still the most important enzyme used for meat tenderizing, although a considerable number of enzymes have been promoted in this area. The simplest way to apply the enzyme is to sprinkle it over slices of raw meat, but papain also may be injected into the carcass and circulated through the veins (Schwimmer, 1981). Several problems are associated with controlling the reaction so excessive degradation does not occur. The reader is referred to the relevant literature on this rather special application of proteases.

Fish meat is much less tough than beef. When proteases are applied to fish muscle, the desired result is generally a thorough degradation of the fish muscle into soluble peptides. The traditional autolysis processes leading to the various fish sauces of Southeast Asia exemplify this application. The reaction may be accelerated by the addition of enzyme preparations (usually papain, bromelain, or bacterial proteases; Beddows and Ardeshir, 1979).

B. New Functional Protein Ingredients

The applications mentioned in the previous section are all typical examples of processing aids, that is, of using a protease to increase the productivity of a process leading to a food product of traditional quality. An entirely different purpose is pursued in the production of new products where quality is the goal. Unconventional protein sources can be modified by protein hydrolysis to improve their functional properties with respect to a particular use in food systems. Thus, an increased solubility and a concomitant improvement of the emulsifying or foaming properties can be achieved with many different food proteins. Such processes are extensively reviewed in *Enzymic Hydrolysis of Food Proteins* (Adler-Nissen, 1986a), so only some general comments will be presented here.

Among the many types of food proteins, the most widely used seem to be soy proteins. They are relatively cheap, are generally available, and have an established record for use in the food industry as functional ingredients. Enzymatic hydrolysates of soy proteins were introduced into the food industry during World War II as egg white replacers in the confectionary industry (Gunther, 1979). However, a broader utilization of soy protein hydrolysates has only occurred recently in the food industry, with the development of a process for making ISSPH.

As the name implies, ISSPH is 100% soluble, even at the isoelectric point of soy proteins, that is, around pH 4.2. The hydrolysate therefore originally was envisioned to be used for the nutritional fortification of soft drinks, an issue that was much debated in the 1970s. In the late 1970s an industrial process was developed (Adler-Nissen, 1978; Olsen and Adler-Nissen, 1979) and fruit juices fortified with ISSPH were marketed (Anonymous, 1982). The commercial development, however, was slow and the original product concept was modified so that products for clinical nutrition came into focus. The rationale for this change was the realization that large groups of hospitalized patients do suffer from an inadequate intake of dietary nitrogen, and that supplementation with extra protein in a soluble form had a positive effect in clinical trials with surgical patients (Hessov and Wara, 1983).

The flowchart for the ISSPH process (Fig. 12) contains some principal steps that are typical for all controlled protein hydrolysis processes:

* preparation of the substrate
* hydrolysis to a defined DH value
* inactivation
* separation
* further processing

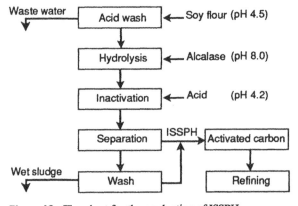

Figure 12 Flowchart for the production of ISSPH.

A closely related process for producing blood cell hydrolysate (BCH), a decolorized protein hydrolysate from the red cell fraction of slaughterhouse blood (Olsen, 1983), also consists of these five steps. In the case of other protein hydrolysis processes, the separation step may be omitted so the entire reaction mixture is recovered as product. Apart from this adaptation, the five steps mentioned must be considered in the design of any new protein hydrolysis process. However, a detailed discussion of the design considerations is beyond the scope of this chapter.

The most prominent functional property of ISSPH is, of course, its solubility at acid pH, but additional studies have revealed that it has other interesting functional properties as well. The peptides of ISSPH also act as emulsifiers (Adler-Nissen, 1988b) and give rise to a sensation of mouthfeel or body when added to low-calorie soft drinks (Adler-Nissen, 1986). The use of ultrafiltration as a separation step may yield modified fractions that exert excellent whipping properties (Olsen and Adler-Nissen, 1981). BCH may be incorporated into ham during the salt curing process. BCH appears to increase the water-binding capacity of the meat, which currently is achieved using polyphosphates.

In conclusion, the current tendency is one of acceptance of protein hydrolysates produced by controlled enzymatic hydrolysis in the food industry as a new class of food ingredients with interesting functional properties. This development has been spurred by two factors: the availability of a range of suitable proteases of food-grade quality from commercial enzyme suppliers and the results gained through the 1970s and 1980s from the studies on designing and controlling protein hydrolysis processes on an industrial scale.

References

Adler-Nissen, J. (1978). Hydrolysis of soy protein. U.S. Patent No. 4,100,024.

Adler-Nissen, J. (1979). Determination of the degree of hydrolysis of food protein hydrolysates by trinitrobenzenesulfonic acid. *J. Agric. Food Chem.* **27**, 1256–1262.

Adler-Nissen, J. (1984). Control of the proteolytic reaction and the level of bitterness in protein hydrolysis processes. *J. Chem. Technol. Biotechnol.* **34B**, 215–222.

Adler-Nissen, J. (1986a). "Enzymic Hydrolysis of Food Proteins." Elsevier Applied Science, London.

Adler-Nissen, J. (1986b). Relationship of structure to taste of peptides and peptide mixtures. *In* "Protein Tailoring and Reagents for Food and Medical Uses" (R. E. Feeney and J. R. Whitaker, eds.), pp. 97–122. Marcel Dekker, New York.

Adler-Nissen, J. (1988a). Bitterness intensity of protein hydrolysates — Chemical and organoleptic characterization. *In* "Frontiers of Flavor" (G. Charalambous, ed.), pp. 63–77. Elsevier, Amsterdam.

Adler-Nissen, J. (1988b). Enzymic modification of proteins — Effect on physico-chemical and functional properties. *In* "Functional Properties of Food Proteins" (R. Lászity and M. Ember-Kárpáti, eds.), pp. 31–39. METE, Budapest.

Adler-Nissen, J., and Eriksen, S. (1986). Peptides as bulking agents. In "The Shelf-Life of Foods and Beverages" (G. Charalambous, ed.), pp. 551–567. Elsevier, Amsterdam.

Adler-Nissen, J., and Olsen, H. S. (1979). The influence of peptide chain length on taste and functional properties of enzymatically modified soy protein. Am. Chem. Soc. Symp. Ser. 92, 125–146.

Adler-Nissen, J., and Olsen, H. S. (1982). Taste and taste evaluation of soy protein hydrolysates. In "Chemistry of Foods and Beverages — Recent Developments" (G. Charalambous and G. Inglett, eds.), pp. 149–169. Academic Press, New York.

Adler-Nissen, J., Eriksen, S., and Olsen, H. S. (1983). Improvement of the functionality of vegetable proteins by controlled enzymatic hydrolysis. Qual. Plant. Plant Foods Hum. Nutr. 32, 411–423.

Anonymous (1982). Soluble Rynkeby protein and protein drinks. Nord. Mejeri-Tidsskr. 9, 308.

Anson, M. L. (1939). The estimation of pepsin, trypsin, papain, and cathepsin with hemoglobin. J. Gen. Physiol. 22, 79–89.

Arai, S., and Fujimaki, M. (1991). Enzymatic modification of proteins with special reference to improving their functional properties. In "Food Enzymology" (P. F. Fox, ed.), Vol. 2, pp. 83–104. Elsevier Applied Science, London.

Arai, S., Watanabe, M., and Hirao, N. (1986). Modification to change physical and functional properties of food proteins. In "Protein Tailoring and Reagents for Food and Medical Uses" (R. E. Feeney and J. R. Whitaker, eds.), pp. 75–95. Marcel Dekker, New York.

Asselin, J., Hébert, J., and Amiot, J. (1989). Effects of in vitro proteolysis on the allergenicity of major whey proteins. J. Food Sci. 54, 1037–1039.

Auld, D. S., and Vallee, B. L. (1987). Carboxypeptidase A. In "Hydrolytic Enzymes" (A. Neuberger and K. Brocklehurst, eds.), pp. 201–255. Elsevier, Amsterdam.

Aunstrup, K. (1980). Proteinases. In "Microbial Enzymes and Bioconversions" (H. H. Rose, ed.), pp. 49–114. Academic Press, London.

Beddows, C. G., and Ardeshir, A. G. (1979). The production of soluble fish protein solution for use in fish sauce manufacture. I. The use of added enzymes. J. Food Technol. 14, 603–612.

Belitz, H.-D., Chen, W., Jugel, H., Treleano, R., Wieser, H., Gasteiger, J., and Marsili, M. (1979). Sweet and bitter compounds: Structure and taste relationship. Am. Chem. Soc. Symp. Ser. 115, 93–131.

Bigelow, C. C., and Channon, M. (1976). Hydrophobicities of amino acids and proteins. In "Handbook of Biochemistry and Molecular Biology" (G. D. Fasman, ed.), 3d Ed., Vol. 1, pp. 209–243. CRC Press, Cleveland.

Birk, Y. (1987). Proteinase inhibitors. In "Hydrolytic Enzymes" (A. Neuberger and K. Brocklehurst, eds.), pp. 257–305. Elsevier, Amsterdam.

Branner-Jørgensen, S., Eigtved, P., and Schneider, P. (1981). Reduced thermostability of modified Mucor miehei rennet. Neth. Milk Dairy J. 35, 361–364.

Brocklehurst, K., Willenbrock, F., and Salih, E. (1987). Cysteine proteinases. In "Hydrolytic Enzymes" (A. Neuberger and K. Brocklehurst, eds.), pp. 39–158. Elsevier, Amsterdam.

Chobert, J.-M., Bertrand-Harb, C., and Nicolas, M.-G. (1988). Solubility and emulsifying properties of caseins and whey proteins modified enzymatically by trypsin. J. Agric. Food Chem. 36, 883–892.

Clegg, K. M. (1978). Dietary enzymic hydrolysates of protein. In "Biochemical Aspects of New Protein Food" (J. Adler-Nissen, B. O. Eggum, L. Munck, and H. S. Olsen, eds.), pp. 109–117. Pergamon Press, Oxford.

Clegg, K. M., Smith, G., and Walker, A. L. (1974). Production of an enzymic hydrolysate of casein on a kilogram scale. J. Food Technol. 9, 425–431.

Creighton, T. E. (1984). "Proteins. Structures and Molecular Properties." Freeman, New York.

Dambmann, C., and Aunstrup, K. (1981). The variety of serine proteases and their industrial significance. In "Proteinases and Their Inhibitors. Structure, Function, and Applied Aspects" (V. Turk and L. Vitale, eds.), pp. 231–244. Pergamon Press, Oxford.

Dixon, M., and Webb, E. C. (1979). "Enzymes," 3d Ed., pp. 874–900. Longman, London.

Emi, S., Myers, D. V., and Iacobucci, G. A. (1976). Purification and properties of the thermostable acid protease of Penicillium duponti. Biochemistry 15, 842–848.

Eriksen, S. (1983). Application of enzymes in soy milk production to improve yield. *J. Food Sci.* **48**, 445–447.

Farrow, R. I., and Reichelt, J. R. (1983). Safe handling. *In* "Industrial Enzymology. The Application of Enzymes in Industry" (T. Godfrey and J. Reichelt, eds.), pp. 157–169. MacMillan, London.

Fastrez, J., and Fersht, A. R. (1973). Demonstration of the acyl-enzyme mechanism for the hydrolysis of peptides and anilides by chymotrypsin. *Biochemistry*, **12**, 2025–2034.

Foltman, B. (1981). Gastric proteinases—Structure, function, evolution, and mechanism of action. *Essays Biochem.* **17**, 52–84.

Fruton, J. S. (1982). Proteinase-catalyzed synthesis of peptide bonds. *Adv. Enzymol.* **53**, 239–306.

Fruton, J. S. (1987). Aspartyl proteinases. *In* "Hydrolytic Enzymes" (A. Neuberger and K. Brocklehurst, eds.), pp. 1–37. Elsevier, Amsterdam.

Fujimaki, M. (1978). Nutritional improvement of food proteins by enzymatic modification, especially by plastein synthesis reaction. *Ann. Nutr. Alim.* **32**, 233–241.

Fujimaki, M., Yamashita, M., Arai, S., and Kato, H. (1970). Plastein reaction. Its application to debittering of proteolyzates. *Agric. Biol. Chem.* **34**, 483–484.

Ginsburg, A., and Carroll, W. R. (1965). Some specific ion effects on the conformation and thermal stability of ribonuclease. *Biochemistry* **4**, 2159–2174.

Godfrey, T. (1983a). Brewing. *In* "Industrial Enzymology. The Application of Enzymes in Industry" (T. Godfrey and J. Reichelt, eds.), pp. 221–259. MacMillan, London.

Godfrey, T. (1983b). Comparison of key characteristics of industrial enzymes by type and source. *In* "Industrial Enzymology. The Application of Enzymes in Industry" (T. Godfrey and J. Reichelt, eds.), pp. 466–569. MacMillan, London.

Godfrey, T., and Reichelt, J. (eds.) (1983). "Industrial Enzymology. The Application of Enzymes in Industry." MacMillan, London.

Gololobov, M. Y., Belikov, V. M., Vitt, S. V., Paskonova, E. A., and Titova, E. F. (1981). Transpeptidation in concentrated solution of peptic hydrolyzate of ovalbumin. *Nahrung* **25**, 961–967.

Gould, B. J. (1975). Enzyme data. *In* "Handbook of Enzyme Biotechnology" (A. Wiseman, ed.), pp. 128–162. Ellis Horwood, Chichester.

Gunther, R. C. (1979). Chemistry and characteristics of enzyme-modified whipping proteins. *J. Am. Oil Chem. Soc.* **56**, 345–349.

Hessov, I., and Wara, P. (1983). Improved nutritional intake after colonic surgery. *Acta Chir. Scand. Suppl.* **516**, 26.

Hvidt, A. (1983). Interactions of water with nonpolar solutes. *Ann. Rev. Biophys. Bioeng.* **12**, 1–20.

Jacobsen, C. F., Léonis, J., Linderstrøm-Lang, K., and Ottesen, M. (1957). The pH-stat and its use in biochemistry. *Meth. Biochem. Anal.* **4**, 171–210.

Jost, R., and Monti, J. C. (1977). Partial enzymatic hydrolysis of whey protein by trypsin. *J. Dairy Sci.* **60**, 1387–1393.

Lalasidis, G. (1978). Four new methods of debittering protein hydrolysates and a fraction of hydrolysates with high content of essential amino acids. *Ann. Nutr. Alim.* **32**, 709–723.

Laskowski, M., Jr., and Kato, I. (1980). Protein inhibitors of proteinases. *Ann. Rev. Biochem.* **49**, 593–626.

Liener, I. E. (1986). Trypsin inhibitors: Concern for human nutrition or not? *J. Nutr.* **116**, 920–923.

Liener, I. E., and Kakade, M. L. (1980). Protease inhibitors. *In* "Toxic Constituents of Plant Foodstuffs" (I. E. Liener, ed.), 2d. Ed., pp. 7–71. Academic Press, New York.

Linderstrøm-Lang, K. (1952). Proteins and enzymes. III. The initial stages in the breakdown of proteins by enzymes. *Lane Medical Lectures*, **6**, 53–72. Stanford University Press, California.

Markussen, J. (1987). "Human Insulin by Tryptic Transpeptidation of Porcine Insulin and Biosynthetic Precursors." MTP Press, Lancaster.

Mietsch, F., Fehér, J., and Halász, A. (1989). Investigation of functional properties of partially hydrolyzed proteins. *Nahrung* **33**, 9–15.

Minagawa, E., Kaminogawa, S., Tsukasaki, F., and Yamauchi, K. (1989). Debittering mechanism in bitter peptides of enzymatic hydrolysates from milk casein by aminopeptidase T. *J. Food Sci.* **54**, 1225–1229.

Mock, W. L., and Zhang, J. Z. (1991). Mechanistically significant diastereoselection in the sulfoximine inhibition of carboxypeptidase A. *J. Biol. Chem.* **266**, 6393–6400.

Moll, M. (1987). Colloidal stability of beer. *In* "Brewing Science" (J. R. A. Pollock, ed.), Vol. 3, pp. 1–327. Academic Press, London.

Monti, J. C., and Jost, R. (1978). Enzymatic solubilization of heat-denatured cheese whey protein. *J. Dairy Sci.* **61**, 1233–1237.

Monti, J. C., and Jost, R. (1979). Papain-catalyzed synthesis of methionine-enriched soy plasteins. Average chain length of the plastein peptides. *J. Agric. Food Chem.* **27**, 1281–1285.

Morihara, K. (1974). Comparative specificity of microbial proteinases. *Adv. Enzymol.* **41**, 179–243.

Neuberger, A., and Brocklehurst, K. (eds.) (1987). "Hydrolytic Enzymes." Elsevier, Amsterdam.

Ney, K. H. (1971). Voraussage der Bitterkeit von Peptiden aus deren Aminosäurezusammensetzung. *Z. Lebensm. Untersuch. Forsch.* **147**, 64–71.

Nozaki, Y., and Tanford, C. (1971). The solubility of amino acids and two glycine peptides in aqueous ethanol and dioxane solutions. Establishment of a hydrophobicity scale. *J. Biol. Chem.* **246**, 2211–2217.

Olsen, H. S. (1983). Herstellung neuer Proteinprodukte aus Schlachttierblut zur Verwendung in Lebensmitteln. *Z. Lebensm. Technol. Verfahrenstechnik* **34**, 406–412.

Olsen, H. S., and Adler-Nissen, J. (1979). Industrial production and applications of a soluble enzymatic hydrolysate of soya protein. *Process. Biochem.* **14**(7), 6–11.

Olsen, H. S., and Adler-Nissen, J. (1981). Application of ultra- and hyperfiltration during production of enzymatically modified proteins. *Am. Chem. Soc. Symp. Ser.* **154**, 133–169.

Ottesen, M., and Svendsen, I. (1970). The subtilisins. *In* Methods in Enzymology (G. E. Perlmann and L. Lorand eds.). **XIX**, 199–215.

Pace, C. N., and Barrett, A. J. (1984). Kinetics of tryptic hydrolysis of the arginine–valine bond in folded and unfolded ribonuclease T_1. *Biochem. J.* **249**, 411–417.

Paulson, J. C. (1989). Glycoproteins: What are the sugar chains for? *Trends Biochem. Sci.* **14**, 272–276.

Perlmann, G. E., and Lorand, L. (eds.) (1970). Proteolytic Enzymes. *In* "Methods in Enzymology," Vol. XIX. Academic Press, New York.

Polgár, L. (1987). Structure and function of serine proteases. *In* "Hydrolytic Enzymes" (A. Neuberger and K. Brocklehurst, eds.), pp. 159–200. Elsevier, Amsterdam.

Polgár, L., and Halász, P. (1982). Current problems in mechanistic studies of serine and cysteine proteinases. *Biochem. J.* **207**, 1–10.

Poulsen, B. U., and Adler-Nissen, J. (1989). "Enzymic hydrolysis of fish protein intended for use in fish feed — screening of enzymes." Presentation at Conference on Pelagic Fish, Torry Research Station, Aberdeen, 27–29 Sept., 1989.

Rackis, J. J., Wolf, W. J., and Baker, E. C. (1986). Protease inhibitors in plant foods. *In* "Nutritional and Toxicological Significance of Enzyme Inhibitors in Food" (M. Friedman, ed.), pp. 299–347. Plenum, New York.

Rebeca, B. D., Peña-Vera, M. T., and Díaz-Castañeda, M. (1991). Production of fish protein hydrolysates with bacterial proteases; Yield and nutritional value. *J. Food Sci.* **56**, 309–314.

Reichelt, J. R. (1983). Baking. *In* "Industrial Enzymology. The Application of Enzymes in Industry" (T. Godfrey and J. Reichelt, eds.), pp. 210–220. MacMillan, London.

Schwimmer, S. (1981). "Source Book of Food Enzymology." AVI, Westport, Connecticut.

Sörensen, S. P. L. (1908). Enzymstudien. I. Über die quantitative Messung proteolytischer Spaltungen. "Die Formoltitrierung." *Biochem. Z.* **7**, 45–101.

Svendsen, I. (1976). Chemical modifications of the subtilisins with special reference to the binding of large substrates. A review. *Carlsberg Res. Commun.* **41**, 237–291.

Svendsen, I., and Breddam, K. (1992). Isolation and amino acid sequence of a glutamic acid specific endopeptidase from *Bacillus licheniformis*. *Eur. J. Biochem.* **204**, 165–171.

Tanford, C. (1962). Contribution of hydrophobic interactions to the stability of the globular conformation of proteins. *J. Am. Chem. Soc.* **84**, 4240–4247.

Thomas, T. D., and Pritchard, G. G. (1987). Proteolytic enzymes of dairy starter cultures. *FEMS Microbiol. Rev.* **46**, 245–268.

Umetsu, H., Matsuoka, H., and Ichishima, E. (1983). Debittering mechanism of bitter peptides from milk casein by wheat carboxypeptidase. *J. Agric. Food Chem.* **31**, 50–53.

Vega, R. E., and Brennan, J. G. (1988). Enzymic hydrolysis of fish offal without added water. *J. Food Engin.* **8**, 201–215.

Visser, S. (1981). Proteolytic enzymes and their action on milk proteins. A review. *Neth. Milk Dairy J.* **35**, 65–88.

Volkin, D. B., and Klibanov, A. M. (1989). Minimizing protein inactivation. *In* "Protein Function: A Practical Approach" (T. E. Creighton, ed.), pp. 1–24. IRL Press, Oxford.

Ward, O. P. (1983). Proteinases. *In* "Microbial Enzymes and Biotechnology" (W. M. Fogarty, ed.), pp. 251–317. Elsevier Applied Science, London.

Warshel, A., Naray-Szabo, G., Sussman, F., and Hwang, J.-K. (1989). How do serine proteases really work? *Biochemistry* **28**, 3629–3637.

Wieser, H., and Belitz, H.-D. (1976). Zusammenhänge zwischen Struktur und Bittergeschmack bei Aminosäuren und Peptiden. II. Peptide und Peptidderivate. *Z. Lebensm. Untersuch. Forsch.* **160**, 383–392.

Yamamoto, A. (1975). Proteolytic enzymes. *In* "Enzymes in Food Processing" (G. Reed, ed.), 2d. Ed., pp. 123–179. Academic Press, New York.

You-Shang, Z. (1983). Enzymatic synthesis of proteins and peptides. *Trends Biochem. Sci.* **8**, 16–17.

CHAPTER 8

Lipases

SVEN ERIK GODTFREDSEN

As were most enzymes applied in food processing, lipases originally were used for preparation of foods in an entirely empirical fashion. The early users of lipases were unaware of the involvement of lipases in traditional, well-established processes for food preparation and processing. This statistic applies to the use of microorganisms for development and enhancement of flavors in cheese manufacturing (Peppler *et al.* 1975; Nasr, 1983; Arnold *et al.*, 1985), in which lipases generated by the applied microorganisms play a key role by liberating flavor precursors or actual flavor compounds from esters present in milk fat.

The early uses of enzyme preparations recognized as lipases were developed mainly to mimic such established uses of lipases while improving their reproducibility and, therefore, the consistency of products made by processes involving these lipases. Preparations of lipases applied in this fashion initially were derived from animal sources and later from industrial fermentation of microorganisms (Nagaoka *et al.*, 1969; Brockerhoff and Jensen, 1974; Borgström and Brockman, 1984; Yeoh *et al.*, 1986; Kouker and Jaeger, 1987; Huge-Jensen *et al.*, 1989). Advances achieved using such lipase preparations, in addition to the reproducibility of food processing, included the possibility of using alternative raw materials for cheese production or of modulating the characteristics of the finished product to meet market requirements.

These early developments in the use of lipases in food processing and manufacturing are quite important from the perspective of the consumer, the manufacturer, and the enzyme producer. More recent developments in lipase technology, however, must be expected to exert an even stronger influence on the food industry. Currently, lipase technologies are so advanced that production of entirely new food ingredients must be considered industrially feasible, as must the production of a wide variety of compounds, such as flavors (Okumura *et al.*, 1979; Iwai *et al.*, 1980) and emulsifiers

(Jensen *et al.*, 1978; Kirk *et al.*, 1989), that are of importance to the food sector.

The number of known microbial species that produce lipases is rather large and is, in fact, still increasing. The most important of these lipase-producing microorganisms are listed in Table I. In Table II, important industrial sources of lipases are given, with the actual commercial producers of the enzymes. Many of these industrial lipases have been developed for uses outside the food industry. Some enzymes were selected and industrialized

TABLE I
Sources of Microbial Lipases

Microorganism	References[a]
Absidia corymbifera	Satyanarayana and Jori, 1981
Absidia hyalospora	(Hirohara *et al.*, 1985)
Achromobacter sp.	(*Mitsuda* et al., 1988)
Achromobacter lipolyticum	Sztajer and Zboinska, 1982; Tahoun *et al.*, 1985
Acinetobacter calcoaceticus	Sztajer and Zboinska, 1982; Fischer and Kleber, 1987
Acinetobacter pseudoalcaligenes	Farin et al., *1986*
Alcaligenes sp.	(Mitsuda *et al.*, 1988)
Alcaligenes denitrificans	[Odera *et al.*, 1986]
Amylomyces rouxii	Koritala *et al.*, 1987
Arthrobacter sp.	(Mitsuda *et al.*, 1988)
Aspergillus awamori	Yokozeki *et al.*, 1982
Aspergillus flavus	Yokozeki *et al.*, 1982; Koritala *et al.*, 1987
Aspergillus fumigatus	Satyanarayana and Jori, 1981
Aspergillus japonicus	[Werdelman and Schmid, 1982]; Vora *et al.*, 1988
Aspergillus niger	Fukumoto *et al.*, 1963; (Iwai *et al.*, 1964); Tsujisaka *et al.*, 1973; [Okumura *et al.*, 1976]; Yokozeki *et al.*, 1982; (Linfield *et al.*, 1984); (Hirohara *et al.*, 1985); (Maerae, 1985); (Macrae and How, 1986); (Vermier *et al.*, 1987); (Kalo, 1988); (Miyazawa *et al.*, 1988); (Sonnet and Antonia, 1988)
Aspergillus oryzae	Yokozeki *et al.*, 1982; Koritala *et al.*, 1987
Bacillus cereus	Tahoun *et al.*, 1985
Bacillus megaterium	(Hirohara *et al.*, 1985)
Bacillus stearothermophilus	El-Hoseiny, 1986; Gowland *et al.*, 1987
Candida antarctica	Mitsuda *et al.*, 1988; (Kirk *et al.*, 1989)
Candida auricularia	Mitsuda *et al.*, 1988
Candida curvata	Tahoun *et al.*, 1985
Candida cylindraceae	(Linfield *et al.*, 1984); (Seino *et al.*, 1984); (Hog *et al.*, 1985); (Koshiro *et al.*, 1985); (Lazar, 1985); *Fujii* et al., *1986;* (Kohr *et al.*, 1986); (Macrae, 1986); (Sih, 1986); (Bucchiarelli *et al.*, 1988); (Miyazawa *et al.*, 1988); (Naumura *et al.*, 1988); (Sonnet and Antonia, 1988); *Thornton* et al., *1988*; (Babayan, 1989)
Candida deformans	Tahoun *et al.*, 1985
Candida foliorum	Mitsuda *et al.*, 1988
Candida humicula	Mitsuda *et al.*, 1988
Candida lipolytica	Tahoun *et al.*, 1985
Candida rugosa	Tahoun *et al.*, 1985
Candida tsukubaensis	Mitsuda *et al.*, 1988
Chaetomium thermophile	Satyanarayana and Jori, 1981
Chromobacterium chocolatum	(Hirohara *et al.*, 1985)

(continues)

TABLE I (continued)

Microorganism	References[a]
Chromobacterium viscosum	(Hog et al., 1985); (Sih, 1986); (Mitsuda et al., 1988)
Corynebacterium acnes	Sztajer and Zboinska, 1982; Tahoun et al., 1985
Flavobacterium arborescens	(Hirohara et al., 1985)
Fusarium oxysporum	[Joshi and Mathur, 1985];(Joshi and Dhar, 1987)
Fusarium solari	Tahoun et al., 1985
Geotrichum candidum	Tsujisaka et al., 1973,[1977]; [Okumura et al., 1976]; Chander and Klostermeyer; 1983; (Sih, 1986); (Vermiere et al., 1987); (Sztajer et al., 1988); Whitesides and Ladner, 1988
Humicula grisea	Adams and Deploey, 1978
Humicula insulens	Adams and Deploey, 1978
Humicula lanuginosa	Arima et al., 1972; Liu et al., 1973 [a],b; Morinaga et al., 1986; Huge-Jensen and Gormsen, 1989b
Lactobacillus sp.	Tahoun et al., 1985
Leishmania donovani	Chandhuri et al., 1986
Malbrancheae pulcella	Ogundero, 1980
Micrococcus freudenreichii	Lawrence et al., 1967
Mucor sp.	Nagoaka and Yamada, 1969
Mucor javanicus	[Ishihara et al., 1975]
Mucor lipolyticus	[Nagaoka and Yamada, 1973]
Mucor miehei	(Hog et al., 1985); (Macrae, 1986); (Sonnet, 1987); (Kalo, 1988); (Sonnet and Antonia, 1988); Thornton et al., 1988; Huge-Jensen et al., 1989a
Mucor pusillus	Ogundero, 1980; Sztajer and Zboinska, 1982
Myxococcus xanthus	Tahoun et al., 1985
Penicillium crustosum	Oi et al., 1967
Penicillium cyclopium	Tsujisaka et al., 1973; [Iwai et al., 1975]; [Okumura et al., 1976]; (Vermiere et al., 1987)
Penicillium roquefortii	Eitenmiller et al., 1970
Phycomyces nitens	(Hog et al., 1985)
Pichia miso	(Ohta et al., 1988)
Propionibacterium acnes	Sztajer and Zboinska, 1982; Tahoun et al., 1985
Propionibacterium granulosum	Sztajer and Zboinska, 1982; Tahoun et al., 1985
Proteus sp.	Tahoun et al., 1985
Pseudomonas sp.	Thornton et al., 1988
Pseudomonas aeruginosa	Sztajer and Zboinska, 1982; (Hamaguchi et al., 1986a,b); [Odera et al., 1986]; [Sugihara et al., 1988]
Pseudomonas fluorescens	Sztajer and Zboinska, 1982; (Hog et al., 1985); (Mitsuda et al., 1988); (Miyazawa et al., 1988); [Roussis et al., 1988]
Pseudomonas fragi	[Kugimiya et al., 1986]
Pseudomonas pseudoalcaligenes	Farin et al., 1986
Pseudomonas stutzeri	Farin et al., 1986
Rhizopus sp.	(Seino et al., 1984)
Rhizopus arrhizus	(Gargouri et al., 1984); (Knox and Cliffe, 1984); (Linfield et al., 1984); (Sih, 1986); Thornton et al., 1988
Rhizopus chinensis	(Kyotani et al., 1988a,b)
Rhizopus delemar	Fukomoto et al., 1964; Tsujisaka et al., 1973; Iwai and Tsujisaka, 1974 a[b]; [Okumura et al., 1976]; (Gargouri et al., 1984); (Hog et al., 1985); (Tatara et al., 1985); (Hamaguchi et al., 1986b), (Vermiere et al., 1987); (Mitsuda et al., 1988); (Sonnet and Antonia, 1988)

(continues)

TABLE I *(continued)*

Microorganism	References[a]
Rhizopus japonicus	(Hog *et al.*, 1985); (Hamaguchi *et al.*, 1986b); (Macrae and How, 1986); (Mitsuda *et al.*, 1988)
Rhizopus microsporus	Tahoun *et al.*, 1985
Rhizopus niveus	(Macrae, 1985); (Sih, 1986)
Rhizopus nodosus	Tahoun *et al.*, 1985
Rhizopus oligosporus	Koritala *et al.*, 1987
Rhizopus oryzae	(Sih, 1986)
Rhodotorula minuta	(Fukui and Tanaka, 1982)
Rhodotorula pilimonae	Tahoun *et al.*, 1985
Saccharomyces fragilis	Lawrence *et al.*, 1967; (Hirohara *et al.*, 1985)
Saccharomyces lipolytica	Tahoun *et al.*, 1985
Schizosaccharomyces pombe	(Hirohara *et al.*, 1985)
Sporotrichum thermophile	Adams and Deploey, 1978
Staphylococcus aureus	Sztajer and Zboinska, 1982; Tahoun *et al.*, 1985
Staphylococcus carnosus	Götz *et al.*, 1985
Streptococcus lactis	Sztajer and Zboinska, 1982; Tahoun *et al.*, 1985
Streptomyces coelicolor	Sztajer and Zboinska, 1982
Streptomyces fradiae	Sztajer and Zboinska, 1982
Streptomyces panayensis	(Hirohara *et al.*, 1985)
Talaromyces thermophilus	Ogundero, 1980
Thielavia minor	Satyanarayana and Jori, 1981
Torula thermophila	Adams and Deploey, 1978

[a] References in parentheses deal mainly with synthesis; references in brackets deal primarily with characterization of the lipase; references in italics deal mainly with detergent applications of lipases.

because of their possible application in the detergent industry whereas others were developed because of their potential as catalysts in organic chemical processing (Tanaka *et al.*, 1981; Okumura *et al.*, 1984; Langrand *et al.*, 1985; Bello *et al.*, 1986; Lin *et al.*, 1987; Glänzer *et al.*, 1988; Gotor *et al.*, 1988; Satoshi *et al.*, 1988; Marples and Roger Evans, 1989). Table III lists major application areas for lipases as well as the most important uses of the enzymes in the food sector. Several discussions of the mode of action of lipases (Desnuelle, 1961; Gargouri *et al.*, 1986), their various uses in and out of the food area (Fullbrook, 1983; Ratledge, 1984; Nielsen, 1985), and their properties (Tsujisaka and Iwai, 1984; Johri *et al.*, 1985; Stuer *et al.*, 1986) have been published, as excellent reviews that discuss most aspects of these enzymes (Brockerhoff and Jensen, 1974; Borgström and Brockman, 1984).

Several lipases have been cloned and expressed in industrial yields in hosts that are particularly amenable to industrial fermentations. Lipases derived from such fermentations are, in some cases, of very high purity, for example, Lipolase®, which was marketed in 1988 as a detergent lipase. The lipases of *Mucor miehei* and *Candida antarctica* also have been cloned and expressed in industry-friendly hosts. The experience gained from lipase

TABLE II
Industrial Sources of Lipases

Lipase	Source	Reference
Achromobacter sp.	Meito Sangyo	Mitsuda *et al.* (1988)
Alcaligenes sp.	Meito	Hirohara *et al.* (1985)
Arthrobacter sp.	Sumitomi	Mitsuda *et al.* (1988)
Aspergillus niger	Amano	Andree *et al.* (1980); Akita *et al.* (1986); Macrae and How (1986); Kalo (1988); Sonnet and Antonia (1988)
	Novo Nordisk	Linfield *et al.* (1984)
	Röhm	Höfelmann *et al.* (1985); Gerlach *et al.* (1988)
Candida cylindraceae	Amano	Sih (1986)
	Enzyme Development	Linfield *et al.* (1984); Sonnet and Antonia (1988)
	Meito	Hog *et al.* (1985); Lazar (1985); Fujii *et al.* (1986); Macrae (1986); Gerlach *et al.* (1988); Thornton *et al.* (1988)
Chromobacterium viscosum	U.S. Biochemicals	Sih (1986)
	Toyo Jozo	Hirohara *et al.* (1985); Hog *et al.* (1985); Mitsuda *et al.* (1988)
Humicula languinosa	Amano	Hirohara *et al.* (1985); Sih (1986)
	Novo Nordisk	Huge-Jensen and Gormsen (1989)
Mucor miehei	Amano	Akita *et al.* (1986); Sonnet (1987); Sonnet and Antonia (1988); Thornton *et al.* (1988)
	Gist Brocades	Sonnet and Antonia (1988)
	Röhm	Gerlach *et al.* (1988)
	Novo Nordisk	Posorske (1984); Hog *et al.* (1985); Kalo (1988); Sonnet and Antonia (1988); Thornton *et al.* (1988); Huge-Jensen *et al.* (1989)
Phycomyces nitens	Takeda	Hog *et al.* (1985)
Pseudomonas sp.	Amano	Hirohara *et al.* (1985); Sih (1986); Thornton *et al.* (1988)
Pseudomonas aeruginosa	Amano	Hamaguchi *et al.* (1986b)
Pseudomonas fluorescens	Amano	Inada *et al.* (1984); Hog *et al.* (1985)
Rhizopus sp.	Serva	Sih (1986)
	Nagase	Andree (1980)
Rhizopus arrhizus	Precibio (France)	Gargouri *et al.* (1984)
	Boehringer-Mannheim	Sih (1986)
	Rapidase	Gerlach *et al.* (1988)
	Gist Brocades	Gerlach *et al.* (1988); Thornton *et al.* (1988)
Rhizopus delemar	Chemical Dynamics	Sih (1986)
	Tanabe	Hog *et al.* (1985); Mitsuda *et al.* (1988), Sonnet and Antonia (1988)
Rhizopus japonicus	Amano	Hamaguchi *et al.* (1986a)
	Nagase	Hamaguchi *et al.* (1986a)
	Osaka Saiken	Hog *et al.* (1985)
Rhizopus niveus	Amano	Sih (1986)
Rhizopus oryzae	Amano	Sih (1986)

manufacturing using genetically engineered strains increases the likelihood that industrially important lipases of the future will be produced in this fashion. The purity and price of lipases manufactured using genetically engineered strains are unlikely to be surpassed by lipases produced using conventional techniques.

For some important applications of lipases in food manufacturing and processing, immobilized rather than free lipases are required. The lipases listed in Table II may be immobilized for such processes, if necessary. Some lipases are manufactured and marketed in an immobilized state that is well suited to application in food-related processes such as modification of fats and oils. The immobilized lipase of *Mucor miehei,* marketed under the trade name Lipozyme®, is an example of an immobilized lipase available on an industrial scale.

Some of the most fascinating developments in the application of lipases in the food industry concern the use of lipases for the modification of

TABLE III
Industrial Application Areas for Microbial Lipases

Industry	Effect	Product
Dairy	Hydrolysis of milk fat (Arnold *et al.,* 1975)	Flavor agents
	Cheese ripening (Peppler *et al.,* 1975; Nasr, 1983)	Cheese
	Modification of butter fat (Kalo, 1988)	Butter
Bakery	Flavor improvement and shelf life prolongation	Bakery products
Beverage	Improved aroma	Beverages
Food dressing	Quality improvement	Mayonnaise, dressings, and whipped toppings
Health food	Transesterification	Health foods
Meat and fish	Flavor development and fat removal	Meat and fish products
Fat and oil	Transesterification	Cocoa butter, margarine
	Hydrolysis	Fatty acids, glycerol, mono- and di-glycerides
Chemical	Enantioselectivity	Chiral building blocks and chemicals
	Synthesis	Chemicals
Pharmaceutical	Transesterification	Specialty lipids
	Hydrolysis	Digestive aids
Cosmetic	Synthesis	Emulsifiers, moisturing agents
Leather	Hydrolysis	Leather products
Paper	Hydrolysis	Paper products
Cleaning	Hydrolysis	Removal of cleaning agents, e.g., surfactants

triglycerides. A key factor in this context is the specificity of the enzymes for the position and the nature of the acyl residues in triglyceride molecules. Some lipases preferentially attack the 1 and 3 position in triglyceride molecules, whereas other lipases exhibit no preference for any of the three acyl residues. In addition, most lipases show some specificity based on the nature of the acyl residue in triglycerides by preferentially attacking unsaturated fatty acids or short chain fatty acids. Some examples of 1,3-specific lipases are the *Mucor miehei* lipase and Lipolase®, which cleave the 1- and 3-ester bonds in triglycerides almost exclusively. Human pancreatic lipase and human lipoprotein lipase likewise exhibit 1,3-specificities.

The possible industrial exploitation of these specificities has been recognized since the late 1960s in the fat and oil industry. A major target has been the production of new triglyceride types with, for example, desirable melting properties by interesterification of readily available triglycerides (Macrae, 1983,1984,1985; Chakrabhaty, 1985; Tatara *et al.*, 1985; Macrae and How, 1986). For comparison, see the example shown in Fig. 1. This technological potential of lipases first was envisioned by Macrae and the Unilever research team (Macrae and Hammond, 1985), as was the possibility of applying lipases in nonaqueous environments for industrial purposes—a requirement for the processes depicted in Fig. 1 since lipases, in the presence of water, will hydrolyze triglycerides rapidly.

One of the crucial issues for the realization of the Unilever process was the development of immobilized preparations of lipases that exhibit the desired specificities. One such product, Lipozyme®, is currently available in industrial quantities. New improved immobilized lipase preparations are being developed. Many discussions have been published that address various aspects of improved biocatalysts for the Unilever process, for example, new

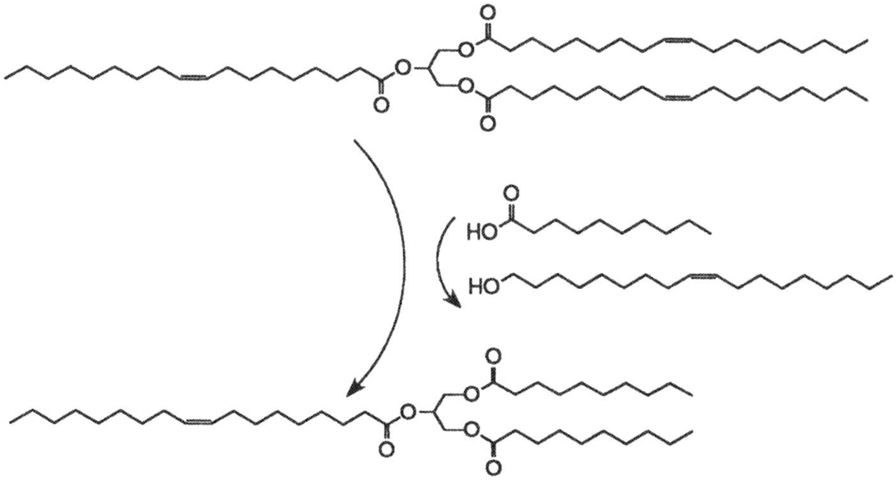

Figure 1 Example of triglyceride with desirable melting properties.

carrier types (Brady *et al.*, 1986; Sztajer *et al.*, 1988), chemically modified lipases soluble in organic solvents (Inada *et al.*, 1984), and lipases entrapped in hydrophobic polymers (Fukui and Tanaka, 1982; Kyotani *et al.*, 1988b). Also, several reviews discuss improvements of the actual process design by using loop reactor systems (Knox and Cliffe, 1984), membrane reactor types (Hog *et al.*, 1985), batch reactor systems (Kyotani *et al.*, 1988a), microemulsion techniques (Babayan, 1989), and packed-bed reactor systems (Macrae, 1983). Packed-bed systems are currently the preferred method for interesterification of triglycerides. Collectively, the technological problems in the use of microbial lipases for triglyceride modification have been solved to the extent that interesterification of triglycerides can be considered an industrially feasible process. These technological advances, as represented by Lipozyme®, must be seen as a breakthrough in industrial enzymology, since there are no previous examples of applications of enzymes in nonaqueous environments that actually generate products in industrial yields (10 tons product/1 kg immobilized biocatalyst). The actual application of the process for triglyceride modification is a strategic issue and will depend on the developments of competing technologies, such as plant genetics, and on the availability of natural materials that match the performance of biosynthesized materials.

The interesterification process is likely to be used first in the preparation of high value materials for use as specialty nutrients or as pharmaceuticals. It is thus conceivable that, with the aid of lipase technologies, it will be possible to make fats rich in essential fatty acids, for example, DHA (4,7,10,13,16,19-all *cis*-docosahexaenoic acid), EPA (eicosapentaenoic acid), arachidonic acid, linoleic acid, or linolenic acid. These fats would exhibit improved food properties by increasing bioavailability of essential fatty acids over naturally occurring materials such as fish oils. Triglyceride molecules that currently are expected to exhibit these highly desirable properties will carry the essential fatty acid in the 2 position and short-chain fatty acids (i.e., C_8 and C_{10} carboxylic acids) at the C-1 and C-3 positions of the triglyceride molecules (Babayan, 1989). These lipids (so-called structured lipids) will be hydrolyzed at an increased rate in the digestive tract because of the increased activity of the pancreatic lipase toward short-chain fatty acids and, thus, will provide an easily absorbable monoglyceride that contains the essential fatty acid. Structured lipids carrying DHA may, in the future, become of prime importance in infant nutrition whereas structured lipids containing EPA may become an important remedy for the general public for prevention of heart diseases.

Structured lipids manufactured with the aid of lipases are likely to become important in connection with total parenteral nutrition (TPN) of hospitalized patients. Currently, triglyceride emulsions based on sunflower oil are used extensively to supply energy efficiently in a TPN regime. Some disadvantages of the currently applied emulsions may be reduced by using structured lipids instead. These molecules are hydrolyzed more rapidly than their nonstructured counterparts by the 1,3-specific lipoprotein lipase that hydrolyzes fat emulsions in blood vessels. Structured lipids will give rise to more hydrophilic fatty acids, which are distributed more easily in the body, whereas the essential fatty acid in the 2 position of the triglyceride molecule will be readily bioavailable as its monoglyceride after removal of the short-chain fatty acids by lipoprotein lipase. Moreover, these short-chain fatty acids can be transported directly to the liver and used there for energy generation. In addition, use of structured lipids will lessen the overload with essential fatty acids encountered by patients in connection with the use of traditional lipid emulsions over extended periods of time. Finally, lipase technologies make possible the design of triglyceride molecules with precise compositions as specified by the demand of the diseased patient. Because of the central involvement of the essential fatty acids as precursors of prostaglandin biosynthesis (Gustafson *et al.*, 1988), they may even be used to regulate the diseased state. Structured lipids must be expected to have an exciting future in the area of TPN and probably will be among the first examples of entirely new food molecules designed on the basis of lipase technologies.

The potential application of lipases to hydrolysis of edible fats and oils (Linfield *et al.*, 1984; Kohr *et al.*, 1986) has been the subject of several studies and has been found to be technologically feasible. However, this technology is subject to competition from other technologies such as steam splitting. Application of the microbial lipases to fat splitting therefore is likely to occur mainly for treatment of very valuable materials such as triglycerides that contain valuable and sensitive polyunsaturated fatty acids, since these are gaining increasing importance in the area of nutrition and as pharmaceutical products, as mentioned earlier.

In the field of synthesis of food ingredients such as emulsifiers and flavor molecules, interest in the application of lipases is increasing. This change is due not only to the superiority of lipases over traditional methods for some esterification reactions but also to the safety of the enzymatic procedures and the natural quality of materials synthesized using biocatalysts. The concept of "green" chemistry and of "green" products ("bioesters") appears to appeal not only to the general public but also to the authorities. Therefore, compounds manufactured with the aid of lipases are likely to gain importance in the future.

The power of lipases in the synthesis of esters of interest to the food industry is illustrated by the process shown in Fig. 2. This process is based on the best currently available lipase for ester synthesis — the lipase produced by *Candida antarctica* (Michiyo, 1986). The example shown (Kirk *et al.*, 1989) illustrates how lipases can be applied for regioselective esterification of polyfunctional alcohols. The process shown is carried out simply by mixing the reactants in the presence of the enzyme and removing water from the reaction mixture over the course of the reaction (Kirk *et al.*, 1989). The process shown is also an example of how new, potentially useful chemicals become abundantly available because of the developments of lipase technologies. The glycoside esters shown are excellent surfactants and emulsifiers and can be prepared at low cost using the lipase-based process. The products are natural, since the process for their synthesis and the raw materials used are also natural.

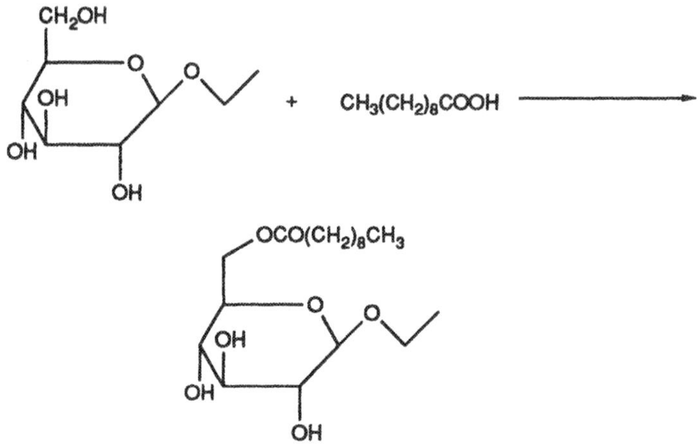

Figure 2 Process of application of lipases to synthesis of esters.

To date, lipases applied in food manufacture and processing have been found in nature mainly as microbial products. Some of these enzymes have been cloned subsequently and expressed in host systems that are, as described earlier, particularly amenable to industrial fermentations. Such processes provide the lipases cost effectively in a pure state that is well suited to the various applications described. These expression systems also lend themselves to the production of engineered lipases in which changes are made in the protein molecule to optimize its reactivity and specificity, according to the needs of the user of the enzymes. A good basis for design of such improved biocatalysts is the structure of the enzymes. Until recently, no triglyceride lipase structures had been solved to high resolution, but at least three structures are currently available. One structure is that of human pancreatic lipase, which was solved by investigators at Hofmann La Roche. Structures of Novo Nordisk lipase derived from *Mucor miehei* and Novo Nordisk Lipolase® also have been solved. Several additional lipase structures are likely to become available in the next few years. Thus, a firm basis for design of improved lipases will become available. Also, new compounds amenable to preparation in lipase-catalyzed processes and useful in foods are likely to be identified and are likely to play an important role in the future in connection with increasing consciousness of and knowledge about the effects of nutritional fats and oils. The application of lipase technologies in the food area is still at its beginning and is expected to become increasingly important.

References

Adams, P. R., and Deploey, J. J. (1978). Enzymes produced by thermophilic fungi. *Mycologia* **70**, 906–910. (1986).

Akita, H., Matsukura, H., and Oishi, T. Lipase catalyzed enantioselective hydrolysis of 2-methyl 3-acetoxy esters. *Tet. Lett.* **27**, 5241–5244.

Andree, H., Müller, W.-R., and Schmid, R. D. (1980). Lipases as detergent components. *J. Appl. Biochem.* **2**, 218–229.

Arima, K., Liu, W.-H., and Beppu, T. (1972). Isolation and identification of the lipolytic and thermophilic fungus. *Agric. Biol. Chem.* **36**, 1913–1917.

Arnold, R. G., Shahani, K. M., and Dwivedi, B. K. (1975). Application of lipolytic enzymes to flavour development in dairy products. *J. Food Sci.* **50**, 1127–1143.

Babayan, V. K. (1989). Medium chain triglycerides. *In* "Dietary Fat Requirements in Health and Development", (J. BeareRogers, ed.) pp. 73–86. American Oil Chemist's Society.

Bello, M., Pievic, M., Adenier, H., Thomas, D., and Legoy, M.-D. (1986). Transestérification, interesterification et synthèse d'esters catalysées par des lipases en microémulsions. *C. R. Acad. Sci. Paris* **303**, 187–192.

Bianchi, D., Cesti, P., and Battistel, E. Anhydrides as acylating agents in lipase-catalysed stereoselective esterification of racemic alcohols. *J. Org. Chem.* **53**, 5531–5534.

Borgström, B., and Brockman, H. L. (ed.) (1984). "Lipases." Elsevier, Amsterdam.

Brady, C. D. *et al.* (1986). "Hydrolysis of Fats." United States Patent No. 4,629,742.

Brockerhoff, H., and Jensen, R. G. (1974). "Lipolytic Enzymes." Academic Press, New York.

Bucciarelli, M., Forni, A., Moretti, I., and Prati, F. (1988). Enzymatic resolution of chiral N-alkyloxaziridine-3,3-dicarboxylic esters. *J. Chem. Soc. Chem. Commun.* 1614–1615.

Chakrabarty, M. M. (1985). Interesterification reaction of glycerides and tehir industrial uses (including a discussion of India's edible oil deficits and possible remedies). *J. Indian Chem. Soc.* **62**, 1–6.

Chander, H., and Klostermeyer, H. (1983). Production of lipase by *Geotrichum candidum*, under various growth conditions. *Milchwissensch.* **38**, 410–412.

Chandhuri, G., Pal, S., and Banerjee, A. B. (1986). Lipase and phospholipases of *Leishmania donovani promastigotes. IRCS Med. Sci* **14**, 1091–1092.

Desnuelle, P. (1961). Pancreatic lipase. *Adv. Enzymol.* **23**, 129–161.

Eitenmiller, R. R., Vakil, J. R., and Shahani, K. M. (1970). Production and properties of *Penicillium roquefortii* lipase. *J. Food Sci.* **35**, 130–133.

El-Hoseiny, M. M. (1986). Production of lipases by certain thermotolerant bacteria. *Egypt. J. Microbiol.* **21**, 81–92.

Farin, F., Labout, J. J. M., and Verschoor, G. J. (1986). "Novel Lipolytic Enzymes and their Use in Detergent Compositions." European Patent Application 0, 218,272.

Fischer, B. E., and Kleber, H.-P. (1987). Isolation and characterization of the extracellular lipase of *Acinotobacter calcoaceticus* 69 V. *J. Basic Microbiol.* **27**, 427–432.

Fujii, T., Tatara, T., and Minagawa, M. (1986). Studies on application of lipolytic enzymes in detergency. I. Effect of lipase from *Candida cylindraceae* on removal of olive oil from cotton fabric. *J. Am. Oil Chem. Soc.* **63**, 796–799.

Fukumoto, J., Iwai, M., and Tsujisaka, Y. (1963). Studies on lipase. I. Purification and crystallisation of a lipase secreted by *Aspergillus niger. J. Gen. Appl. Microbiol.* **9**, 353–361.

Fukomoto, J., Iwai, M., and Tsujisaka, Y. (1964). Studies on lipase. IV. Purification and properties of a lipase secreted by *Rhizopus delemar. J. Gen. Appl. Microbiol.* **10**, 257–265.

Fukui, S., and Tanaka, A. (1982). Bioconversion of lipophilic or water insoluble compounds by immobilized biocatalysts in organic solvent systems. *In* "Enzyme Engineering" (I. Chibata, S. Fukui, and L. B. Wingard, Jr., ed.), Vol. 6, pp. 191–200. Plenum Press, New York.

Fullbrook, P. D. (1983). The use of enzymes in the processing of oilseeds. *J. Am. Oil Chem. Soc.* **60**, 476–478.

Gargouri, Y., Julien, R., Pieroni, G., Verger, R., and Sarda, L. (1984). Studies on the inhibition of pancreatic and microbial lipases by soybean proteins. *J. Lipid Res.* **25**, 1214–1221.

Gargouri, Y., Piéroni, G., Rivière, C., Sarda, L., and Verger, R. (1986). Inhibition of pancreatic and microbial lipases by proteins: Kinetic and binding studies. *In* "Enzymes of Lipid Metabolism II," (L. Freysz, ed.). NATO ASI Series A, Vol. 116, pp. 23–27.

Gerlach, D., Missel, C., and Schreier, P. (1988). Screening for Lipases for the enantiomeric resolution of R,S-octanol by esterification in organic medium. *Z. Lebensm. Unters. Forsch.* **186**, 315–318.

Glänzer, B. I., Faber, K., and Griengl, H. (1988). Microbial resolution of O-acetylpantoyl lactone. *Enz. Microb. Technol.* **10**, 689–690.

Götz, F., Popp, F., Korn, E., and Schleifer, K. H. (1985). Complete nucleotide sequence of the lipase gene from *Staphylococcus hyicus* cloned in *Staphylococcus carnosus. Nuc. Acid. Res.* **13**, 5895–5906.

Gotor, V., Brieva, R., and Rebolledo, F. (1988). A simple procedure for the preparation of chirale amides. *Tet. Lett.* **29**, 6973–6974.

Gowland, P., Kernick, M., and Sundaram, T. K. (1987). Thermophilic bacterial isolates producing lipase. *FEMS Microbiol. Lett.* **48**, 339–343.

Gustafson, C., Franzén, L., and Tagesson, C. (1988). Phospholipase activation and arachidonic acid release in isolated epithelial cells. *Scand. J. Gastroenterol.* **23**, 413–421.

Hamaguchi, S., Ohashi, T., and Watanabe, K. (1986a). Lipase-catalyzed stereoselective hydrolysis of 2-acyloxy-3-chloropropyl *p*-toluenesulfonate. *Agric. Biol. Chem.* **50**, 375–380.

Hamaguchi S., Ohashi, T., and Watanabe, K. (1986b). Lipase-catalyzed stereoselective hydrolysis of 2-acyl and 1-*p*-tolylsulfonyl substituted propanediol and butanediol. *Agric. Biol. Chem.* **50**, 1629–1632.

Hirohara, H., Mitsuda, S., Ando, E., and Komaki, R. (1985). Enzymatic preparation of optically active alcohols related to synthetic pyrethroids insecticides. *Stud. Org. Chem.* **22**, 119–134.

Höfelmann, M., Hartmann, J., Zink, A., and Schreier, P. (1985). Isolation, purification and characterization of lipase isoenzymes from a technical *Aspergillus niger* enzyme. *J. Food Sci.* **50**, 1721–1725.

Hoq, M. M., Tagami, H., Yamane, T., and Shimizu, S. (1985). Some characteristics of continuous glyceride synthesis by lipase in a microporous hydrophobic membrane reactor. *Agric. Biol. Chem.* **49**, 335–342.

Huge-Jensen, I. B., and Gormsen, E. (1989). "Enzymatic Detergent Additive." United States Patent No. 4,810,414.

Huge-Jensen, B., Boel, E., Thim, L., Christensen, M., Andreasen, F., and Christensen, T. (1989). Triglyceride lipases from fungi. Paper presented at 15th Nordisk Lipid Symposium. Rebild, Denmark.

Inada, Y., Nishimura, H., Takahashi, K., Yoshimoto, T., Saha, A. R., and Saito, Y. (1984) Ester synthesis catalyzed by polyethylene glycol modified lipase in benzene. *Biochem. Biophys. Res. Commun.* **122**, 845–850.

Ishihara, H., Okuyama, H., Ikezawa, H., and Tejima, S. (1975). Studies of lipase from *Mucor javanicus. Biochem. Biophys. Acta* **388**, 413–422.

Iwai, M., and Tsujisaka, Y. (1974a). The purification and properties of three kinds of lipases from *Rhizopus delemer. Agric. Biol. Chem.* **38**, 1241–1247.

Iwai, M., and Tsujisaka, Y. (1974b). Interconversion of two lipases from *Rhizopus delemar. Agric. Biol. Chem.* **38**, 1249–1254.

Iwai, M., Tsujisaka, Y., and Fukumoto, J. (1964). Studies on lipase (II). Hydrolysis and esterifying action of crystalline lipase of *Aspergillus niger. J. Gen. Appl. Microbiol.* **10**, 13–22.

Iwai, M., Okumura, S., and Tsujisaka, Y. (1975). The comparison of the properties of two lipases from *Penicillium cyclopium* Westring. *Agric. Biol. Chem.* **39**, 1063–1070.

Iwai, M., Okumura, S., and Tsujisaka, Y. (1980). Synthesis of terpene alcohol esters by lipase. *Agric. Biol. Chem.* **44**, 2731–2732.

Jensen, R. G., Gerrior, S. A., Hagerty, M. M., and McMahon, K. E. (1978). Preparation of acylglycerols and phospholipids with the aid of lipolytic enzymes. *J. Am. Oil Chem. Soc.* **55**, 422–427.

Johri, B. N., Jain, S., and Chouhan, S. (1985). Enzymes from thermophilic fungi: Proteases and lipases. *Proc. Indian Acad. Sci. (Plant Sci.)* **94**, 175–196.

Joshi, S., and Dhar, D. N. (1987). Specificity of fungal lipase in hydrolytic cleavage of oil. *Acta Microbiol. Hungarica* **34**, 111–114.

Joshi, S., and Mathur, J. M. S. (1985). Specificity of lipase isolated from *Fusarium oxysporum. Ind. J. Microbiol.* **25**, 76–78.

Kalo, P. (1988). Modification of butter fat by interesterifications catalyzed by *Aspergillus niger* and *Mucor miehei* lipases. *Meijeritieteellinen Aikakauskirja* **46**, 36–47.

Kirk, O., Björkling, F., and Godtfredsen, S. E. (1989). A highly selective enzme catalysed esterification of simple glucosides. *J. Chem. Soc. Chem. Commun.* 934–935.

Knox, T., and Cliffe, K. R. (1984). Synthesis of long chain esters in a loop reactor system using a fungal cell bound enzyme. *Proc. Biochem.* **19**, 188–192.

Kohr, H. T., Tan, N. H., and Chau, C. L. (1986). Lipase-catalyzed hydrolysis of palm oil. *J. Am. Oil Chem. Soc.* **63**, 538–540.

Koritala, S., Hesseltine, C. W., Pryde, E. H., and Mounts, T. L. (1987). Biochemical modification of fats by microorganisms: A preliminary survey. *J. Am. Oil Chem. Soc.* **64**, 509–513.

Koshiro, S., Sonomoto, K., Tanaka, A., and Fukui, S. (1985). Stereoselective esterification of *d,l*-menthol by polyurethane-entrapped lipase in organic solvent. *J. Biotech.* **2**, 47–57.

Kouker, G., and Jaeger, K.-E. (1987). Specific and sensitive plate assay for bacterial lipases. *Appl. Environ. Microbiol.* **53**, 211–213.

Kugimiya, W., Otani, Y., Hashimoto, Y., and Takagi, Y. (1986). Molecular cloning and nucleotide sequence of the lipase gene from *Pseudomonas fragi. Biochem. Biophys. Res. Commun.* **141**, 185–190.

Kyotani, S., Fukuda, H., Morikawa, H., and Yamane, T. (1988a). Kinetic studies on the interesterification of oils and fats using dried cells of fungus. *J. Ferment. Technol.* **66**, 71–83.

Kyotani, S., Fukuda, H., Nojima, Y., and Yamane, T. (1988b). Interesterification of fats and oils by immobilized fungus at constant water concentration. *J. Ferment. Technol.* **66**, 567–575.

Langrand, G., Secchi, M., Buono, G., and Baratti, J. (1985). Triantaphylides: Lipase-catalyzed ester formation in organic solvents. An easy preparative resolution of α-substituted cyclohexanols. *Tet. Lett.* **26**, 1875–1860.

Lawrence, R. C., Fryer, T. F., and Reiter, B. (1967). The production and characterization of lipases from a *Micrococcus* and *Pseudomonas. J. Gen. Microbiol.* **48**, 401–418.

Lazar, G. (1985). Estersyntese mit lipasen. *Fette Seif. Anstrichm.* **87**, 394–400.

Lin, J. T., Yamazaki, T., and Kitazume, T. (1987). A microbially based approach for the preparation of chirale molecules possessing the trifluoromethyl group. *J. Org. Chem.* **52**, 3211–3217.

Linfield, W. M., Barauskas, R. A., Sivieri, L., Serota, S., and Stevenson, R. S., Sr. (1984). Enzymatic fat hydrolysis and synthesis. *J. Am. Oil Chem. Soc.* **61**, 191–195.

Liu, W.-H., Beppu, T., and Arima, K. (1973a). Purification and general properties of the lipase of the thermophilic fungus *Humicula lanuginosa* S-38. *Agric. Biol. Chem.* **37**, 157–163.

Liu, W.-H., Beppu, T., and Arima, K. (1973b). Effect of various inhibitors on lipase action of thermophilic fungus *Humicula lanuginosa* S-38. *Agric. Biol. Chem.* **37**, 2487–2492.

Macrae, A. R. (1983). Lipase-catalyzed interesterification of oils and fats. *J. Am. Oil Chem. Soc.* **60** 291–294.

Macrae, A. R. (1984). Microbial lipases as catalysts for the interesterification of oils and fats. *Am. Oil Chem. Soc. Monogr.* **11**, 189–199.

Macrae, A. R. (1985). Enzyme-catalysed modification of oils and fats. *Phil. Trans. R. Soc. Lond. B* **310**, 227–233.

Macrae, A. R. (1986). "Process for the Preparation of Esters." European Patent Application. 0, 274,798.

Macrae, A. R., and Hammond, R. C. (1985). Present and future application of lipases. *Biotech. Gen. Eng. Rev.* **3**, 193–217.

Macrae, A. R., and How, P. (1986). "Rearrangement Process." United States Patent No. 4,719,178.

Marples, B. A., and Roger-Evans, M. (1989). Enantioselective lipase-catalysed hydrolysis of esters of epoxy secondary alcohols: An alternative sharpless oxidation. *Tet. Lett.* **30**, 261–264.

Michiyo, I. (1986). "Positionally Non-Specific Lipase from *Candida species*. A Method for Producing It and a DNA Process for Producing It." PCT Patent Application No. WO 88/02775.

Mitsuda, S., Umemura, T., and Hirohara, H. (1988). Preparation of an optically pure secondary alcohol of synthetic pyrethroids using microbial lipases. *Appl. Microbiol. Biotechnol.* **29**, 310–315.

Miyazawa, T., Takitani, T., Ueji, S., Yamada, T., and Kuwata, T. (1988). Optical resolution of unusual amino acids by lipase-catalysed hydrolysis. *J. Chem. Soc. Chem. Commun.* **1988**, 1214–1215.

Morinaga, T., Kanda, S., and Nomi, R. (1986). Lipase production of a new thermophilic fungus, *Humicula lanuginosa* var. *catenulata. J. Ferment. Technol.* **64**, 451–453.

Nagaoka, K., and Yamada, Y. (1969). Studies on *Mucor lipases. Agric. Biol. Chem.* **33**, 986–993.

Nagaoka, K., and Yamada, Y. (1973). Purification of *Mucor* lipases and their properties. *Agric. Biol. Chem.* **37**, 2791–2796.

Nagaoka, K., Yamada, Y., and Koaze, Y. (1969). Studies on *Mucor* lipases. Part I. Production of lipases from a newly isolated *Mucor* sp. *Agric. Biol. Chem.* **33**, 299–305.

Nasr, M. (1983). Acceleration of Romi cheese ripening by addition of fungal esterase lipase powder. *Egypt. J. Dairy Sci.* **11**, 309–315.

Naumura, K., Takahashi, N., and Chikamatsu H. (1988). Enzyme-catalysed asymmetric hydrolysis of *meso*-substrate. The facile synthesis of both enantiomers of *cis*-2,5-disubstituted tetrahydrofuran derivatives. *Chem. Lett.* **1988**, 1717–1720.

Nielsen, T. (1985). Industrial application possibilities of lipase. *Fette Seif. Anstrichm.* **87**, 15–19.

Odera, M., Takeuchi, K., and Tho-E, A. (1986). Molecular cloning of lipase genes from *Alcaligenes denitrificans* and their expression in *Escherichia coli. J. Ferment. Technol.* **64**, 363–371.

Ogundero, V. W. (1980). Lipase activity of thermophilic fungi from mouldy groundnuts in Nigeria. *Mycologia* **72**, 118–126.

Ohta, H., Kimura, Y., and Sugano, Y. (1988). Kinetic resolution of ketone cyanohydrin acetates with a microbial enzyme. *Tet. Lett.* **29**, 6957–6960.

Oi, S., Sawada, A., and Satomura, Y. (1967). Purification and properties of two types of *Penicillium* lipase, I and II, and conversion of types I and II under various modification conditions. *Agric. Biol. Chem.* **31**, 1357–1366.

Okumura, S., Iwai, M., and Tsujisaka, Y. (1976). Positional specificity of four kinds of microbial lipases. *Agric. Biol. Chem.* **40**, 655–660.

Okumura, S., Iwai, M., and Tsujisaka, Y. (1979). Synthesis of various kinds of esters by four microbial lipases. *Biochem. Biophys. Acta* **575**, 156–165.

Okumura, S., Iwai, M., and Tominaga, Y. (1984). Synthesis of ester oligomer by *Aspergillus niger* lipase. *Agric. Biol. Chem.* **48**, 2805–2808.

Peppler, H., Dooley, J. G., and Huang, H. T. (1975). Flavour development in Fortina and Romano cheese by fungal esterase. *J. Dairy Sci.* **59**, 859–862.

Posorske, L. H. (1984). Industrial-scale application of enzymes to the fats and oil industry. *J. Am. Oil Chem. Soc.* **61**, 1758–1760.

Ratledge, C. (1984). Biotechnology as applied to the oil and fats industry. *Fette Seif. Anstrichm.* **86**, 379–389.

Roussis, I. G., Karabalis, I., Papadopoulou, C., and Drainas, C. (1988). Some properties of extracellular lipase from a *Pseudomonas fluorescens* strain. *Lebensm. Wiss. Technol.* **21**, 188–194.

Satoshi, M., Umemura, T., and Hirohara, H. (1988). Preparation of an optically pure secondary alcohol of synthetic pyrethroids using microbial lipases. *Appl. Microbiol. Biotechnol.* **29**, 310–315.

Satyanarayana, T., and Jori, B. N. (1981). Lipolytic activity of thermophilic fungi of paddy straw compost. *Curr. Sci.* **50**, 680–682.

Seino, H., Uchibori, T., Nishitani, T., and Inamasu, S. (1984). Enzymatic synthesis of carbohydrate esters of fatty acid (I). Esterification of sucrose, glucose, fructose and sorbitol. *J. Am. Oil Chem. Soc.* **61**, 1761–1765.

Sih, C. J. (1986). "Process for Preparing Optically Active 3-Acylthio-2-methylpropionic Acid Derivatives." PCT Patent Application No. WO 87/05328.

Sonnet, P. E. (1987). Kinetic resolution of aliphatic alcohols with a fungal lipase from *Mucor miehei. J. Org. Chem.* **52**, 3477–3479.

Sonnet, P., and Antonia, E. (1988). Synthesis and evaluation of pseudolipids to characterize lipase selectivities. *J. Agric. Food Chem.* **36**, 856–862.

Stuer, W., Jaeger, K. E., and Winkler, U. K. (1986). Purification of extracellular lipase from *Pseudomonas aeruginosa. J. Bacteriol.* **168**, 1070–1074.

Sugihara, A., Shimada, Y., and Tominaga, Y. (1988). Enhanced stability of a microbial lipase immobilized to a novel synthetic amine polymer. *Agric. Biol. Chem.* **52**, 1589–1590.

Sztajer, H., and Zboinska, E. (1982). Microbial lipases in biotechnology. *Acta Biotechnol.* **8**, 169–175.

Sztajer, H., Maliszewska, I., and Wieczorek, J. (1988). Production of exogenous lipases by bacteria, fungi, and actinomycetes. *Enz. Microb. Technol.* **10**, 492–497.

Tahoun, M. K., El-Kady, M., and Wahba, A. (1985). Glyceride synthesis by an intracellular lipase from *Aspergillus niger. Microbios Lett.* **28**, 133–139.

Tanaka, T., Ono, E., Ishihara, M., Yamanaka, S., and Takinami, K. (1981). Enzymatic acyl exchange of triglycerides in *n*-hexane. *Agric. Biol. Chem.* **45**, 2387–2389.

Tatara, T., Fujii, T., Kawase, T., and Minagawa, M. (1985). Studies on application of lipolytic enzymes in detergency II. Evaluation of adaptability of various kinds of lipases in practical laundry conditions. *J. Am. Oil Chem Soc.* **62**, 1053–1058.

Thornton, J., Howard, S. P., and Buckley, J. T. (1988). Molecular cloning of a phospholipid acyltransferase from *Aeromonas hydrophila.* Sequence homologies with lecithin-cholesterol acyltransferase and other lipases. *Biochem. Biophys. Acta* **959**, 153–159.

Tsujisaka, Y., and Iwai, M. (1984). Comparative study of microbial lipases. *Kagaku Kogyo (Osaka)* **58**, 60–69.

Tsujisaka, Y., Iwai, M., and Tominaga, Y. (1973). Purification, crystallization and some properties of lipase from *Geotrichum candidum* link. *Agric. Biol. Chem.* **37**, 1457–1464.

Tsujisaka, Y., Okumura, S., and Iwai, M. (1977). Glyceride synthesis by four kinds of microbial lipases. *Biochem. Biophys. Acta* **489**, 415–422.

Vermiere, A., Pille, S., Himpe, J., and Vandamme, E. (1987). Screening and production of microbial lipases. *Med. Fac. Landbouww. Rijksuniv. Gent.* **52**, 1853–1861.

Vora, K. A., Bhandara, S. S., Pradhan, R. S. Amin, A. R., and Modi, V. V. (1988). Characterization of extracellular lipase produced by *Aspergillus japonicus* in response to *Calotropis gigentea* latex. *Biotech. Appl. Biochem.* **10**, 465–472.

Werdelmann, B. W., and Schmid, R. D. (1982). The biotechnology of fats—A challange and an opportunity. *Fett. Seif. Anstrichm.* **84**, 436–443.

Whitesides, G. H., and Ladner, W. (1988). "Method for Making Chiral Epoxy Alcohols." United States Patent No. 4,732,853.

Yeoh, H. H., Wong, F. M., and Lim, G. (1986). Screening for fungal lipases using chromogenic lipid substrates. *Mycologia* **78**, 298–300.

Yokozeki, K., Tanaka, T., Yamanaka, S., Takinami, T., Hirose, Y., Sonomoto, K., Tanaka, A., and Fukui, S. (1982). Ester exchange of triglyceride by entrapped lipase in organic solvent. *Enz. Eng.* **6**, 151–152.

CHAPTER 9

Oxidoreductases

FRANK E. HAMMER

I. Polyphenol Oxidase

A. *General Characteristics*

Polyphenol oxidase has been given two entries in the International Union of Biochemistry (IUB) classification; as EC 1.14.18.1, monophenol monooxygenase, and EC 1.10.3.1, catechol oxidase. Common trivial names for monophenol monoxygenase are tyrosinase, phenolase, monophenol oxidase and cresolase. Common trivial names for catechol oxidase are diphenol oxidase, *o*-diphenolase, phenolase, polyphenol oxidase, and tyrosinase (Enzyme Nomenclature, 1984). Tyrosinase was the first name used for polyphenol oxidase because tyrosine was the first substrate studied. The mammalian enzyme has limited specificity, acting only on tyrosine and dihyroxyphenylalanine (DOPA). The enzyme complexes from higher plants and fungi are virtually nonspecific and oxidize a wide variety of monophenolic and *o*-diphenolic compounds (Scott, 1975). In this chapter, the enzyme will be referred to as polyphenol oxidase. Depending on its biological source, it may have monophenol monooxygenase and/or catechol oxidase activities. Monophenol monoxygenase (tyrosinase) can oxidize both tyrosine and DOPA to melanin, whereas catechol oxidase can convert only DOPA to melanin (Robb, 1984). Tyrosinase oxidizes monohydroxy phenols to *o*-dihydroxy phenols and catechol oxidase dehydrogenates *o*-dihydroxyphenols to *o*-quinones (Scott, 1975). In some publications, tyrosinase is described to carry out both the hydroxylation and the dehydrogenase reaction (Robb, 1984; Kahn and Andrawis, 1985). Robb (1984) gives the reactions as sequential: hydroxylation followed by dehydrogenation, with the final product of *o*-quinone.

Polyphenol oxidases can catalyze many reactions that involve phenolic compounds. When the monophenol *p*-cresol is the substrate, it is oxidized to the diphenol 4-methyl catechol. The oxygen for the hydroxylation comes from O_2, not from water. Thus, the enzyme acts as a monooxygenase in this reaction. Hydroxylation of monophenols is usually referred to as cresolase activity because *p*-cresol is frequently the substrate. When the *o*-diphenol catechol is the substrate, it is dehydrogenated to *o*-benzoquinone.

Polyphenol oxidase is a copper-containing enzyme widely distributed in plants, animals, and humans. In this chapter, we will discuss the enzyme as it occurs in plants. Many reasons have been put forward for the lack of pigment formation in the intact cell, despite the coexistence of polyphenol oxidase and its substrates in the same organ and the same cell. Some of these reasons are separation of substrate and enzyme, substrate presence as precursor, enzyme inactivity, enzyme complexing with an inhibitor, and reduction of quinones formed (Robb, 1984).

Polyphenol oxidase is little understood in its role in higher plants. It is a plastidic enzyme that is nuclear encoded, but is not active until incorporated into the plastid. Apparently, it also exists free in the cytoplasm in degenerating or senescent tissues such as ripe or ripening fruit. Cytochemical techniques to determine the localization of polyphenol oxidase established that the enzyme is found in root plastids, potato tuber amyloplasts, hypocotyl plastids, epidermal plastids, carrot tissue culture plastids, leukoplasts of *Aegopodium*, apical plastids, etioplasts, chromoplasts, and chloroplasts of many different species (Vaughn, 1984). The function of polyphenol oxidase in plants is an enigma because, although it is localized in the plastid, most of the phenolic compounds are in the vacuole, a cellular location not juxtaposed to the plastid. The plastidic location of the enzyme might be necessary to provide a strong reducing environment to prevent the oxidation of *o*-diphenols to *o*-diquinones (Vaughn, 1984). The plastidic polyphenol oxidase also may be necessary to provide an electron donor for monophenol oxygenase activity. Polyphenol oxidase and chlorogenic acid have been implicated in resistance of the tomato plant. These enzymes are known to form orthoquinones in damaged plant tissue. Orthoquinones are known to alkylate amine and sulfhydryl groups of proteins and amino acids. This alkylation alters solubility, digestibility, and infectivity for some pathogenic viruses. The effects of orthoquinone alkylation on an important microbial insecticide, *Bacillus thuringiensis kurstaki* (Btk), were tested. When Btk was incubated with polyphenol oxidase and chlorogenic acid and fed to *Heliothis zea* larvae, there was a more toxic response than without the phytochemicals. Less dramatic but similar results were obtained with only polyphenol oxidase. Digestibility experiments indicated that alkylation enhanced solubilization and/or proteolysis of the crystal protein *in vivo* (Ludlum, 1991).

To determine the physiological importance of an enzyme, reliable assay methods are imperative. Unfortunately, in many studies of polyphenol oxidase, attempts have not been made to differentiate it from peroxidase (Kahn, 1985). In the presence of very small concentrations of hydrogen peroxide, peroxidase will form quinones from phenolic compounds. Peroxidase activity generally can be eliminated or minimized by adding catalase or alcohols to the assay solution. Catalase (EC 1.11.1.6) will remove the hydrogen peroxide and the alcohol will be oxidized preferentially to its aldehyde. Kahn (1985) discusses the use of tropolone (2,4,6-cycloheptatriene-1-one) to differentiate between tyrosinase and peroxidase activities. Tropolone is an effective copper chelator and, thus, inhibits tyrosinase. Inhibition (50%) occurs at 100, 3, and $0.4 \times 10^{-6} M$ tropolone when 4-methyl catechol, dopamine, and D,L-DOPA, respectively, are the substrates. In addition, tropolone can serve as a substrate for peroxidase. The resulting product is yellow and has an absorbance peak at 418 nm. The oxidation of tropolone occurs only when both peroxidase and hydrogen peroxide are present.

A commonly used polyphenol oxidase assay is the measurement of the formation of quinones and quinone polymers spectrophotometrically at 470–500 nm. Another method is the measurement of the decrease of O_2 by polarographic methods. A common dilemma for all polyphenol oxidase assay methods is the low affinity of the enzyme for oxygen. Consequently aeration of the assay solution is necessary for reliable results.

B. Of Mushroom

Mushroom polyphenol oxidase usually is referred to as tyrosinase. Tyrosinases catalyze the hydroxylation of monophenols, for example, tyrosine, to o-diphenols and convert o-diphenols to o-quinones. The mushroom enzyme can use tyrosine, catechol, or DOPA as a substrate (Flurkey, 1989).

Many physical characteristics have been determined for mushroom tyrosinases, including diffusion coefficients, sedimentation coefficients, frictional ratios, Stokes radii, isoelectric points, and apparent molecular weights. Mushroom tyrosinase appears to be composed of two heavy chains of 43–45 kDa each and two light chains of about 13 kDa each. The heavy chains contain the catalytic sites. Different isoenzymes apparently have different heavy chains. The role of the light chains is unknown. Although there is abundant information on the physical and catalytic properties of mushroom tyrosinase, little is known about the enzyme in developing mushrooms (Flurkey, 1989).

Tyrosinase activity was monitored in four stages of mushroom development: small pins (0–0.5 cm), large pins (0.5–1 cm), immature mushrooms, and mature mushrooms (classified by cap size, gill development, and veil covering). Using catechol as the substrate, there was no apparent trend in enzyme activity during development. In contrast, DOPA oxidase activity appeared to decrease with a concomitant increase in latent tyrosinase activity. The investigators concluded that tyrosinase activity and the isoenzyme content (1) are complex, (2) differ in preharvest development, (3) vary from tissue to tissue, and (4) are composed of active and latent forms. No new isoenzyme forms appear to be made during development (Flurkey and Ingebrigtsen, 1989).

Other workers investigated the effect of ascorbic acid, sodium bisulfite, and thiol compounds on mushroom polyphenol oxidase (Golan-Goldhirsch, 1984). Ascorbic acid and sodium bisulfite had little effect on initial reaction velocity whereas dithiothreitol and glutathione decreased activity by 35%. By spectrophotometry, the I_{50} values were: dithiothreitol, 0.06 mM; glutathione, 0.17 mM; sodium bisulfite, 0.20 mM; and ascorbic acid, 0.24 mM. At 0.1 mM dithiothreitol, a complete loss of activity was seen after 70 min at 25°C, while sodium bisulfite, glutathione, and ascorbic acid caused 50% inactivation after 28, 106, and 130 min, respectively, at 5 mM.

C. Of Grape

Enzymatic browning of musts and wines depends mainly on the oxidation of endogenous phenols by grape polyphenol oxiedases that are released during wine making. Catecholase and cresolase activities of grape polyphenol oxidase were investigated during development and maturation of Monastrell and Airien grape berries. Cresolase and catecholase were extracted from the grapes by placing the flesh in 100 mM phosphate buffer at pH 7.3 with 10 mM sodium ascorbate. The mixture was homogenized, centrifuged, and extracted with Triton X-100. The supernatant was subjected to ammonium sulfate precipitation and, after dialysis, was used as the enzyme source. Phenolic compounds were extracted from the grape with 80% ethanol and determined spectrophotometrically at 480 and 520 nm.

Catecholase activity measured below pH 5.0 increased during maturation, whereas its activity measured above pH 5.0 decreased during maturation, indicating the presence of two catecholases in Monastrell grapes. These two activities disappeared after further maturation, revealing the presence of a third enzyme. These changes in catecholase activity appear to be related to the formation of anthocyanins, because there was no shift in the optimum pH of catecholase activity in the white grape Airen, nor was there any change in the polyphenol oxidases separated by gel electrophoresis during the 7-week study (Sanchez-Ferrer, 1989). The catecholase activity of both cultivars when measured at pH 4.0 increased continuously during the 7-week study. In contrast, the cresolase activity of both cultivars increased much less. In an earlier study, partially purified polyphenol oxidase from Arien grapes also had both cresolase and catecholase activities. Catecholase had a pH optimum of 3.5–4.5 and was relatively heat stable. The apparent K_m for 4-methylcatechol was 9.5 mM. Cresolase activity had a lag phase, but the presence of o-diphenol abolished it. The apparent K_m for p-cresol was 0.35 mM and the activation constant for o-diphenol was 1.75 μM (Valero, 1988).

Catecholase and cresolase activities of purified Monastrell grape polyphenol oxidase were characterized (Sanchez-Ferrer, 1988). The polyphenol oxidase was purified 126-fold by ammonium sulfate fractionation. The partially purified enzyme had both catecholase and cresolase activity. The former had temperature and pH optima of 20–40°C and 3.5–5.0, respectively. The apparent K_m for 4-methyl catechol was 9 mM. Cresolase activity showed a lag phase, which was decreased by a higher enzyme concentration and increased when the substrate concentration was increased. Cresolase activity rose with a pH change from 3.5 to 7.5 without attaining a defined maximum activity. The apparent K_m for p-cresol was evaluated by Hanes plot and a value of 0.5 mM was observed.

Polyphenol oxidase activities of Koshu grapes were affected by storage temperature. Polyphenol oxidase activity of the filtered homogenate of grapes stored at −20°C was twice as high as when the grapes were stored at 4°C (Nakamura, 1985). The best harvest time for Koshu grapes, judging from soluble polyphenol oxidase activity in their juice could be determined because of the relationship between soluble polyphenol oxidase activity and Brix and total acid. There was no correlation between the insoluble fraction of polyphenol oxidase and Brix or total acid (Nakanishi, 1987). The existence of tasteless or, in Japanese, ajinashi berries among Koshu grapes is believed to be due to a virus infection. In the ripe stage of the grapes, the polyphenol oxidase activity and total phenolic content of the ajinashi grape were about the same as those of the uninfected grape. The number of active bands of polyphenol oxidase in ripe ajinashi berries on electrophoresis was smaller than in normal berries. The intensity of the bands also indicated less polyphenol oxidase in the infected berries (Nakamura, 1984). An *in vitro* oxidation system was used to determine the extent of enzymatic oxidation on seven grape seed phenolics: gallic acid, (+)-catechin, (−)-catechin, and four procyanidins. There was a considerable depletion of the phenolics. In the absence of polyphenol oxidase, the decrease of phenolics was much less (Oszmianski, 1985). There was a significant correlation between red grape browning and polyphenol oxidase content, which accounts for 74% of the browning. Polyphenol oxidase only accounted for 16% of the browning in white grapes. Browning of white grapes was concluded to be caused by other factors. There was no correlation between browning of red grapes and hydroxycinnamic acid content (Romeyer, 1985).

D. Of Pear

The extraction of polyphenol oxidase from plant sources is complicated by the presence of endogenous phenolic compounds. These are oxidized to the corresponding quinones or semiquinone radicals by the enzyme. In turn, these phenolic compounds react with the polyphenol oxidase, changing its properties. Research was conducted to investigate different methods of extracting polyphenol oxidase from d'Anjou pears to determine the most efficient method of binding the phenolic compounds during extraction (Smith, 1985). A water extract of the pear tissue at pH 4.2 and 4°C lost 60% of its polyphenol oxidase activity 7 hr after extraction. Visible browning of the extract occurred within 30 min and absorbance at 410 nm increased for 10 hr, indicating the formation of polymerized phenolic compounds. This increase in absorbance at 410 nm was accompanied by a decrease in absorbance at 324 and 280 nm. To prevent these changes, the investigators prepared a buffered extract of acteone powders, and extracts using the phenol adsorbants polyvinylpolypyrrolidone (PVPP), Amberlite XAD-4, and the ionexchange resins BioRad AG 1-X8 and AG 2-X8. Electrophoresis of the water extract revealed three regions of polyphenol oxidase activity with 11 distinct bands of activity. In later experiments, when the phenolic compounds were removed with XAD-4, the number of apparent isoenzymes was decreased to three. Thus, endogenous phenolic substrates can react with protein to produce additional isoenzymes that are not present in the intact tissue. In the pH range 2.0–6.5, the purified polyphenol oxidase was most stable at pH 4.0 at 4°C. The enzyme was unstable at pH 6.5.

d'Anjou pear polyphenol oxidase was extracted in the presence of a phenolic binder, AG 2-X8, and the nonionic detergent Triton X-100 (Wisseman, 1985). After extraction, the enzyme was extensively pruified by chromatography on Phenyl Sepharose CL-4B, DEAE–cellulose, and hydroxyapatite columns. An interesting discovery was that sharp peaks and good resolution were obtained only after the columns were run at room temperature. When the columns were run at 4°C, the resolution was poor. Hydrophobic chromatography, using Phenyl Sepharose CL-4B resin, provided a rapid partial purification of pear polyphenol oxidase. The enzyme was eluted as a single major peak and clearly separated from the bulk or the other 280-nm absorbing material. The purification was 24-fold. Occasionally the yield was close to 200%, indicating the removal of an inhibitor. After DEAE chromatography, three peaks appeared with purifications of 30-, 12-, and 7-fold. The isoenzymes corresponding to fractions 1 and 2 were purified further using hydroxyapatite chromatography. The isoenzyme corresponding to peak 1 was purified further to 148-fold and the one corresponding to peak 2 was purified further to 103-fold. The DEAE peak was only purified 7-fold, probably because of its high protein concentration. The stability of the purified polyphenol oxidases was excellent. The enzyme of peak 1 lost 15% of its activity when stored at 4°C over 15 days at pH 7.0, but the remaining activity remained at that level for 120 days.

The effect of dimethylsulfoxide (DMSO) on enzyme activity was studied using the peak 1 fraction. DMSO concentrations below 50% significantly increased the enzyme activity, with a maximum of 180% occurring at 5% DMSO. At 80% DMSO, the activity decreased to about 30%. The compatibility of pear polyphenol oxidase suggests that the enzyme is active in a hydrophobic environment. The purified polyphenol oxidase is being used currently to study to the mode of inhibition by sulfite.

E. Of Kiwi Fruit

The polyphenol oxidase of kiwifruit was purified partially by ammonium sulfate fractionation, dialysis, and chromatography on a DEAE–cellulose column (Park, 1985). The cresolase fraction was in the first peak and the catecholase appeared in the fourth peak when eluted from the DEAE–cellulose column. Six peaks absorbed at 280 nm. The catecholase was shown to have four isoenzymes by polyacrylamide electrophoresis and enzyme staining. These four isoenzymes (FA1P–FA4P) were characterized further. The estimated molecular masses were 15, 20, 25, and 40 kDa. The FA1P fraction was less heat resistant than the FA4P fraction. The K_m values were 50 mM and 8.7 mM for FA1P and FA4P, respectively, when (+)-catechin was the substrate. The activation energy on (+)-catechin was 4 kcal/mol for FA1P and 7 kcal/mol for FA4P. The pH optima were not too different; pH 6.8 for FA1P and pH 7.3 for FA4P when catechol was the substrate. When (+)-catechin was used, the pH optima were 7.5 and 8.0. Experiments to determine the active site of kiwifruit polyphenol oxidase were done using Rose Bengal. The results indicated that histidine might be associated with the active site. Inhibition studies using NaCN, reduced and oxidized glutathione, ethylenediaminetetraacetic acid (EDTA), sodium diethyldithiocarbamate, and quinic acid were performed. Reduced glutathione diminished FA4P activity to 9%, but had no effect on FA1P. EDTA had little effect on the activity of any of the four fractions. Sodium diethyldithiocarbamate decreased the activity of fraction FA4P to 11% and the activity of fraction FA1P to 91%. Quinic acid had little effect.

F. Of Sago Palm

Sago palm has a higher productivity of starch than either wheat or corn. In addition, it can grow on tropical acidic marshy ground that cannot be used as farmland. The plant is also somewhat salt resistant. However sago starch is not exploited fully because of its brownish color, which is caused by browning of the pith. To prevent browning and improve starch quality, the mechanism of browning was studied and found to be the oxidation of polyphenols by polyphenol oxidases in the pith. The oxidation of D-catechin and D,L-epicatechin in the pith by polyphenol oxidase was responsible for the browning. Two polyphenol oxidases were extracted and partially purified from sago palm pith by hydroxyapatite chromatography, DEAE–cellulose chromatography, and gel filtration (Okamoto, 1988). The purification procedure appeared to remove essential copper from both enzymes, because 0.1 mM $CuSO_4$ increased the relative activity to about 300% for both enzymes. Also of interest is that 0.9 mM $ZnSO_4$ increased the relative activity of both enzymes to about 140%. The two enzymes showed similar behavior with inhibitors. Among them, KCN and Na-diethyldithiocarbamate were the most potent inhibitors. Reducing agents such as Na_2SO_3 and $NaHSO_3$ were also potent inhibitors, as might be expected. The optimum pH for both enzymes was 6.5; there was little activity below pH 4 and above pH 9. One of the enzymes was surprisingly stable at pH 10. About 90% of its activity remained after 18 hr at 4°C. The other enzyme was more acid stable, losing only about 10% of its activity at pH 3. Both enzymes exhibited similar thermal denaturation. The activities began to decrease precipitously at about 50°C and were almost 0 at about 80°C. 4-Methyl catechol was the best substrate for both enzymes, having a relative activity of 925% for one of the enzymes and 168% for the other when D-catechin was set at 100%. In general, both enzymes hardly oxidized monohydroxy or m- or p-dihydroxy phenols. They oxidized trihydroxy phenols to some degree. The activities of both enzymes were inhibited by an excess of D-catechin. Michaelis–Menton's formula could not be applied for either enzyme at more than 15 mM. Both these enzyme fractions had molecular masses of about 40 kDa.

G. Of Tea

Partially purified polyphenol oxidase from fresh tea leaves oxidizes amino acids to aldehydes in the presence of (+)-catechin or other diphenols. These aldehydes contribute to development of aroma in tea. Formation of the aldehydes was directly proportional to the concentration of the amino acids, polyphenol oxidase, and (+)-catechin. However, when (+)-catechin concentration was greater than 0.3 mM, production of aldehydes was reduced, possibly because of substrate saturation of polyphenol oxidase. As expected, diethyldithiocarbamate inhibited formation of aldehydes. The inhibition was reversed by Cu^{2+}. Diethyldithiocarbamate inhibited polyphenol oxidase noncompetetively (Srivastava, 1986). Green tea leaves and crushed, torn, and curled tea (CCT) had activities of 10.9 polyphenol oxidase units per g dry weight. Fermented tea had no detectable activity and black tea had little activity. All three polyphenol oxidase assays were done using substrates present in tea. When the assays were repeated using pyragallol as the substrate, polyphenol oxidase activities were 18.0 for green tea leaves, 11.2 for CCT tea, 5.4 for fermented tea, and 0.9 for black tea, indicting that polyphenol oxidase was still present in fermented tea and, in a lesser quantity, still present in black tea (Matheis, 1987). Depression in polyphenol oxidase during withering of tea leaf *(Camellia sinsensis)* affected the oxidative condensation of tea flavanols by forming theaflavins and thearubigins. These compounds are associated with the briskness, brightness, and body of tea liquors (Ullah, 1984). The addition of pectic substances during processing was shown to reduce the content of theaflavin and thearubigin. This reduction may have been related to the reduction of polyphenol oxidase content (Dev Choudhury, 1984). Oxidation of black tea components theaflavin, theaflavin monogallates, and theaflavin digallates by polyphenol oxidase was tested in the presence of (−)-epicatechin in a model fermentation system under aerobic conditions. Oxidation of theaflavin monogallates was more rapid than that of theaflavin digallate (Bajaj, 1987). The variation of polyphenol oxidase activity, total polyphenol content, and catechins on the quality of black tea (*Camellia sinensis* L. O Kuntze) was studied with respect to different clones and shoot components. There was a wide variation of polyphenol oxidase in the six different clones. Optimum fermentation time and polyphenol oxidase activity of different clones showed a hyperbolic relationship. A good but nonlinear relationship was found between polyphenol oxidase activity of fresh tea shoots of different clones and the quantity of theaflavins of corresponding black teas. Formation of theaflavins during fermentation of different shoot components and the quality of black tea agreed well with polyphenol oxidase activity (Thanaraj and Seshadri, 1990). In another study, clonal tea was found to be better than seedling tea. One of the differences noted between clonal and seedling leaves was that polyphenol oxidase activity was higher in the former. As a result, the theaflavin content was higher in black tea made using the clone; therefore, the browning potential is greater in the clonal tea (Van Lelyveld, 1986). When 0.1% of a microbial polyphenol oxidase was added to black tea during fermentation, the fermentation time was reduced. The color of the infusion was improved, and theaflavin and thearubigin concentrations were increased significantly (Kato, 1987). Polyphenol oxidase and peroxidase activities of tea associated microflora were evaluated. Ten bacterial and seven fungal isolates were screened for these activities. In general, bacterial strains were more promising than fungal isolates. Polyphenol oxidase activity was much higher than peroxidase in the three most potent *Bacillus* strains (Pradhan, 1990).

H. In Cocoa Processing

One of the least understood factors in the quality of cocoa products is the presence of polyphenol oxidase in cocoa beans. In chocolate production, polyphenol oxidase is one component responsible for the formation of flavor precursors, beginning in the oxidative phase of the fermentation and continuing into the drying phase. Among the changes affecting flavor are reduction in the bitterness and astrigency that result from polymerized polyphenol–protein interactions. An increase in oxygen penetration into the cocoa bean mass during drying induces maximum oxidation of (−)-epicatechin and procyanidins, causing the production of melanin and melanoproteins that are responsible for the color in brown chocolate. In an attempt to understand the role of the enzyme in chocolate flavor and color development more fully, polyphenol oxidase was extracted from cocoa beans, then isolated and enriched. Crude polyphenol oxidase extract was prepared by grinding frozen beans and extracting them with buffer at pH 8.0. Ammonium sulfate at 65% saturation precipitated the polyphenol oxidase. The enzyme was enriched by high performance liquid chromatography (HPLC) using a hydrophobic resin resulting in a 6.5-fold purification with a 25% recovery. There appear to be multiple isoforms of polyphenol oxidase and no contaminating enzymes in the purified preparation. This rapid purification technique should allow for easier study of the properties of cocoa polyphenol oxidase (Wong, 1990).

I. Of Green Pepper

Green pepper, an important article of commerce, is obtained from unripe but fully developed berries of *Piper nigrum.* The fresh berries become black in a short time unless preserved. This blackening was found to be due to the enzyme-catalyzed oxidation of 3,4-dihydroxyphenylethanol glycoside. The enzyme described was polyphenol oxidase, which was found to oxidize the aglycone of the glycoside. Damage to skin tissues resulted in a very rapid blackening due to oxidation (Variyar, 1988). Among five commerical varieties of *P. nigrum* studied, there was great variation in the quantities of phenolics and polyphenol oxidase-oxidizable phenolics. However, no correlation was observed between the total or oxidizable phenolics of a particular variety and the extent of blackening during sun-drying. This result might be explained by the findings that sun-drying resulted in a 75% decrease in total phenolic content and a complete loss of polyphenol oxidase. A new compound, 3,4-dihydroxy-6-(*N*-ethylamino)benzamide, was identified in pepper as an efficient substrate for pepper polyphenol oxidase (Bandyopadhyay, 1990).

J. Of Strawberry

The red color in strawberries comes from two main anthocyanin pigments: pelargonidin-3-glucoside (PGN) and cyanidin-3-glucoside (CYN). Anthocyanins are very unstable. Color loss can occur after thawing of frozen fruit and during processing of strawberry products. The presence of an active polyphenol oxidase system can cause the loss of anthocyanins, but the fate of the pigment is unknown. Model systems containing PGN, CYN, D-catechin (CAT), and polyphenol oxidase were prepared. It was concluded that strawberry polyphenol oxidase played a role in the degradation of anthocyanin pigments. Direct oxidation of these pigments did not seem to be the main pathway of decolorization, although polyphenol oxidase could oxidize cyanin at a very slow rate. The reaction of polyphenol oxidase with CAT, a flavonoid found in strawberries, occurred at a rapid rate with the formation of a polymer absorbing at 390 nm. Quinones and other intermediates formed during the oxidation of CAT by polyphenol oxidase may be responsible for the destruction of anthocyanin, either by oxidation or copolymerization (Wesche-Ebeling, 1990). In an earlier study by the same group, polyphenol oxidase was extracted from strawberries using a hydrophobic Phenyl Sepharose resin. Two polyphenol oxidase fractions were isolated, both of which had very high activity with D-catechin. Studies of model systems containing CAT, CYN, or PGN indicated that anthocyanin pigments were destroyed either by direct oxidation by quinones formed from the D-catechin resulting from polyphenol oxidase action or by copolymerization into tannin formed by D-catechin – quinone polymerization (Wesche-Ebeling, 1984).

K. Of Potato

Dopachrome is the first visible pigment produced during adventitious enzymatic browning of plant tissue. Homogenized raw potato was used as a source of polyphenol oxidase and its substrate, DOPA. Dopachrome formation was found to be inhibited by sodium benzoate, reduced temperature, and acid pH. Results indicate that the benzoate competes with DOPA (Moon, 1985). The effect of isopropyl-N-(3-chlorophenyl) carbamate (CIPC) on total nitrogen content and polyphenol oxidase activity was evaluated on potatoes in India. CIPC treatment increased total nitrogen. Polyphenol oxidase activity was reduced by CIPC treatment at room temperature but not at colder temperatures (Randhawa, 1986). The extent of enzymatic browning at the cut surfaces of potatoes correlated with total phenolic compounds, tyrosine, and, to a lesser extent, polyphenol oxidase activity. Browning in the Atlantic cultivars was almost eliminated by a water dip. Russett Burbank was much more susceptible to browning than the Atlantic cultivars (Sapers, 1989). The effects of soil applications of sodium molybdate on Katahdin potato quality was investigated. The parameters investigated were polyphenol oxidase activity, total phenolic content, chlorogenic acid level, tyrosine, and the amount of ascorbic acid. Soil applications of high levels of sodium molybdate increased ascorbic acid but decreased polyphenol oxidase activity, enzymatic discoloration, chlorogenic acid, and tyrosine. Lower levels of sodium molybdate did not affect these parameters (Munshi, 1988). One of the problems in attempting to determine polyphenol oxidase activity in potato is the inhibition of polyphenol oxidase by endogenous phenols. Endogenous phenols were removed from potato tuber homogenates by polyvinylpyrrolidone (PVP), extensive dialysis, ammonium sulfate or acteone precipitation, or Sephadex G-25 chromatography. G-25 gel filtration was superior in removing endogenous phenols. When the potato homogenate was treated in this manner, polyphenol oxidase activity was retained (Hsu, 1988).

L. Of Pomegranate

Husk scald incidence on Wonderful pomegranate fruit during storage was related to the amount of o-dihydroxyphenols extracted from the husk and was controlled by inhibiting their oxidation by polyphenol oxidase. Postharvest measures included dipping the fruit in boiling water for 2 min and in antioxidant solutions. The most effective control of husk scald was storage of late-harvested fruit in a low-oxygen environment (2% oxygen at 2°C). However, this treatment resulted in the accumulation of ethanol, which caused development of off-flavors (Ben-Arie, 1986).

M. Of Persimmon

Seasonal changes of polyphenol oxidase activity in the flesh of the Japanese persimmon were measured from July to October. Results showed that polyphenol oxidase activity increased rapidly in two cultivars (Fuyu and Sangokuichi) and there was flesh darkening. In the other two cultivars (Yotsuyasaijo and Hagakushi), polyphenol oxidase activity remained relatively low during fruit development and no flesh darkening was observed (Taira, 1987). Ethanol vapor treatment of attached intact Japanese persimmon quickly removed the astringency and induced flesh darkening within 3 weeks. A crude polyphenol oxidase preparation as an acetone powder from persimmon mesocarp was much more active when extracted at room temperature than at 4°C. Of the five most common substrates, 4-methyl catechol was oxidized most rapidly by the enzyme. Polyphenol oxidase activity increased sharply just after the decrease in astringency and coincided with the onset of darkening at the calyx end. High polyphenol oxidase activity was maintained during progressive darkening of the entire flesh (Sugiura, 1985).

N. Of Olive

A positive correlation between the rate of browning of the crude homogenate of the fruit and the polyphenol oxidase content of the acetone powder extracted from the fruit was seen in five olive (*Olea europaea*) varieties. An inhibitor was ruled out and it was concluded that the rate of olive browning was directly proportional to polyphenol oxidase activity (Sciancalepore and Longone, 1985). Although there were significant differences between the varieties in *o*-diphenol content and polyphenol oxidase activity, the browning could be correlated to these parameters and not to peroxidase activity (Sciancalepore, 1985).

O. Of Beet

Endogenous polyphenol oxidase activity is associated with the "black ring" defect in canned beet. This black ring resulted after 5–10 min exposure of beet roots to live steam. A previous study implicated peroxidase and polyphenol oxidase activity as the cause (Parkin, 1990). This study produced evidence that polyphenol oxidase was the primary cause of discoloration. When beet roots were dipped in 5 mM cysteine, discoloration was inhibited;

when they were dipped in 5 mM dithiothreitol, it was prevented. Both sulfhydryl agents are known inhibitors of polyphenol oxidase. Discoloration also was prevented by 1 mM cyanide, a known inhibitor of metalloproteins, and 1 mM tropolone, a copper chelator and peroxidase substrate. Beet slices dipped in a solution containing both tropolone and hydrogen peroxide discolored to a similar extent as the 0.1% sodium chloride control, indicating that, in the presence of peroxide, peroxidase also can contribute to the discoloration of thermally processed beet roots (Im, 1990).

II. Peroxidase

A. General Characteristics

Peroxidases are distributed ubiquitously in both the plant and animal kingdoms. Perhaps the one most widely known to enzymologists is horseradish peroxidase (EC 1.11.1.7), because of its wide use as an indicator enzyme in standard spectrophotometric and immunoassay techniques. All peroxidases act on hydrogen peroxide as an electron acceptor and oxidize a multitude of donor compounds, many times to colored end products. Horseradish peroxidase has a heme prosthetic group and is a hemoprotein. NADH peroxidase (EC 1.11.1.1) is a flavoprotein. Glutathione peroxidase (EC 1.11.1.9) is a selenium-containing protein. Chloride peroxidase (EC 1.11.1.10) is probably a heme–thiolate protein.

B. Studies on Horseradish Peroxidase

Thermal inactivation of horseradish peroxidase in buffer was studied by differential scanning calorimetry and residual activity determination. For both experimental systems, the apparent reaction order for thermal inactivation was 1.5. Since the four isoenzymes varied in heat resistance, the 1.5 order probably was caused by the heterogenity of the enzyme. Sucrose at concentrations of 40% or greater stabilized the enzyme against thermal inactivation. At 10–20% sucrose, the thermal stability was reduced. At equimolar concentrations of glucose, fructose, or lactose, horseradish peroxidase was inactivated more rapidly than in the presence of sucrose (Chang, 1988). The thermal stability of immobilized horseradish peroxidase was tested in water–organic solvent systems. Water–dioxane reduces and water–DMSO increases thermal stability. Results also indicated that horseradish peroxidase is stabilized against thermal inactivation by immobilization (D'Angiuro, 1987). Aminated horseradish peroxidase was inactivated by interaction with S-Sepharose. Peroxidase that was aminated by periodate oxidation and reductive amination was purified by cation-exchange chromatography on S-

Sepharose. Instead of the single peak of aminated enzyme, two distinct peaks were eluted from the column. Evaluation of the protein in each of the peaks showed that the first peak was similar in spectral properties and activity to the native enzyme. The second peak had a 3-fold reduction in extinction in the Soret region at 404 nm and had no activity (Allen, 1991).

The gene encoding horseradish peroxidase has been cloned and sequenced, so the gene is commercially available (British Biotechnology, 1989). A synthetic gene encoding horseradish peroxidase isoenzyme C was expressed in *Escherichia coli*. The nonglycosylated recombinant enzyme was made in inclusion bodies in an insoluble inactive form containing only traces of heme. The peroxidase was solubilized and the conditions under which it folded to produce active enzyme were determined. The folding was shown to be critically dependent on the concentrations of urea, Ca^{2+}, and heme, and on oxidation by oxidized glutathione. The purified peroxidase had about half the activity of the enzyme from horseradish root. Glycosylation is not essential for correct folding and activity. The overall yield of enzyme was only 2–3%, but could be increased by reprocessing material that precipitated during folding (Smith, 1990). More recently, a cDNA clone encoding a neutral horseradish peroxidase was isolated and characterized. The cDNA contained 1378 nucleotides excluding the poly(A) tail and the deduced protein contained 327 amino acids, including a 28-amino-acid leader sequence. The predicted amino acid sequence is 9 amino acids shorter than the major isoenzyme belonging to the horseradish peroxidase C group (HRP-C). The sequence of the neutral isoenzyme shows 53.7% identity with HRP-C but appears to have three fewer glycosylation sites. The described cDNA clone was concluded to encode a neutral horseradish peroxidase that belongs to a new, not previously described horseradish peroxidase group (Bartonek-Roxa, 1991). Genomic DNAs encoding the two peroxidase isoenzymes prxC2 and prxC3 were cloned and sequenced. Genes *prx*C2 and *prx*C3 encoded 347 and 349 amino acid residues, respectively, including putative signal sequences at the N termini (Fujiyama, 1990). Properties of four monoclonal antibodies to horseradish peroxidase were investigated. The antibodies were shown to be directed to different epitopes on the polypeptide chain (Kim, 1991). A review on the structure and kinetic properties of horseradish peroxidase has been published (Dunford, 1991).

C. Other Analytical Peroxidases

Peroxidase from *Coriolus versicolor* is now being used in analytical procedures as a substitute for horseradish peroxidase. The peroxidases from *C. versicolor* are excreted into the medium only when the carbon source is limited and veratryl alcohol is added to the medium as an enzyme inducer. In a medium with limited nitrogen, the excretion of enzyme is not dependent on veratryl alcohol. The optimum pH of the peroxidase is 4.0, and it is immunologically similar to horseradish peroxidase (Morohoshi, 1990). A European Patent

Application described a peroxidase derived from the basidiomycete *Coprinus cinereus* (Asami, 1986) that also can substitute for horseradish peroxidase. A peroxidase also can be made efficiently from the callus tissue of wasibi (Monma, 1990). A review on *N*- and *O*-demethylations catalyzed by peroxidases has been published (Meunier, 1991).

D. Blanching of Foods

1. General Considerations

Peroxidases damage most foods if they are not inactivated before further processing. Therefore, the inactivation and characteristics of peroxidases have been studied extensively in many fruits and vegetables, as well as in pasta and dough. Peroxidases in foods cause the formation of undesirable end products unless they are completely or at least partially inactivated. However, tomato and potato tissue cell walls apparently are not degraded by either endogenous or added peroxidase, although the application of peroxidase substrates such as indoleacetic acid, NADH, KI, and Na_2SO_3 led to an increase in weight loss of ripening tomato tissue. Exogenous peroxidase added to potato slices did not lead to tissue degradation, but added pectinase caused cell wall degradation (Araujo, 1985). Blanching usually is employed to preserve most vegetables and even some fruits before further processing. Other methods to inactivate peroxidases in various food products have been studied also. Blanching, however, is still the most widely used method to inactivate peroxidases and other enzymes. Blanching of 16 vegetable varieties was done in water at $70-100°C$ for various times. Peroxidase and lipoxygenase activities were determined before and after the various blanching regimes. Inactivation of the two enzymes in beans, asparagus, zucchini, and carrots showed a biphasic pattern. In all vegetables except French beans, lipoxygenase had the higher heat stability. Heat resistance was lower for soluble than for ionically bound peroxidase (Pizzocaro, 1988). A thermokinetic model of peroxidase inactivation in corn-on-the-cob during blanching was developed (Luna, 1986). Using this thermokinetic model, the average peroxidase activity retention in the kernel and outer-cob during blanching–cooling was predicted (Garrote, 1987). Flavor changes that occur in unblanched frozen vegetables have been suggested to be related to enzyme activity. The activities of palmitoyl-CoA hydrolase, α-oxidase, lipoxygenase, peroxidase, and catalase were determined in 22 fresh vegetable types. Protein and ascorbic acid were also determined. All vegetable types had palmitoyl-CoA dehydrogenase and lipoxygenase activity. Palmitoyl-CoA dehydrogenase activity ranged from 0.05 (onion and beetroot) to 3.63 (potato) mmol/min/100 g wet weight. Lipoxygenase activity ranged from 1.0 (parsley) to 44.9 (peas). α-Oxidation activity was not detected in 9 vegetable types but ranged from 0.3 (crisp-headed lettuce) to 63.6 (kale). All vegetables tested except red pepper had peroxidase activity; levels ranged from 2 (onion) to 2656 (Brussels sprouts). All vegetables except red pepper, snap pea, and tomato had catalase activity; levels ranged from 2 (cucumber) to 403 (peas). Linear regression analysis failed to find any correlation between any of the enzyme activities or between enzyme activities and ascorbic acid concentration. Palmitoyl-CoA dehydrogenase was suggested to be useful as a blanching indicator (Baardseth, 1987). The temperature resistance of peroxidase was studied in cabbage, leeks, carrot, celery, spinach, squash, potatoes, onions, and green beans. Peroxidase activities of cabbage and green beans were very high. In contrast, onions had very little activity. Blanching of green beans, potatoes, and squash at 75°C was not adequate for inactivating peroxidase. Inactivation also was affected by the type of vegetable and the size of the vegetable pieces (Mueftuegil, 1985).

2. Blanching of Asparagus

The thermal stability of peroxidase in fresh asparagus tips was investigated by heating at 50, 60, and 70°C. Peroxidase inactivation followed a biphasic curve and the activation energy for the partially purified enzyme was 41.9 kcal/mol (Ganthavorn, 1991). Asparagus spears were blanched at 88°C for 1–4.5 min and stored at −18°C. Hexanal, ascorbic acid, free sulfhydryl, and peroxidase activity were monitored during storage. Blanching for 2 min reduced peroxidase activity by 98%. No hexanal was detectable in product blanched for 2 min or more. Ascorbate stability during storage increased with blanching time, but there was no direct relationship with peroxidase loss. Free sulfhydryl level was sensitive to heating and also decreased during storage of unblanched asparagus. However, there appeared to be a stable fraction of free sulfhydryls (Ganthavorn, 1988).

3. Blanching of Cauliflower and Cabbage

The residual peroxidase activity was determined in cauliflower after blanching and sulfur dioxide treatment. The organoleptic qualities were assessed after the cauliflower was stored for up to 1 yr in sealed polyethylene bags at −35°C. The quality of the blanched product correlated inversely with the peroxidase activity. Blanched cauliflower following a brief dip in metabisulfite solution prior to freezing gave a superior product, even when stored for 1 yr. The sulfur dioxide disappeared after 3 min of cooking in boiling water (Ramaswamy, 1989). The heat stability of purified spring cabbage peroxidase isoenzymes was determined. The anionic isoenzyme (pI 3.7) was relatively heat stable. The cationic isoenzyme (pI 9.9) was inactivated by heat more readily. Neither isoenzyme showed regeneration after heat inactivation (McLellan, 1987a).

E. Reduction of Peroxidase Activity by Other Methods

Reduction of peroxidase activity was determined in tomato, carrot, or eggplant extracts using the naturally occurring antioxidants quercetin, rutinic acid, chlorogenic acid, and α-tocopherol, alone or with heat treatment. Regeneration of enzyme at room temperature and after frozen storage ($-18°C$ for 4 wk) was tested also. Combining antioxidant (125 mg%) and heat (2 min at 75°C) resulted in almost complete inhibition of vegetable peroxidases. Regeneration of peroxidase in the extracts that were treated with heat and antioxidant was less than in extracts treated with heat or antioxidant alone (Hemeda, 1988,1991). These same antioxidants without heat were tested as inhibitors of peroxidase in tomato, carrot, or eggplant extracts. Four concentrations of each were tested: 125, 250, 500, and 1000 mg%. Peroxidase activity of carrot and eggplant was inhibited more effectively than that of tomato by all four antioxidants. α-Tocopherol was the least effective and chlorogenic acid was the most effective, followed by quercetin and rutinic acid (Hemeda, 1990). Gibberellic acid has been tested as a postharvest treatment for mango fruit to reduce peroxidase activity. Gibberellic acid at the rate of $100-200$ mg/liter significantly delayed ripening, reduced peroxidase and amylase activities, and retarded the loss of weight, chlorophyll, and ascorbic acid content (Khader, 1988).

F. Of Apple

Peroxidases have been associated with flavor deterioration of fruit juice during storage. Cox's Orange Pippin apples frequently are used in the production of apple juice. The heat stability and number of peroxidase isoenzymes in the peel and pulp were determined. Soluble and ionically bound peroxidases were extracted from the peel and pulp. Most activity was in the soluble fraction of the peel. Soluble peroxidases were cationic and anionic. The ionically bound peroxidases were solely cationic. Five isoenzymes were found by isoelectric focusing. The soluble peroxidase in pulp was most heat stable. A limited regeneration of activity was noticed after heating in all instances. pH optima were $5-6$. A commercial fresh apple juice contained slight peroxidase activity when assayed at pH 6.0 (Moulding, 1987). Apple polyphenol oxidase was inactivated by 98, 63, and 34% of the initial activity after 20 min at 25°C at adjusted pH values of 2.00, 2.25, and 2.50, respectively (Zemel, 1990). Apple peel peroxidase was purified 30-fold with 80% yield using hydrophobic chromatography followed by affinity chromatography on concanavalin A-substituted agarose. Isoelectric focusing showed the presence of at least four isoenzymes in the purified extract. The pI values ranged from $4.6-9.8$. The bulk or 99% of apple peel peroxidase activity was due to the cationic forms. Kinetic studies with guaiacol, chlorogenic acid, and (+)-catechin indicated that peroxidase followed Michaelian kinetics with a ping-pong mechanism and that it was inhibited by high substrate concentration. Apple peel peroxidases also were able to act on the main polyphenolic compounds of this fruit: catechins, hydroxycinnamate derivatives, and flavonol compounds. Apple peel peroxidase is believed to be involved in the enzymatic browning of apple fruit (Richard, 1989).

G. Of Orange

Flavedo peroxidase from Valencia Late oranges was purified by ammonium sulfate fractionation, gel filtration, and ion-exchange chromatography. The anionic fraction consisted of an apparently homogeneous isoenzyme with a pI of 2.5. An anionic fraction contained three isoenzymes with pIs from 9.5 to 10.5. Both fractions had similar specific activities with benzidine and indoleacetic acid; the anionic fraction had a considerably higher specific activity with eugenol and guaiacol (Catala, 1987).

H. Of Bean

Peroxidase activity was determined in haricot seed prior to germination, 5 days after germination, after 50 days of development, and in the mature succulent green bean. Activity increased during each of the stages until maturity, when it decreased from 190 to 55 mmol per mg protein per min. Blanching denatured 98% of the peroxidase, but 20% was restored when the seed was stored for 12 months at $-20°C$. Peroxidase was regenerated dramatically when the seed was held at room temperature. Freeze drying also inactivated the enzyme, but air-drying showed limited inactivation (Kermasha *et al.*, 1988). Haricot seed peroxidase was extracted and partially purified by ammonium sulfate fractionation. Optimum pH for activity was 5.4. When 4 mM hydrogen peroxide was added, the activity increased about 92-fold. At higher concentration of hydrogen peroxide, the activity was decreased. Carbonyl compounds resulting from peroxidase action were isolated as dinitrophenylhydrazones and purified by preparative thin layer chromatography. Subsequent analysis showed that acetone was the principle carbonyl compound (Kermasha and Metche, 1988).

Black beans *(Phaseolus vulgaris)* were studied to characterize the soluble peroxidase and to determine the effect of heat pretreatment and storage conditions on hardening of the bean and inactivation of enzyme. The enzyme had a pH optimum of $6.4-6.8$, and optimal temperature of $40-45°C$, and an activation energy of 1.19 kJ/mol. There was strong dependence on water activity and little sensitivity to Ca^{2+} ions. Thermal inactivation of peroxidase was facilitated with increasing moisture. When the peroxidase was inactivated by heat at $102°C$, there was a reduction of hardening. When the temperature was raised to $105°C$, there was a further reduction of hardening. Thus, heating only to inactivate peroxidase was not adequate to inactivate the hardening mechanism (Rivera, 1989).

I. Of Grape

Soluble and ionically bound peroxidase was determined in homogenized Ohane grapes. The soluble fraction contained 87% of the activity. Heat inactivation was nonlinear; about 90% of the activity was destroyed after 10 min at 80°C. There was no significant regeneration of activity. Using isoelectric focusing, six peroxidase isoenzymes with pIs from 3.5 to 9.8 were detected (Robinson, 1989). Peroxidase was extracted from grape berries and purified 124-fold by ammonium sulfate fractionation and column chromatography. Four anionic isoenzymes were detected. Estimated molecular mass was about 40 kDa. pH and temperature optima were 5.5 and 40°C. Partial heat inactivation occurred at 65°C in 2 min; complete inactivation was achieved at 85°C in 5 min. K_m was 5.5 and 16 mM for guiacol and hydrogen peroxide, respectively (Sciancalepore *et al.*, 1985).

J. Of Brussels Sprouts

Using gel filtration and ion-exchange chromatography, four peroxidase isoenzymes were isolated from extracts of Brussels sprouts. The isoenzymes differed in their substrate specificities and heat stability properties. Three showed biphasic thermal inactivation, whereas the fourth was relatively more heat labile and was inactivated in a more linear manner with time. Regeneration was not observed (McLellan, 1987b).

K. Of Barley

An endosperm-specific barley peroxidase was cloned, characterized, and expressed. A barley peroxidase (BP 1) was purified from mature barley grains. Using antibodies to this enzyme, a cDNA clone was isolated from a cDNA expression library. BP 1 is less than 50% identical to other sequenced plant peroxidases. Analyses of RNA and protein from aleurone, endosperm, and embryo tissue indicated maximal expression 15 days after flowering. High levels of enzyme were found only in the endosperm. BP 1 is not expressed in leaves (Rasmussen, 1991).

L. In Spices

Peroxidases are believed to contribute to the damage of spices by oxidizing malonaldehyde, a product of lipid oxidation. Malonaldehyde in the presence of Mn(II), dissolved molecular oxygen, and horseradish peroxidase is oxidized with consumption of O_2 in an apparent oxygenase reaction catalyzed by the peroxidase. However, results show that the peroxidase functions as a normal peroxidase. A reaction sequence is initiated by the autoxidation of malonaldehyde, generating a free radical that reacts with molecular oxygen, forming hydrogen peroxide that is acted on by the peroxidase to form more free radicals, causing more malonaldehyde oxidation, and additional oxygen consumption (MacDonald, 1989).

M. Of Potato

The cDNA of a highly anionic peroxidase from potato was cloned and sequenced. This peroxidase was strongly suggested to be involved in deposition of the aromatic domain of suberin. The deduced amino acid sequence of this anionic peroxidase showed considerable homology to other peroxidases. This anionic peroxidase was barely detectable 2 days after wounding and reached a maximal level 8 days after wounding. Using the cDNA for the anionic peroxidase as a probe, the mRNA for the enzyme was shown to be induced in suberizing potato. The maximum level was reached in 3 days (Roberts, 1988).

Many of the 1000 potato varieties cultivated throughout the world have similar organoleptic features, posing commercial problems for identification of different potato varieties. It has been proposed that peroxidase zymograms of tuber sap obtained after polyacrylamide gel electrophoresis at pH 7.9 could be used to identify the 50 potato varieties marketed in Spain. The peroxidase pattern obtained after electrofocusing in a pH 3.5–9.5 gradient could be used if the results are in doubt (Nieto, 1990). Yams also can be identified using isoelectric focusing of peroxidase and acid phosphatase isoenzymes. This technique was suitable for the positive identification of shoot culture materials of four yam species. This method was also suitable to identify somatic hybrids following protoplast fusion and the regeneration of hybrid plants from microcalluses (Twyford, 1990).

N. Of Cucumber

In cucumber, as in many other plants, peroxidases are a major ethylene-induced protein. Ethylene-induced peroxidases do not appear to play a major role in cell wall lignification, chlorophyll degradation, or disease resistance. In normal plants, an increase in peroxidase is associated with aging tissue. Since ethylene promotes aging, the induction of peroxidase is thought to be the consequence of aging (Abeles, 1990). From a cDNA library of ethylene-treated cucumber cotyledons, two cDNA clones encoding putative peroxidases were isolated using a synthetic DNA probe based on a partial amino acid sequence of a 33 kDa cationic peroxidase that was known to be induced by ethylene. DNA sequencing indicated that the two clones were derived from two closely related RNAs related to published plant peroxidase sequences. An increase in the level of mRNA is evident after 3 hr ethylene exposure and plateaus after 15 hr (Morgens, 1990). Antibodies to the pI-9 and pI-4 isoenzymes were used in a radial immunodiffusion assay to show that ethylene induced similar peroxidases in other cultivars of *Cucumis sativis* other species of *Cucumis* and other genera of Cucurbitaceae. Examination of ethylene-induced peroxidases by isoelectric focusing demonstrated the presence of a series of other peroxidases, most slightly acidic, with an isoelectric focusing pH of ~6. The data suggest that the induction of peroxidase isoenzymes during ethylene-induced senescence is a common response in this family of plants. It also appears that the acidic and basic peroxidases are highly conserved (Abeles, 1989). Peroxidase activity in fresh cucumber fruit was highest in the skin, followed by pericarp and, finally, carpel tissues. Its optimum pH was about 7. Using hydrogen peroxide with benzidine, guaiacol, p-phenylenediamine, o-phenylenediamine, o-dianisidine, or 3-amino-9-ethylcarbazole as substrate, more anodic isoenzymes were seen in the skin than in the pericarp or carpel. There was no reaction using these substrates in the absence of hydrogen peroxide. The high thermal stability of cucumber peroxidase is believed to play a role in the darkening of processed cucumber products (Miller, 1990). Mechanical stress also stimulates peroxidase activity in cucumber. Moderate and severe mechanical stress caused the appearance of a new slow-migrating peroxidase isoenzyme immediately after treatment. This enzyme disappeared after 24 hr of storage, then reappeared after 48 hr. Severe stress also stimulated the appearance of two additional moderate-migrating peroxidases 24 hr after treatment. Using peroxidase as an indicator, mechanical stress appears to induce accelerated aging of cucumber (Miller, 1989). Peroxidase activity was determined in tissues and brines of pickling cucumbers. Peroxidase activity was relatively unaffected by the first 3 days of brining, but decreased during fermentation and storage, and after processing. $CaCl_2$ reduced the loss of activity in tissue and brine during fermentation and storage. No peroxidase, catalase, or lipoxygenase activity was detected in cucumber extracts fermented by four different lactic acid bacteria (Buescher, 1987). Peroxidase activities were determined in commercially obtained pickle products. Large variations were found between products and sources of a given product. All sources of refrigerated dills, fresh pack dills, and spears had activity. Spear tissues located in the center and bottom portions of jars had the highest peroxidase activity (Buescher, 1986).

O. Of Pea

A simple method was developed to identify peroxidase isoenzymes in crude pea seed extract without the need for purification. Analytical thin layer isoelectric focusing on polyacrylamide gel resulted in high resolution of the isoenzymes. o-Dianisidine : H_2O_2 and 3-amino-9-ethylcarbazole : H_2O_2 were sensitive and specific sensing reagents. In all three cultivars, 18 isoenzymes were detected that appeared consistently (Lee, 1988). Residual peroxidase activity was determined in untreated and blanched peas to control the blanching of deep-frozen peas. Enzyme extracts were prepared by homogenizing samples in phosphate buffer at pH 6 and 4°C. Activity was measured using o-dianisidine : H_2O_2. These tests allow the rapid characterization of enzyme extracts relative to pH and conditions of blanching (Omran, 1989). Peroxidase isoenzymes were purified from green peas by ion-exchange chromatography. Three isoenzymes, one neutral and two cationic, were identified and characterized. A further study attempted to optimize blanching conditions by examining the effect of various blanching treatments on the activity of peroxidase, polyphenol oxidase, lipoxygenase, and catalase and by monitoring quality changes during storage at $-23°C$. Enzyme activity was concluded not to be a reliable indicator for the adequacy of blanching (Halpin, 1989). The combined effects on peroxidase activity of crude extracts of green peas of heat treatment and various antioxidants were evaluated. The main conclusions were that a combination of antioxidant [butylated hydroxyanisole (BHA), butylated hydroxytoluene (BHT), or α-tocopherol] and heat treatment was more effective in inactivating peroxidase than heat treatment alone. In the presence of antioxidant, the heat treatment time could be reduced, permitting better retention of heat sensitive nutrients and organoleptic qualities (Lee, 1989).

P. Of Tomato

Peroxidase in tomato extract was measured using 2,2'-azino-bis-(3-ethylbenzthiazol-6-sulfonic acid) (ABTS). This substrate is much superior to guaiacol because there was only one oxidation product and the molar extinction coefficient of the oxidation product was 31,100, six times higher than that of tetraguaiacol. Peroxidase activity measured with ABTS had a pH optimum of 3.3 when measured in tomato extract. The measurement could be completed in a much shorter time: 30 sec compared with 3 min using guaiacol (Arnao, 1990). In suberizing tomato *(Lycopersicon esculentum)* fruit, the mRNA showed induction and reached a maximum in 3 days. The data suggest that the induction of peroxidase by wounding is preceded by transcriptional activation of the peroxidase gene or by increased stabilization of the mRNA (Roberts, 1988). Activities of acidic and basic peroxidases from tomato fruit were determined at six ripening stages, from green to red-ripe fruits. The acidic and basic isoenzymes reached a maximum activity during the climacteric, at the pink stage. The relative increase of the basic peroxidase was much more pronounced (Rothan, 1989). Tomatoes were harvested at three stages of ripeness and three peroxidase isoenzymes were isolated. One was soluble and two were ionically bound. All three had molecular masses of about 46 kDa and optimal pH for activity of ~ 6 (Fils, 1985).

Q. Of Paprika

The relationship between pigment content, peroxidase activity, and sugar composition was determined in paprika *(Capsicum annuum* L.). The pods were ripened and treated afterward with ethylene. The skin was analyzed for glucose, fructose, sucrose, peroxidase, and pigment concentration. Results indicated that sugar content could be related to color and that the method of drying had a marked effect on all the characteristics. Peroxidase activity decreased as the pods turned red and decreased about 50% when dried at room temperature. Freeze drying produced inferior paprika that had a higher peroxidase content. Whether this was a causal relationship was not determined (Vamos-Vigyazo, 1985a). Analysis of samples taken from different points of the band driers used for large scale processing of paprika showed that paprika is damaged by current production methods. Heat damage caused the loss of pigment as well as a considerable loss of peroxidase activity and sugar content. The loss of color is believed to be a nonenzymatic process and caused by, among other factors, Maillard browning (Vamos-Vigyazo, 1985b). There were six isoenzyme bands in red pepper extracts and eight in green fruits (Park, 1990).

R. Of Peanut

Native cationic and anionic peroxidases from peanut cultures lost activity and became more susceptible to proteolysis when they were deglycosylated or their carbohydrates were oxidized with periodate. In the presence of tunicamycin, which inhibits glycosylation, the nonglycosylated form of the peroxidases was secreted. The oligosaccharides in peanut peroxidases may not be essential for their secretion but they are important in maintaining catalytically active conformation and stability (Hu, 1989). Cationic and anionic peanut peroxidases were purified to homogeneity, as shown by two-dimensional electrophoresis. The two isoenzymes appeared to be similar and only showed a small difference in their heme absorption maxima and specific activity. Antisera cross-reacted with both isoenzymes (Hu, 1990). The two major isoenzymes of peanut cationic peroxidase found in the medium of cultured cells were characterized (Van Huystee, 1990). A monoclonal antibody for use in a solid phase enzyme-linked immunosorbent assay (ELISA) specific for an anionic peroxidase from peanut suspension cell medium was developed. The antibody had high affinity and specificity for the anionic peroxidase, had only weak interaction with α-amylase, and had virtually no reactivity with other enzymes (Xu, 1990). An extensive review on the molecular aspects and biological functions of peanut peroxidases has been published (Van Huystee, 1991).

S. In Canola

In some instances, for example, in canola, peroxidase activity is desirable because it is directly related to degreening. Peroxidase bleaches chlorophyll in the presence of hydrogen peroxide and 2,4-dichlorophenol. Peroxidase activity in isolated thylakoids from degreening canola seeds was demonstrated. Activity is inhibited initially and then stimulated by sublethal freezing. Therefore, inhibition of peroxidase activity following sublethal freezing may be responsible for the degreening of seeds. The economic value of canola is increased when there are fewer green seeds (Johnson-Flanagan, 1990).

III. Lactoperoxidase

A. General Characteristics

Lactoperoxidase is a hemoprotein that catalyzes the oxidation of thiocyanate and iodide ions in the presence of hydrogen peroxide to produce highly reactive oxidizing agents. These products have a broad spectrum of antimicrobial effects. They can kill viruses, gram positive and gram negative bacteria, fungi, and probably also mycoplasmas and parasites. The molecular components that are oxidized in the cells are sulfhydryl groups, NADH and NADPH, and, under some conditions, aromatic amino acids. The major products responsible for these effects are the hypothiocyanate ion (OSCN⁻), hypothiocyanous acid (HOSCN), and iodine (I₂) (Pruitt, 1985). Lactoperoxidase is found in milk, saliva, tears, cervical mucous, and nasal glands. Human milk does not appear to contain lactoperoxidase (Tenovuo, 1985). However, a more recent study confirmed the presence of lactoperoxidase in human colostrum. The human enzyme resembles bovine lactoperoxidase B in immunoreactive properties and has an M_r of 80,000. The enzyme is a glycoprotein containing about 10% carbohydrate (Langbakk, 1989).

The lactoperoxidase in mammalian milk has a heme very tightly bound to a single polypeptide chain of M_r 78,000. Studies of the prosthetic group were carried out using the proton homonuclear Overhauser effect, a technique useful in studying paramagnetic proteins. Data indicate the presence of arginine and histidine residues in the distal pocket with stereochemistry resembling those in cytochrome C peroxidase and horseradish peroxidase (Thanabal, 1989). The interaction of thiocyanate ion and lactoperoxidase was studied using ^1H and ^{15}N nuclear magnetic resonance. The results were consistent with specific binding of SCN⁻ to the enzyme approximately 0.72 nm from the ferric ion in the heme group in the vicinity of the porphyrin periphery. Cyanide ion bound to the ferric ion at the sixth position and did not inhibit SCN⁻ binding to the enzyme. Iodide inhibited the binding of SCN⁻ and appeared to bind at the same site (Modi, 1989).

B. In Milk

In food processing, the most widely studied lactoperoxidase system is in bovine milk and milk products. The lactoperoxidase system is a potent bacteriostatic or bacteriocidal system that has been used to preserve raw milk without refrigeration. It inhibits cell multiplication, lactic acid production in milk, and oxygen uptake by resting cells. This system was studied by ^{15}N NMR and optical spectroscopy at different concentrations of thiocyanate (SCN^-) and H_2O_2, and different pH values in an attempt to identify the chemical species responsible for antimicrobial activity. Conditions giving optimal activity, that is, maximum inhibition of oxygen uptake by *Streptococcus cremoris,* were pH < 6.0 and equimolar concentrations of SCN^- and H_2O_2 (Modi, 1991). The use of equimolar concentrations of SCN^- and H_2O_2, in this instance, 0.25 mM, for optimum effect was also found by Wang (1987). The inhibitory power of the lactoperoxidase system also was investigated in milk on *Listeria monocytogenes* and *Staphylcoccus aureus.* The naturally occurring lactoperoxidase in milk was activated by the addition of 1.2 or 2.4 mM SCN^- and 0.3 or 0.6 mM H_2O_2 for *S. aureus* and *L. monocytogenes* studies, respectively. The lactoperoxidase system increased the predicted time to reach half the maximum attainable cfu/ml by 326 hr for *L. monocytogenes* at 10°C and 36 hr for *S. aureus,* also at 10°C (Kamau, 1990). Another study in milk indicated that the lactoperoxidase should be supplied from an exogenous source and that 0.4 mM H_2O_2 was necessary for activation (Mijacevic, 1989). A lactoperoxidase system consisting of lactoperoxidase (0.37 U/ml), KSCN (0.3 mM), and H_2O_2 (0.3 mM) delayed but did not prevent the growth of *L. monocytogenes* at 20°C in milk (Siragusa, 1989). In another study, it was recommended that the lactoperoxidase system in milk be reactivated periodically by adjusting SCN^- and H_2O_2 concentrations to 0.25 mM during storage of raw milk at 4°C prior to pasteurization (Martinez, 1988). Using the lactoperoxidase system in milk activated by the addition of 0.2 mM SCN^- and H_2O_2, it was found that the growth of different *Salmonella* serotypes decreased markedly in milk acidified to pH 5.3 (Wray, 1987). The lactoperoxidase system, when used in milk acidified to pH 5.5, caused a reduction of *Campylobacter jejuni* (Ewais, 1985). A glucose–glucose oxidase-activated lactoperoxidase system delayed the onset of exponential growth of *Salmonella typhimurium* in infant formula milk (Earnshaw, 1990). The preservative properties of the lactoperoxidase system were evaluated in another study on infant formula. The formula was reconstituted with lake water that contained a mixed flora population and incubated at 30°C for 72 hr. The lactoperoxidase system was activated by glucose–glucose oxidase. In the control sample, proliferation of *Enterococcus* and *Pseudomosas* spp. and *Enterobacteriaceae* occurred in 24 hr. In the presence of the lactoperoxidase system, these bacteria failed to grow in the first 24 hr. However, yeasts proliferated and *Rhodotorula glutinis* was the dominant species after 48 hr (Banks, 1985).

C. In Milk Products

The lactoperoxidase system also has been tested to reduce the numbers of viable microorganisms in milk products. Cottage cheese previously inoculated with *Pseudomonas fragi, P. fluorescens, E. coli,* and *S. typhimurium* was preserved by the lactoperoxidase system. The system was activated by H_2O_2 generated by glucose–glucose oxidase. Detectable levels of the test microorganisms were not present during the 21-day test period (Earnshaw, 1989). In French soft cheese, the lactoperoxidase system applied to the cheese surface led to the elimination of viable *L. monocytogenes* from cheeses previously inoculated with the organism (Denis, 1989). The manufacture of a pickled soft cheese (Domiati type) was not affected when the lactoperoxidase system was used to treat the milk prior to the making of cheese. The processing time was reduced by the lactoperoxidase system, which avoided the direct addition of NaCl to the milk; because the whey was unsalted it could be utilized more readily (Hefnawy, 1986). Cow, buffalo, and mixed milks were treated with SCN^- and H_2O_2 or heat treated at 90°C for 5 min, then refrigerated for 2 days or held at 25°C for 1 day before being inoculated with a 2% yogurt culture and incubated at 40°C for 48 hr. Rate of increase of acidity and decrease of pH was similar for all three treatments. During refrigerated storage for 7 days, there were no significant differences in chemical composition of the three yogurts (Mehanna, 1988). Some studies on the effects of the lactoperoxidase system on starter cultures used in the manufacture of cheeses also have been done. Thirteen strains of thermophilic lactic acid bacteria commonly used in the manufacture of cheese were tested for their sensitivity to the lactoperoxidase system by continuous pH monitoring. Inhibition was expressed as the difference in time to reach pH 5 between control and test treatments. *Lactobacillus acidophilus* and three strains of *L. bulgaricus* were inhibited in the presence of SCN^- and lactoperoxidase, which indicated their ability to produce H_2O_2. *Lactobacillus helveticus, Streptococcus thermophilus*, and one strain of *L. lactis* required an external source of H_2O_2 to cause inhibition. *Streptococcus faecium* and one strain each of *L. lactis* and *L. bulgaricus* were resistant to the lactoperoxidase system (Guirguis, 1987). In another study, a delay of 4.5 hr in coagulation was observed when a thermophilic cheese starter was inoculated into milk 2 hr after activation of the lactoperoxidase system. When milk was inoculated 6 hr after activation, there was no increase in the coagulation time but less acid was produced (Valdez, 1988).

IV. Catalase

A. General Characteristics

Catalase (EC 1.11.1.6) is a hemoprotein that catalyzes the dismutation of hydrogen peroxide into dioxygen and water. Several organic substances, especially ethanol, can act as hydrogen donors; ethanol is oxidized to acetaldehyde. Catalase isoenzymes with enhanced peroxidase activity were isolated from barley and mutant maize (Havir, 1990). A manganese catalase from *Lactobacillus lactis* was crystallized and its structure studied (Baldwin, 1991). In plants, the major function of catalase is believed to be the decomposition of potentially harmful hydrogen peroxide (Inamine, 1989). When seedlings were submitted to increasing water deficit, peroxidase and catalase activities were preserved or increased as the availability of water to plant tissue was decreased. In most species tested, the two hydrogen peroxide scavenging abilities vary in a complex way in relation to the degree of water stress (Badiani, 1990). The reader is referred to the excellent treatment of catalase by Scott (1975) for a review.

B. In Various Fruits and Vegetables

Catalase from the pericarp of green tomato fruit was purified to homogeneity as judged by silver staining of SDS gels. The M_r of the holoenzyme was 225 kDa and that of each subunit polypeptide 55 kDa. Catalase activity was optimum at pH 6.0–6.5; half-maximal activity was observed at pH 4.5 and 8.0. Tomato pericarp catalase was four times as sensitive to inactivation by silver nitrate as beef liver catalase. Tomato catalase also was more sensitive to protein-modifying agents (Inamine, 1989).

Catalase from potato tubers was purified to homogeneity. The purified enzyme had a low specific activity, about 3000 U per mg protein, and a marked tendency to form aggregates. SDS–polyacrylamide gel electrophoresis showed a single 56-kDa peptide. The apparent M_r of the holoenzyme was estimated to be 224 kDa. There was some homology between the N-terminal sequences of potato *(Ipomea batatas)* and bovine liver catalase (Beaumont, 1990). There were time dependent increases of catalase, superoxide dismutase, and α-tocopherol when potato tubers were stored for 40 wk at various temperatures. The increase of these antioxidants is believed to be in response to activated oxygen production (Spychalla, 1990).

Catalase also occurs in spices. 34 spices were tested for catalase activity. Catalase content was highest in basil, marjoram, and thyme. It was lowest in turmeric and cloves (Gerhardt, 1983).

C. In Milk

Catalases from milk have been isolated and characterized. The separation and purification of catalase from bovine milk has been performed rarely and much still remains to be understood concerning its structure and mechanism of action. A study by Ito (1990) answered some of these questions. This group of investigators made a catalase antibody in rabbit using crystalline bovine liver catalase as the antigen. This IgG antibody was used to prepare an affinity matrix that was able to adsorb bovine milk catalase. The purified protein was electrophoretically homogeneous. Approximately 7 mg catalase was obtained from 10 liter of milk. The enzyme had a specific activity of 328 U per mg protein, compared with a specific activity of 0.13 U per mg protein in bovine milk. Catalase distribution in goat milk was determined (Ito, 1982). Catalase isoenzymes from goat milk were isolated and purified. About 20% of the catalase was found in the cream and about 72% in the skim milk. Recovery was 17% in buttermilk and 55% in whey. The catalase from these two sources was purified 156- and 56-fold, respectively, with yields of 6% and 26%, respectively. On polyacrylamide gel electrophoresis, two bands were formed that had catalase activity. The apparent isoenzymes had different pH optima (Ito, 1984).

V. Sulfhydryl Oxidase

A. General Characteristics

Sulfhydryl oxidase catalyzes the oxidation of thiols to their corresponding disulfides using molecular oxygen as the electron acceptor. The reaction requires that two moles of thiol are consumed for each mole of oxygen. A FAD-activated sulfhydryl oxidase was isolated and identified from whey obtained from bovine skim milk (Kiermeier, 1970). The enzyme also has been identified in the spores of the fungus *Myrothecium werrucaria* (Mandels, 1956). More recently, the enzyme was isolated and characterized from *Aspergillus sojae* (Katkochin, 1986) and *Aspergillus niger* (de la Motte, 1987; Hammer, 1990).

B. Mammalian

Sulfydryl oxidase from bovine milk was isolated and characterized (Janolino and Swaisgood, 1975). Cream was separated from the skim milk by centrifugation. Chymosin was added to the skim milk to precipitate the casein and the precipitate was separated from the whey by centrifugation. The whey was saturated to 50% with ammonium sulfate and the precipitate collected by centrifugation. The precipitate was dissolved in phosphate buffer at pH 7.0 and dialyzed against the same buffer. The dialyzed enzyme was concentrated to about 3% protein and allowed to stand overnight at 4°C. The sedimented enzyme was centrifuged and the precipitate was dissolved in buffer, dialyzed against deionized water, and lyophilized. The sulfhydryl oxidase was purified further by gel filtration and differential centrifugation. The enzyme was purified about 3200-fold with a 41% recovery. The specific activity was 104 μmol oxygen consumed/min/mg N.

The reaction catalyzed by bovine milk sulfhydryl oxidase involves oxidation of sulfhydryl groups to the disulfide in the presence of molecular oxygen. The sulfhydryl groups in cysteine, glutathione, or even protein can be oxidized. Glutathione was chosen as the substrate because it is more stable than cysteine and reacts more quickly than protein-bound sulfhydryl groups. This enzyme can be assayed by three techniques: oxygen consumption, disappearance of sulfhydryl groups, and hydrogen peroxide production. The oxygen consumption method usually is used because of its greater sensitivity and ease of use. The enzyme exhibited a sharp peak at 35°C and had only 10% activity at the relatively low temperature of 50°C. It had an optimal pH of 7.0 and showed very little activity at pH 4.5 and 8.5. The K_m for glutathione was 9.0×10^{-5} at 35°C and pH 7.0.

Sulfhydryl oxidase is a glycoprotein (Janolino and Swaisgood, 1975) containing about 11% carbohydrate, consisting of 7.2% total hexose, 2.9% N-acetyl hexosamine, 0.4% sialic acid, and 0.44% fucose. Atomic absorption spectroscopy revealed the presence of 0.030% iron and 0.011% copper. Apparently, 1 gram-atom of iron may be required for two enzyme subunits, each having a molecular mass of about 45 kDa, as determined by gel electrophoresis. The molecular mass may be inaccurate since it was difficult to determine because of the tendency of the enzyme to form aggregates. As expected, 0.1 mM EDTA inhibited its activity almost 100%. When the EDTA-treated enzyme was assayed in the presence of 1 µM iron, the activity was restored almost completely. Cupric ion restored its activity about 50% and manganese ion about 40%. Cobalt, zinc, and molybdenum ions had little effect.

Only recently have data elucidating the physiological role of sulfhydryl oxidase become available. Evidence has shown that sulfhydryl oxidase may be involved in the interconversion of xanthine dehydrogenase to xanthine oxidase (Clare, 1981). Sulfhydryl oxidase was demonstrated to catalyze the refolding of reductively denatured ribonuclease (Janolino and Swaisgood, 1975; de la Motte, 1987; Hammer, 1990). Molecular oxygen did not appear to be responsible for the catalytic effect.

Chromatographically purified sulfhydryl oxidase was used to examine the refolding of reductively denatured ribonuclease. It was demonstrated that no readily diffusible oxygen species was responsible for the reaction (Clare, 1981). Reductively denatured ribonuclease A (0.040 µmol) was renatured in about 4 hr using sulfhydryl oxidase (22 µg protein) at pH 7.0. Controls using boiled enzyme, ferrous sulfate, or no enzyme required 24 hr to renature ribonuclease A completely.

The possible interaction of sulfhydryl oxidase with xanthine oxidase was first revealed when both enzymes were present in the same void volume from a BioGel A-150 M column (Clare, 1981). This result probably indicated that the enzymes were bound to each other because the xanthine oxidase has a much higher molecular mass (303 kDa) than sulfhydryl oxidase (90 kDa). A purified xanthine oxidase preparation from bovine milk that had been converted to the dehydrogenase by dithiothreitol was converted to xanthine oxidase by sulfhydryl oxidase. However, the xanthine oxidase lost 50% of its activity in about 4 hr when it was incubated with sulfhydryl oxidase. The assay method for xanthine dehydrogenase used NAD^+ as the cofactor; the oxidase was assayed in the absence of NAD^+ and the presence of oxygen. However, it was not demonstrated that the xanthine oxidase produced hydrogen peroxide and superoxide anion.

The rate of oxidation of reduced glutathione by sulfhydryl oxidase was increased 5-fold with horseradish peroxidase (Swaisgood, 1980). The degree of enhancement was directly proportional to the amount of peroxidase. Further, the enhancement was not affected by excess catalase. Therefore, it appears that some sort of direct interaction between sulfhydryl oxidase and peroxidase may occur because, on initiation of the reaction in the presence of peroxidase, the absorption of the peroxidase increased at 417 nm. The wavelength of the increase in the absorbance spectrum indicates that the species formed corresponds to HRP II, the one-electron reduced form of oxidized horseradish peroxidase. These observations are consistent with a direct transfer of an intermediate species of oxygen, presumably a hydroperoxy group, from sulfhydryl oxidase to horseradish peroxidase. Sulfhydryl oxidase activity also increased with the addition of peptide disulfides and oxytocin to assay mixtures. These results were interpreted as being caused by substrate binding to a second site on the enzyme. In the presence of oxidized glutathione, the expected substrate inhibition in the presence of reduced glutathione does not occur. The increased activity of the enzyme in the presence of some disulfides may be due to competitive binding at an allosteric binding site for the substrate. Sulfhydryl oxidase was isolated from bovine milk in a form sufficiently pure for sequencing. Although precautions were taken during the electrophoretic elution to prevent reactions that cause N-terminal blocking, negative results when sequence analysis was attempted indicated that the N terminus is blocked (Janolino, 1990).

C. Microbial

Sulfhydryl oxidase from *Aspergillus sojae* was purified from about 500 U/liter in the fermetation broth to about 2000 U/g (Katkochin, 1986). The enzyme was produced by aerobically culturing the microorganism in a media containing cerelose, inorganic nitrogen, and essential minerals. During growth, the pH was prevented from falling below 3.5 by the addition of alkali, the preferred pH being 4.5–6.0. When a suitable biomass was attained, the culture was harvested, the cells were removed from the whole broth by centrifugation, and the cell-free broth was filtered using filter aid. The broth was concentrated by ultrafiltration followed by further concentration by diafiltration with a membrane with a 10,000 molecular weight cut-off. When the intracellular and extracellular sulfhydryl oxidases were recovered, the overall recovery was 40%. The sulfhydryl oxidase preparation contained an acid and a neutral proteinase. These were removed by chromatography using BioRad P100 resin.

The activity profile of sulfhydryl oxidase from *A. sojae* appears to be similar to that of the bovine milk enzyme. In both instances, there is conversion of two disulfide groups in the presence of molecular oxygen to the corresponding disulfide with the production of hydrogen peroxide. With reduced glutathione set at 100% in an oxygen depletion assay, only L-cysteine ethyl ester showed appreciable activity of about 20%. Five other sulfhydryl compounds had activities of less than 10% compared with reduced glutathione. The enzyme exhibited a pH optimum of 6.5, a temperature optimum of 42°C, and a very sharp decrease in activity above this temperature. The enzyme showed relatively good stability to about 40°C. Above this temperature, the enzyme activity decreased sharply and at 60°C there was virtually no activity. Unlike the bovine milk enzyme, the sulfhydryl oxidase from *A. sojae* does not appear to require metal cofactors because it is not inhibited by EDTA and cyanide.

The sulfhydryl oxidase from *A. niger* was characterized and found to be a flavin enzyme (de la Motte and Wagner, 1987; Hammer, 1990). Stoichiometric measurements showed a relationship of 1.7 mol flavin/mol protein. This sulfhydryl oxidase is of interest because it requires two reducing substrate molecules. Most flavin oxidases derive two reducing equivalents from one substrate molecule.

The procedure for the isolation and purification of *A. niger* sulfhydryl oxidase involved methanol precipitation, ammonium sulfate precipitation, gel filtration, acetone precipitation, hydroxyapatite chromatography, and DEAE fractionation. The specific activity increased from 7.5 to 115 units/mg protein. Recovery was about 36%.

Sedimentation equilibrium experiments indicated a molecular mass of about 106 kDa and SDS-polyacrylamide gel electrophoresis indicated two subunits, each with a molecular mass of 53 kDa. The enzyme was a glycoprotein containing about 20% hexose and 2% aminohexose. The amino acid profile was unremarkable. The flavin component could be separated from the protein by boiling for 5 min. The apoenzyme, when treated with 5 M guanidine chloride, could be reconstituted to about 60% of the original activity when incubated with excess FAD. Incubation with excess FMN regained no activity.

The rate of reactivation of reductively denatured ribonuclease A was enhanced by sulfhydryl oxidase, demonstrating that the enzyme could oxidize sulfhydryl groups associated with protein molecules. Glutathione substrate had the lowest K_m (0.3). Other thiol-containing substrates—cysteine, dithiothreitol, and 2-mercaptoethanol—had much higher K_m values: 43, 66, and 340, respectively. There was no activity with substrates of other oxidases. The substrates tested were D-glucose, L-lysine, D-lysine, benzylamine, xanthine, putrescine, catechol, NADH, and NADPH. With GSH as the substrate, pH optimum was between 5.0 and 5.5. There was no activity at pH 2.5 and very little at pH 10.0.

VI. Glucose Oxidase

A. Failure in the Food Industry

Glucose oxidase (EC 1.1.3.4) has not lived up to its promise as an antioxidant in the food industry. It was modified by treatment with succinic anhydride to be much more acid stable. In fact, it had a half-life over 24 hr at pH 2.4 at room temperature (Wagner, 1986). One of its major applications was seen to be the stabilization of orange soft drinks against oxidative deterioration. This application failed because of the reversible inhibition of the glucose oxidase by the dyes in the orange soft drink below pH 3.0. There is considerable precedent for the formation of complexes between various chemical agents and the bound flavin moieties of flavoproteins (Wagner, 1986). Another important reason for the lack of success in the food industry has been its inactivation by hydrogen peroxide, one of its products, when the enzyme is complexed with glucose (Malikkides, 1982).

B. Of Aspergillus niger

The production of glucose oxidase in *A. niger* was enhanced by treatment with mutagens (Markwell, 1989). A number of mutants of *A. niger* affected in glucose oxidase expression are described. The overproducing mutants could be classified into seven groups (Swart, 1990). Conidia of *A. niger* strain G 13 were subjected to four stage mutagenization using different combinations of mutagens. Each stage was treated with uv radiation and nitrosomethylurea, ethyleneimine, acryflavin, or N-methyl-N'-nitrosoguanidine. In all, 400 of 5208 colonies isolated after mutagenization had higher glucose oxidase activity. The 9 most active showed an activity increase of almost 30% (Fiedurek, 1990).

The formation and location of the enzyme was studied also in *A. niger* that was pregrown under citric acid-producing conditions. Glucose oxidase could be induced by shifting the pH of the medium from 1.7 to 5.5. The induction required either glucose or glucose-6-phosphate. Rapid secretion into the medium of oxidase produced in this fashion by a mechanism other than autolysis was surprising, because in *A. niger* glucose oxidase is usually an intracellular enzyme and must be recovered from the mycelium by grinding or lysis. Histochemical evidence for the peroxisomal location of glucose oxidase in *A. niger* would lead one to propose an intracellular origin of gluconic acid (Van Dijken, 1980). When O_2-enriched air was used to increase the dissolved oxygen concentration of an *A. niger* fermentation, the glucose oxidase activity increased about 3-fold. At the same time, the production of gluconic acid increased about 2-fold. The enzyme was produced in both endocellular and exocellular forms (Traeger, 1991).

Glucose oxidase from *A. niger* was cloned from cDNA and genomic libraries using oligonucleotide probes. Two yeast expression plasmids containing the entire glucose oxidase coding sequence, the untranslated region, and about 2 kb flanking gt10 sequences were transformed into yeast. One plasmid contained a yeast alcohol dehydrogenase II–glyceraldehyde-3-phosphate dehydrogenase promoter, glucose oxidase presequence, glucose oxidase coding sequence, and glyceraldehyde-3-phosphate dehydrogenase terminator inserted into the *Bam* HI site of the yeast–*E. coli* shuttle vector pAB24. The second plasmid had the yeast α-factor leader substituted for the glucose oxidase presequence. The transformed yeast was grown in yeast extract peptone medium containing 4% glucose for 144 hr in shaker flasks at 300 rpm. When grown under derepressing conditions, transformants of these plasmids in yeast secrete large amounts of active glucose oxidase into the medium. Comparison of the yeast-derived glucose oxidase with purified A. niger glucose oxidase by polyacrylamide gel electrophoresis showed that the yeast glucose oxidase migrated more slowly than the A. niger enzyme. When the yeast enzyme was treated with endoglycosidase H, it migrated with about the same mobility as the A. niger glucose oxidase. Thus, it appears that the yeast enzyme has more extensive N-linked glycosylation. The yeast-derived enzyme had a turnover number of $17,000-20,000$ min^{-1} compared with $16,200$ min^{-1} for the wild-type enzyme. Other kinetic parameters for both the yeast-derived and wild-type enzymes were similar. The hyperglycosylated yeast enzyme may be more thermostable than the wild-type glucose oxidase (Frederick, 1990). The primary structure of an O-linked sugar chain derived from the glucose oxidase of A. niger was examined. Mannitol appeared to be the only O-linked monosaccharide detected. The O-linked sugar chains were mainly present as single mannose residues. These short O-glycosidically linked oligosaccharide units contained mannosyl-serine and mannosyl-threonine linkages (Takegawa, 1991). The A. niger glucose oxidase gene was also cloned into A. niger, Aspergillus nidulans, and Saccharomyces cerevisiae. Self-cloning of the gene into A. niger increased enzyme production; cloning into A. nidulans conferred a novel capacity to produce glucose oxidase. Cloning the gene into S. cerevisiae used a plasmid in which the enzyme gene was preceded by a yeast secretion signal. This effected the high-level production of extracellular glucose oxidase in the transformed yeast (Whittington, 1990). Monoclonal antibodies to glucose oxidase from A. niger were made using apoenzyme as the antigen. Five of these antibodies, all of the IgG1 subisotype, were characterized. Glucose oxidase lost its immunlogical reactivity with the antibodies under direct covalent immobilization. The carbohydrate moiety of the enzyme is not immunogenic. None of the antibodies had any detectable effect on the catalytic properties of the enzyme and all five antibodies bound the nonnative enzyme coated on ELISA plates in preference to the enzyme in solution (Wimalasena, 1990).

C. *Of* Penicillium

Aglyco- and glycoglucose oxidases from *Penicillium amagasakiense* were compared. *P. amagasakiense* secreted the aglyco version when grown in the presence of 20 mg/liter of tunicamycin. The carbohydrate content of the purified aglyco version was 2.9% and the molecular mass of it and its subunit were 130 kDa and 70 kDa, respectively. The glycosylated version had 20.4% carbohydrate and molecular mass of 150 kDa and 70 kDa. Both versions showed remarkably similar properties. For the aglyco and glyco versions, K_M values of

3.4 and 2.7 mM, V_{max} values of 320 and 270 U/mg protein, and pH optima of 5.5 and 6.0 were observed, respectively. Isoelectric focusing also showed similar patterns. Thermostability was also similar. It was concluded that the carbohydrate moiety plays no significant role in the functioning of glucose oxidase from *P. amagasakiense* (Kim and Schmid, 1991). Glucose oxidase from *P. amagasakiense* was purified by fast protein liquid chromatography in combination with ion-exchange chromatography. Elution conditions were optimized to permit the simultaneous purification and separation of the glucose oxidase isoforms. Elution was performed with a mixed pH and salt gradient. Three peaks, each consisting of 1–2 isoforms, were resolved with a very flat linear gradient in 40 ml elution buffer. Three more peaks were eluted at 10, 30, and 100% elution buffer (Kalisz, 1990). Glucose oxidase apoenzyme from *Penicillium chrysogenum* was purified by chromatography on Con A–Sepharose 4B (Krysteva, 1990).

D. Of Talaromyces flavus

Talaromyces flavus produces a glucose oxidase that may be involved in the control of the fungal plant pathogen *Verticillium dahliae*. The extracellular glucose oxidase was isolated and purified and found to be a glycoprotein. Amino acid analysis indicated that it was similar to other fungal glucose oxidases. It was stable from pH 3.0 to 7 and the optimum pH for activity was 5.0 (Kim *et al.*, 1990).

VII. Pyranose Oxidase

Pyranose oxidase or glucose-2-oxidase (EC 1.1.3.10) oxidizes the C-2 atoms of D-glucose, D-xylose, D-sorbose, and D-glucono-1,5-lactone to form their respective pyranosones. Hydrogen peroxide is formed as a by-product. It was studied in the 1980s as a more economical way to produce D-fructose from D-glucose. The production of D-fructose from D-glucose by glucose isomerase suffers from a drawback because the isomerization reaction provides an equilibrium mixture typically containing only 42–50% D-fructose. Pyranose oxidase typically oxidizes 95% of D-glucose to glucosone. D-Fructose can be prepared from the glucosone using catalytic hydrogenation (Horwath, 1986). Pyranose oxidase free from glucosone-utilizing impurities was obtained from strains of *Coriolus versicolor, Lenzites betulinus,* and *Polyporus obtusus*. The enzyme can be used to make pure glucosone for use in making high purity food sugars, for example, fructose, mannitol, and sorbitol. When pyranose oxidase is prepared from *P. obtusus,* it is a tetramer with a molecular mass of about 200 kDa with a subunit mass of about 69 kDa. The enzyme is a flavo-protein with a sedimentation coefficient of 11.4. Its relative activities toward glucose, sorbose, xylose, and δ-gluconolactone substrates are

100:80:48:12, respectively, at pH 5.0. The enzyme can be used to make D-glucosone from D-glucose, D-xylosone from D-xylose, 5-keto-D-fructose from L-sorbose, or a mixture of 2-keto-D-gluconic acid and D-isoacscorbic acid from δ-D-gluconolactone (Koths, 1986a,b,c). Pyranose oxidase can be stabilized against heat inactivation partially by amidinating it with either dimethyl adipimidate or dimethyl suberimidate. The stabilized enzyme has about three times the heat stability of the native enzyme at 65°C (Shaked, 1986). When the organisms are exposed to a mutagenic agent such as uv radiation, higher amounts of pyranose oxidase can be produced (Ring, 1986). A new basidiomycete that produces pyranose oxidase has been isolated. Pyranose oxidase also can be used for the quantitative determination of glucose without the need for mutarotase, which might be needed by glucose oxidase if the reaction is expected to go to completion in a reasonable period of time. The enzyme also is useful for assaying 1,5-anhydroglucitol, a marker for the diagnosis of diabetes (Izumi, 1990).

VIII. Xanthine Oxidase

A. General Characteristics

Xanthine oxidase (EC 1.1.3.22) is an iron–molybdenum flavoprotein that can oxidize xanthine, hypoxanthine, some other purines and pterins, and aldehydes. It is unique among oxidases because it can produce superoxide anion (O_2^- as well as H_2O_2. Superoxide dismutase dismutates two molecules of O_2^- to H_2O_2 and water.

B. In Milk

Xanthine oxidase is found in milk and liver. This section addresses xanthine oxidase in milk and dairy products primarily. About 200 commercially processed dairy products were assayed polarographically for xanthine oxidase activity. Average activity of 22 of 25 fluid milks ranged from 4.5 to 139.6 mU per liter O_2 g^{-1} hr^{-1}. Most of the dried milks and evaporated milk products and all the butters had no activity. Average activity of ice creams and yogurts ranged from 0 to 40.3 mU; 50% of the 24 brands of creams had activity from 4.2 to 405 mU. About two-thirds of 111 brands and varieties of cheeses had activities ranging from 0 to 420 mU. Activity of fresh raw milk increased 3.7-fold in storage at 4°C and declined when stored for longer periods (Zikakis, 1980a). An earlier study showed that raw milks had an average xanthine oxidase activity of 110 mU/ml. Homogenization and pasteurization inactivated more than 50% of the enzyme and, when packaged, averaged 34 mU/ml. Raw creams had activities from 153 to 319 mU/ml (Cerbulis, 1977). Much of the xanthine oxidase in bovine milk is localized in the lipid membrane of fat globules. Low speed agitation (5000 rpm for 20 sec) of washed fat globules added to skim milk released xanthine oxidase (Bhavadasan, 1982). Effects of cold storage (5°C for 24 hr) and heat treatment (60°C for 5 min) on free and membrane xanthine oxidase were studied. Both treatments increased total xanthine oxidase activity in milk. In buttermilk, activity of membrane xanthine oxidase increased and free xanthine oxidase decreased with both treatments (Bhavadasan, 1980). When bovine milk was heated at 70°C for 30 sec, xanthine oxidase activity generally increased, 3- to 4-fold in some samples. Pasteurization at 85 or 90°C, however, almost completely inactivated the enzyme (Zmarlicki, 1978).

Xanthine oxidase was isolated and purified from raw bovine milk. The milk was treated with 1–2% Triton X-100, followed by a two-step fractionation using ammonium sulfate. The enzyme was then purified by chromatography on Sephadex G-75, G-100, and Sepharose 6B columns. Final purification was by chromatography on a DEAE–Sephadex A-50 column eluted with a continuous salt gradient in 0.005–0.1 M phosphate buffer at pH 8.6. The purified enzyme exhibited a single peak on gel chromatography or polyacrylamide electrophoresis. The enzyme had an average protein to flavin ratio of 2.0:4.1 (Zikakis, 1976,1979a,1980b). The effect of isotopic substitution of hydrogen-8 of xanthine (with ^{2}H and ^{3}H) on the rate of oxidation by bovine xanthine oxidase was measured. Results suggested the presence of a rate-limiting step prior to the irreversible C–H bond cleavage step (Ardenne, 1990). Xanthine oxidase was converted to xanthine dehydrogenase by 1% 2-mercaptoethanol. This conversion was inhibited partially by 6 M urea. When purified from fresh bovine milk, the enzyme had a specific activity of 3.59 IU/mg, contained 14.8% protein N, had no lipid, was hydrophobic in nature, and had lysine as the N-terminal residue. It was indicated that xanthine oxidase could not be the source of milk riboflavin (Cheng, 1984). Xanthine oxidase purified from bovine milk had a flavin protein spectral absorbance ratio of 5.0, indicating that it was a highly purified protein. Blood serum from a rabbit hyperimmunized against this enzyme fraction exhibited two precipitation lines. The second antigen was identified as bovine IgG. Commercial preparations of xanthine oxidase also contained IgG as a contaminant (Clare, 1991). When xanthine oxidase was prepared from raw bovine milk in the presence of dithioerythritol, 94% of its xanthine oxidase was found as a dehydrogenase type. The enzyme was converted to an oxidase type when dithioerythritol was removed (Nakamura, 1982). Xanthine dehydrogenase also can be converted to the oxidase form with sulfhydryl oxidase (Clare, 1981). When xanthine oxidase was purified by limited proteolysis, structural alterations results in diminished stability, suboptimal substrate binding, and reduced catalytic efficiency (Zikakis, 1979b). Freshly prepared enzyme isolated from the fat globule membrane using deoxycholate yielded one zone on SDS-polyacrylamide gel electrophoresis corresponding to a M_r of 153 kDa. After storage at 4°C for 30 days, there were three zones. It was suggested that an endogenous milk proteinase, possibly of membrane origin, was causing this breakdown. After exposure of the enzyme to trypsin the enzyme retained full activity, although selective cleavage of the enzyme was occurring, as shown by gel electrophoresis. However, after 24 hr of digestion, the enzyme retained only 24% of its activity and the only protein present had an M_r of 92 kDa (Mangino, 1976).

The influence of various milk proteins on lipid oxidation in a related model system was assessed with purified milk metalloproteins on the oxidation of triglyceride emulsions. Xanthine oxidase had little effect on lipid oxidation in the absence of added metals, but was strongly prooxidative in the presence of 10 mM Cu^{2+} (Allen, 1982).

IX. Lipoxygenase

A. General Characteristics

Plant lipoxygenases have been known for over 50 years. Several have been purified and mechanistic details have been elucidated. The first mammalian lipoxygenase, 12-lipoxygenase from human platelets, was described in 1974 (Rokach, 1989). Lipoxygenases play an important role in the stability of various processed foods. All lipoxygenases use molecular oxygen (dioxygen) to catalyze the oxidation of polyunsaturated fatty acids and initially form fatty acid peroxy free radicals that remove a hydrogen from another unsaturated fatty acid molecule to produce hydroperoxides. The hydroperoxides are converted into acids, aldehydes or ketones, and other compounds (Matheis, 1987a). Lipoxygenases also can undergo an anaerobic reaction. Soybean lipoxygenase-1 can use the hydroperoxide product as its second substrate. Although the mechanism of the reaction is simple, a complex pattern of products is formed: fatty acid dimers, oxodienoic acids, and n-pentane (Vliegenthart, 1982). Lipoxygenases also show the phenomenon of self-inactivation. The instabilities of the lipoxygenases of pea seeds and wheat were confirmed in recent studies. In the kinetic study of the reaction of soybean lipoxygenase with arachidonic acid as substrate, a self-catalyzed destruction of the enzyme was observed that was involved in the interaction of the enzyme with the hydroperoxy fatty acid in only a minor way (Schewe, 1986). Several types of lipoxygenases are classified by the Enzyme Commission. The lipoxygenase that oxidizes linoleate to 13-hydroperoxyoctadeca-9,11-dienoate has been assigned the Enzyme Commission number 1.13.11.12. Other lipoxygenases that have been studied are arachidonate 12-lipoxygenase (EC 1.13.11.31), arachidonate 15-lipoxygenase (EC 1.13.11.33), and arachidonate 5-lipoxygenase (EC 1.13.11.34). Many investigators have suggested that the physiological roles of lipoxygenases in plants might be lipid oxidation with respect to seed aging and death, regulation of metabolism in germinating seeds, production of plant hormones, and inhibitory effects on pathogenic organisms. These effects are not clearly understood. Defense against insect attack was studied for the lipid oxidation products of soybean lipoxygenases. Bean bugs were repelled by products of linoleic acid oxidation such as linoleic acid monohydroperoxides and hexanal. The insects also preferred lipoxygenase null seeds more than normal soybean seeds. The oxidation products also repelled two species of leaf beetles that usually do not feed on soybean seeds. Although the repellant powers of the lipoxygenase products were not great, the results suggested a defensive role for lipoxygenases (Mohri, 1990). More recently, lipoxygenase was implicated in carotenoid metabolism and synthesis of abscisic acid in *Lycopersicon* and *Phaseolus* seedlings. *In vitro*, lipoxygenase cleaved neoxanthin and violaxanthin, a widely distributed carotenoid pigment formed in plants from zeaxanthin, to small ($\leq C_{13}$) fragments. Perhaps *in vivo*, any apocarotenoids formed by the specific cleavage of 9'-*cis*-neoxanthin during abscisic acid synthesis are metabolized rapidly by lipoxygenase or similar enzymes (Parry, 1991).

B. Of Soybean

In the plant kingdom, the most widely studied lipoxygenase is soybean lipoxygenase-1. This enzyme is a metalloenzyme with a nonheme iron as its prosthetic group. The iron is bound directly to the polypeptide backbone and is not readily removed by iron chelators such as bipyridyl and O-phenanthroline. The M_r of the enzyme is 98.5 kDa (Vliegenthart, 1982). Soybean isoenzymes lipoxygenase-1, -2a, -2b, and -2c were examined spectroscopically for the presence of covalently bound pyrroloquinone quinone (PQQ) after derivatization by phenylhydrazine, 2,4-dinitrophenylhydrazine, and 3-methyl-2-benzothiazolinone. None of these compounds showed evidence for PQQ, contrary to a previous report (Michaud-Soret, 1990). Alkyl free radicals from the β-scission of fatty acid alkoxyl radicals were detected by spin trapping in a soybean lipoxygenase system in borate buffer and linoleic acid, linolenic acid, or arachidonic acid (Chamulitrat, 1990).

Soybean lipoxygenase has been used since the 1930s to bleach flour to produce white bread crumb. Improvement of dough-forming properties and baking performance of wheat flour using soybean lipoxygenase to oxidize free lipids in the flour is well known. When the free lipids are removed from the flour, this effect is diminished greatly (Matheis, 1987b).

The stability of foods made from soybeans are affected by their lipoxygenase content. In addition to the oxidation of polyunsaturated fatty acids, soybean lipoxygenase has been reported to cooxidize fat-soluble vitamins. Although vitamins A and D do not occur naturally in plant tissues, they may be present in foods containing animal products. Lipoxygenase-catalyzed oxidation of linoleic acid can be accompanied by considerable losses of vitamins A, D_2, D_3, and E at neutral pH (Gordon, 1989). Cooxidation of β-carotene was also reported at pH 7.4 (Barimalaa, 1988). Chlorophyll A was bleached by soybean lipoxygenase L1. It was concluded that linoleinyl radicals were responsible (Abbas, 1988).

Among the methods studied to prevent the deleterious effect of lipoxygenase on soy-containing foods are inhibitors, inactivators, and heat denaturation. Dithiothreitol, EDTA, $CaCl_2$ and ascorbic acid decreased lipoxygenase activity but did not eliminate it (Monma, 1990). Lipoxygenase activity in soy milk was decreased by propyl gallate, which inhibited the enzyme more than BHA, BHT, and ascorbic acid. Citric acid and ascorbic acid enhanced this inhibition by propyl gallate (Vijayvaragiya, 1991). Turmeronol A and turmeronol B, phenolic sesquiterpene ketones derived from turmeric, inhibited soybean lipoxygenase at IC_{50} values of 16 and 9 μM, respectively. These compounds prevented the autooxidation of linolenic acid at a concentration of about 200 ppm (Imai, 1990). The prior incubation of soybean lipoxygenase with arachidonic acid resulted in a gradual decrease of activity. Preincubation with linoleic acid or 11,14,17-eicosatrienoic acid had no effect on activity. A lipoxygenation product of arachidonic acid, 15(S)-hydroperoxyeicosatetraenoic acid, irreversibly inhibited soybean lipoxygenase in a time-dependent manner (Kim, 1990). Inactivation of soybean lipoxygenase at acid pH was evaluated by grinding whole beans in water acidified to pH 3.0. Lipoxygenase was found to be inactivated completely and irreversibly at or below this pH. The acid treatment slightly reduced the functional properties of soy proteins produced from the beans. Protein solubility and extractibility were reduced but emulsifying ability was not affected. Emulsion stability was decreased and viscosity was increased. Water binding capacity also decreased (Ali, 1989). Various acids, without heat, were used to inactivate soybean lipoxygenase with little effect on protein solubility. An ancillary benefit of the acid treatment was the reduction of urease activity to acceptable levels (Che Man, 1989a). Lipoxygenase-1 was stable when preincubated at pH 3.2–9.2, but was irreversibly denatured at or below pH 3.0 (Asbi, 1989). The prevention of the "beany" and bitter flavor in soy full-fat flour, caused by lipoxygenase, during milling reportedly was prevented by a simple blanch-dry process. Soybeans were dehulled and blanched in water at 80–99°C. Water blanching for 3 min at 99°C was necessary when the cotyledons were subsequently sun dried (Seth, 1988). Other workers found that the presence of soybean solids stabilizes lipoxygenase against heat inactivation (Ostergaard, 1988).

Breeding programs to eliminate lipoxygenases in soybean seeds is another technique to attempt to reduce the detrimental effects of lipoxygenases in processed food products. Lipoxygenase isoenzymes are localized in the cytoplasm of cotyledons of germinating soybean seeds. Lipoxygenase appears to be absent from the mitochondria, glyoxysomes, protein bodies, lipid bodies, and other organelles (Song, 1990). A simple test to detect lipoxygenase isoenzymes using the preferred substrate in a coupled cooxidation and destruction of β-carotene was devised to assay for the three isoenzymes, LOX1, LOX2, and LOX3, in soybean seeds. A simple test is necessary to determine the progress of genetic manipulations (Ecochard, 1990).

Lipoxygenase-free soybean seeds were used to study accelerated aging compared with seeds containing lipoxygenase. When stored at 40°C, 100%

relative humidity, for 1–10 days, germination data indicated that the absence of lipoxygenase did not increase resistance of seeds to accelerated aging (Wang, 1990). The storage stability of lipoxygenase-free full-fat soy flours were compared with those with lipoxygenase (19–21 U per mg) activity. Free fatty acids, peroxide value, lipoxygenase activity, N solubility index, and β-carotene content were evaluated. Lipoxygenase-free flour stored better than flour with lipoxygenase (Seth, 1989). The role played by lipoxygenase in flavor quality of soybean oil was tested by comparing oil processed from special soybeans lacking lipoxygenase-1 with oil from normal beans. No significant differences were found in flavor quality or stability based on total volatiles. 2,4-Decadienal levels were also similar. Factors other than lipoxygenase-1 appear to affect the food quality of soybean oils (Frankel, 1988).

Soybean lipoxygenase may have some important applications in genetic engineering. The regulatory sequence of the lipoxygenase gene is expressed vigorously in soybean germination. This sequence may be useful for directing high levels of expression of heterologous genes in transgenic plants (Shibata, 1990). The preparative synthesis of linoleic acid 13(S)-hydroperoxide using soybean lipoxygenase-1 has been reported (Iacazio, 1990). Immobilized lipoxygenase of soybean was used to prepare 13-hydroxy-γ-linolenic acid. The enzyme was immobilized on chitosan; product yield may reach 70% at pH 9.0 (Nakahara, 1990).

C. Of Potato

Potato arachidonate 5-lipoxygenase was purified and characterized (Reddanna, 1990; Shimizu, 1990). Potato tuber lipoxygenase preparations converted α-linolenic acid not only to 9(S)-hydroperoxy10(E),12(Z),15(Z)-octadecatrienoic acid but also to more polar metabolites (Grechkin, 1991). Potato 5-lipoxygenase also appeared to have peroxidase-like activity. It catalyzed the reduction of 12(S)-hydroperoxy-9(Z),11(E)-octadecadienoic acid in the presence of vitamin E (Cucurou, 1991). Hydroxyperoxy acids functioned as suicide substrates and irreversibly inactivated potato lipoxygenase. Preincubation of lipoxygenase in a hemoglobin solution abolished this inhibitory effect. The results indicate that substrates that can be converted to an epoxide intermediate act as suicide substrates (Kim, 1989).

The thermal inactivation kinetics of two lipoxygenase isoenzymes were studied in solution. Each of the two isoenzymes followed first-order kinetics with different inactivation rate constants. However, the apparent inactivation kinetics of crude and reconstituted lipoxygenases showed second-order kinetics. This was explained by the presence of two isoenzymes with different thermal properties (Park, 1988).

With the purpose of exploring the reasons for enhanced destruction of carotenoids in irradiated potato tubers during storage at 15°C, changes in carotenoids and lipoxygenase activity were measured during storage at various temperatures. Irradiation enhanced the disappearance of carotenoids in potatoes stored at 15 and 20°C and reduced its formation at 4 and 25–30°C. Irradiation at a sprout-inhibiting dose of 100 Gy caused an immediate decrease in lipoxygenase activity as well as in its capacity to cooxidize β-carotene. The reduced lipoxygenase activity of irradiated tubers was seen throughout their storage, regardless of the temperature. Irradiated tubers also had decreased protein content and increased levels of peptides and amino acids. There appeared to be no relationship between lipoxygenase activity and carotenoid destruction in irradiated potato tubers (Bhushan, 1990).

D. Of Tomato

A microsomal membrane-associated lipoxygenase from tomato was characterized. It appears to use the free fatty acids released from phospholipids. The microsomal enzyme is active over a pH range of 4.5–8.0, comprises over 38% of the total (microsomal plus soluble) lipoxygenase activity in the tissue, has an apparent K_m of 0.52 mM, and a V_{max} of 0.186. The membranous enzyme also cross-reacts with polyclonal antibodies raised against soybean lipoxygenase-1 and has an apparent molecular mass of 100 kDa. The reaction oxidizing linoleic acid was inhibited strongly by N-propyl gallate (Todd, 1990). Lipoxygenase was extracted from tomato by an ammonium sulfate technique. Crude and partially purified preparations were stable for 1 wk at 40°C and pH 7. With linoleic acid as the substrate, the enzyme had a K_m of $0.37 \times 10^{-5} M$ and a pH optimum of 4.0–4.5. The effects of ammonium sulfate and EDTA on the enzyme suggested that natural inhibitors are separated during purification and trace metals are needed for activity (Daood, 1988). A large amount of 5-lipoxygenase can be extracted from the callus of tomato. The callus is induced by culturing the cut tomato stem in a medium containing nutrients, hormones, and vitamins. 5-Lipoxygenase may be useful in the production of leukotrienes (Hatamoto, 1990).

E. Of Maize

Maize seeds, after 5 days of germination, contain at least two lipoxygenase isoenzymes, L1 and L2. These were extracted from acetate-buffered acetone powder containing 0.1% Brij 99, a nonionic detergent. Both isoenzymes had a broad pH optimum: 6.0–8.2 for L1 and 7–9 for L2. The pI of L1 was 6.40; isoelectric focusing of L2 resulted in two peaks with pI values of 5.55 and 5.70. When L1 and L2 were heated at various temperatures for 5 min, 50% residual activity remained at 45 and 65°C, respectively. Substrate specificity was tested using linoleic, α- and γ-linolenic, and arachidonic acids. Using initial velocity determinations, linoleic acid was the preferred substrate for both isoenzymes. However, apparent K_m and V_{max} values showed that linolenic acids were the best substrates for L1 whereas linoleic acid was best for L2 (Poca, 1990).

F. Of Strawberry

In strawberry homogenates, lipoxygenase oxidized unsaturated fatty acids in water but not in chloroform/methanol. The differences between ripe and unripe fruit can be detected by measuring the lipoxygenase content. 10-Hydroxyoctadeca-8,12-dienoic acid and 10-hydroxyoctadeca-8-enoic acid were detected in large quantities in the aqueous homogenates. These compounds are toxic to fungi and may be involved in protecting the plant from pathogenic fungi (Gorst-Allman, 1988).

G. Of Canola Seed

Lipoxygenase from canola seed was purified using ammonium sulfate at 20–50% saturation followed by conventional anion-exchange chromatography using DEAE–cellulose. Only a single active fraction appeared. The pH for optimal activity was 7.5; the apparent K_m was 2×10^{-4} when linoleic acid was the substrate. The partially purified enzyme had a considerably higher activity with free linoleic acid than with linoleic acid esters. The enzyme also showed a high reactivity with the lipids extracted from canola (Khalyfa, 1990). Rapeseed free of erucic acid has both lipoxygenase-1 and -2 activities; there was great variability of the lipoxygenase activities of different rapeseed samples (Meshehdani, 1990).

H. Of Bean

Two lipoxygenase isoenzymes from French bean pericarp were separated and characterized. The two isoenzymes differed in molecular weight, pI, apparent K_m, and lag phase. Both could be characterized as lipoxygenase-2 because of their acidic optimum pH, simultaneous production of carbonyl compounds and hydroperoxides in the presence of oxygen, aerobic chlorophyll bleaching ability, and inhibition by chlorophyll a. In the presence of preformed hydroperoxides, both isoenzymes were inhibited competitively by chlorophyll a (Abbas, 1989). The effect of cyanide on green bean lipoxygenase, in both the crude and purified hydrophobic forms, was studied to determine if this method could be used to reduce the blanching treatments used in the vegetable industry. The results indicated that cyanide, probably in the undissociated acid form, is a weak competitive inhibitor of green bean lipoxygenase. It is possible that other nontoxic N-containing compounds also inhibit lipoxygenase (Adams, 1989). Green bean lipoxygenase was purified in an attempt to determine its role in food quality deterioration. Lipoxygenase was extracted from fresh and unblanched frozen beans using buffer and subsequently fractionated with ammonium sulfate. The crude enzyme was purified by hydrophobic interaction chromatography followed by isoelectric focusing, ion-exchange chromatography, and gel filteration chromatography. The results suggest three forms of green bean lipoxygenase that can be separated by virtue of their ionic charge and hydrophobicity (Adams and Ongley, 1989).

The combined effect of microwave heat treatment and irradiation on lipoxygenase in fresh and dry faba beans was studied. Faba beans were irradiated at 500 krad, then microwaved at 190°C (1500 Watt, 2540 MHz) for various lengths of time. When dry beans were irradiated for 4 min followed by microwaving for 2 min, the lipoxygenase activity was destroyed completely. The inactivation of lipoxygenase was faster in fresh beans because of their higher moisture content (Al-Fayadh, 1988).

I. Of Other Plants

Lipoxygenases from other plant sources have been studied also. The enzyme was purified from wheat kernels using ammonium sulfate precipitation and DEAE–Sephadex A-50 anion-exchange chromatography. Arachidonic acid was converted mainly to $5\Delta8$-hydroperoxy-6(E),8(Z),11(Z),14(Z)-eicosatetraenoic acid. Moreover, evidence was obtained for leukotriene A4 synthase activity of the wheat lipoxygenase (Heydeck, 1991).

γ-Irradiation at doses of 4 and 6 kGy reduced the lipoxygenase activity of intact buckwheat seeds (Henderson, 1990).

Avocado lipoxygenase was purified to near homogeneity by affinity chromatography. The pH and temperature optimum were 7.1 and 36°C. The K_m for linoleic acid was 7.2×10^{-2} mM and V_{max} was 432 mU per hr per mg protein. Epicatechin was a competitive inhibitor with a K_i of 9.0×10^{-5} mM (Marcus, 1988).

The thermal inactivation of asparagus lipoxygenase was studied by Ganthavorn (1991). Inactivation followed first-order kinetics and activation energies for thermal denaturation of partially purified enzyme was 47.5 kcal/mol.

Lipoxygenase heterogenity in two species of pea was demonstrated by antibody testing. At least seven different polypeptides were identified; five of these were identified as lipoxygenase precursors. Limited N-terminal sequence data indicated further heterogeneity when compared with sequences predicted from cDNAs (Domoney, 1990). Cloning and sequencing of two cDNAs from mRNA of maturing pea seeds allowed the deduction of the complete amino acid sequence of a lipoxygenase polypeptide that is very similar to that of soybean lipoxygenase-2. The predicted molecular mass of this polypeptide is about 97 kDa, and its sequence permits comparisons of lipoxygenase-2 and -3 isoforms from pea and soybean (Ealing, 1989).

J. Of Fish

Lipoxygenases are also widely distributed in the animal kingdom. In teleost fish, arachidonic 12-lipoxygenase is the predominant lipoxygenase. An arachidonic acid 15-lipoxygenase was found in gill tissue of teleost fish and is believed to help control the quality and characteristics of volatile compounds that are implicated in fish odor and flavor. Total activity of this enzyme following hydroxyapatite chromatography was significantly higher than in the crude tissue preparation, suggesting that inhibitors were removed. The enzyme was active toward polyunsaturated fatty acids present in the tissue, producing hydroxylated products from fatty acids with 18-, 20-, and 22-carbon chain lengths at Carbons 13, 15, and 17. The enzyme had a molecular mass of 70 kDa (German, 1990). The arachidonic acid 12-lipoxygenase from trout gill tissue effectively bleaches β-carotene in conjunction with the peroxidation of different polyunsaturated fatty acid substrates. The maximum velocity of bleaching was higher for trout lipoxygenase than for arachidonic acid 15-lipoxygenase from soybeans. This difference may reflect the presence of oxidative cofactors in the trout gill lipoxygenase preparation (Stone, 1989). Trout gill arachidonic acid 12-lipoxygenase can initiate oxidation of polyunsaturated fatty acids to produce acyl peroxides. Using a gill homogenate as a model system, the ability of gill lipoxygenase depended on concentration of the enzyme and substrate and length of time. The major volatile compounds produced from the oxidation of arachidonic and eicosapentaenoic acids were 1-octen-3-ol, 2-octenol, 2-nonanal, 2-nonadienal, 1,5-octadiene-3-ol, and 2,5-octadien-1-ol. Formation of oxidative flavors was related to the activity of the lipoxygenase and was inhibited by esculetin and BHA (Hsieh, 1989). The arachidonic 12-lipoxygenase of rainbow trout gill tissue was active at pH 7–9 and was inactivated rapidly at temperatures higher than 40°C. Inhibition studies showed that inhibition was generally noncompetitive and that the most potent inhibitors were fisetin and quercetin (Hseih, 1988a). The presence and relative activities of arachidonic acid 12-lipoxygenase were investigated in 15 species of freshwater fish. All the species contained lipoxygenases capable of initiating the production of off-flavors by the formation of highly unstable hydroperoxides, which appear to be the direct precursors of the off-flavors. Gill lipoxygenase of rainbow trout was reactive with arachidonic, eicosopentaenoic, and docosahexaenoic acids, but had low reactivity with linoleic acid. Optimum activity was at pH 7.5 and inactivation occurred at temperatures higher than 40°C. No lipoxygenase activity was found in muscle, but activity was found in the skin of all the species, although levels varied among species. Glutathione enhanced the stability of gill lipoxygenase, indicating that thiol groups are involved in the activity of the enzyme (Hseih, 1988b). Lipoxygenase activity also was detected in sardine skin. A crude lipoxygenase from sardine was partially characterized. Its substrate specificity was elucidated; the enzyme was the most active on α-linolenic acid (113%) when linoleic acid was set at 100% (Mohri, 1990).

K. Of Chicken Meat

Lipoxygenase-type enzymes were shown to be present in chick muscle. Examination of the oxidation products of [^{14}C]-arachidonic acid revealed the presence of 15-lipoxygenase. The enzyme was purified partially by affinity chromatography on linoleoyl–aminoethyl Sepharose. The enzyme was stable for 12 months at $-20°C$. Lipoxygenases were suggested to play a role in some of the oxidative changes occurring in fatty acids while chicken meat is in frozen storage (Grossman, 1988).

X. Dehydrogenases

A. General Characteristics

In the Enzyme Nomenclature (1984) listing of oxidoreductases that act on the CH-OH group of donors (EC 1.1), 199 dehydrogenases using NAD$^+$ or NADP$^+$ as acceptors (EC 1.1.1) and 3 using cytochrome as acceptor (EC 1.1.2) are listed. These oxidoreductases are categorized mostly as dehydrogenases, although 37 are categorized as reductases. This class of enzyme is very important to the metabolic functions of plants and animals but, in foods or food processing, only a few might be considered important, that is, recently mentioned in the literature. These enzymes are alcohol dehydrogenases, shikimate dehydrogenases, malate dehydrogenases, isocitrate dehydrogenases, and lactate dehydrogenases.

B. In Various Plants

In peach, 18 dehydrogenase isoenzyme systems were examined. Among these were 3 isocitrate dehydrogenases (EC 1.1.1.41), 3 malate dehydrogenases (EC 1.1.1.37), and 2 shikimate dehydrogenases (EC 1.1.1.25) (Mowrey, 1990).

A shikimate dehydrogenase from cucumber was extracted and purified 7-fold by precipitation with ammonium sulfate and elution from columns of Sephadex G-25, DEAE–cellulose, and hydroxyapatite. Two activity bands were detected on polyacrylamide gel electrophoresis at the last purification step. The enzyme was inhibited competitively by protocatechuic acid. Heat inactivation at $50-55\,^{\circ}C$ was biphasic. NADPH or shikimic acid protected the enzyme against heat inactivation (Lourenco, 1991).

An NAD-dependent alcohol dehydrogenase was isolated and purified from parsley leaves. The M_r of the enzyme was 37 kDa by SDS-polyacrylamide gel electrophoresis and 68 kDa by Sephadex G-100. The enzyme was assumed to consist of two equal-sized subunits. The pH optimum was 8.9; K_m and V_{max} values for acetaldehyde were 0.6 mM and 279 U/mg, respectively (Liang, 1990).

The nucleotide sequence of an alcohol dehydrogenase gene in octoploid strawberry was deduced (Wolyn, 1990).

Studies at the International Rice Research Institute on the N-assimilating isoenzymes in rice showed that the shikimate dehydrogenase high-mobility alloenzyme (Sdh-12) was associated with higher protein accumulation. The Sdh-12 homozygotes had a significantly higher accumulation of oryzenin. Since oryzenin is the major component of seed storage proteins in rice, the Sdh-12 alloenzyme can be an effective marker in screening for higher seed protein content among different cultivars (Shenoy, 1990).

A rapid and economical method was developed to ascertain hybrid purity of tomato by testing for two variants of alcohol dehydrogenase. In the cultivated tomato, two electrophoretic variants of alcohol dehydrogenase are encoded by two different alleles that are present in the desired F_1 hybrid, open-pollinated varieties, other inbred lines, or commercial hybrids. One of the alcohol dehydrogenases is encoded by an allele in the F_1 hybrid. Therefore, testing for this enzyme will give an indication of the purity of the hybrid (Van den Berg, 1991).

A watermelon mitochondrial malate dehydrogenase (EC 1.1.1.37) cDNA was sequenced and cloned (Gietl, 1990).

In barley, alcohol dehydrogenase and lactate dehydrogenase were induced by anaerobiosis in both aleurone layers and roots. Under aerobic conditions, developing seeds accumulate alcohol dehydrogenase activity that survives seed drying and dehydration. Activity of lactate dehydrogenase also increases during seed germination, but the activity in dry or rehydrated seeds is very low. Developmental expression of alcohol and lactate dehydrogenase was monitored from 0 to 24 days postgermination. Neither activity was induced to any extent in germinating seeds; however, both enzymes were highly induced by anoxia in root tissue during development. Based on gel electrophoresis results, this increase in activity was due to the differential expression of different *Adh* and *Ldh* genes in root tissue. The increase in activity seems to be due to the *de novo* synthesis of these two proteins (Good, 1989). The alcohol dehydrogenase gene family of barley was cloned and analyzed to address the question of gene homology between barley and maize systems directly. Alleles of each of the three loci specifying alcohol dehydrogenase in barley were isolated by molecular cloning and analyzed. Using the criteria of DNA and amino acid homology with the maize *Adh* genes, two of the genes (*Adh2* and *Adh3*) have been designated as maize *Adh2*-like and the third was identified as an *Adh1* mutant allele (Trick, 1988).

C. Meat Lactate Dehydrogenase

Lactate dehydrogenase may be important in meat research. There is a relationship between the lactate dehydrogenase isoenzyme pattern and different types of muscle fibers. Differences exist in muscle lactate dehydrogenase activity and in isoenzymes among different species of animal (Hamm, 1990,1991).

References

Abbas, J., Rouet-Mayer, M. A., Tremolieres, A., and Philippon, J. (1988). Bleaching of chlorophyll A by soy lipoxygenase 1. Effect of various O2 conditions. *Sci. Alim.* **8**, 83–96.

Abbas, J., Rouet-Mayer, M. A., and Lauriere, C. (1989). Lipoxygenase isoenzymes of French bean pericarp: Separation, characterization and chlorophyll-bleaching ability. *Phytochem.* **28**, 1019–1024.

Abeles, F. B., Biles, C. L., and Dunn, L. J. (1989). Hormonal regulation and distribution of peroxidase isoenzymes in the Cucurbitaceae. *Plant Physiol.* **91**, 1609–1612.

Abeles, F. B., Biles, C. L., and Dunn, L. J. (1990). Induction of peroxidase as a response to environmental stimuli. *Monogr. Br. Soc. Plant Growth Regul.* **20**, 199–215.

Adams, J. B. (1989). Inhibition of green bean lipoxygenase by cyanide. *Food Chem.* **31**, 243–250.

Adams, J. B., and Ongley, M. H. (1989). The behaviour of green bean lipoxygenase on chromatography and isoelectric focusing. *Food Chem.* **31**, 57–71.

Al-Fayadh, M. H., and Al-Baldawi, A. M. H. (1988). Combined effect of microwave heat treatment and irradiation on trypsin inhibitors and lipoxygenase in fresh and dry faba beans. *Iraqui J. Agric. Sci.* **6**, 137–146.

Allen, J. C., and Wrieden, W. L. (1982). Influence of milk proteins on lipid oxidation in aqueous emulsion. II. Lactoperoxidase, lactoferrin, superoxide dismutase and xanthine oxidase. *J. Dairy Res.* **49**, 249–263.

Allen, M. P., Choo, S. H., Li, T. M., and Parrish, R. F. (1991). Inactivation of animated horseradish peroxidase by interaction with S-Sepharose. *Anal. Biochem.* **192**, 453–457.

Araujo, P. J. (1985). Cell wall degradation of tomato and potato tissue as influenced by oxidative metabolism and acid conditions. *Diss. Abs. Inter.* **46**, 1–150.

Ardenne, S. S., and Edmondson, D. E. (1990). Kinetic isotope effect studies on milk xanthine oxidase and on chicken liver xanthine dehydrogenase. *Biochemistry* **29**, 9046–9052.

Arnao, M. B., Casas, J. L., del Rio, J. A., Canovas, F. G., Acosta, M., and Sabater, F. (1990). Use of 2,2'-azino-bis-(3-ethylbenzthiazol-6-sulphonic acid) (ABTS) for the measurement of peroxidase activity in tomato fruit extracts. *Rev. Agro. Tecnol. Aliment.* **30**, 333–340.

Asami, S., Amano, N., Amachi, T., and Yoshizumi, H. (1986). Process for producing peroxidase. European Patent Office No. 0179486.

Asbi, B. A., Wei, L. S., and Steinberg, M. P. (1989). Effect of pH on the kinetics of soybean lipoxygenase-1. *J. Food Sci.* **54**, 1594–1595.

Baardseth, P., and Slinde, E. (1987). Enzymes and off-flavours: palmitoyl-CoA hydrolase, lipoxygenase, α-oxidation, peroxidase, catalase activities and ascorbic acid content in different vegetables. *Norweg. J. Agric. Sci.* **1**, 111–117.

Badiani, M., De Biasi, M. G., Colognola, M., and Artema, F. (1990). Catalase, peroxidase and superoxide dismutase activities in seedlings submitted to increasing water deficit. *Agrochim.* **34**, 90–102.

Bajaj, K. L., Anan, T., Tsushida, T., and Ikegaya, K. (1987). Effects of (−)-epicatechin on oxidation of theaflavins by polyphenol oxidase from tea leaves. *Agric. Biol. Chem.* **51**, 1767–1772.

Baldwin, E. (1991). Crystallization and structural studies of *Lactobacillus plantarum* manganese catalase. *Diss. Abs. Inter.* **51**, 1–339.

Banks, J. G., and Board, R. G. (1985). Preservation by the lactoperoxidase system (LP-S) of a contaminated milk formula. *Lett. Appl. Microbiol.* **1**, 81–85.

Barimalaa, I. S., and Goedon, M. H. (1988). Cooxidation of β-carotene by soybean lipoxygenase. *J. Agric. Food Chem.* **36**, 685–687.

Bartonek-Roxa, E., Eriksson, H., and Mattiasson, B. (1991). The cDNA sequence of a neutral horseradish peroxidase. *Biochim. Biophys. Acta* **1088**, 245–250.

Beaumont, F., Jouve, H. M., Gagnon, J., Gaillard, J., and Pelmont, J. (1990). Purification and properties of a catalase from potato tubers *(Solanum tuberosum)*. *Plant Sci.* **72**, 19–26.

Ben-Arie, R., and Or, E. (1986). The development and control of husk scald on Wonderful pomegranate fruit during storage. *J. Am. Soc. Hort. Sci.* **111**, 395–399.

Beumer, R. R., Noomen, A., Marjis, J. A., and Kampelmacher, E. H. (1985). Antibacterial action of the lactoperoxidase system on *Campylobacter jejuni* in cow's milk. *Neth. Milk Dairy J.* **39**, 107–114.

Bhavadasan, M. K., and Ganguli, N. C. (1980). Free and membrane-bound xanthine oxidase in bovine milk during cooling and heating. *J. Dairy Sci.* **63**, 362–367.

Bhavadasan, M. K., Abraham, M. J., and Ganguli, N. C. (1982). Influence of agitation on milk lipolysis and release of membrane-bound xanthine oxidase. *J. Dairy Sci.* **65**, 1692–1695.

Bhushan, B., and Thomas, P. (1990). Effects of gamma irradiation and storage temperature on lipoxygenase activity and carotenoid disappearance in potato tubers. *J. Agric. Food Chem.* **38**, 1586–1590.

bin Ali, A. (1989). Kinetics of acid inactivation of soybean lipoxygenase and its effects on the functional properties of soy protein. *Diss. Abs. Inter.* **50**, 1–162.

British Biotechnology Limited Catalogue (1989).

Buescher, R. W., and McGuire, C. (1986). Peroxidase activities in cucumber pickle products. *J. Food Sci.* **51**, 1079–1080.

Buescher, R. W., McGuire, C., and Skulman, B. (1987). Catalase, lipoxygenase, and peroxidase activities in cucumber pickles as affected by fermentation, processing, and calcium chloride. *J. Food Sci.* **52**, 228–229.

Catala, C., and Chamarro, J. (1987). Partial purification and properties of anionic and cationic flavedo peroxidase isoenzymes from Valencia Late oranges. *Rev. Agro. Tecnol. Aliment.* **27**, 509–518.

Cerbulis, J., and Farrell, H. M., Jr. (1977). Xanthine oxidase activity in dairy products. *J. Dairy Sci.* **60**, 170–184.

Chamulitrat, W., and Mason, R. P. (1990). Alkyl free radicals from the β-scission of fatty acid alkoxyl radicals as detected by spin trapping in a lipoxygenase system. *Arch. Biochem. Biophys.* **282**, 65–69.

Chang, B. S., Park, K. H., and Lund, D. B. (1988). Thermal inactivation kinetics of horseradish peroxidase. *J. Food Sci.* **53**, 920–923.

Che Man, Y. B., Wei, L. S., and Nelson, A. I. (1989b). Acid inactivation of soybean lipoxygenase with retention of protein solubility. *J. Food Sci.* **54**, 963–967.

Che Man, Y. B. (1989a). Acid inactivation of soybean lipoxygenase and its effects on protein solubility. *Diss. Abs. Inter.* **49**, 3523–3524.

Cheng, G. S. G. (1984). Studies on xanthine oxidase in cow's milk. *Diss. Abs. Inter.* **44**, 1–163.

Clare, D. A., and Lecce, J. G. (1991). Copurification of bovine milk xanthine oxidase and immunoglobin. *Arch. Biochem. Biophys.* **286**, 233–237.

Cucurou, C., Battioni, J. P., Daniel, R., and Mansuy, D. (1991). Peroxidase-like activity of lipoxygenases: different substrate specificity of potato 5-lipoxygenase and soybean 15-lipoxygenase and particular affinity of vitamin E derivatives for the 15-lipoxygenase. *Biochim. Biophys. Acta* **1081**, 99–105.

D'Angiuro, L., Galliani, S., and Cremonesi, P. (1987). Thermal stability of immobilized horseradish peroxidase (HRP) in water-organic solvent systems. "Biocatalysis in Organic Media." Elsevier, Amsterdam.

Daood, H., and Biacs, P. A. (1988). Some properties of tomato lipoxygenase. *Acta Aliment.* **17**, 53–65.

de la Motte, R. S., and Wagner, F. (1987). *Aspergillus niger* glucose oxidase. *Biochemistry* **26**, 7363–7371.

Denis, F., and Ramet, J. P. (1989). Antibacterial activity of the lactoperoxidase system on *Listeria monocytogenes* in trypticase soy broth, UHT milk and French soft cheese. *J. Food Protect.* **52,** 706–711.

Dev Choudhury, M. N., and Bajaj, K. L. (1984). Chemical nature of pectic substances in tea shoots and their effect on polyphenol oxidase activity. *Two and a bud.* **31,** 59–65.

Domoney, C., Firmin, J. L., Sidebottom, C., Ealing, P. M., Slabas, A., and Casey, R. (1990). Lipoxygenase heterogeneity in *Pisum sativum. Planta* **181,** 35–43.

Dunford, H. B. (1991). Horseradish peroxidase: structure and kinetic properties. *Peroxidases Chem. Biol.* **2,** 1–24.

Ealing, P. M., and Casey, R. (1989). The cDNA cloning of a pea *(Pisum sativum)* seed lipoxygenase. Sequence comparisons of the two major pea seed lipoxygenase isoforms. *Biochem. J.* **264,** 929–932.

Earnshaw, R. G., Banks, J. G., Defrise, D., and Francotte, C. (1989). The preservation of Cottage cheese by an activated lactoperoxidase system. *Food Microbiol.* **6,** 285–288.

Earnshaw, R. G., Banks, J. G., Francotte, C., and Defrise, D. (1990). Inhibition of *Salmonella typhimurium* and *Escherichia coli* in an infant milk formula by an activated lactoperoxidase system. *J. Food Protect.* **53,** 170–172.

Ecochard, R., Maltese, S., Dayde, J., and Planchon, C. (1990). Detection of lipoxygenase isoenzymes in soybean seeds *(Glycine max. L. Merrill):* use of a simple test for breeding purposes. *Oleagineux* **45,** 333–336.

Fiederuk, J., and Ilczuk, Z. (1990). Intensification of glucose oxidase synthesis with *Aspergillus niger* by the way of multistage mutagenization. *Acta Biotechnol.* **10,** 371–376.

Fils, B., Sauvage, F. X., and Nicolas, J. (1985). Tomato peroxidases purification and some properties. *Sci. Aliment.* **5,** 217–232.

Flurkey, W. H. (1989). Polypeptide composition and amino-terminal sequence of broad bean polyphenoloxidase. *Plant Physiol.* **91,** 481–483.

Flurkey, W. H., and Ingebrigtsen, J. (1989). Polyphenol oxidase activity and enzymatic browning in mushrooms. Am. Chem. Soc. Symposium Series; Quality factors in fruits and vegetables: *Chem. Technol.* **405,** 44–54.

Frankel, E. N., Warner, K., and Klein, B. P. (1988). Flavor and oxidative stability of oil processed from null lipoxygenase-1 soybeans. *J. Am. Oil Chem. Soc.* **65,** 147–150.

Frederick, K. R., Tung, J., Emerick, R. S., Masiarz, F. R., Chamberlin, S. H., Vasavada, A., Rosenberg, S., Chakraborty, S., Schopter, L. M., and Massey, V. (1990). Glucose oxidase from *Aspergillus niger.* Cloning, gene sequence, secretion from *Saccharomyces cerevisiae,* and kinetic analysis of a yeast derived enzyme. *J. Biol. Chem.* **265,** 3793.

Fujiyama, K. I., and Takano, M. (1990). Genomic DNA structure of two new horseradish peroxidase-encoding genes. *Gene* **89,** 163–169.

Ganthavorn, C., and Powers, J. R. (1988). Changes in peroxidase activity, hexanal, ascorbic acid and free sulfhydryl in blanched asparagus during frozen storage. *J. Food Sci.* **53,** 1403–1405.

Ganthavorn, C., Nagel, C. W., and Powers, J. R. (1991). Thermal inactivation of asparagus lipoxygenase and peroxidase. *J. Food Sci.* **56,** 47–49.

Garrote, R. L., Luna, J. A., Silva, E. R., and Bertone, R. A. (1987). Prediction of residual peroxidase activity in the blanching-cooling of corn-on-the-cob and its relation to off-flavor development in frozen storage. *J. Food Sci.* **52,** 232–233.

Gerhardt, U., and Naumann, W. (1983). Occurrence of enzymes, especially catalase, in spices. *Fleischerei* **34,** 508–510.

German, J. B., and Creveling, R. K. (1990). Identification and characterization of a 15-lipoxygenase from fish gills. *J. Agric. Food Chem.* **38,** 2144–2147.

Gietl, C., Lehnerer, M., and Olsen, O. (1990). Mitochondrial malate dehydrogenase (MDH, EC 1.1.1.37) from watermelon *(Citrullus vulgaris):* sequence of cDNA clones and primary structure of the higher-plant precursor protein. *Plant Mol. Biol.* **14,** 1019–1030.

Golan-Goldhirsch, A., and Whitaker, J. R. (1984). Effect of ascorbic acid, sodium bisulfite, and thiol compounds on mushroom polyphenol oxidase. *J. Agric. Food Chem.* **32,** 1003–1009.

Good, A. G., and Crosby, W. L. (1989). Induction of alcohol dehydrogenase and lactate dehydrogenase in hypoxically induced barley. *Plant Physiol.* **90,** 860–866.

Gordon, M. H., and Barimala, I. S. (1989). Co-oxidation of fat soluble vitamins by soybean lipoxygenase. *Food Chem.* **32**, 31–37.

Gorst-Allman, C. P., and Spiteller, G. (1988). Investigation of lipoxygenase-like activity in strawberry homogenates. *Z. Leben. Unter. Forsch.* **187**, 330–333.

Grechkin, A. N., Kuramshin, R. A., Safonova, E. Y., Efremov, Y. I., Latypov, S. K., Ilyasov, A. V., and Tarchevskii, I. A. (1991). Double hydroperoxidation of α-linolenic acid by potato tuber lipoxygenase. *Biochim. Biophys. Acta* **1081**, 79–84.

Grossman, S., Bergman, M., and Sklan, D. (1988). Lipoxygenase in chicken muscle. *J. Agric. Food Chem.* **36**, 1268–1270.

Guirguis, N., and Hickey, M. W. (1987). Factors affecting the performance of thermophilic starters. II. Sensitivity to the lactoperoxidase system. *Austr. J. Dairy Technol.* **42**, 14–16.

Halpin, B. E. (1989). Purification and characterization of peroxidase isoenzymes from green peas *(Pisum sativum)* and relation of enzyme activity to quality of frozen peas. *Diss. Abs. Inter.* **49**, 1–163.

Hamm, R. (1990). Lactate dehydrogenase and its importance in meat research. II. Physiological function of the isoenzymes, the structure of lactate dehydrogenase, and distribution and binding of lactate dehydrogenase in muscle cells. *Fleischwirtsch.* **70**, 1336–1339.

Hamm, R. (1991). Lactate dehydrogenase and its importance in meat research. III. Dependence of lactate dehydrogenase on muscle fibre type and muscle development; effects of animal-specific factors; the importance of muscle activity and innervation. *Fleischwirtsch.* **71**, 102–106.

Hammer, F. E., de la Motte, R. S., Ray, L., Scott, D., and Wagner, F. W. (1990). Microbial sulfyhydryl oxidase and method; culture of *Aspergillus*, oxidation of glutathione. U.S. Patent No. 4894340.

Hatamoto, H., Fujita, T., Murata, K., and Takato, S. (1990). Culturing of tomato callus and 5-lipoxygenase extraction from the callus. *Jpn. Kokai Tokyo Koho* 02,291,261.

Havir, E. (1990). Plant catalases: demonstration and characteristics of a catalase isoenzyme with enhanced peroxidatic activity. *Plant Biol.* **10**, 271–283.

Hefnawy, S. A., Ewais, S. M., and Abd-El-Salam, M. H. (1986). Manufacture of pickled soft cheese from milk preserved by the lactoperoxidase system. *Egypt. J. Dairy Sci.* **14**, 219–223.

Hemeda, H. M. (1988). Effect of α-tocopherol and some flavonoid compounds on inactivation of peroxidase (purified and unpurified) of selected vegetables. *Diss. Abs. Inter.* **49**, 1–134.

Hemeda, H. M., and Klein, B. P. (1990). Effects of naturally occurring antioxidants on peroxidase activity of vegetable extracts. *J. Food Sci.* **55**, 184–185.

Hemeda, H. M., and Klein, B. P. (1991). Inactivation and regeneration of peroxidase activity in vegetable extracts treated with antioxidants. *J. Food Sci.* **56**, 68–71.

Henderson, H. M., Eskin, N. A. M., and Borsa, J. (1990). Buckwheat lipoxygenase: inactivation by gamma-irradiation. *Food Chem.* **38**, 97–103.

Heydeck, D., Wiesner, R., Kuehn, H., and Schiewe, T. (1991). On the reaction of wheat lipoxygenase with arachidonic acid and its oxygenated derivatives. *Biomed. Biochim. Acta* **50**, 11–15.

Horwath, R. O. (1986). Method for screening microorganisms for the production of glucose-2-oxidase. U.S. Patent No. 4568638.

Hsieh, R. J. (1988). The properties of gill lipoxygenase and the generation of volatile flavor compounds in fish tissue. *Diss. Abs. Inter.* **49**, 1–341.

Hsieh, R. J., and Kinsella, J. E. (1989). Lipoxygenase generation of specific volatile flavor carbonyl compounds in fish tissues. *J. Agric. Food Chem.* **37**, 279–286.

Hsieh, R. J., German, J. B., and Kinsella, J. E. (1988). Lipoxygenase in fish tissue: some properties of the 12-lipoxygenase from trout gill. *J. Agric. Food Chem.* **36**, 680–685.

Hsu, A. F., Thomas, C. E., and Brauer, D. (1988). Evaluation of several methods for the estimation of the total activity of potato polyphenol oxidase. *J. Food Sci.* **53**, 1743–1745.

Hu, C., and van Huystee, R. B. (1989). Role of carbohydrate moieties in peanut *(Arachis hypogaea)* peroxidases. *Biochem. J.* **263**, 129–135.

Hu, C., Krol, M., and Van Huystee, R. B. (1990). Comparison of anionic with cationic peroxidase from cultured peanut cells. *Plant Cell Tiss. Org. Cult.* **22**, 65–70.

Iacazio, G., Langrand, G., Baratti, J., Buono, G., and Triantaphylides, C. (1990). Preparative, enzymatic synthesis of linoleic acid (13S)-hydroperoxide using soybean lipoxygenase-1. *J. Org. Chem.* **55**, 1690–1691.

Im, J-S., Parkin, K. L., and Von Elbe, J. H. (1990). Endogenous polyphenoloxidase activity associated with the "black ring" defect in canned beet *(Beta vulgaris L.)* root slices. *J. Food Sci.* **55**, 1042–1045.

Imai, S., Morikiyo, M., Furihata, K., Hayakawa, Y., and Seto, H. (1990). Turmeronol A and turmeronol B, new inhibitors of soybean lipoxygenase. *Agric. Biol. Chem.* **54**, 2367–2371.

Inamine, G. S., and Baker, J. E. (1989). A catalase from tomato fruit. *Phytochem.* **28**, 345–348.

Ito, O., and Akuzawa, R. (1982). Distribution of catalase in goats milk. *Bull. Nippon Vet. Zootech. Coll.* **31**, 198–202.

Ito, O., and Akuzawa, R. (1984). Isoenzymes of catalase in goats milk. *Jpn. J. Zootech. Sci.* **55**, 220–226.

Ito, O., Kamata, S., Kaki-Ichi, N., Suzuki, Y., Hayashi, M., and Uchida, K. (1990). Purification of milk catalase by immunoaffinity chromatography. *J. Food Sci.* **55**, 1172–1173.

Izumi, Y., Furuya, Y., and Yamada, H. (1990). Isolation of a new pyranose oxidase producing *Basidomycete*. *Agric. Biol. Chem.* **54**, 799–801.

Janolino, V. G., and Swaisgood, H. E. (1975). Isolation and characterization of sulfhydryl oxidase from bovine milk. *J. Biol. Chem.* **250**, 2532–2538.

Janolino, V. G., Morrison-Rowe, S. J., and Swaisgood, H. E. (1990). Confirmation of a blocked amino terminus of sulfhydryl oxidase. *J. Dairy Sci.* **73**, 2287–2291.

Johnson-Flanagan, A. M., and McLachlan, G. (1990). Peroxidase-mediated chlorophyll bleaching in degreening canola *(Brassica napus)* seeds and its inhibition by sublethal freezing. *Physiol. Plant.* **80**, 453–459.

Kahn, V. (1985). Tropolone—a compound that can aid in differentiating between tyrosinase and peroxidase. *Phytochem.* **24**, 915–920.

Kahn, V., and Andrawis, A. (1985). Inhibition of mushroom tyrosinase by tropolone. *Phytochem.* **24**, 905–908.

Kalisz, H. M., Hendle, J., and Schmid, R. (1990). Purification of the glycoprotein glucose oxidase from *Penicillium amagasakiense* by high-performance liquid chromatography. *J. Chromatogr.* **521**, 245–250.

Kamau, D. N., Doores, S., and Pruitt, K. M. (1990). Antibacterial activity of the lactoperoxidase system against *Listeria monocytogenes* and *Staphylococcus aureus* in milk. *J. Food Protect.* **53**, 1010–1014.

Katkocin, D., Miller, C. A., and Starnes, R. L. (1986). Microbial sulfhydryl oxidase. U.S. Patent No. 4632905.

Kato, M., Okamoto, J., Omori, M., Obata, Y., Saijo, R., and Takeo, T. (1987). Effect of microbial polyphenol oxidase on the colour of black tea infusions. *J. Agric. Soc. Japan* **61**, 599–601.

Kermasha, S., and Metche, M. (1988). Studies on seed peroxidase (from) *Phaseolus vulgaris* cv. haricot. *J. Food Sci.* **53**, 247–249.

Kermasha, S., Alli, I., and Metche, M. (1988). Changes in peroxidase activity during the development and processing of *Phaseolus vulgaris* cv., haricot seed. *J. Food Sci.* **53**, 1753–1755.

Khader, S. E. S. A., Singh, B. P., and Khan, S. A. (1988). Effect of GA3 as a post-harvest treatment of mango fruit on ripening, amylase and peroxidase activity and quality during storage. *Sci. Hort.* **36**, 261–266.

Khalyfa, A., Kermasha, S., and Alli, I. (1990). Partial purification and characterization of lipoxygenase of canola seed *(Brassica napus var. Westar)*. *J. Agric. Food Chem.* **38**, 2003–2008.

Kermeier, F., and Ranfft, K. (1970). Some characteristics of milk sulphydryl oxidase. *Z. Lebensm. Unters. Forsch.* **143**, 11–15.

Kim, B. B., Khegai, L. A., Cherednikova, T. V., Gavrilova, E. M., and Egorov, A. M. (1991). Interaction of monoclonal antibodies with horseradish peroxidase. *Bioorg. Khim.* **17**, 35–41.

Kim, J. M., and Schmid, R. (1991). Comparison of *Penicillium amagasakiense* glucose oxidase purified as glyco- and aglyco-proteins. *FEMS Microbiol. Lett.* **78**, 221–225.

Kim, K. K., Fravel, D. R., and Papavizas, G. C. (1990). Production, purification, and properties of glucose oxidase from the bicontrol fungus *Talaromyces flavus*. *Can. J. Microbiol.* **36**, 199–205.

Kim, M. R., and Sok, D. E. (1990). Involvement of the unstable eicosanoids in self-inactivation of soybean lipoxygenase during incubation with arachidonic acid. *Han'guk Saenghwa Hakhoechi* **23**, 478–485.

Kim, M. R., Kim, S. H., and Sok, D. E. (1989). Inactivation of potato lipoxygenase by hydroperoxy acids as suicide substrates. *Biochem. Biophys. Res. Commun.* **164**, 1384–1390.

Koths, K. E., and Halenbeck, R. F. (1986). *P. obtusus* pyranose-2-oxidase enzyme used for converting pyranose compounds to pyranosone compounds, e.g. glucosone. U.S. Patent No. 4569913.

Koths, K. E., Halenbeck, R. F., and Ring, D. B. (1986a). Pure pyranose-2-oxidase enzyme preparation obtained by culture of *Coriolus versicolor* or *Lenzites betulinus*. U.S. Patent No. 4568645.

Koths, K. E., Halenbeck, R. F., and Fermandes, P. M. (1986b). Pyranosone, especially glucosone, production from pyranose using pyranose-2-oxidase enzyme free of pyranosone-utilising enzymes. U.S. Patent No. 4569910.

Krysteva, M., Papukchieva, S., and Zlateva, T. (1990). Affinity chromatography purification of apoglucose oxidase from *Penicillium chrysogenum*. *Dokl. Bolg. Akad. Nauk.* **43**, 69–72.

Langbakk, B., and Flatmark, T. (1989). Lactoperoxidase from human colostrum. *Biochem. J.* **259**, 627–631.

Lee, H. C., and Klein, B. P. (1989a). A simple method of identifying peroxidase isoenzymes from crude pea seed extracts. *Food Chem.* **29**, 275–282.

Lee, H. C., and Klein, B. P. (1989b). Evaluation of combined effects of heat treatment and antioxidant on peroxidase activity of crude extract of green peas. *Food Chem.* **32**, 151–161.

Liang, Z. Q., Hayase, F., Nishimura, T., and Kato, H. (1990). Purification and characterization of NAD-dependent alcohol dehydrogenase and NADH-dependent 2-oxoaldehyde reductase from parsley. *Agric. Biol. Chem.* **54**, 2727–2729.

Lourenco, E. J., Silva, G. M. L., and Neves, V. A. (1991). Purification and properties of shikimate dehydrogenase from cucumber *(Cucumis sativus L.) J. Agric. Food Chem.* **39**, 458–462.

Luna, J. A., Garrote, R. L., and Bressan, J. A. (1986). Thermo-kinetic modeling of peroxidase inactivation during blanching-cooling of corn on the cob. *J. Food Sci.* **51**, 141–145.

MacDonald, I. D., and Dunford, H. B. (1989). Mechanism of horseradish peroxidase-catalyzed oxidation of malonaldehyde. *Arch. Biochem. Biophys.* **272**, 185–193.

McLellan, K. M., and Robinson, D. S. (1987a). The heat stability of purified spring cabbage peroxidase isoenzymes. *Food Chem.* **26**, 97–107.

McLellan, K. M., and Robinson, D. S. (1987b). Purification and heat stability of Brussels sprout peroxidase isoenzymes. *Food Chem.* **23**, 305–319.

Malikkides, C. O., and Weiland, R. H. (1982). On the mechanism of immobilized glucose oxidase deactivation by hydrogen peroxide. *Biotechnol. Bioeng.* **24**, 2419–2438.

Mandels, G. R. (1956). Properties and surface location of a sulfhydryl oxidizing enzyme in fungus spores. *J. Bacteriol.* **72**, 230–234.

Mangino, M. E. (1976). Characterization of xanthine oxidase in the fat globule membrane. *Diss. Abs. Inter.* **37**, 2748–2749.

Marcus, L., Prusky, D., and Jacoby, B. (1988). Purification and characterization of avocado lipoxygenase. *Phytochem.* **26**, 323–327.

Markwell, J., Frakes, L. G., Brott, E. C., Osterman, J., and Wagner, F. W. (1989). *Aspergillus niger* mutants with increased glucose-oxidase production. *Appl. Microbiol. Biotech.* **30**, 166–169.

Martinez, C. E., Mendoza, P. G., Alacron, F. J., and Garcia, H. S. (1988). Reactivation of the lactoperoxidase system during raw milk storage and its effect on the characteristics of pasteurized milk. *J. Food Protect.* **51**, 558–561.

Matheis, G., and Whitaker, J. R. (1987). A review: enzymatic cross-linking of proteins applicable to foods. *J. Food Biochem.* **11**, 309–327.

Matheis, G., Vizthum, O. G., and Weder, J. K. P. (1987). On the polarographic determination of polyphenol oxidase activity in different teas. *Chem. Mikrobiol. Tech. Leben.* **11**, 1–4.

Mehanna, N. M., and Hefnawy, S. A. (1988). Effect of thiocyanate-lactoperoxidase-hydrogen peroxide system on the manufacture and properties of yoghurt. *Egypt. J. Dairy Sci.* **16**, 55–63.

Meshehdani, T., Pokorny, J., Davidek, J., and Panek, J. (1990). The lipoxygenase activity of rapeseed. *Nahrung* **34**, 727–734.

Meunier, B. (1991). N- and O-demethylations catalyzed by peroxidases. *Peroxidases Chem. Biol.* **2**, 201–217.

Michaud-Soret, I., Daniel, R., Chopard, C., Mansuy, D., Cucurou, C., Ullrich, V., and Chottard, J. C. (1990). Soybean lipoxygenases-1, -2a, -2b and -2c do not contain PQQ. *Biochem. Biophys. Res. Commun.* **172**, 1122–1128.

Mijacevic, Z., Otenhajmer, I., and Ivanovic, D. (1989). Antimicrobial effect of the lactoperoxidase thiocyanate-hydrogen peroxide system in milk. *Mljekarstvo* **39**, 199–204.

Miller, A. R., and Kelley, T. J. (1989). Mechanical stress stimulates peroxidase activity in cucumber fruit. *HortSci.* **24**, 650–652.

Miller, A. R., Kelley, T. J., and Mujer, C. V. (1990). Anodic peroxidase isoenzymes and polyphenol oxidase activity from cucumber fruit: tissue and substrate specificity. *Phytochem.* **29**, 705–709.

Modi, S., Behere, D., and Mitra, S. (1989). Binding of thiocyanate to lactoperoxidase: 1H and 15N nuclear magnetic resonance studies. *Biochemistry* **28**, 4689–4694.

Modi, S., Deodhar, S. S., Behere, D. V., and Mitra, S. (1991). Lactoperoxidase-catalyzed oxidation of thiocyanate by hydrogen peroxide: nitrogen-15 nuclear magnetic resonance and optical spectra studies. *Biochemistry* **30**, 118–124.

Mohri, S., Cho, S-Y., Endo, Y., and Fujimoto, K. (1990a). Lipoxygenase activity in sardine skin. *Agric. Biol. Chem.* **54**, 1889–1891.

Mohri, S., Endo, Y., Matsuda, K., Kitamura, K., and Fujimoto, K. (1990b). Physiological effects of soybean seed lipoxygenases on insects. *Agric. Biol. Chem.* **54**, 2265–2270.

Monma, T., and Fukunaga, Y. (1990). Peroxidase and its manufacture with wasabi callu culture. *Jpn. Kokai Tokyo Koho* 02,276,574.

Monma, M., Sugimoto, T., Hashizume, K., and Saio, K. (1990). Study on soybean lipoxygenase. IV. Effect of several lipoxygenase inhibitors on lipoxygenase activities in soybean homogenate. *Nippon Shokuhin Kogyo Gakkaishi* **37**, 625–627.

Moon, Y. S., and Kim, M. S. L. (1985). Inhibiting pattern of dopachrome formation as influenced by sodium benzoate in raw potato tubers. *Korean J. Food Sci. Tech.* **17**, 232–236.

Morgens, P. H., Callahan, A. M., Dunn, L. J., and Abeles, F. B. (1990). Isolation and sequencing of cDNA clones encoding ethylene-induced putative peroxidases from cucumber cotyledons. *Plant Mol. Biol.* **14**, 715–725.

Morohoshi, N., Tamura, R., Katayama, Y., and Morimoto, M. (1990). Properties of phenol oxidases secreted by *Coriolus versicolor*. Characteristics of extracellular and intracellular laccase and peroxidases. *Tokyo Noko Daigot Noga. Ensh. Hokoku* **27**, 47–56.

Moulding, P. H., Grant, H. F., McLellan, K. M., and Robinson, D. S. (1987). Heat stability of soluble and ionically bound peroxidases extracted from apples. *Internat. J. Food Sci. Technol.* **22**, 391–397.

Mowrey, B. D., Werner, D. J., and Byrne, D. H. (1990). Inheritance of isocitrate dehydrogenase, malate dehydrogenase, and shikimate dehydrogenase in peach and peach almond hybrids. *J. Am. Soc. Hort. Sci.* **115**, 312–319.

Mueftuegil, N. (1985). The peroxidase enzyme activity of some vegetables and its resistance to heat. *J. Food Sci. Agric.* **36**, 877–880.

Munshi, C. B., and Mondy, N. I. (1988). Effect of soil applications of sodium molybdate on the quality of potatoes: polyphenol oxidase activity, enzymatic discoloration, phenols, and ascorbic acid. *J. Agric. Food Chem.* **36**, 919–922.

Nakahara, H., Suzuki, O., and Yokochi, T. (1990). 13-Hydroperoxy-gamma-linolenic acid preparation with immobilized lipoxygenase of soybean. *Kokai Tokyo Koho* 02,207,792.

Nakamura, K., Amano, Y., and Kagami, M. (1984). Comparison of polyphenol oxidase in 'normal' and 'ajinashi' berries of Koshu grapes. *J. Inst. Enol. Vitic.* **19**, 7–12.

Nakamura, K., Amano, Y., and Kagami, M. (1985). Effect of storage temperature on polyphenol oxidase activities of Koshu grapes. *J. Inst. Enol. Vitic.* **20**, 17–20.

Nakanishi, K., Makino, S., and Yokotsuka, K. (1987). Changes in polyphenol oxidase activity and must components during fruit maturation. *J. Inst. Enol. Vitic.* **22**, 1–9.

Nieto, A. R., Sancho, A. C., Barros, M. V., and Gorge, J. L. (1990). Peroxidase zymograms at constant and gradient pH electrophoresis as an analytical test in the identification of potato varieties. *J. Agric. Food Chem.* **38**, 2148–2145.

Nomenclature Committee of IUB, Enzyme Nomenclature (1984). Academic Press, Orlando, Florida.

Okamoto, A., Imagawa, H., and Arai, Y. (1988). Partial purification and some properties of polyphenoloxidases from sago palm. *Agric. Biol. Chem.* **52**, 2215–2222.

Omran, H., Buehler, K. D., Tian, C., and Gierschner, K. (1989). Determination of lipoxygenase and peroxidase activities in untreated and blanched peas. *Ind. Obst Gemuese* **74**, 275–281.

Ostergaard, A., and Adler-Nissen, J. (1988). Thermal denaturation of lipoxygenase from soya beans in the presence of its natural substrate. *Lebens. Wissen. Tech.* **21**, 8–12.

Oszmianski, J., Sapis, J.-C., and Macheix, J.-J. (1985). Changes in grape seed phenols as affected by enzymic and chemical oxidation *in vitro. J. Food Sci.* **50**, 1505–1506.

Park, E. Y., and Luh, B. S. (1985). Polyphenol oxidase of kiwifruit. *J. Food Sci.* **50**, 678–684.

Park, K. H., Kim, Y. M., and Lee, C. W. (1988). Thermal inactivation kinetics of potato tuber lipoxygenase. *J. Agric. Food Chem.* **36**, 1012–1015.

Park, W. M., and Kim, S. H. (1990). Comparison of peroxidase, polyphenoloxidase and protein in tissues of red pepper fruit *(Capsicum annuum L.)* between immature and mature stage. *J. Korean Soc. Hort. Sci.* **31**, 199–206.

Parkin, K. L., and Im, J.-S. (1990). Chemical and physical changes in beet *(Beta vulgaris L.)* root tissue during simulated processing—relevance to the "black ring" defect in canned beets. *J. Food Sci.* **55**, 1039–1041.

Parry, A. D., and Horgan, R. (1991). Carotenoid metabolism and the biosynthesis of abscisic acid. *Phytochem.* **30**, 815–821.

Pizzocaro, F., Ricci, R., and Zanetti, L. (1988). New aspects of blanching of vegetables before freezing. I. Peroxidase and lipoxygenase activity. *Ind. Aliment.* **27**, 993–998.

Poca, E., Rabinovitch-Chable, H., Cook-Moreau, J., Pages, M., and Rigaud, M. (1990). Lipoxygenases from *Zea mays L.* Purification and physicochemical characteristics. *Biochim. Biophys. Acta* **1045**, 107–114.

Pradhan, R., and Paul, A. K. (1990). Polyphenol oxidase and peroxidase activities of tea-associated microflora. *Environ. Ecol.* **8**, 643–645.

Ramaswamy, H. S., and Ranganna, S. (1989). Residual peroxidase activity as influenced by blanching, SO2 treatment and freezing of cauliflowers. *J. Sci. Food Agric.* **47**, 376–382.

Randhawa, K. S., Khurana, D. S., and Bajaj, K. L. (1986). Effect of CIPC (isopropyl-N-(3-chlorophenyl)carbamate) on total N content and polyphenol oxidase activity in relation to processing of potatoes. *Qual. Plant.* **36**, 207–212.

Rasmussen, S. K., Welinder, K. G., and Hejgaard, J. (1991). cDNA cloning, characterization and expression of an endosperm-specific barley peroxidase. *Plant. Mol. Biol.* **16**, 317–327.

Reddanna, P., Whelan, J., Maddipati, K. R., and Reddy, C. C. (1990). Purification of arachidonate 5-lipoxygenase from potato tubers. *Meth. Enzymol.* **187**, 268–277.

Richard, F., and Nicolas, J. (1989). Purification of apple peel peroxidase: studies of some properties and specificity in relation to phenolic compounds. *Sci. Aliment.* **9**, 335–350.

Ring, D. B. (1986). *P. obtusus, C. versicolor* and *L. betulinus* strains which produce high levels of pyranose-2-oxidase for pyranosone production. U.S. Patent No. 4569915.

Rivera, J. A., Hohlberg, A. I., Aguilera, J. M., Plahk, L. C., and Stanley, D. W. (1989). Hard-to-cook defect in black beans—peroxidase characterization and effect of heat pretreatment and storage conditions of enzyme inactivation. *Can. Inst. Food Sci. Technol. J.* **22**, 270–275.

Robb, D. A. (1980). Tyrosinase. II, Copper Proteins and Copper Enzymes. Tyrosinases. "Copper Proteins and Copper Enzymes" (R. Lontie, ed.) Vol. II, pp. 207–234. CRC Press, Boca Raton, Florida.

Roberts, E., Kutchan, T., and Kolattukudy, P. E. (1988). Cloning and sequencing of cDNA for a highly anionic peroxidase from potato and the induction of its mRNA in suberizing potato tubers and tomato fruits. *Plant Mol. Biol.* **11**, 15–26.

Robinson, D. S., Bretherick, M. R., and Donnelly, J. K. (1989). The heat stability and isoenzyme composition of peroxidases in Ohane grapes. *Int. J. Food Sci. Technol.* **24**, 613–618.

Rokach, J. (1989). Leukotrienes and lipoxygenases. *Bioactive Mol.* **11**, 1–199.

Romeyer, F. M., Sapis, J.-C., and Macheix, J.-J. (1985). Hydroxycinnamic esters and browning potential in mature berries of some grape varieties. *J. Sci. Food Agric.* **36**, 728–732.

Rothan, C., and Nicolas, J. (1989). Changes in acidic and basic peroxidase activities during tomato fruit ripening. *HortSci.* **24**, 340–342.

Sanchez-Ferrer, A., Bru, R., and Cabanes, J. (1988). Characterization of catecholase and cresolase activities of Monastrell grape polyphenol oxidase. *Phytochem.* **27**, 319–321.

Sanchez-Ferrer, A., Bru, R., and Valero, E. (1989). Changes in pH-dependent grape polyphenoloxidase activity during maturation. *J. Agric. Food Chem.* **37**, 1242–1245.

Sapers, G. M., Douglas, F. W., Bilyk, A., Hsu, A. F., Dower, H. W., Garzarella, L., and Kozemple, M. (1989). Enzymatic browning in Atlantic potatoes and related cultivars. *J. Food Sci.* **54**, 362–365.

Schewe, T., Rapoport, S. M., and Kuhn, H. (1986). Enzymology and physiology of reticulocyte lipoxygenase: comparison with other lipoxygenases. *Adv. Enzymol.* **58**, 191.

Sciancalepore, V. (1985). Enzymatic browning in five olive varieties. *J. Food Sci.* **50**, 1194–1195.

Sciancalepore, V., and Longone, V. (1984). Polyphenol oxidase activity and browning in green olives. *J. Agric. Food Chem.* **32**, 320–321.

Sciancalepore, V., Longone, V., and Alviti, F. S. (1985). Partial purification and some properties of peroxidase from Malvasia grapes. *Am. J. Enol. Vitic.* **36**, 105–110.

Seth, K. K., and Nirankar, N. (1988). A simple blanch-dry process for lipoxygenase inactivation in soybean cotyledons. *Int. J. Food Sci. Tech.* **23**, 275–279.

Seth, K. K., and Nirankar, N. (1989). Storage stability of lipoxygenase-free full-fat soyflour packed in polyethylenes. *Int. J. Food Sci. Tech.* **24**, 559–565.

Shaked, Z., and Wolfe, S. N. (1987). Catalase and pyranose-2-oxidase used for glucosone production from glucose stabilised against glucosone inactivation by crosslinking or reacting with amidinating agent respectively. U.S. Patent No. 4650758.

Shenoy, V. V., Seshu, D. V., and Sachan, J. K. S. (1990). Shikimate dehydrogenase-12 allozyme as a marker for high seed protein content in rice. *Crop Sci.* **30**, 937–940.

Shibata, D. (1990). Cloning of regulatory sequences of the lipoxygenase gene of soybean. *Jpn. Kokai Tokyo Koho* 02,100,679.

Shimizu, T., Nonda, Z., Miki, I., Seyama, Y., Izumi, T., Raadmark, O., and Samuelsson, B. (1990). Potato arachidonate 5-lipoxygenase: purification, characterization, and preparation of 5(S)-hydroperoxyeicosatetraenoic acid. *Meth. Enzymol.* **187**, 296–306.

Siragusa, G. R., and Johnson, M. G. (1989). Inhibition of *Listeria monocytogenes* growth by the lactoperoxidase-thiocyanate-H2O2 antimicrobial system. *Appl. Environ. Microbiol.* **55**, 2802–2805.

Smith, A. T., Santama, N., Dacey, S., Edwards, M., Bray, R. C., Thorneley, R. N. F., and Burke, J. E. (1990). Expression of a synthetic gene for horseradish peroxidase C in *Escherichia coli* and folding and activation of the recombinant enzyme with calcium and heme. *J. Biol. Chem.* **265**, 13335–13343.

Smith, D. M., and Montgomery, M. W. (1984). Improved methods for the extraction of polyphenol oxidase from d'Anjou pears. *Phytochem.* **24**, 901–904.

Song, Y., Love, M. H., and Murphy, P. (1990). Subcellular localization of lipoxygenase-1 and -2 in germinating soybean seeds and seedlings. *J. Am. Oil Chem. Soc.* **67**, 961–965.

Spychalla, J. P., and Desborough, S. L. (1990). Superoxide dismutase, catalase, and α-tocopherol content of stored potato tubers. *Plant Physiol.* **94**, 1214–1218.

Srivastava, R. A. K. (1986). Polyphenol oxidase activity in the development of acquired aroma in tea *(Thea sinensis var. Assamica L.)* *Curr. Sci.* **55**, 284–287.

Stone, R. A., and Kinsella, J. E. (1989). Bleaching of β-carotene by trout gill lipoxygenase in the presence of polyunsaturated fatty acid substrates. *J. Agric. Food Chem.* **37**, 866–868.

Swaisgood, H. E., and Horton, R. (1980). Sulfhydryl oxidase: oxidation of sulfhydryl groups and the formation of three-dimensional structure in proteins. *Ciba Found. Symp.* **72**, 205–222.

Swart, K., Van De Vondervoort, P. J. I., Witteveen, C. F. B., and Visser, J. (1990). Genetic localization of a series of genes affecting glucose oxidase levels in *Aspergillus niger*. *Curr. Genet.* **18**, 435–439.

Taira, S., Sugiura, A., and Tomana, T. (1987). Relationship between natural flesh darkening and polyphenol oxidase activity in Japanese persimmon *(Diospyros kaki Thunb.)* fruits. *J. Japan. Soc. Food Sci. Tech.* **34**, 612–615.

Takegawa, K., Fujiwara, K., Iwahara, S., and Yamomoto, K. (1991). Primary structure of an O-linked sugar chain derived from glucose oxidase of *Aspergillus niger. Agric. Biol. Chem.* **55**, 883–884.

Thanabal, V., and Mar, G. N. (1989). A nuclear Overhauser effect investigation of the molecular and electronic structure of the heme crevice in lactoperoxidase. *Biochemistry* **28**, 7038–7044.

Thanaraj, S. N. S., and Seshadri, R. (1990). Influence of polyphenol oxidase activity and polyphenol content of tea shoot on quality of black tea. *J. Sci. Food Agric.* **51**, 57–69.

Todd, J. F., Paliyath, G., and Thompson, J. E. (1990). Characteristics of a membrane-associated lipoxygenase in tomato fruit. *Plant. Physiol.* **94**, 1225–1232.

Traeger, M., Qazi, G. N., Onken, U., and Chopra, C. L. (1991). Contribution of endo- and exocellular glucose oxidase to gluconic acid production at increased dissolved oxygen concentrations. *J. Chem. Technol. Biotechnol.* **50**, 1–11.

Trick, M., Dennis, E. S., Edwards, K. J. R., and Peacock, W. (1988). Molecular analysis of the alcohol dehydrogenase gene family of barley. *Plant Mol. Biol.* **11**, 147–160.

Twyford, C. T., Viana, A. M., James, A. C., and Mantell, S. H. (1990). Characterization of species and vegetative clones of *Dioscorea* food yams using isoelectric focussing of peroxidase and acid phosphatase isoenzymes. *Tropical Agric.* **67**, 337–341.

Ullah, M. R., Gogoi, N., and Baruah, D. (1984). The effect of withering on fermentation of tea leaf and development of liquor characters of black teas. *J. Sci. Food Agric.* **35**, 1142–1147.

Valdez, G. F., de Bibi, W., and Bachmann, M. R. (1988). Antibacterial effect of the lactoperoxidase/thiocyanate/hydrogen peroxide (LP) system on the activity of thermophilic starter culture. *Milchwissensch.* **43**, 350–352.

Valero, E., Varon, R., and Garcia-Carmona, F. (1988). Characterization of polyphenol oxidase from Arien grapes. *J. Food Sci.* **53**, 1482–1485.

Vamos-Vigyazo, L., Polacsek-Racz, M., Schmidt, K., Joo-Farkas, I., Pauli, M. P., Horvath, G., Kiss, K., and Horvath, L. (1985a). Relationship between pigment content, peroxidase activity and sugar composition of red pepper *(Capsicum annuum L.)* I. Influence of cultivar, drying method and a ripening accelerator. *Acta Aliment.* **14**, 173–189.

Vamos-Vigyazo, L., Polacsek-Racz, M., Kampis, A., Pauli, M. P., and Horvath, G. (1985b). Relationship between pigment content, peroxidase activity and sugar composition of red pepper *(Capsicum annuum L.)* II. Changes occurring the industrial drying process. *Acta Aliment.* **14**, 191–200.

Van den Berg, B. M. (1991). A rapid and economical method for hybrid purity testing of tomato *(Lycopersicon esculentum)* F1 hybrids using ultrathin-layer isoelectric focusing of alcohol dehydrogenase variants from seeds. *Electrophoresis* **12**, 64–69.

Van Huystee, R. B. (1991). Molecular aspects and biological functions of peanut peroxidases. *Peroxidases Chem. Biol.* **2**, 155–170.

Van Huystee, R. B., Hu, C., and Sesto, P. A. (1990). Comparisons between cationic and anionic peanut peroxidases as glycoproteins. *Prog. Clin. Biol. Res.* **344**, 315–325.

Van Lelyveld, L. J., and De Rooster, K. (1986). Browning potential of tea clones and seedlings. *J. Hort. Sci.* **61**, 545–547.

Variyar, P. S., Pendharkar, M. B., Banerjee, A., and Bandyopadhyay, C. (1988). Blackening in green pepper berries. *Phytochem.* **27**, 715–717.

Vaughn, K. C., and Duke, S. O. (1984). Function of polyphenol oxidase in higher plants. *Physiol. Plant.* **60**, 106.

Vijayvaragiya, R. R., and Pai, J. S. (1991). Lowering of lipoxygenase activity in soy milk preparation by propyl gallate. *Food Chem.* **41**, 63–67.

Vliegenthart, J. F. G., and Veldink, G. A. (1982). Lipoxygenases. *In* "Free Radicals in Biology" (W. A. Pryor, ed.). Vol. V, pp. 29–64. Academic Press, Inc.

Wang, J., Fujimoto, K., Miyazawa, T., Endo, Y., and Kitamura, K. (1990). Sensitivities of lipoxygenase-lacking soybean seeds to accelerated aging and their chemiluminescence levels. *Phytochem.* **29**, 3739–3742.

Wang, P., Lin, C., Wu, K., and Lu, Y. (1987). Preservation of fresh milk by its natural lactoperoxidase system. *Sci. Agric. Sin.* **20**, 76–81.

Wesche-Ebeling, P. A. E. (1984). Purification of strawberry polyphenol oxidase and its role in anthocyanin degradation. *Diss. Abs. Int.* **44**, 1–180.

Wesche-Ebeling, P., and Montgomery, M. W. (1990). Strawberry polyphenol oxidase: its role in anthocyanin degradation. *J. Food Sci.* **55**, 731–734.

Whittington, H., Kerry-Williams, S., Bidgood, K., Dodsworth, N., Peberdy, J., Dodson, M., Hinchliffe, E., and Ballance, D. J. (1990). Expression of the *Aspergillus niger* glucose oxidase in *A. niger*, *A. nidulans* and *Saccharomyces cerevisiae*. *Curr. Genet.* **28**, 531–536.

Wimalasena, R. L., De Alwis, W. U., Wilson, G. S. (1990). Influence of screening procedures on properties of monoclonal antibodies to glucose oxidase. *Anal. Chim. Acta* **241**, 105–113.

Wissemann, K. W., and Montgomery, M. W. (1985). Purification of d'Anjou pear *(Pyrus communis L.)* polyphenol oxidase. *Plant Physiol.* **78**, 256–262.

Wolyn, D. J., and Jelenkovic, G. (1990). Nucleotide sequence of an alcohol dehydrogenase (ADH, EC 1.1.1.1) gene in octoploid strawberry *(Fragaria Ananassa Duch.)*. *Plant Mol. Biol.* **14**, 855–857.

Wong, M. K., Dimick, P. S., and Hammerstedt, R. H. (1990). Extraction and high performance liquid chromatographic enrichment of polyphenol oxidase from *Theobromo cacao* seeds. *J. Food Sci.* **55**, 1108–1111.

Wray, C., and McLaren, I. (1987). A note on the effect of the lactoperoxidase systems on salmonellas *in vitro* and *in vivo*. *J. Appl. Bacteriol.* **62**, 115–118.

Xu, Y., Hu, C., and Van Huystee, R. E. (1990). Quantitative determination of a peanut anionic peroxidase by specific monoclonal antibodies. *J. Exp. Bot.* **41**, 1479–1488.

Zemel, G. P., Sims, C. A., Marshall, M. R., and Balaban, M. (1990). Low pH inactivation of polyphenoloxidase in apple juice. *J. Food Sci.* **55**, 562–563.

Zikakis, J. P. (1979). Preparation of high purity xanthine oxidase from bovine milk. U.S. Patent No. 4172763.

Zikakis, J. P. (1980). Xanthine oxidase. U.S. Patent No. 4238566.

Zikakis, J. P., and Biasotto, N. O. (1976). An improved isolation and purification method of xanthine oxidase from bovine raw milk. *Abs. Papers* **172**, BIOL 140.

Zikakis, J. P., and Wooters, S. C. (1980). Activity of xanthine oxidase in dairy products. *J. Dairy Sci.* **63**, 893–904.

Zikakis, J. P., and Silver, M. R. (1979). Bovine milk xanthine oxidase: the effect of limited proteolysis on kinetics and structure. *Abs. Papers* **178**, ACS AGFD.

Zmarlicki, S., Salek, A., and Kopec, A. (1978). Thermal inactivation of xanthine oxidase in cow's milk. *Int. Dairy Cong.* **XX**, 317–320.

CHAPTER 10

Applications of Oxidoreductases

THOMAS SZALKUCKI

Oxidoreductases are not currently large volume products for enzyme manufacturers, but they are being used in interesting specialized applications in food and associated industries. Laboratory and small scale production processes for applications other than those covered in this chapter (Barker, 1986, Enzyme applications in the food industry; Bottomley *et al.*, 1989, decolorizing annatto whey using peroxidase; Suwa *et al.*, 1989, deoxygenation of foods and beverages with lactate monoxygenase) indicate many more possible uses for oxidoreductases, but to date, these applications are not known to be of commercial significance.

Because of their high degree of specificity, oxidoreductases are used extensively as analytical tools for quantitative and qualitative analysis. These enzymes have been used in colorimetric test strips, in test kits with enzymes in solution, and immobilized in membranes for "instant" read out sensors. Most of these test methods were developed originally for medical diagnostic use, but now are used frequently as analytical tools by the food industry.

I. Alcohol Oxidase

Alcohol oxidase catalyzes the oxidation of short chain linear aliphatic alcohols to the corresponding aldehyde and H_2O_2. Because the reaction rate varies with the alcohol (Provesta Enzymes, 1987a), the approximate alcohol composition of the system must be understood before using the enzyme. Although the reaction catalyzed by alcohol oxidase should make the enzyme ideal for the deoxygenation of alcoholic beverages such as wine, wine coolers, and beer, the pH optimum for the enzyme is 7.5–8.0 with an operating range of 6.0–9.5 (Provesta Enzymes, 1978b). Alcohol oxidase also is inhibited reversibly by 5–10% ethanol or methanol (Provesta Enzymes, 1987c). With its high pH optimum and substrate inhibition, alcohol oxidase solubilized in these relatively acidic beverages is not an effective deoxygenation system, but could be used in these products if contained behind a selectively permeable cap liner.

A 1990 U.S. patent describes a multienzyme system used to minimize or prevent the formation of off-odors in fish oil emulsions. Alcohol oxidase is one of the enzymes in this system designed to limit the formation of alcohols and aldehydes that contribute off-odors and -flavors (Antrim and Taylor, 1990).

Despite the fact that alcohol oxidase is not currently being used directly in food, it is being used extensively as an analytical tool for the determination of alcohol concentration. Methodologies have been established to use this reaction for the quantification of alcohol with free and immobilized enzyme. Two interesting food analysis procedures have been developed using alcohol oxidase. The first of these uses an alcohol oxidase enzyme electrode to determine the amount of aspartame in beverages. The procedure involves the addition of α-chymosin to cleave methanol from aspartame; methanol then is measured by the enzyme electrode (Smith *et al.*, 1989). The second method uses alcohol oxidase to determine the methyl ester content of pectins by measuring the quantity of methanol released from the pectin (Klavons and Bennett, 1986).

II. Catalase

In 1988, the use of H_2O_2 in North America grew by 15%, the highest rate of growth of any commodity chemical (Anonymous, 1989a). Although much of this increase was for nonfood use (mostly paper and pulp), H_2O_2 producers saw considerable growth in small but growing niches. The growth of the market is caused predominantly by the perceived environmental safety of the product and by new technologies such as waste treatment and aseptic packaging (Ainsworth, 1989). The use of catalase for removing residual H_2O_2 also has shown significant growth, despite the fact that many of these applications do not require the removal of excess peroxide.

The U.S. Food and Drug Administration (FDA) has affirmed the use of H_2O_2 as a direct food substance with GRAS status (Generally Recognized as Safe) [Code of Federal Regulations (21 CFR) 184.1366, 1989)]. The primary functional uses of H_2O_2 in the food industry are categorized by the FDA (Table I) as an antimicrobial agent, an oxidizing and reducing agent (as defined in 21 CFR), or as a bleaching agent. The FDA regulations stipulate functional uses of H_2O_2 for each food listed. However, the actual functional use is sometimes not exactly what appears in the regulations. For example,

TABLE I
FDA Approved Uses for $H_2O_2^a$

Food	Maximum treatment level in food (%)	Function
Milk intended for use during the cheese-making process as permitted in the appropriate standards of identity	0.05	Antimicrobial agent
Whey, during the preparation of modified whey by electrodialysis methods	0.04	Antimicrobial agent
Colored (annatto) cheese whey	0.05	Bleaching agent
Wine, dried egg whites, and dried egg yolks	Amount sufficient for the purpose	Oxidizing and reducing agent
Tripe, beef feet, herring, and instant tea	Amount sufficient for the purpose	Bleaching agent
Starch	0.15	Antimicrobial agent; removal of SO_2 from starch slurry
Corn syrup	0.15	Reduction of SO_2 levels in finished syrup
Wine vinegar	Amount sufficient for the purpose	Reduction of SO_2 in wine prior to fermentation
Emulsifier containing fatty acid esters	1.25	Bleaching agent

^{a} Adapted from the Code of Federal Regulations (1989).

two of the U.S. Department of Agriculture (USDA) approved methods for pasteurization of egg white, the Armour and the Standard Brands methods, use a combination of heat and H_2O_2 to achieve the desired microbial reduction. 21 CFR lists the use of H_2O_2 in egg products as an oxidizing and reducing agent, not as an antimicrobial agent.

FDA regulations also stipulate that residual H_2O_2 be removed from the product by appropriate physical and chemical means. Although this section does not define an actual method for removal of residual peroxide, many of the regulations dealing with specific products still suggest catalase as the enzyme to be used to catalyze the decomposition of H_2O_2. Catalase does provide a relatively simple and cost effective means of removing residual H_2O_2.

Two commonly used sources for catalase sold in the United States are bovine liver and *Aspergillus niger*. Each type of catalase has unique benefits that makes it useful in different situations. Beef liver catalase catalyzes a "burst effect" of rapid H_2O_2 decomposition but has a fairly limited pH range (Chapter 8). Fungal catalase, on the other hand, functions over a wide pH range (pH 2–10) at a slower rate than beef liver catalase, but decomposes a greater quantity of H_2O_2 over time.

A. H_2O_2/Catalase in Cheese Milk

Pasteurization (72°C for 16 sec) is one of the methods used to ensure the microbiological safety of milk. When milk is being used for cheese production, this heat treatment adversely affects the flavor quality of the finished cheese. An alternative method permitted by the FDA is the use of up to 0.05% H_2O_2 for the chemical sterilization of milk for Cheddar (21 CFR 133.113), Colby (21 CFR 133.118), washed curd, soaked curd (21 CFR 133.136), granular and stirred curd (21 CFR 133.144), or Swiss and Emmenthaler cheese production. Sufficient catalase to eliminate residual peroxide is required when using H_2O_2. Although the FDA regulations permit the use of up to 0.05% H_2O_2, the chemical generally is used at a lower level in conjunction with HTST (high temperature short time) heat treatment to destroy the bacteria in the cheese milk effectively.

If H_2O_2/HTST treatment is used for cheese milk, the H_2O_2 should be diluted to 5% with water before addition to the milk, then metered into the milk just before passage through HTST equipment (Kosikowski, 1977). The milk is then cooled to the proper temperature for the desired cheese type and pumped into the cheese vat. The residual H_2O_2 is decomposed by catalase added to the treated milk [approximately 4.2 ml beef liver catalase (100 Keil Units/ml) or 3.9 ml fungal catalase (1000 Baker Units/ml) per 1000 lb milk with 0.02% H_2O_2 added; Miles Laboratories, 1984; Finnsugar, bulletin no. 100]. The catalase can be added soon after the milk begins to flow into the cheese vat or it can be mixed in after the vat has been filled. About 20 min after catalase addition, the cheese milk should be tested for residual H_2O_2. It is important to achieve a negative test for residual peroxide to prevent the destruction of the starter culture that will be added. After achieving the negative H_2O_2 test, catalase manufacturers generally suggest a 10 min waiting period before starter culture addition to insure that no residual H_2O_2 remains.

The major benefit attributed to the use of H_2O_2/catalase treatment of cheese milk is the improved flavor of the finished product. The flavors of the cheeses made from H_2O_2/catalase-treated cheese milk are closer to that of cheese made with raw milk without the microbial problems associated with the use of raw milk. Care must be taken when using H_2O_2, since excessive use leads to a chemical off-flavor and a soft pasty cheese.

B. H_2O_2/Catalase in Eggs

As mentioned earlier, two of the pasteurization methods listed by USDA use a combination of H_2O_2 and heat (USDA). The amount of heat required to achieve proper pasteurization of albumen by these methods is lower than that required by other approved methods. USDA regulations stipulate that, when using these H_2O_2 methods, procedures must be followed to insure that the residual peroxide is dissipated completely (negative test by appropriate method) before final packaging. One procedure uses catalase whereas the other states that "methods other than catalase addition are used."

As early as 1901 (Loew, 1901), egg white has been shown to have catalase activity. Ball and Cotterill (1971a,b) characterized and isolated this activity, concluding that catalase is a native component of egg white. They also stated that the amount of catalase activity varies from egg to egg. Because of this variation, the time required for the native catalase to dissipate the residual peroxide can vary. To minimize the hold time and eliminate the variability in the time required for a negative test, large scale producers of egg albumen add additional catalase to the product after heat treatment. Based on a 0.05% H_2O_2 addition to the albumen, a dose of 0.005% (wt/wt) catalase (fungal catalase, 1000 Baker Units/ml) is recommended as a starting dose. The level of addition can be varied based on temperature and time constraints. Higher temperatures increase the rate of reaction but decrease the total amount of H_2O_2 decomposed by a specific quantity of catalase.

C. H_2O_2/Catalase in Whey and Other Products

The FDA permits the use of benzoyl peroxide or H_2O_2 as a bleaching agent for colored (annatto) cheese whey. Both chemicals also are allowed as antimicrobial agents, but H_2O_2 is limited to use in the preparation of modified whey by electrodialysis.

Although benzoyl peroxide is easier to work with, some companies prefer to use H_2O_2/catalase to avoid residual benzoate in their product. Both beef liver and fungal catalase are used by the whey industry to dissipate residual H_2O_2, but the fungal product is preferred when the pH is lower than 5.0 or if the product must be kosher. The FDA limits for H_2O_2 in this application are 0.04% as an antimicrobial and 0.05% as a bleaching agent. The procedure for removal of the residual peroxide with catalase is similar to that used for cheese milk with the exception that, as the pH of the whey decreases, the dose of beef liver catalase must be increased.

As seen in Table I, the FDA permits the use of H_2O_2 in products and procedures other than cheese milk, eggs, and whey. When using catalase to dissipate residual peroxide in these systems, the procedure for cheese milk is a good starting point. Appropriate modifications to the procedure should be made based on time, temperature, and pH of the product and process.

III. Glucose Oxidase/Catalase

Most of the glucose oxidase/catalase used in the food industry is a semipurified enzyme preparation obtained from *Aspergillus niger* that has glucose oxidase (GO) and catalase as the predominant activities. The catalase in glucose oxidase preparations is present to prevent the buildup of H_2O_2 from the oxidation of glucose. The glucose oxidase to catalase ratios of most commercial preparations used in food applications range from 5:1 to 1:1 [Titrimetric units of glucose oxidase (Underkofler, 1957); Baker Units of catalase (Scott and Hammer, 1960)]. The addition of glucose oxidase/catalase to products is based on the glucose oxidase units (GO U) of the product. The reactions catalyzed by glucose oxidase/catalase permit it to be used to scavenge glucose or oxygen or to produce gluconic acid. Glucose oxidase without catalase can be used to produce H_2O_2.

"Glucose oxidase was first identified as of potential commercial interest in 1943 by Dwight Baker. He developed the first methods of commercial type production as he sought a means of coping with the oxidative changes in beer due to the high air level in bottled beer, then averaging 5–6 ml per 12 oz bottle" (Scott, 1975). The laboratory-scale trials of glucose oxidase/catalase in beer were successful (Ohlmeyer, 1957; Reinke et al., 1963; Zetalaki, 1964) but production-scale use never became very large. Having sufficient glucose in bottled beer is a problem and the bottling procedures for beer were improved, reducing the need for an oxygen scavenger.

A. Desugaring of Egg Components

The desugaring of eggs prior to spray drying is one of the best known and earliest successful uses of glucose oxidase/catalase (Baldwin et al., 1953). In this application, glucose oxidase/catalase is used to oxidize the glucose present in eggs, thereby preventing Maillard browning in the finished dry product. Powdered egg producers still remove glucose from eggs prior to spray drying, but microbial culture is now the most common procedure for desugaring. The enzymatic procedure continues to be used for some products, since it is preferred for egg products that are components of finished products with low flavor profiles. In these systems, the flavors from the culture process can interfere with the flavor of the finished product.

Glucose oxidase/catalase can be used to desugar liquid whole eggs or the separated yolk and white (albumen). A procedure for the desugaring of albumen will be presented as a guide for the enzymatic process. Pasteurization can be performed before or after desugaring, but the desugaring process should be performed at a temperature of 45–50°F. The pH of albumen is adjusted to 6.8–7.0 to optimize the performance of the enzyme system. The pH adjustment is made with the slow addition (to prevent localized precipitation) of citric acid (approximately 1 lb/1000 lb albumen). After acidification, about 600 ml of 35% H_2O_2/1000 lb albumen is added. Once the peroxide is blended well into the albumen, add 75,000 GO U/1000 lb. This dose is based on a 16-hr desugaring process. A higher or lower dose can be used to shorten or lengthen the procedure. To prevent a rate reduction caused by O_2 limitation, more H_2O_2 should be added to the albumen during the desugaring. Once the foam subsides after enzyme addition, 35% H_2O_2 is added at a rate of 5 ml/min/1000 lb liquid egg white. Peroxide addition is continued at this rate until foam begins to build up (about 2.5 hr). The rate of addition is reduced to 2.5 ml/min/1000 lb. Again, when foam begins to build, the peroxide addition is reduced by half. This procedure of incremental reduction in the rate of peroxide addition is followed until the desired level of glucose is achieved (0.1% glucose on a dry basis). Once the desugaring is complete, other component additions can be made prior to drying, based on the desired end product.

The desugaring procedure for egg yolk is similar to the procedure for albumen but there is no acid addition and the process is complete in about 4 hr. The procedure for whole egg does not require acidification and the enzyme dose is increased to about 100,000 GO U/1000 lb. At this level, the process takes about 6 hr.

Removal of glucose for the minimization of Maillard browning also is performed commercially for other food ingredients, using procedures similar to those used for egg desugaring. In these applications, the enzymatic process is preferred over fermentation because of the minimal change in the flavor of the treated product. Two glucose oxidase/catalase desugaring processes described in the literature are glucose reduction in potatoes for minimization of browning (Low et al., 1989) and conversion of glucose to gluconic acid in grape juice for the production of a low alcohol wine (Villettaz, 1986).

B. Deoxygenation of Beverages and Other Applications

Glucose oxidase/catalase has been used for many years to deoxygenate beverages, but commercial use is limited currently to certain specific applications. As mentioned earlier, the initial commercial purification procedures for glucose oxidase/catalase came from the desire to deoxygenate bottles of beer. Successful use of glucose oxidase/catalase to deoxygenate wine was also reported by researchers (Yang, 1955; McLeod and Ough, 1970) but large scale use never occurred. One of the primary causes of these unutilized successes is the inability of fungal catalase to function catalatically in the presence of ethanol (Temple and Ough, 1975). Without catalase activity, glucose oxidase still deoxygenates the product but is, in reality, an effective H_2O_2 producer until either glucose or O_2 is depleted or the enzyme is inactivated. Some organoleptic benefit still can be realized with the use of glucose oxidase/catalase in ethanol-containing beverages if other compounds such as SO_2 or ascorbic acid can scavenge the peroxide formed by glucose oxidase and the oxygen permeation rate into the finished package over time is very low. Currently, there is minimal commercial use of glucose oxidase/catalase in ethanol containing beverages.

One of the best markets for glucose oxidase/catalase for deoxygenation of beverages continues to be the production of lemon or orange soft drinks sold in countries in which the product is exposed to light during distribution. In these situations, the use of the traditional oxygen scavengers and antioxidants, ascorbic and erythorbic acids, would bleach the beverages when exposed to light. To avoid a "white" citrus-flavored soft drink, glucose oxidase/catalase is used instead of ascorbic or erythorbic acid to scavenge the oxygen in these products. The use level of the enzyme in this application is 0.25–0.5 GO U/fluid oz beverage. Depending on the temperature, pH, and initial dissolved oxygen content, this dose can remove the dissolved oxygen within 3 hr. The headspace oxygen in the sealed bottled or canned beverage is also scavenged over time, but its rate of removal depends on the rate of diffusion into the beverage.

Soft drinks sold in the United States do not have the extreme light exposure problems prevalent in some other countries with less refined sales and distribution systems. Therefore, ascorbic and erythorbic acids can be used in citrus-flavored soft drinks without bleaching the product. Another reason for the low level of glucose oxidase/catalase use by the United States soft drink industry is that the FD&C dyes (Red #40, Yellow #5, and Yellow #6) used as colorants in orange and lemon beverages inhibit glucose oxidase. Ironically, one of the markets targeted for glucose oxidase/catalase use because of the susceptibility of its flavors to oxidation never materialized because of the dyes used in those beverages.

The commercial use of glucose oxidase/catalase in juices with relatively high ascorbic acid content (e.g., orange or grapefruit) currently is limited to deoxygenation treatment before pasteurization. In these products, using the same quantity of enzyme used in soft drinks, the dissolved O_2 is removed (from O_2 saturation to <0.5 ppm at room temperature) in 30 min as opposed to 3 hr for beverages without ascorbic acid. The deoxygenation in these products is rapid because H_2O_2 from the glucose oxidase-catalyzed reaction oxidizes the ascorbic acid rather than being decomposed by catalase to water and $\frac{1}{2} O_2$. Although this system gives a relatively rapid deoxygenation of high ascorbic content juices, the oxidation of ascorbic acid by H_2O_2 can cause problems in long term storage, especially if the packaging material allows O_2 to diffuse into the product over time. The sterile-filtered addition of glucose oxidase/catalase to aseptically packaged orange juice actually brought about more rapid browning of the product than the control during storage because of slow oxygen permeation into the package and the enzymatically formed H_2O_2 oxidization of ascorbic acid. Despite the failure of glucose oxidase/catalase in aseptically packaged juice, the enzyme system works well as a pre-pasteurization treatment for juice and juice drinks. In this application, O_2 is removed from the juice prior to heating, improving the flavor of the finished product. The enzyme system is inactivated by the heat treatment, preventing the storage problem.

Work conducted by I. Sagi and C. H. Mannheim (1988, personal communication) at Technion, Israel Institute of Technology, has shown that the glucose oxidase/catalase system can be used to increase the shelf life of fresh-squeezed nonpasteurized orange juice. Using the glucose oxidase/catalase system to create an anaerobic environment rapidly after squeezing, the growth of spoilage organisms with concomitant off-flavor production was delayed. Currently, there is no commercial use of this process.

The use of glucose oxidase/catalase is not limited to the use of liquid enzyme for deoxygenation of liquid products. With proper pH, temperature, and substrate concentration conditions, freeze-dried glucose oxidase/catalase can be used to deoxygenate packages of nonliquid food products with a water activity as low as 0.65. A commercial application of this system is the deoxygenation of packages of shredded cheese to inhibit mold growth (Qualcepts, 1986). Use of the dry enzyme system eliminates the spraying of mold inhibitor and the drying of shredded cheese. Instead, the enzyme and anticaking agents are added directly to the cheese coming from the shredder, and the product is packaged immediately. Other applications of this system that become commercialized will be of interest.

Glucose oxidase/catalase has been used commercially since the early 1950s. Although there have been many applications of this enzyme system since that time, only a few of them have been commercially successful. Some of these failures were due to unforeseen problems with H_2O_2 formed by the reaction, whereas others were due to the economics or materials available at that time. Only a few of the many glucose oxidase/catalase applications found in the literature have been presented, since these are known to be practiced commercially.

III. Lactoperoxidase

Lactoperoxidase (LP) is the enzyme component in one of the naturally occurring antimicrobial systems found in bovine milk. For the LP system to exhibit antimicrobial action, lactoperoxidase, thiocyanate, and hydrogen peroxide must be present. Lactoperoxidase catalyzes the oxidation of thiocyanate by H_2O_2 to hypothiocyanous acid and hypothiocyanate. The strong antimicrobial effect is due to the ability of these products to oxidize sulfhydryl groups and reduced nicotinamide nucleotides. The bacteriostatic and bacteriocidal effectiveness of the LP system depend not only on the organism but also on the product conditions (Pruitt and Reiter, 1985; Oleofina, no date; Borch et al., 1989). Several reports provide varying data on the effectiveness of the LP system on *Listeria monocytogenes* (Denis and Ramet, 1989; Earnshaw and Banks, 1989; Siragusa and Johnson, 1989; Kamau et al., 1990).

A. Preservation of Raw Milk

Currently, it appears that one of the first commercial uses of the LP system in foods will be for the preservation of raw milk when refrigeration is not possible. This application relies on the lactoperoxidase naturally found in milk, supplementing it with 14 mg NaSCN/liter milk (due to the variable levels of thiocyanate in milk) and 30 mg sodium percarbonate/liter milk (to supply the H_2O_2) to activate the LP system. The use of this antimicrobial system extends the amount of time that the initial microbiological quality of the milk can be maintained without refrigeration. Depending on the initial quality and the ambient temperatures, these additions can extend the time before spoilage occurs by 7–26 hr, thereby extending the range that raw milk can be transported from collection centers to the central dairy. The LP system is not a substitute for good sanitary procedures or pasteurization; milk treated with the LP system still must undergo pasteurization. This additional treatment insures not only microbiological quality but also complete removal of any residual concentrations of active oxidation products of the LP system. Successful field trials of this system have been conducted in many countries where refrigerated transport is currently not available. The International Dairy Federation prepared a code of practice for the preservation of raw milk by the LP system (International Dairy Federation, 1988) that was submitted to the Joint FAO/WHO Codex Committee on Food Hygiene. This committee advanced the guidelines for review at the next session of the joint FAO/WHO Committee of Government Experts on the Code of Principles Concerning Milk and Milk Products, with the recommendation of rapid adoption by the Codex Commission at its meeting in July 1991 (Anonymous, 1989b).

Much of the development of the LP system focused on preservation of raw milk where refrigeration is not available during transportation, but research results reported by Kamau & co-workers (1991) demonstrate the potential of this system for extending the shelf life of milk transported, treated, and held in a more typical manner. Their results show that the addition of thiocyanate and H_2O_2 to raw milk, followed by pasteurization, significantly extended shelf life at $10°C$ compared with untreated milk (Kamau et al., 1991).

B. Preservation of Other Products

Use of the LP system to preserve raw milk uses the naturally occurring lactoperoxidase in milk. Although the LP system is not currently approved for food use, tests are being conducted in other food products such as cottage cheese and ice cream. Lactoperoxidase for use in these products is purified from nonpasteurized milk or sweet whey. Depending on the product in which the LP system is to be used, thiocyanate and H_2O_2 must be added also. Thiocyanate does occur naturally at levels as high as 600 ppm in plants and vegetables, particularly cruciferous vegetables (e.g., cabbage, cauliflower), but most products being investigated require the addition of sodium or potassium thiocyanate (currently not acceptable food additives). H_2O_2 for the reaction can be formed with the addition of glucose oxidase and glucose. Results of these tests indicate significant extension of shelf life using the LP system, but more acceptable methods of thiocyanate enrichment need to be found to help attain regulatory approval.

IV. Sulfhydryl Oxidase

Sulfhydryl oxidase catalyzes the oxidation of free sulfhydryls, forming a disulfide bond and H_2O_2. This enzyme has been isolated from a variety of sources such as bovine milk, bovine kidney, *Aspergillus niger*, *Aspergillus sojae*, and other microbial species. In several publications, Swaisgood and co-workers have reported the purification and characterization of sulfhydryl oxidase from bovine milk (see Chapter 8).

A. UHT Milk Treatment

Bovine milk sulfhydryl oxidase was used initially to treat UHT (ultra-high temperature) milk for the elimination of the "cooked" flavor (Swaisgood, 1977,1978). In this application, the sulfhydryl oxidase catalyzes the oxidation of the thiols responsible for the cooked flavor, thereby eliminating the off-flavor. Swaisgood *et al.* (1987) report the use of an immobilized sulfhydryl oxidase column for the continuous treatment of UHT milk. Starnes *et al.* (1986) isolated sulfhydryl oxidase from *A. sojae* and demonstrated its ability to remove the "burnt" flavor of UHT milk, but there is no commercial use of either sulfhydryl oxidase in milk at this time. Perhaps if a higher value product could be produced with this technology, sulfhydryl oxidase could be commercialized.

B. Dough Strengthening

The use of sulfhydryl oxidase for the oxidation of sulfhydryl groups with the formation of disulfide bonds has been speculated as a method of strengthening dough in baking applications. Using the bovine milk enzyme, Kaufman and Fennema (1987) evaluated the effectiveness of sulfhydryl oxidase to strengthen wheat flour dough. Although several parameters were measured, the investigators were unable to detect significant differences. A European patent (Haarasilta *et al.*, 1989) demonstrates the effectiveness of sulfhydryl oxidase from *A. niger*, in conjunction with other enzymes, to strengthen dough and improve the quality of baked products made with weak flour. Further testing is necessary to demonstrate commercial utility of this process.

References

Ainsworth, S. J. (1989). Hydrogen peroxide producers look to new markets for growth. *Chem. Eng. News* Sept. 11, 1989.

Anonymous (1989a). Peroxide growth in double digits. *Chem. Market. Rep.* June 25, 5.

Anonymous (1989b). Codex panel to survey expert advice on Listeria. *Food Chem. News* **31**(40), 6–8.

Antrim, R. L., and Taylor, J. B. (1990). U.S. Patent No. 4,961,939.

Baldwin, R. R., Cambell, H. A., Theissen, R., and Lorant, G. J. (1953). The use of glucose oxidase in processing foods with special emphasis on desugaring of egg whites. *Food Technol.* **7**, 275–282.

Ball, H. R., and Cotterill, O. J. (1971a). Egg white catalase: 1. Catalatic reaction. *Poultry Sci.* **50**(2), 435–446.

Ball, H. R., and Cotterill, O. J. (1971b). Egg white catalase: 2. Active component. *Poultry Sci.* **50**(2), 446–452.

Barker, S. A. (1986). Extending enzyme applications in the food industry. *In* "Chemical Aspects of Food Enzymes" (A. T. Andrews, ed.). pp. 137–148. University of Reading, Reading, England.

Borch, E., Wallentin, C., Rosen, M., and Bjorck, L. (1989). Anti-bacterial effect of the lactoperoxidase/thiocyanate/hydrogen peroxide system against strains of *Campylobacter* isolated from poultry. *J. Food Protect.* **52**(9), 638–641.

Bottomley, R. C., Colvin, R. D., and Blanton, M. van. (1989). U.S. Patent No. 4,888,184.

Code of Federal Regulations (1989). U.S. Code 21.

Denis, E., and Ramet, J. P. (1989). Antibacterial activity of the lactoperoxidase system on *Listeria monocytogenes*. *Microbiol. Alim. Nut.* **7**, 25–30.

Earnshaw, R. G., and Banks, J. G. (1989). A note on the inhibition of *Listeria monocytogenes* NCTC 11994 in milk by an activated lactoperoxidase system. *Lett. Appl. Microbiol.* **8**, 203–205.

Finnsugar (Bulletin No. 100). "Fermcolase® in Cheesemaking for the Catalytic Decomposition of Hydrogen Peroxide." Finnsugar Bioproducts, Schaumburg, Illinois.

Haarasilta, S., Pullinen, T., Valisanen, S., and Tammersalo, I. (1989). European Patent No. 338452.

International Dairy Foundation (1988). "Code of Practice for the Preservation of Raw Milk by the Lactoperoxidase System." Bull. No. 234. International Dairy Federation, Brussels.

Kamau, D. N., Doores, S., and Pruitt, K. M. (1990). Enhanced thermal destruction of *Listeria monocytogenes* and *Staphylococcus aureus* by the lactoperoxidase system. *Appl. Environ. Microbiol.* **56**(9), 2711–2716.

Kamau, D. N. Doores, S., and Pruitt, K. M. (1991). Activation of the lactoperoxidase system prior to pasteurization for shelf-life extension of milk. *Milchwissensch.* **46**(4), 213–214.

Kaufman, S. P., and Fennema, O. (1987). Evaluation of sulfhydryl oxidase as a strengthening agent for wheat flour dough. *Cereal Chem.* **64**(3), 172–176.

Klavons, J. A., and Bennett, R. D. (1986). Determination of methanol using alcohol oxidase and its application to methyl ester content of pectins. *J. Agric. Food Chem.* **34**(4), 173–175.

Kosikowski, F. V. (1977). Control of spoilage bacteria in cheese milk. *In* "Cheese and Fermented Milk Foods," pp. 292–303. Kosikowski and Associates, Brooktondale, New York.

Loew, O. (1901). "Catalase. A New Enzyme of General Occurrence, with Special Reference to the Tobacco Plant." Report No. 68. U.S. Department of Agriculture.

Low, N., Jiang, Z., Ooraikul, B., Dokhani, S., and Palcic, M. M. (1989). Reduction of glucose content in potatoes with glucose oxidase. *J. Food Sci.* **54**(1), 118–121.

McLeod, R., and Ough, C. S. (1970). Some recent studies with glucose oxidase in wine. *Am. J. Enol. Vitic.* **21**(2), 54–60.

Miles Laboratories (1984). "Catalase L, Application Information." Miles Laboratories, Elkhart, Indiana.

Ohlmeyer, D. W. (1957). Use of glucose oxidase to stabilize beer. *Food Technol.* **11**(10), 503–507.

Oleofina. The lactoperoxidase system: A new food preservation system. Synfina-Oleofina, Oleofina Division, Brussels.

Provesta Enzymes (1987a). "Alcohol Oxidase RL-100 and RL-150 Substrates and Inhibitors." Technical Bulletin No. 2. Provesta Corporation, Bartlesville, Oklahoma.

Provesta Enzymes (1987b). "Alcohol Oxidase RL-100." Technical Data Sheet. Provesta Corporation, Bartlesville, Oklahoma.

Provesta Enzymes (1987c). "Biosynthesis of Hydrogen Peroxide Using Alcohol Oxidase Enzyme." Technical Bulletin No. 4. Provesta Corporation, Bartlesville, Oklahoma.

Pruitt, K. M., and Reiter, B. (1985). Biochemistry of peroxidase system: Antimicrobial effects. In "The Lactoperoxidase System: Chemistry and Biological Significance" (K. M. Pruitt and D. Tenuovo, ed.), pp. 144–178. Marcel Dekker, New York.

Qualcepts (1986). "FloAm® 200." Qualcepts Nutrients, Minneapolis, Minnesota.

Reinke, B. C., Hoag, L. E., and Kincaid, C. M. (1963). *Proc. Am. Soc. Brew. Chem.*, 175.

Scott, D. (1975). Applications of glucose oxidase. In "Enzymes in Food Processing" (G. Reed, ed.). pp. 519–547. Academic Press, New York.

Scott, D., and Hammer, F. (1960). Assay of commercial catalase. *Enzymologia* **22**, 194.

Siragusa, G. R., and Johnson, M. G. (1989). Inhibition of *Listeria monocytogenes* growth by the lactoperoxidase–thiocyanate–H_2O_2 antimicrobial system. *Appl. Environ. Microbiol.* **55(11)**, 2802–2805.

Smith, V. J., Green, R. A., and Hopkins, T. R. (1989). Determination of aspartame in beverages using an alcohol oxidase enzyme electrode. *J. Assoc. Off. Anal. Chem.* **72(1)**, 30–33.

Starnes, R. L., Katkocin, D. M., Miller, C. A., and Strobel, R. J. Jr. (1986). U.S. Patent No. 4,632,905.

Suwa, Y., Kobayashi, T., Ishigouoka, H., Ono, K., and Ono, M. (1989). U.S. Patent No. 4,867,990.

Swaisgood, H. E. (1977). U.S. Patent No. 4,053,644.

Swaisgood, H. E. (1978). U.S. Patent No. 4,087,328.

Temple, D., and Ough, C. S. (1975). Inhibition of catalase activity in wines. *Am. J. Enol. Vitic.* **26(2)**, 92–96.

Underkofler, L. A. (1957). Properties and applications of the fungal enzyme glucose oxidase. In "Proceedings of the International Symposia on Enzyme Chemistry." pp. 486–490.

United States Department of Agriculture (date). "Egg Inspectors Handbook," AMS-PY Instruction No. 910. U.S. Government Printing Office, Washington, D.C.

Villettaz, J.-C. (1986). European Patent Application 0,194,043,A1.

Yang, H. Y. (1955). *Food Res.* **20**, 42.

Zetalaki, Z. (1964). *Elemenz. Impar.* **18(8)**, 178.

CHAPTER 11

Milling and Baking

BRUNO G. SPROESSLER

I. Introduction

Enzymes are involved in the formation of starch, gluten, hemicelluloses, and lipids during ripening and are the key to dough fermentation and bread making, since they degrade these polymers. The activities of enzymes in the flour used for bread making depend on the climate, the type of grain, and the harvesting conditions. Dry-harvested grain usually contains too low a level of enzyme activity. During germination, the activities greatly increase. A humid climate during the harvest favors germination and can lead to grain with too high an enzyme content, particularly of amylase. Such sprouted wheat cannot be used for bread making.

In times when bread was made by hand, the baker was still able to make good bread from flour poor in enzymes. He could, for example, prolong the action of the enzymes by extending the dough fermentation time. In modern manufacturing processes, the dough must be adapted to the baking process, which is done by adding enzyme preparations. Flours that are poor in enzymes are adjusted at the mill to a standardized activity, by adding malted flour or fungal enzymes. The ideal enzyme quantity for each manufacturing process is added by the bread-maker. In many countries, enzyme activity is reached by adding a combination of bread improvers containing enzymes. In the United States, this state can be achieved by adding enzyme tablets that are dissolved in the baking water. The use of enzymes is also of particular importance in modern baking because tremendously varied products, such as bread, rolls, cookies, and crackers, are made from the same flour. In addition, the demands for machinability of the doughs and constant quality of the baked goods are very high. The baked items must fit into prefabricated packaging containers and fill these optimally.

The typical constituents of wheat flour are shown in Fig. 1. They can vary

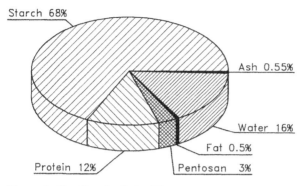

Figure I   Constituents of wheat flour.

to a limited extent. Starch and protein are the main components. The ash content depends on the milling conditions. For example, ash content of 0.55% multiplied by 1000 leads to the flour type 550. The starch granules are partly damaged during milling. The proteins and pentosans are partly soluble and partly insoluble in water. Using enzymes, the properties of the polymer substances starch, protein, and pentosan can be modified. The controlled use of enzymes presupposes methods that determine the activity of the enzymes indigenous to the flour. Determination of the enzyme activities has been proved, in itself, to be insufficient, since no correlation exists with practical efficiency. Therefore, practical rheological methods are required for examining the dough, by means of which the enzyme activities can be evaluated. Moreover, the development of enzymes for flour treatment demands knowledge of manufacturing methods and of the practical needs of the manufacturer. Whereas, in the past, malt was the only substance available for the enzymatic treatment of flour, current microbiological techniques have made available a whole variety of fungal and bacterial enzymes with which the machinability of the flours can be influenced in a controlled manner. Even today, the enzymes are selected largely on an empirical basis. The baking test still finally determines the quality of the enzyme products. Many imponderables still remain in this complex interaction of dough constituents and enzymes with widely varying activities. However, many test results help fit more and more pieces into the "jigsaw puzzle" of baked goods manufacturing.

II. Enzymes in Flour

Flour contains many enzymes of which α- and β-amylases, proteinases, peptidases, hemicellulases, and oxidases are technically most important. These enzymes are normally present at low levels in ungerminated wheat. The low moisture content keeps them inactive during the storage of wheat and flour.

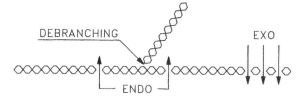

Figure 2 Strategy of enzymatic degradation of natural polymers.

On germination, the activity of most flour enzymes increases to such an extent that sprouted flour is unsuitable for baking.

The enzymes of the greatest technical importance are hydrolases, which act on natural macromolecules and degrade them ultimately to the monomers. This degradation is accomplished effectively by the combined action of endo- and exoenzymes shown in simplified form in Fig. 2. For branched macromolecules, there is normally also a debranching enzyme, for example, pullulanase in the case of starch enzymes. The endoenzymes change the rheological properties of the macromolecule by splitting a few bonds, whereas the exoenzymes are more important in producing sugars and amino acids, which serve as food for the yeast during dough fermentation, as flavor precursors, and as crust-forming compounds.

A. Amylases

1. α-Amylases

Up to 22 multiple forms of α-amylase have been detected in wheat (Marchylo *et al.*, 1980). Some multiple forms are a result of the genetic diversity of cereal grains. Certain multiple forms could also be artifacts of the isolation procedure. Perhaps some of the isoenzymes bind to their inhibitors. α-Amylase can be detected in immature wheat. The amounts decrease with further maturation and desiccation. On germination of the grain, the amylase activity increases several hundredfold over a 4–5 day period (Kruger and Lineback, 1987). The α-amylases of malted wheat have a pH optimum of about 5.5, pI values between 6 and 6.2, and molecular masses of 42–42.5 kDa measured by SDS gel electrophoresis (Tkachuk and Kruger, 1974). Isoelectric focusing also showed that germinated wheat and rye contain multiple forms of α-amylases that could be divided into low (pI < 5.5) and high (pI > 5.8) pI groups (MacGregor *et al.*, 1988). Zawistoswka *et al.* (1988) separated the two groups of α-amylases by immobilized metal affinity chromatography. The α-amylases of immature wheat also exist in multiple form, with a lower pI between 4.6 and 5.1 and higher molecular mass between 52 and 54 kDa. These enzymes are more heat labile but more stable under acid conditions than the germinated α-amylases. Germinated α-amylases adsorb to starch granules (Kruger and Marchylo, 1985). They are preferentially adsorbed to small granules, which they degrade.

An endogenous α-amylase inhibitor was detected in various cereal grains by immunochemical methods (Weselake *et al.*, 1985).

2. β-Amylases

β-Amylases belong to the exoenzyme class (Fig. 1). They split alternate α-(1,4)-D-glucosidic linkages in starch or dextrins stepwise from the nonreducing end. The action stops in the region of α-(1,6)-D-glucosidic linkages. Branched starch can be degraded to maltose and β-limited branched dextrin, which is used in the Sandstedt, Kneen, and Blish (SKB) method (Sandstedt *et al.*, 1939) for determination of α-amylase activity. Multiple forms of β-amylase result from genetic variation, but also from aggregation by disulfide bonds. Up to 80% of the β-amylases are attached to glutenin and are insoluble in water. On germination, the soluble form increases but the total amount does not change significantly. By treatment with papain, insoluble β-amylase can be solubilized to five distinct species with pI values identical to species present in germinated wheat (Kruger, 1979). The multiple forms have molecular masses of about 64.2 kDa, but differ slightly in their pH optima (4.5–5.5) and their pI values (4.1–4.9). β-Amylases are heat labile as is wheat α-amylase, but they are more acid stable. They do not require calcium for stability, but can be inactivated with sulfhydryl reagents.

3. Methods For Determining α- and β-Amylases in Flour

The amount of amylases must be known before proper compensation of enzyme deficiencies by addition of malted flour or a microbial amylase preparation can occur. It is difficult to determine α- and β-amylase separately because of their combined effect on starch degradation. The methods used, therefore, more or less determine both activities. We can assume that, in flour, β-amylase activity is always present in sufficient amounts (Uhlig and Spröessler, 1972). The limiting activity is generally α-amylase. The maltose value [American Association of Cereal Chemists (AACC), 1983, Method 22-15] measures, by the ferricyanide method, the sugars liberated from 5 g flour in 46 ml buffer at pH 4.7 by the action of α- and β-amylase within 1 hr at 30°C. Another practice-oriented method is the determination of the gassing power (AACC, 1983, Method 22-14), in which gas production is measured after 6 hr in a fermenting dough. Both methods depend not only on enzyme activity, but also on the number of damaged starch granules in the flour, which are attacked much more easily enzymatically.

The Hagberg falling number (AACC, 1983, Method 56-81 A) and the Brabender amylograph method (AACC, 1983, Method 22-10) are sensitive to heat-stable α-amylase. In both methods, flour is the substrate for the enzyme and, at the same time, the enzyme preparation. The flour–water suspension is heated. At about 60°C, the starch gelatinizes and the viscosity increases.

With increasing α-amylase activity of the flour, more gelatinized starch is broken down and the viscosity of the starch paste decreases. These two methods are adapted to the heat-stable wheat α-amylase, which can act on gelatinized starch at temperatures up to 90°C. Both methods are not able to determine the activity of fungal α-amylase supplements, because heat labile fungal α-amylase is inactivated at about 65°C and therefore can act on gelatinized starch only slightly. However, both methods are influenced by the properties of the starch. The falling number value of flours with nearly the same α-amylase activity increases with the rising gelatinizing temperature of the starch (Sproessler and Schmerr, 1986). A modified falling number method has been described by Perten (1984) and Sproessler (1982). The flour is mixed with gelatinized starch and the falling time is measured at 30°C. With this method, all amylases can be determined.

According to our experience, the nephelometric method is most suitable for determining both cereal and fungal α-amylase. The enzymes are extracted from the flour and 0.2 ml extract is added to a cloudy substrate suspension of amylopectin or, better, of β-limited dextrins. The turbidity of the suspension decreases according to the α-amylase concentration. The decrease can, for example, be determined with the Model 191 Grain Analyzer (GAA; Perkin–Elmer Corporation, Oak Brook II) within 2 min of incubation at 37°C. By extracting the enzyme, the influence of flour components is diminished and the enzyme activity can be determined on a defined substrate (O'Connell et al., 1980; Kruger and Tipples, 1981; Fretzdorff and Weipert, 1983).

B. Proteases and Peptidases

1. Flour Proteins

The proteins that contribute to dough formation can be divided into glutenins and gliadins, according to their different solubility in a 70% ethanol solution. Both protein fractions are very complex. The rheological properties of flour are determined by the quantity and ratio of gliadin to glutenin (Belitz et al., 1987). At an increasing proportion of gliadin, the resistance to extension decreases and the extensibility increases. In the presence of reducing agents and anionic detergents such as sodium dodecyl sulfate and 2-mercaptoethanol, the glutenins can be dissolved (Bietz and Wall, 1972) and separated chromatographically. The high molecular weight (HMW) subunits have sizes of 95,000–140,000 daltons. The baking properties correlate with the pattern of HMW subunits (Hübner and Wall, 1976; Ng and Bushuk, 1988).

The gluten is strengthened by oxidative polymerization of the proteins. A reduction of the disulfide bonds weakens the gluten. Proteinases weaken the gluten structure by splitting the protein chains. Peptidases form amino acids and can improve the flavor. Both types of enzymes are present in flour.

2. Proteases

Fox and Mulvihill (1982) have compiled the results of research on proteases and peptidases in flour. Most endoproteases are sulfhydryl proteases. The carboxypeptidases of flour contain serine at the active site. Aminopeptidases still have not been detected definitely (Stauffer, 1987). The protease activity is distributed irregularly throughout the cereal grain. The activity increases about 15-fold during germination. Therefore, malt contains protease as well as amylase. Too high a protease activity in relation to the amylase can, therefore, be a hindrance during bread making. Sulfhydryl compounds such as cysteine and glutathione increase the protease activity; oxidants such as bromate and iodacetic acid are inhibitors. Only some of the proteases can be extracted with water.

The proteolytic activity of the extract was a maximum at pH 3.8, whereas maximal activity for whole flour was obtained at pH 4.4. The extractable enzymes are less heat stable. The thermostable nonextractable proteases are complexed with the proteins of the flour and, thus, protected against thermal denaturation (Drapon and Godon, 1987). The absolute number of proteases in the flour is not yet known. Certainly, not all are determined genetically but also arise by autodigestion, microbial infection, or infestation by insects.

3. Peptidases

Flour extracts contain free amino acids that have been formed by the combined action of proteases and peptidases. A peptidase with a molecular mass of 59,000 daltons and a pH optimum of 8.6 – 9.0 has been described by Kruger (1971). The enzyme does not break down hemoglobin or gelatin. Two carboxypeptidases with pH optima of about 4.2 were isolated by Preston and Kruger (1976). They release amino acids from the carboxylic end of hemoglobin and gelatin.

4. Measuring Activity of Proteases in Cereal Products

Analytical determinations of activity are carried out with protein substrates that do not occur in flour. Therefore, the values obtained fail to address the practical effect of the proteases on the baking properties. Rheological measurements are required to judge their effectiveness. Cereal chemists use the method of analytical determination recommended by the AACC (1983, Method 22-60). Its principle is to measure the amount of hydrolyzed protein that cannot be precipitated by trichloracetic acid after enzymolysis under well-defined conditions. This determination can be made by measuring absorption at 275 nm or by using the Folin reagent. Hemoglobin is used as the preferred substrate because of its uniformity and solubility. The activities are expressed conventionally in "hemoglobin units" (HU) per gram. Wheat flour has activities of about 1 – 3 HU and malted wheat flour of about 25 – 50 HU per gram.

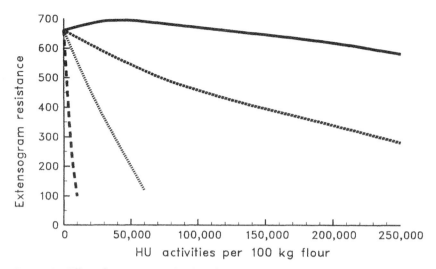

Figure 3 Effect of proteases on dough resistance. The doughs prepared with proteases were fermented (135 min at 28°C). Bacterial protease (VERON P®), — — —; papain, - - -; fungal protease 2 (VERON PS®), - - -; fungal protease 1, ——. Resistance was measured with the Brabender Extensograph.

Many other substrates such as casein, gelatin, and gluten have been used, but none of them have been satisfactory. The action of enzymes on these substrates does not correlate with their performance in bread making (Fig. 3).

C. Nonstarch Polysaccharidases

Our knowledge of the indigenous cellulases, β-glucanases, hemicellulases, and pentosanases is still fragmentary. Drews (1969) had stressed the importance of these enzymes and that of soluble pentosans in the making of rye bread. Nonstarch polysaccharides comprise a series of substances consisting of cellulose, pentosans (hemicellulose), β-glucans, mannans, and gluco- and galactomannans. These polysaccharides form complexes with proteins and ferulic acid. They influence the rheological properties of the dough by forming complexes, by their high capacity to bind water, and by their ability to swell. These complex structures can be influenced by enzymes that hydrolyze polysaccharides, but also by proteases and oxidases. Current knowledge of the nonstarch polysaccharides is reviewed by Meuser and Sukow (1986).

Wheat flour contains approximately 2–3% pentosans, 25% of which are soluble in water. Water-soluble, enzyme-extractable, and total pentosans were estimated in whole wheat and laboratory-milled products by Hashimoto et al. (1987). The enzyme-extractable pentosans were estimated after treatment with enzymes from *Trichoderma viride*. Total pentosans were determined after hydrolysis with 2N HCl at 100°C by the orcinol – HCl method. Wheat flours contain 0.7–0.83% soluble pentosans, 1.66–1.86% enzyme-extractable pentosans, and 1.5–2.12% total pentosans. In rye flour, about 5% pentosans are found, 40% of which are soluble (Meuser et al., 1985).

The water-soluble pentosans contain arabinoxylans and arabinogalactans. Arabinoxylans have an elongated structure of one long chain of β-(1,4)-linked D-xylopyranose units. Every second or third xylose unit has a side chain consisting of L-arabinose. The arabinogalactans consist of a β-(1,3)- and β-(1,6)-linked galactose chain with individual arabinose side chains. These chains are bound covalently with a peptide and thus form a highly branched structure (Strahm et al., 1981). The high solubility of arabinogalactan is attributed to the large degree of branching within the molecule.

Pentosans contribute to the physical properties of dough mostly by nonspecific interaction with gluten proteins. The insoluble pentosans—the so-called tailings—consist chiefly of arabinoxylans that oxidatively form an insoluble network, perhaps via ferulic acid (Geissman and Neukom, 1973). In conjunction with a glycoprotein, a gel forms by means of oxidation that is made of 25% protein, 75% arabinoxylan, and small amounts of ferulic acid (Neukom, 1976). Proteases cause this gel to become fluid. Using proteases, therefore, can alter the properties. The improvement of flour quality by means of oxidases surely results not only from the oxidation of the thiol groups in the gluten protein and the unsaturated fatty acid of the flour lipids, but also from the oxidative modification of the glycoproteins.

The pentosans and glycoproteins absorb a great deal of water and cause gel formation. They are not denatured by heat and do not retrograde in the cold. They thus prolong the shelf life (Casier et al., 1973; Kim and D'Appolonia, 1977) and are able to retain gas in the dough (Hoseney, 1984). These properties make it possible to manufacture bread from rye flours.

In keeping with their higher pentosan proportion and their major contribution to dough formation, pentosanases and β-glucanases or hemicellulases play a particularly important role in the treatment of rye flour. Bruemmer (1972) disclosed the relative effects on the viscosity of rye flour suspensions exhibited by various enzymes such as pentosanase and α-amylase. Rye dough treated with pentosanases gave softer doughs and increased bread volume (Weipert, 1972). Also in treatment of wheat flour, especially of whole wheat flour, hemicellulases or pentosanases can change the swelling and water absorption properties of the dough and improve bread quality (ter Haseborg and Himmelstein, 1988). They improve dough tolerance and ovenspring, bread volume, shape, and texture (Moonen, 1990). During germination of wheat, endo-(1,4)-β-xylanase and endo-(1,3),(1,4)-glucanase activity is very low for the first 3 days but increases rapidly on the fourth and fifth day. Extracts from wheat germinated for 5 days reduce the viscosity of arabinoxylan from rye (Cordner and Henry, 1989).

Gaines and Finney (1989) added several commercial enzyme preparations with protease, cellulase, and β-glucanase to cookie doughs. Cellulase preparations from *Trichoderma reesei* were effective in maintaining stability of dough consistency and cookie size when doughs were held for 2 hr before extrusion.

D. Oxidases

1. Lipoxygenase

Wheat grain contains a relatively high level of lipoxygenase activity in the germ and a low level in the endosperm (Rohrlich et al., 1959). Several isoenzymes have been detected electrophoretically in various types of wheat (Guss et al., 1967). During storage of the wheat, the activity decreases. The temperatures that occur when mixed feed is pelletized are sufficient to inactivate the lipoxygenase almost completely (van Ceumern and Hartfiel, 1984). The lipoxygenases indigenous to the flour only oxidize free linoleic and linolenic acids.

This type-1 lipoxygenase only forms 9-hydroperoxide (Graveland, 1972; Grosch et al., 1976). Soy also contains type-2 lipoxygenases. These enzymes oxidize all lipids containing linoleic acid. During this reaction, peroxy radicals arise that cooxidize the carotenoids of the flour (Weber and Grosch, 1976; Schieberle et al., 1981) and improve the strength of the dough and the baking properties of wheat flour (Kieffer and Grosch, 1980).

The type-2 lipoxygenase of soy has its optimal activity at pH 6.5 and loses its activity more rapidly during storage than the technologically inactive type-1 enzyme that has a pH optimum of 9. Soy preparations used to improve flours therefore should be evaluated by means of activity determination at pH 6.5. Flour improvement using type-2 lipoxygenase probably occurs via the oxidation of low molecular weight sulfhydryl compounds and the cross-linking of protein molecules of the gluten by radicals (Grosch et al., 1976). Flour improvement is not achieved in flours from which the lipids have been extracted but it is improved by the addition of an emulsion of linoleic acid.

2. Peroxidases

Wheat flour contains peroxidases (Fretzdorf, 1980) that presumably can cross-link phenolic constituents such as ferulic and vanillic acid (see Section II, C). However, the pH optimum of the wheat peroxidase is about 4.5 (Kieffer et al., 1981). In the pH range 5–6 of wheat doughs, the enzyme is barely effective, which may be the reason no correlation has been found between the peroxidase content of wheat flours and dough properties (Kieffer et al., 1981). Horseradish peroxidase still has approximately 80% of its optimal activity at the pH of the dough. The addition of horseradish peroxidase in combination with H_2O_2 or pyrocatechol leads to strengthening of the dough. The replacement of pyrocatechol by ferulic or caffeic acid provides similar results. H_2O_2 can be substituted by the H_2O_2-producing system glucose oxidase and glucose. Phenol oxidase from mushroom shows a comparable effect (Nishiyama *et al.*, 1979).

E. Lipases and Esterases

Lipase activity is found mainly in the germ and aleurone layers. Not more than one-third of the total lipase of wheat occurs in patent flours. The distinction between lipases, which only hydrolyze emulsified esters, and true esterases, which act on esters in solution, has not always been clear. In most studies on wheat lipase, esterolytic rather than lipolytic activity has been measured (Fox and Mulvihill, 1982). The activity can be assayed on triacetin or tributyrin. Isoelectric focusing demonstrated the presence of 17 esterase fractions in a crude wheat extract (Cubadda and Quattrucci, 1974). All fractions were inhibited completely by diisopropyl fluorophosphate (DFP), but not by EDTA or *p*-chloromercuric benzoate. Lipases need Ca^{2+} and bile salts for optimal activity with olive oil as substrate (Caillat and Drapon, 1970).

The storage of grain under adverse conditions results in an increase in free fatty acids, indicating grain deterioration during storage. Yeast has a much higher level of lipolytic enzymes than flour. The influence of these lipases and esterases on the functional properties of flour is not well known. All these enzymes are inactivated during baking (Reed and Thorn, 1971). The function of lipids and lipases is described by Nierle *et al.* (1981) and Mohsen *et al.* (1986). Protein/lipid interactions in the dough have been reported by Bushuk (1986), Chung *et al.* (1982), and Zawistowska *et al.* (1984).

III. Use in the Baking Industry

A. Wheat Bread and Rolls

As early as 3500 B.C., flat types of bread were baked in ovens. Because of the sour dough occasionally used even then, the flat bread rose in the oven and also stayed fresh somewhat longer. It is not known whether the Egyptians used malt as a bread improver. At any rate, they were familiar with the procedure for obtaining malt by moistening barley and allowing it to germinate. In an eulogy by Dinias (360 B.C.), Greek bread was described as beautifully white with an admirable taste that was varied by adding salt, oil, cheese, or sesame. Bread is still made in a similar way today. However, industrially manufactured yeast and the improved knowledge of the use of flour additives make is possible to obtain a great variety of baked goods of consistently high quality.

1. Dough Process

Basically, baked goods are prepared using the following steps. The ingredients flour, water, yeast, and additions are kneaded to a dough. The dough processing includes fermentation and leavening, dividing, molding, and shaping.

In the straight dough process, which is used mostly in Europe, the dough is fermented 1–3 hr. In the sponge and dough process consisting of two steps, a sponge is prepared from about 60% of the total flour and water quantity, yeast (2.5%), and enzymes. Mixing is done for 1 min at low speed and 3 min at high speed. The fermentation time is 3–5 hr at about 30°C. For dough preparation, the remaining 40% flour and water, salt (2%), sugar (8%), and shortenings are added and mixed. The mixture is then fermented (20 min), divided, proofed for about 55 min at 42°C, and baked (18 min at 230°C). No-time dough is prepared by immediately dividing the dough after mixing without a fermentation time and proofing for about 55 min at 42°C. This short-time process requires fungal amylase and proteinase, which stabilize the dough, in addition to oxidizing agents such as potassium bromate and ascorbic acid. The Chorleywood Bread Process is a special no-time dough process used in the United Kingdom that has the particular feature of using a vacuum during dough mixing.

2. Dough Development and Baking

Flour doughs contain 60–70% water, depending on the water-binding capacity of the proteins, pentosans, and damaged starch. During mixing, water is bound initially to insoluble protein (glutenin). A water content of about 30–35% free water dissolves soluble flour components such as gliadins. The gluten is the structure maker; gliadin acts as a plasticizer. The well-developed gluten network is filled with starch granules, admixed with pentosans and lipids. Microscopic investigations support the significance of the gluten network, filled with starch present as native, partially, or wholly gelatinized granules (Wassermann and Doerfner, 1974; Pomeranz et al., 1984). Air bubbles are introduced during mixing and CO_2 bubbles by yeast fermentation. The yeast converts sugar in a sequence of 12 enzymatic reactions to carbon dioxide gas (CO_2). Theoretically, 1 g glucose yields 0.4 g, that is, 249 ml CO_2. In practice, this value is lowered by side reactions that yield glycerol, organic acids, alcohol (Reed and Peppler, 1973), and precursors for flavor development.

The CO_2 gas expands the dough, provided that the protein network is able to retain the gas and to expand without rupture. The baking properties also depend on the distribution of the gas and the size and stability of the gas bubbles, which can be influenced by kneading and baking improvers (Sluimer, 1990). During the baking process, the starch gelatinizes. Proteins are denatured at the same time and build a rigid structure. Water is released and picked up by the starch. In the heat of the oven, a dark crust encloses a

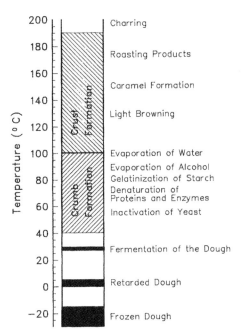

Figure 4 Changes in dough and baking affected by temperature.

fine, tender crumb. Figure 4 summarizes the changes during baking. These changes take place in both the crust and the crumb. The browning reaction of the crust involves caramelization of sugars and interaction between sugars and amino acids or peptides. At the same time, flavor and taste compounds are formed. Both browning and flavor forming are influenced by enzymatically produced sugars and peptides during dough fermentation and early baking. At early stages of baking, the increase in temperature enhances enzymatic activity and growth of yeast. At about 50–60°C, the yeast is killed. Above that temperature, starch gelatinization, protein denaturation, enzyme inactivation, and alcohol and water evaporation take place. The crumb temperature inside the loaf does not exceed 95–100°C. The temperatures in the crust are much higher. The taste of the baked bread depends on dough composition, fermentation time, addition of components such as enzymes, and the baking process. Bread flavor is created by the combined effects of many substances. The whole breadmaking process, including crumb and crust formation and flavor development, can be influenced by amylases, proteinases, pentosanases, β-glucanases, hemicellulases, lipases, and oxidases.

3. Effects of Amylase Supplementation on Bread Quality

A certain number of starch granules are ruptured when wheat is milled into flour. This damaged starch binds water and can be attacked in the dough by amylases and converted into dextrins, which are then hydrolyzed by β-amylase to maltose or by amyloglucosidase to glucose. These reactions constantly supply the yeast with fermentable sugar. A uniform yeast fermentation and gas production in the dough is reached. The addition of sucrose or glucose results in gas production at the wrong stage of the baking process.

Amylases also improve the gas retention properties of the dough through starch modification. The addition of fungal amylase leads to doughs with better machinability and bread with a finer crumb.

The crust color is intensified by means of cereal amylase, but also by fungal amylase. The higher reducing sugar content increases Maillard browning and also gives more flavor. The combined addition of fungal amylase and amyloglucosidase increases the amount of reducing sugars, particularly in sourdough (Nagargoje et al., 1984). The importance of amylases in bread making has been reviewed by Reed and Thorn (1971), Fox and Mulvihill (1982), Grampp (1982), Harinder et al. (1983), Linko and Linko (1986), Drapon and Godon (1987), Peppler and Reed, (1987), Cauvain and Chamberlain (1989), and Uhlig et al. (1988).

Microbiological amylase products usually are standardized to SKB units/g (Sandstedt et al., 1939). The highly active amylase concentrates are diluted with starch or maltodextrin in such a way that they can be distributed evenly in practice in the 100 ppm range. The activity of the malted flour may vary from type to type. It is not possible to dose different exogenous amylases based on SKB activity, as is shown in Table I.

With flours poor in maltose, fungal amylase can be added up to 150,000 SKB per kg flour without producing a negative effect. In contrast, malted flour may be used only up to approximately 15,000 SKB units and bacterial amylases only to about 1000 SKB per 100 kg flour (Sproessler, 1972). The low dosage tolerance is related to the thermal stability of the enzymes. Table II shows the results of a straight-dough baking test. Amylases increase the loaf volume. Fungal amylases additionally improve the crumb score. The comparison of different fungal amylase preparations at the same SKB level gave different results (Table II). The highest volume and the best dough tolerance with a very fine crumb were obtained using a fungal amylase preparation (VERON ST). This fungal amylase product not only contains different fungal amylases, but also contains β-glucanase and hemicellulase activity (Section II, C).

a. Thermal stability of α-amylases from various sources Differences in the heat stability of the amylases are important because of the decisive effect of

TABLE I
Dosage and Inactivation of Different Exogenous Amylases

Commercial product	Concentration (SKB units/g)	Dosage per 100 kg flour (SKB units)	Inactivation temperature (°C)
Fungal α-amylase	500–20,000	25,000–150,000	65
Bacterial α-amylase	50–5,000	Maximum 1,000	80
Malted flour	40–80	8,000–15,000	70

TABLE II
Straight-Dough Baking Test Using Different Amylase Preparations

Preparation	Dosage (g/100 kg)	α-Amylase (SKB/100 kg)	Loaf volume (ml)	Crumb grain	Texture score
Control			1200	Fine	Fair
Malted wheat	100	6,000	1300	Fine	Good
	200	12,000	1350	Coarse	Fair
Fungal amylase A	15	25,000	1370	Fine	Good
Fungal amylase B (VERON ST)	15	25,000	1480	Very fine	Very good
Bacterial amylase	10	1,000	1290	Fine	Fair
Bacterial amylase	30	3,000	1330	Coarse	Very poor

enzyme action during baking when the starch begins to gelatinize at 60–65°C and become highly susceptible to hydrolysis. The temperature in the center of a loaf of white pan bread rises by 4–6°C/min during baking and reaches 90–95°C. Figure 5 shows the baking temperature and baking time of rolls and 1-kg white pan bread. In the case of rolls, even heat-stable amylases have only a short time to act on gelatinized starch before they are inactivated by heat. Fungal amylase is inactivated at 75°C within 10 min (Fig. 6). Malt and bacterial amylases are more heat stable and have an adequate time to act on gelatinized starch. This difference explains the high yield of soluble dextrins in the crumb. The thermostability of α-amylases from cereal, fungal, and bacterial sources in the yeast dough systems was investigated by Asp *et al.* (1985). In this model experiment with small rolls, bacterial α-amylase activity increased during fermentation, proofing, and especially during the last 4 min of baking. An excess of heat-stable amylases results in a sticky crumb. Such loaves cannot be sliced and bagged directly

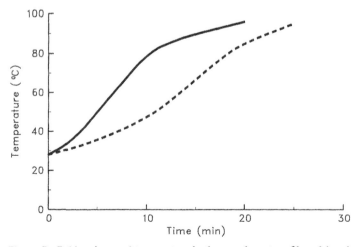

Figure 5 Baking time and temperature in the crumb center of bread (– – –) and rolls (———).

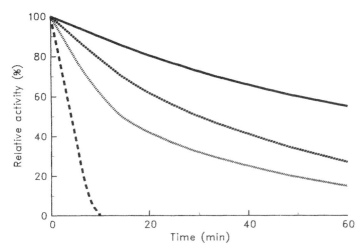

Figure 6 Heat stability of different amylases at 75°C in a 1% enzyme solution in water, pH 5.8, heated for different lengths of time. Enzymes tested were bacterial amylase (——), malt amylase (---), VERON F 25® (- - -), and fungal amylase (— — —).

after baking. Flour of partly sprouted grain or an overdose of malt or bacterial amylase can cause such a serious problem. However, such enzymes can improve the shelf life of the bread if they are used in a controlled manner.

4. Proteases and Dough Consistency

During the preparation of dough, the proteins of wheat flour form a network. A simplified model of a gluten network with disulfide bridges is shown in Fig. 7. Each square represents an amino acid molecule. The gluten structure gets stronger the more sulfur bridges are present. Bromate and dehydroascorbic acid strengthen the dough in this manner. The proteases break down the network by splitting peptide bonds. Therefore, the addition of proteases to a strong dough improves the elasticity and texture of the gluten and the handling properties of the dough. The gas formed by the yeast is retained better. The gluten structure is softened and expands under the increasing gas pressure. Thus it is possible to obtain good dough rise, even with strong flours, and, therefore, make bread with high volume and good crumb structure. Fungal proteases are used in large quantities, particularly in the United States. About two-thirds of the white bread in the United States is produced with the addition of proteases. Protease enzymes relevant to the baking industry have been reviewed by Lyons (1982). The reason fungal proteases are particularly suitable for bread making is that they soften the gluten structure of the dough specifically, and only to a certain extent (Sproessler, 1981). Unintentional overdosing does not lead to decomposition of the structure, as can happen easily when bacterial protease or papain is used. Figure 8 shows that the extension resistance of doughs measured in the

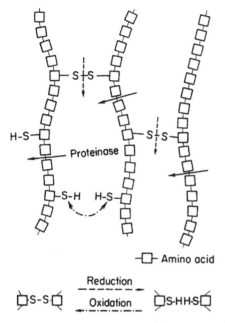

-□- Amino acid

$$\square\text{S-S}\square \quad \xleftarrow{\text{Reduction}} \quad \square\text{S-H H-S}\square$$
$$\xrightarrow{\text{Oxidation}}$$

Figure 7 A simplified model of gluten network and possibilities for modification.

extensograph drops rapidly because of the action of papain and bacterial proteases. The gentle action of the fungal proteases is, according to the most recent tests (Belitz and Kieffer, 1989), attributed to the fact that the fungal proteases only have limited access to the compounds in the dough complex

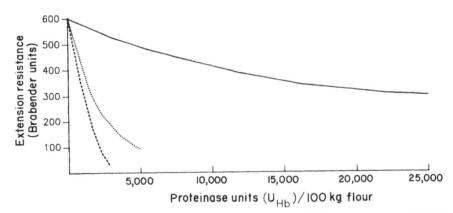

Figure 8 Extension resistance of doughs after treatment with different proteinases: VERON PS®, ——; papain (COROLASE L 10®) ---; bacterial proteinase (VERON P®), — — —. Dough resistance was measured with the Brabender extensograph.

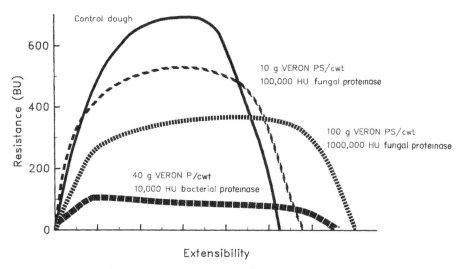

Figure 9 Effect of fungal proteinase (VERON PS®) and bacterial protease (VERON P®) on Brabender extensographs of flour dough.

that they will break down. This limited access also may explain why many fungal proteases in the dough show almost no effect, although they are excellent at breaking down proteins such as hemoglobin and glutenin in solution. The practical efficiency of various fungal proteases therefore cannot be determined analytically, for example, by the HU method, but only by means of methods similar to practical conditions. Suitable equipment is the Brabender extensograph or the Chopin alveograph. These methods are based on rheological measurement of the dough. The extension resistance and the extensibility of doughs thus can be determined. Extensograph curves of flour dough treated with fungal and bacterial proteases are shown in Fig. 9. Added proteases must act within the pH range of the dough (pH 6–5.5). Alkaline or acid proteases are therefore less suitable. Further, it must be considered that flour contains protease inhibitors that affect the added protease products in different ways. Thus Gabor *et al.* (1982) found that the bacterial protease thermitase and brewery enzymes can be inhibited by a serine protease inhibitor derived from wheat; VERON P®, on the other hand, is not influenced by this inhibitor.

5. Bread Staling

Extended shelf life has become of crucial importance to the industrial baker. At least 5% of all bread produced in the United States is lost as a result of staling. We know that retrogradation of starch plays a major role, but gluten, lipids, and pentosans are also factors contributory to staling. Enzymes cannot change the quantity of these polymers but can change their proper-

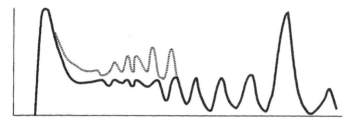

Figure 10 Gel chromatographs of bread crumb extracts on Biogel P2-400 mesh (column: 200 × 2.5 cm, 65°C). carbohydrates in the eluate are detected with orcinol sulfuric acid at 420 nm. Control extract, ——; baked with VERON F 25®, ---.

ties. Bacterial amylase delays staling by increasing hydrolysis of gelatinized starch to dextrins. Dextrins of a particular size interfere with cross-links between starch and protein (Martin and Hoseney, 1991). The heat-stable bacterial amylase can act on the starch gelatinized at approximately 65°C for a longer time during baking. Thus, more starch is degraded and its retrogradation is delayed. It is possible to detect more dextrins in bread crumb extract with a degree of polymerization of 6 or higher (Fig. 10). The properties of the starch in the bread crumb change. Bread crumb from bread made with bacterial amylase, that has been homogenized with water, shows lower gelatinization maxima on heating in the amylogram (Sproessler, 1985; Fig. 11). It is very difficult to control the level of bacterial α-amylase. An excess of

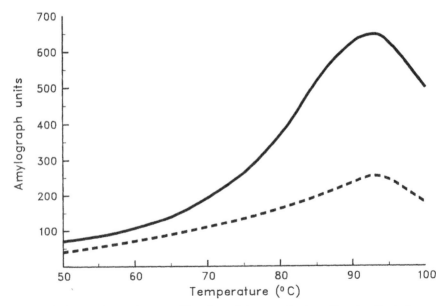

Figure 11 Brabender amylograms of bread crumb homogenization with water. Control extract, ——; baked with VERON F 25®, ---.

enzyme leads to a soft gummy crumb. Good results are obtained with 3 SKB
α-amylase units per kg flour (Rubentaler *et al.*, 1965). Rye bread made with
the addition of bacterial amylase was kept fresh for 6 months (Schulz and
Uhlig, 1972). The dosage used for this purpose was about 100 times that for
white bread. At the low pH value of the rye doughs ($\sim$4.2), both the activity
and the heat stability of the bacterial amylase are reduced greatly. Attempts
were made to achieve the same effect as the bacterial amylase without its side
effects by stabilizing the fungal amylase in sucrose syrup (Cole, 1981). "Intermediate stability" amylases with thermostability properties between those
of traditional fungal and bacterial enzymes have been described by Hebeda *et
al.* (1990). The debranching enzyme pullulanase improves the crumb structure when used in combination with bacterial amylase (Anonymous, 1986).

Fungal enzymes can be used in combination with emulsifiers without the
risk of overdose to prolong the shelf life of bread. By carefully balancing
amylase, proteinase, pentosanases, and β-glucanases, and by combining them
with sodium stearyl lactylate and monoglycerides, a higher volume and an
improved crumb structure are obtained and shelf life can be prolonged by
3–7 days (Rohm Tech, 1986; Fig. 12).

The influence of monoglycerides, proteins, and pentosans and their
effects have been reviewed by Kulp and Ponte (1981). Staleness of bread
involves several sensory perceptions such as odor, flavor, mouth feel, and
firmness of the crumb. All these attributes are used by a taste panel to judge
the freshness of bread.

The firming of the crumb during storage can be measured by instruments

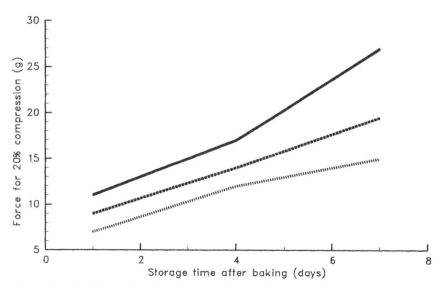

Figure 12 Shelf life studies of bread baked with emulsifier and enzymes. Breads were stored in
bags at 28°C. Crumb softness was measured with a compressimeter. Control bread, ——; with
bacterial amylase (5 g/cwt), – – – –; with VERON ESL® super (5 g/cwt), ---.

such as the Baker Compressimeter and the Instron testing instrument (Kamel *et al.*, 1984; Redlinger *et al.*, 1985). These instruments measure the force required to depress a plunger to a given depth into the bread crumb. The changes occurring during bread staling have been reviewed by Pomeranz (1987). Despite considerable work in this field, bread staling is of interest because many questions remain unanswered (Eliasson and Lijunger, 1988; Siljestroem *et al.*, 1988).

The storage temperature has a strong influence on staling. Bread should be stored at room temperature or deep frozen, but should not be kept in the refrigerator. The retrogradation of the starch is lower at room temperature. Thus, it is best to store bread at room temperature under conditions that minimize moisture loss. At $-22°C$, bread can be stored properly for $15-20$ days, but the flavor and aroma deteriorate thereafter. Freezing maintains bread at the freshest level at the time of freezing, but cannot improve the quality of stale bread.

Retrogradation is basically irreversible, but stale bread can be improved to a limited extent by reheating. The starch crystallinity can be measured by X-ray diffraction studies (Dragsdorf and Varriano Marston, 1980) or by differential thermal scanning (Czuchajowska and Pomeranz, 1989).

6. Frozen and Retarded Doughs

Consumers increasingly expect an oven-fresh assortment of baked goods. The frozen dough process puts deep freezing to work to preserve dough pieces at temperatures below their freezing point. The fermentation stops and the action of the enzymes ceases. In the retarded dough process, dough pieces undergo intermediate storage at temperatures between -5 and $+8°C$. The required storage time can be adjusted by the storage temperature and the yeast content. Fermentation is retarded, but not interrupted completely. Before baking, the dough pieces are warmed again and proofed.

Correct freezing, flours with low thermostable cereal α-amylase content, and freezing-tolerant yeast are most important. Because of its low heat stability, a fungal amylase shows no effect on gelatinized starch. Because of the prolonged dough processing times, the swelling phenomena are amplified and result in decreased water retention in the baked goods. Special fungal enzymes with amylase and balanced side activities improve dough stability and fermentation tolerance, and increase ovenspring and loaf volume (Himmelstein, 1985; ter Haseborg, 1985).

In addition to fungal enzyme treatment, it is recommended that flours with low natural enzyme content be used for frozen or retarded dough. Frozen dough must be fermented after thawing. Instead of freezing the dough, one can store deep frozen prebaked rolls. The rolls are baked insufficiently to gelatinize the starch, form a rigid structure, and destroy microorganisms. Completing the baking process immediately before consumption insures a fresh product with crisp crust. Prebaked rolls are ready for use after 10 min baking time.

7. High-Fiber Bread

Although per capita consumption of bread has dropped in the United States, specialty breads such as whole-wheat, multigrain, and other high-fiber breads are produced increasingly (Miller, 1981). Dietary fibers have different physiological as well as functional effects. They are not digested by human digestive enzymes and therefore are not absorbed nutritionally in the small or large intestine. They can be determined analytically by the AACC Method 32-10 (AACC, 1983). In fiber-rich dough, the percentage of regular wheat flour is reduced. The swelling and water absorption properties of the dough are changed, which can result in slack and sticky or excessively stiff doughs and leads to bread of inferior quality. The particle size of the added whole wheat and multigrain and the relative proportions of soluble and insoluble fiber are important, causing relatively faster or slower water absorption.

Negative effects of the hemicelluloses, β-glucans, and pentosans in the supplement can be offset by enzyme preparations containing hemicellulase or pentosanase (ter Haseborg and Himmelstein, 1988). Straight-dough bread formulations containing fiber-rich materials can be influenced by enzyme preparations. VERON HE®, developed for rye flour treatment, hydrolyzes high-polymer insoluble pentosans, breaking them down into fast-swelling fragments. VERON ST® is designed to have a stabilizing effect on slack and sticky dough. Figure 13 shows whole wheat bread prepared with and without enzyme preparations.

B. Treatment of Rye Flour

The use of rye increases the number of different types of bread in Europe. It is also possible to manufacture a variety of rye–wheat mixed breads. Rye bread consisting entirely or partly of rye flour is made using sour dough or

Control	VERON ST	VERON HE
	(15 g/100 lbs of flour)	(15 g/100 lbs of flour)

Figure 13 Whole wheat bread with and without enzyme preparations.

acidifiers. A sour dough is acidified and leavened by its microorganisms. Sour dough processing leavens the dough. Leavening of a sour dough by yeast frequently is augmented by adding baker's yeast to the final dough. Sour dough processing also develops flavor. Acidification of the dough results from the action of lactic acid bacteria. Such acidification is essential for baking rye bread. A high ratio of lactic acid to acetic acid is desirable. Acidification is retarded considerably by adding salt to the rye sour doughs. Sour doughs use a starter culture in the initial stage. The starter cultures can be bought or taken from a ripe sour dough. The culture also can be produced by spontaneous fermentation initiated by microorganisms indigenous to the flour (Pomeranz, 1987). The addition of fungal amylase improves the fermentation of the sour doughs. The low pH value of rye doughs influences the activity of the flour enzymes. The amylases indigenous to the flour become less active as the degree of acidification increases. By adding acid ingredients, even rye rich in amylase can be made into bread. In contrast to wheat dough, the pentosans rather than the proteins play the decisive role in dough formation. Rye flour contains more soluble and insoluble pentosans. They absorb many times their own weight in water, but produce porous and well-leavened baked products only with difficulty. The water bound by the pentosans is transferred to the starch during heating in the oven. Amounts and properties of starch are of major significance in rye bread making. The starch forms the sheetlike component. The gums, probably in cooperation with protein, form fibrils (Pomeranz, 1987). Rye starch gelatinizes at $50-70°C$. Thus, even the heat-labile fungal amylase can act, until inactivated at $65°C$, on the gelatinized readily accessible starch during baking. Therefore, the addition of fungal amylase can be detected both in the amylogram and by means of the falling number. Both methods serve to characterize the quality of the rye flour and serve as a basis for treatment with amylases.

In the amylogram, the gelatinization temperature of the starch as well as the quantity is important for the evaluation. Good baking properties are obtained above 200 Amylogram Units and $63°C$ gelatinization temperature. The sooner the starch gelatinizes, the greater its degree of enzymatic degradation. This characteristic can lead to a damp crumb and a weakening of crumb elasticity. The gelatinization temperature therefore also should be taken into account (Bolling, 1989). The amylogram can be lowered not only by means of amylases, but also by means of pentosanases, indicating that the pentosans also contribute greatly to the viscosity at increased temperature. In contrast to proteins, pentosans do not denature on heating. Rye flour from a dry harvest contains few enzymes, and high molecular weight starch and pentosans. The doughs are tough and subject to poststiffening. The bread has low volume and the crumb is dry and tends to crack. The addition of pentosanase solves these problems. The insoluble high molecular weight pentosans already are transformed partially into swellable fractions during kneading. Therefore the doughs do not become tougher during fermentation. The loaves are larger, the crumb is softer, and the bread stays fresh longer.

C. Use of Enzymes in Cookie and Cracker Production

For cookie and cracker making, flours poor in protein with a weak gluten structure are most suitable. The dough from such flours does not shrink. Such flours are very difficult to obtain in uniform quality. By adding suitable proteases, flours rich in protein can be used also. Particularly suitable for this purpose are bacterial proteases and papain (Sproessler, 1987). These enzymes hydrolyze the gluten structure to such an extent that the dough loses its elastic properties, is easy to roll out, keeps its shape, and does not shrink. The energy required for kneading is clearly reduced. Proteases replace sodium sulfite, which is no longer allowed in many countries and which softens the gluten structure by reducing the disulfide bridges. The incorporation of proteases permits baked goods of superior organoleptic quality to be obtained (Guèrivière and Bussièr, 1974). The enzyme dosage depends on the flour grade, recipe, and method of dough preparation. When using protein-rich flours and for recipes with a high fat and sugar content, high doses of enzymes are used. Intensive kneading or a higher water content promote the contact between the glutenin substrate and the enzyme. A higher dough temperature, for example, 40°C, accelerates the enzyme reaction. Both measures help reduce the enzyme quantity used. Gaines and Finney (1989) measured the effects of different enzymes on cookie spread and cookie dough consistency.

D. Wafer Production

When protein-rich flour is used in wafer making, the batter becomes thick and the protein partially forms lumps, which clogs the sieves and pouring nozzles. In contrast to cookie and cracker doughs, wafer dough contains a high proportion of water which promotes the effect of the enzyme. However, the batter is poured onto the wafer plates and baked immediately after the dough has been prepared. Bacterial protease and papain are suitable for wafer making because of their rapid hydrolysis of the gluten. However, it has been shown that these products first promote thickening and only liquefy the dough in the desired manner after a rest of 10–20 min.

Special enzyme products function without dough rests. They can be used without altering the manufacturing process, since the dough becomes liquid in the mixer and does not stiffen subsequently. Therefore, it is possible to obtain the advantages described by Drechsel and Ruttloff (1975) without reservation. It becomes possible to manufacture wafers from any type of flour. The homogeneous batters are more liquid. Thus, pumpable doughs with a higher flour content can be made, meaning less water must be evaporated during baking, leading to a decrease in energy use. If the flour content remains unchanged, the water evaporates more rapidly from the thin-bodied dough, leading to a reduction in baking time and an increase in production capacity.

IV. Conclusion

Flour contains enzymes. Their type and quantity, however, are not optimally suited to the needs of baked goods manufacturing. This deficiency was recognized long ago and compensated for by the addition of malt flour. Microbiological enzymes such as fungal amylases, fungal proteases, pentosanases, hemicellulases, and β-glucanases have since then taken the place of malt flour.

In the new generation of microbiologically produced enzymes, special products of bread, rolls, cookies, crackers, and wafers, retarded dough processing, and shelf-life extension figure ever more prominently in research efforts. This trend is likely to continue.

References

American Association of Cereal Chemists (1983). "Approved Methods of the American Association of Cereal Chemists," 8th Ed. American Association of Cereal Chemists, St. Paul, Minnesota.

Anonymous (1986). Bread antistaling composition and method. *Res. Discl.* **268**, 494–495.

Asp, E. H., Midness, L. T., Bakshi, A. S., and Smith, L. B. (1985). Change in activity of cereal, fungal, or bacterial alpha-amylase in yeast dough during fermentation and baking. *Food Technol. Australia* **37**, 198–201.

Belitz, H. D., Kim, J.-J., Kieffer, R., Seilmeier, W., Werbeck, U. and Wieser, H. (1987). Separation and characterization of reduced glutenins from different wheat varieties and importance of the gliadin/glutenin ratio for the strength of gluten. *In* "Proceedings of the 3rd International Workshop on Gluten Proteins" (R. Lásztity and F. Bédés, ed.), pp. 189–205. World Scientific, Singapore.

Bietz, J. A., and Wall, J. S. (1972). Wheat gluten subunits. Molecular weights determined by sodium dodecyl sulfate–polyacrylamide gel electrophoresis. *Cereal Chem.* **49**, 416–430.

Bolling, H. (1989). Development of wheat and rye quality during the last 25 years. *Brot Backwaren* **37**, 314–327 *(in German)*.

Bushuk, W. (1986). Protein–lipid and protein–carbohydrate interactions in flour–water mixtures. *In* "Chemistry and Physics of Baking" (J. M. V. Blanshard, P. F. Frazier, and T. Galliard, eds.). pp. 147–154. Royal Society of Chemistry, London.

Caillat, J. M., and Drapon, R. (1970). Influence des sels biliaires et des ions calcium sur l'hydrolyse des triglycerides en emulsion par la lipase de gemmule de blé germe. *Bull. Soc. Chim. Biol.* **52**, 59–73.

Casier, J. P. J., De Paepe, G., and Bruemmer, J. M. (1973). Effect of water-insoluble wheat and rye pentosans on the baking properties of wheat meal and other raw materials. *Getreide Mehl Brot* **27**, 36–44.

Cauvain, S. P., and Chamberlain, N. (1989). The bread improving effect of fungal alpha-amylase. *J. Cereal Sci.* **8**, 239–248.

Chung, O. K., Pomeranz, Y., and Finney, K. F. (1982). Relation of polar lipid content to mixing requirement and loaf volume potential of hard red winter wheat flour. *Cereal Chem.* **59**, 14–20.

Codner, A. M., and Henry, J. (1989). Carbohydrate-degrading enzymes in germinating wheat. *Cereal Chem.* **66**, 435–439.

Cole, M. S. (1981). Effect of proteinases as flour supplement. *Getreide Mehl Brot* **35**, 60–62.

Czuchajowska, Z. and Pomeranz, Y. (1989). Differential scanning calorimetry, water activity, and moisture contents in crumb center and near-crust zones of bread during storage. *Cereal Chem.* **66**, 305–309.

Drapon, R., and Godon, B. (1987). Role of enzymes in baking. *In* "Enzymes and Their Role in Cereal Technology" (J. E. Kruger, F. E. Kniger, D. Lineback, and C. E. Stauffer, eds.), pp. 290–304. American Association of Cereal Chemists, St. Paul, Minnesota.

Dragsdorf, R. D., and Varriano-Marston, E. (1980). Bread staling: X-Ray diffraction studies on bread supplemented with alpha-amylase from different sources. *Cereal Chem.* **57**, 310–314.

Drechsel, W., and Ruttloff, H. (1975). Use of proteinases for wafer production. *Bäcker Konditor* **7**, 212–213 *(in German)*.

Drews, E. (1969). Amylograms with respect to some quality criteria of rye and its mill products. *Brot Gebäck* **23**, 165–170 *(in German)*.

Eliasson, A.-C., and Lijunger, G. (1988). Interactions between amylopectin and lipid additives during retrogradation in a model system. *J. Sci. Food Agric.* **44**, 353–361.

Fox, P. F., and Mulvihill, D. M. (1982). Enzymes in wheat, flour, and bread. *In* "Advances in Cereal Science and Technology" (V. Y. Pomeranz, ed.), pp. 107–156. American Association of Cereal Chemists, St. Paul, Minnesota.

Frazier, P. J., Brimblecombe, F. A., Daniels, N. W. R., and Eggitt, P. W. R. (1977). The effect of lipoxygenase action on the mechanical development of doughs from fat-extracted and reconstituted wheat flours. *J. Sci. Food Agric.* **28**, 247–254.

Fretzdorff, B. (1980). Determination of peroxidase activity in cereals. *Z. Lebensm. Unters. Forsch.* **170**, 187–193 *(in German)*.

Fretzdorff, B., and Weipert, D. (1983). A nephelometric method for the measurement of alpha-amylase activity in cereals. *Z. Lebensm. Unters. Forsch.* **177**, 167–172 *(in German)*.

Gabor, R., Teufel, A., and Ruttloff, H. (1982). Mode of action of various microbial protease preparations on wheat gluten and wheat flour. *Nahrung* **26**, 37–46 *(in German)*.

Gaines, C. S., and Finney, P. L. (1989). Effects of selected commercial enzymes on cookie spread and cookie dough consistency. *Cereal Chem.* **66**, 73–78.

Geissmann, T., and Neukom, H. (1973). Ferulic acid as a constituent of the water-insoluble pentosans of wheat flour. *Cereal Chem.* **50**, 414–416.

Grampp, E. G. (1982). Modification of certain foodstuffs by enzymes. *Process Biochem.* **17**, 2–4, 6, 12.

Graveland, A. (1973). Enzymic oxidation of linolenic acid in aqueous wheat flour suspensions. *Lipids* **8**, 606–611.

Grosch, W., Laskawy, G., and Weber, F. (1976). Formation of volatile carbonyl compounds and cooxidation of beta-carotene by lipoxygenase from wheat, potato, flax, and beans. *J. Agric. Food Chem.* **24**, 456–459.

Guèrivière, J.-F., and Bussière, G. (1974). The use of proteases in the cereal baking industries. *Ann. Technol. Agric.* **23**, 257–268 *(in French)*.

Guss, P. L., Richardson, T., and Stahmann, M. A. (1967). Oxidation reduction enzymes of wheat. III. Isoenzymes of lipoxidase in wheat fractions and soybean. *Cereal Chem.* **44**, 607–610.

ter Haseborg, E. (1985). Will enzyme treatment of flour help the frozen dough process and dough retarding? Singapore Institute of Food Science and Technology *Newslett.* **5**, 14–18.

ter Haseborg, E. (1988). Enzyme use in the milling industry. *Alimenta* **27**, 2–10 *(in German)*.

ter Haseborg, E., and Himmelstein, A. (1988). Quality problems with high-fiber breads solved by use of hemicellulase enzymes. *Cereal Foods World* 419–422.

Harinder, K., Maninder, K., and Bains, G. S. (1983). Effect of cereal, fungal, and bacterial alpha-amylase on the rheological and breadmaking properties of medium-protein wheats. *Nahrung* **27**, 609–618.

Hashimoto, S., Shogren, M. D., and Pomeranz, Y. (1987). Cereal pentosans: Their estimation and significance. I. Pentosans in wheat and milled wheat products. *Cereal Chem.* **64**, 30–34.

Hebeda, R. E., Bowles, L. K., and Teague, W. M. (1990). Developments in enzymes for retarding staling of baked goods. *Cereal Foods World* **35**, 453–457.

Himmelstein, A. (1985). Enzyme treatment of flour—Will it help frozen and retarded doughs? *Baker's Dig.* **58**, 8, 11–12.

Hoseney, R. C. (1984). Functional properties of pentosans in baked foods. *Food Technol.* **38,** 114–117.

Huebner, F. R., and Wall, J. S. (1976). Fractionation quantitative differences in glutenin from wheat varieties varying in baking quality. *Cereal Chem.* **59,** 258.

Kamel, B. S., Wachnuik, S., and Hoover, J. R. (1984). Comparison of the Baker Compressimeter and the Instron in measuring firmness of bread containing various surfactants. *Cereal Foods World* **29,** 159–161.

Kieffer, R., and Grosch, W. (1980). Improvement of dough and baking properties of wheat flour by type II lipoxygenase from soybeans. *Z. Lebensm. Unters. Forsch.* **170,** 258–261 *(in German)*.

Kieffer, R., Matheis, G., Hofmann, H. W., und Belitz, H.-D. (1981). Improvement of baking properties of wheat flours by addition of horse radish peroxidase, hydrogen peroxide, and phenols. *Z. Lebensm. Unters. Forsch.* **173,** 376–379 *(in German)*.

Kim, S. K., and D'Appolonia, B. L. (1977). The role of wheat flour constituents in bread staling. *Baker's Dig.* **51,** 38–42, 44, 57.

Kruger, J. E. (1971). Purification and some properties of malted wheat BAPA-ase. *Cereal Chem.* **48,** 512–522.

Kruger, J. E. (1979). Modification of wheat beta-amylase by proteolytic enzymes. *Cereal Chem.* **56,** 298–302.

Kruger, J. E., and Marchylo, B. A. (1985). A comparison of the catalysis of starch components by isoenzymes from the two major groups of germinated wheat alpha-amylase. *Cereal Chem.* **62,** 11–18.

Kruger, J. E., and Lineback, D. R. (1987). Carbohydrate-degrading enzymes in cereals. *In* "Enzymes and Their Role in Cereal Technology" (E. Kruger *et al.*, eds.), pp. 117–139. American Association of Cereal Chemists, St. Paul, Minnesota.

Kruger, J. E., and Tipples, K. H. (1981). Modified procedure for use of the Perkin–Elmer model 191 grain amylase analyzer in determining low levels of alpha-amylase in wheats and flours. *Cereal Chem.* **58,** 271–274.

Kulp, K., and Ponte, J. G. (1981). Staling of white pan bread: Fundamental causes. *CRC Crit. Rev. Food Sci. Nutr.* **15,** 1–48.

Linko, Y. Y., and Linko, P. (1987). Enzymes in baking. *In* "Chemistry and Physics of Baking" (J. M. V. Blanshard, P. F. Frazier, and T. Galliard, eds.), pp. 105–116. Royal Society of Chemistry, London.

Lyons, T. P. (1982). Proteinase enzymes relevant to the baking industry. *Biochem. Soc. Trans.* **10,** 287–290.

MacGregor, A. W., Marchylo, A., and Kruger, J. E. (1988). Multiple alpha-amylase components in germinated cereal grains determined by isoelectric focusing and chromatofocusing. *Cereal Chem.* **65,** 326–333.

Marchylo, B., LaCroix, L. J., and Kruger, J. E. (1980). alpha-Amylase isoenzymes in Canadian wheat cultivars during kernel growth and maturation. *Can. J. Plant Sci.* **60,** 433–443.

Martin, M. L., and Hoseney, R. C. (1991). A mechanism of bread firming. II. Role of starch hydrolyzing enzymes. *Cereal Chem.* **68,** 503–507.

Meuser, F., and Sukow, P. (1985). AIF-Forschungsbericht Nr. 5144.

Meuser, F., and Sukow, P. (1986). Non-starch polysaccharides. *In* "Chemistry and Physics of Baking" (J. M. V. Blanshard, P. F. Frazier, and T. Galliard, eds.), pp. 42–61. Royal Society of Chemistry, London.

Miller, B. S. (ed.) (1981). Variety breads in the United States. Proceedings of a symposium presented at the AACC 65th annual meetings. American Association of Cereal Chemists, St. Paul, Minnesota.

Mohsen, S. W., Alian, A. M., Attia, R., and El-Azhary, T. (1986). Specificity of lipase produced by *Rhizopus delemar* and its utilization in bread making. *Egypt. J. Food Sci.* **14,** 175–182.

Moonen, H. (1990). Synergism between enzymes group and other functional ingredients in baking. Paper presented at "Using Enzymes Technology in Bakery Foods." St. Louis, Missouri, Nov. 8–9.

Nagargoje, K. M., Maninder, K., and Bains, G. S. (1984). Interaction of amyloglucosidase and alpha-amylase supplements in bread making. *Nahrung* **28,** 837–849.

Neukom, H. (1976). Chemistry and properties of the non-starchy polysaccharides (NSP) of wheat flour. *Lebensm. Wiss. Technol.* **9**, 143–148.

Ng, P. K. W., and Bushuk, W. (1988). Statistical relationship between high molecular weight subunits of glutenin and bread making quality of Canadian-grown wheats. *Cereal Chem.* **65**, 408–413.

Nierle, W., and El Baya, A. W. (1981). Wheat lipids-function and effect in flour processing. *Fette Seifen Anstrichm.* **83**, 391–395.

Nishiyama, J., Kuninori, T., and Matsumoto, H. (1979). Effect of mushroom extract on dough-reaction of sulfhydryl-gluten with mushroom-oxidized phenols. *J. Ferm. Technol.* **57**, 387–394.

O'Connell, B. T., Rubenthaler, G. L., and Murbach, N. L. (1980). Evaluation of a nephelometric method for determining cereal alpha-amylase. *Cereal Chem.* **57**, 411–415.

Peppler, H. J., and Reed, G. (1987). Enzymes in food and feed processing. *In* "Biotechnology" (J. F. Kennedy, ed.), Vol. 7a, pp. 547–603. VCH, Weinheim, Germany.

Perten, H. (1984). A modified falling-number method suitable for measuring both cereal and fungal alpha-amylase activity. *Cereal Chem.* **61**, 108–111.

Pomeranz, Y. (1987). "Modern Cereal Science and Technology." VCH, Weinheim, Germany.

Pomeranz, Y., Meyer, D., and Seibel, W. (1984). Wheat, wheat rye, and rye dough and bread studied by scanning electron microscopy. *Cereal Chem.* **61**, 53–59.

Preston, K. R., and Kruger, J. E. (1976). Purification and properties of two proteolytic enzymes with carboxypeptidase activity in germinated wheat. *Plant. Physiol.* **58**, 516–520.

Redlinger, P. A., Setzer, C. S., and Dayton, A. D. (1985). Measurements of bread firmness using the Instron Universal Testing Instrument: Differences resulting from test conditions. *Cereal Chem.* **62**, 223–226.

Reed, G., and Peppler, H. J. (1973). "Yeast Technology." AVI, Westport, Connecticut.

Reed, G., and Thorn, J. A. (1971). Enzymes. *In* "Wheat: Chemistry and Technology" (Y. Pomeranz, ed.), 2d Ed., pp. 453–491. American Association of Cereal Chemists, St. Paul, Minnesota.

Rohm Tech (1986). Product showcase. Enzyme preparation. *Cereal Foods World* **31**, 259–260.

Rohrlich, M., Ziehmann, G., and Lenschau, R. (1959). The application of the electrochemical measuring process according to Tödt for the determination of the lipoxidase and the catalase in grain. *Getreide Mehl* **9**, 13–18 *(in German)*.

Rubenthaler, G., Finney, K. F., and Pomeranz, Y. (1965). Effects on loaf volume and bread characteristics of alpha-amylases from cereal, fungal, and bacterial sources. *Food Technol.* **19**, 239–241.

Sandstedt, R. M., Kneen, E., and Blish, M. J. (1939). A standardized Wohlgemuth procedure for α-amylase activity. *Cereal Chem.* **16**, 712.

Schieberle, P., Grosch, W., Kexel, H., and Schmidt, H. L. (1981). A study of oxygen isotope scrambling in the enzymic and nonenzymic oxidation of linoleic acid. *Biochim. Biophys. Acta* **666**, 322–326.

Schulz, A., and Uhlig, H. (1972). Possibilities of addition of alpha-amylases to bakery products with special consideration of bacterial amylase. *Getreide Mehl Brot* **26**, 215–221 *(in German)*.

Siljestroem, M., Bjoerck, I., Eliasson, A. C., Loenner, C., Nyman, M., and Asp, N.-G. (1988). Effects on polysaccharides during baking and storage of bread—In vitro and in vivo studies. *Cereal Chem.* **65**, 1–8.

Sluimer, P. (1990). Baking properties from the technical point of view. *Getreide Mehl Brot* **44**, 80–84 *(in German)*.

Sproessler, B. (1977). Enzymatic treatment of strong wheat flours and rye flours with high amylogram values. *Mühle* **114**, 235–236 *(in German)*.

Sproessler, B. (1981). Effect of proteinases as flour supplement. *Getreide Mehl Brot* **35**, 60–62 *(in German)*.

Sproessler, B. (1982). New analytical methods for measuring enzymes in the flour. *Mühle* **119**, 425–430 *(in German)*.

Sproessler, B. (1985). Enzyme activity and effectiveness. *Mühle* **122**, 406–407 *(in German)*.

Sproessler, B. (1987). Use of enzymes in biscuit, cracker and wafer production. *Gordian* **9**, 168–171 *(in German)*.

Sproessler, B., and Schmerr, H. (1986). Comparison of analytical methods for the measurement of enzymes in flours. *Mühle* **123**, 465 *(in German)*.

Sproessler, B., and Uhlig, H. (1972). Effectiveness of microbial amylases. *Getreide Mehl Brot* **26**, 210–215 *(in German)*.

Stauffer, C. E. (1987). Proteases, peptidases, and inhibitors. *In* "Enzymes and Their Role in Cereal Technology" (J. E. Kruger, D. Lineback, and C. E. Stauffer, eds.), pp. 201–251. American Association of Cereal Chemists, St. Paul, Minnesota.

Strahm, A., Amado, R., and Neukom, H. (1981). Structure of arabinogalactan and arabinogalactan-peptides. *Phytochem.* **20**, 1061.

Tkachuk, R., and Kruger, J. E. (1974). Wheat α-amylases. II. Physical characterization. *Cereal Chem.* **51**, 508.

Uhlig, H., and Sproessler, B. (1971). Application of proteases and amylases in the flour mill. *Mühle* **108**, 226–228 *(in German)*.

Uhlig, H., and Sproessler, B. (1972). The natural enzymes of cereals and their supplementation with microbiological preparations. *Mühle* **109**, 221–223 *(in German)*.

Uhlig, H., Sproessler, B., Plainer, H., and Taeger, T. (1988). Application of enzyme preparations in engineering. *In* "Biotechnology-Focus 1" (R. K. Finn and P. Praeve, ed.), pp. 263–297. Carl Hanser Verlag, Vienna.

van Ceumern, S., and Hartfiel, W. (1984). Activity of lipoxygenase in cereals and possibilities of enzyme inhibition. *Fette Seifen Anstrichm.* **86**, 204–208.

Wassermann, L., and Doerfner, H. H. (1974). Rasterelectron microscopy of baked goods. *Getreide Mehl Brot* **28**, 324–328.

Weber, F., and Grosch, W. (1976). Co-oxydation of a carotenoid by the enzyme lipoxygenase: Influence on the formation of linoleic acid hydroperoxides. *Z. Lebensm. Unters. Forsch.* **161**, 223–230.

Weipert, D. (1972). Rheology of rye dough. *Getreide Mehl Brot* **26**, 275–280 *(in German)*.

Weselake, R. J., Mac Gregor, A. W., and Hill, R. D. (1985). Endogenous alpha-amylase inhibitor in various cereals. *Cereal Chem.* **62**, 120–123.

Zawistowska, U., Bekes, F., and Bushuk, W. (1984). Wheat intercultivar variations in lipid content composition, and distribution and their relation to baking quality. *Cereal Chem.* **61**, 527–531.

Zawistowska, U., Sangster, K., Zawistowski, J., Langstaff, J., and Friesen, A., D. (1988). Immobilized metal affinity chromatography of wheat α-amylases. *Cereal Chem.* **65**, 413–416.

CHAPTER 12

Starches, Sugars, and Syrups

RONALD E. HEBEDA

I. Introduction

A. Background

Corn is the largest and most important crop grown in the United States, as evidenced by an annual production rate of 8–9 billion bushels. Total corn utilization in the United States has averaged 7.3 billion bushels per year from 1980/81 to 1990/91. In the 1990/91 crop year, 60% of the corn was processed into domestic feed, 23% was exported, and the remaining 16% was used for food and industrial purposes [Corn Refiners Association (CRA), 1991].

The corn refining industry is the largest single consumer of corn in the food and industrial catagory. The 24 corn refining plants operating in the United States in 1990 used nearly 1 billion bushels of corn for the production of starch, sweeteners, syrups, and alcohol (CRA, 1991). Assuming a yield of 115 bushels per acre, 8.7 million acres, or almost 14,000 square miles of land devoted to corn production are required annually to satisfy the needs of the corn refining industry.

Corn use by the corn refining industry has grown significantly since the mid 1970s. For instance, from the 1975/76 to the 1990/91 crop years, corn used for starch and sweetener production increased by 49 and 185%, respectively (Table I). Currently, corn usage for these two products averages 9–10% of total utilization (CRA, 1991).

Corn is nearly the sole source of starch, syrups, and sweeteners in the United States, because of the abundant supply, low cost, and ease of storage that insures a continuous source of raw material throughout the year. Elsewhere in the world, other crops such as sorghum, wheat, rice, potato, tapioca, arrowroot, and sago are used in addition to corn for the production of starch

TABLE I
Corn Utilization for Starch and Sweeteners[a]

Crop year	U.S. corn utilization (million bu)			Starch/sweeteners (% of total)
	Total	Starch	Sweeteners[b]	
75/76	5767	116	207	5.6
80/81	7283	120	348	6.4
81/82	6975	130	368	7.1
82/83	7249	127	403	7.3
83/84	6693	147	445	8.8
84/85	7032	143	497	9.1
85/86	6494	152	516	10.3
86/87	7385	155	524	9.2
87/88	7757	167	546	9.2
88/89	7260	164	558	9.9
89/90	8113	172	576	9.2
90/91	8020	173	590	9.5

[a] Adapted from Corn Refiners Association, 1988, 1991.
[b] Includes dextrose, high fructose corn syrups, and traditional corn syrups.

and starch-based sweeteners. The selection of raw material generally is based on availability, cost, and by-product value. However, refined starch products all exhibit essentially the same properties, regardless of raw material source.

B. Starch Production

Starch is produced in the corn refining industry by a process in which aqueous streams are used to facilitate separation of the individual components present in the corn kernal. Consequently, the process is called "wet-milling," as opposed to the "dry-milling" process that separates corn fractions by milling, aspiration, and screening.

In the wet-milling process, shelled corn is first steeped, or soaked, in a sulfur dioxide solution at the proper temperature and pH to initiate a lactic acid fermentation. The corn absorbs water and is softened to facilitate subsequent separation into the various components. The steeped corn is broken by coarse grinding, the germ is separated by centrifugal hydroclones, and the fiber is recovered by screening. The remaining material consists of gluten (a high protein fraction) and starch. Gluten is removed by centrifugation and a starch stream containing 3–5% protein is recovered and further purified by washing with water through a series of hydroclones. The relatively pure starch slurry (about 45% w/w solids) is processed to a dry product or used as substrate for production of sweeteners and syrups.

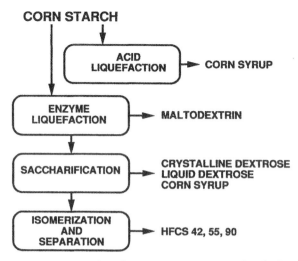

Figure I Production of dextrose, corn syrups, and maltodextrin from starch.

C. Sweetener and Syrup Production

Corn sweeteners and syrups are produced by enzyme or acid hydrolysis of starch (Fig. 1). Degree of hydrolysis often is expressed in terms of dextrose equivalents (DE). DE is defined as the percentage of reducing sugars present, calculated as dextrose (D-glucose) on a dry solids basis. Starch DE is essentially 0 since very few reducing end groups are present in the polysaccharide. Complete hydrolysis of starch yields dextrose, that is, a 100 DE product. Therefore, an intermediate degree of hydrolysis yields a product exhibiting a DE between 0 and 100. DE is measured by a variety of techniques, including oxidation/reduction reactions, osmometry, and liquid chromatography (Delheye and Moreels, 1988).

Initial hydrolysis (liquefaction) is conducted with a thermostable bacterial α-amylase to produce a soluble low-viscosity hydrolysate of $10-15$ DE. The liquefied hydrolysate can be processed directly to a maltodextrin[1] product or used as the substrate for conversion (saccharification) to dextrose with the enzyme glucoamylase. The dextrose hydrolysate is processed to crystalline dextrose,[2] liquid dextrose, or a high dextrose corn syrup. In addition, a portion of the dextrose can be isomerized to fructose with a glucose isomerase enzyme to produce a high fructose corn syrup (HFCS).

A wide variety of corn syrups[3] can be made by acid or enzyme hydrolysis. Direct acid hydrolysis of corn starch yields straight acid corn syrups. Enzy-

---

[1] Maltodextrin is defined as a hydrolysis product of less than 20 DE.
[2] Dextrose is defined as a product of ≥ 99.5 DE.
[3] Corn syrups are defined as products of $20-99.4$ DE.

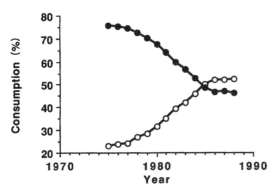

Figure 2   Sweetener consumption in the United States. Refined sugar, ●; corn sweeteners, ○.

matic saccharification of acid hydrolysate or of low DE enzyme hydrolysate produces acid–enzyme or enzyme–enzyme corn syrups, respectively.

D. Sweetener and Syrup Consumption

Consumption of corn sweeteners is shown in Fig. 2 and compared with data for refined sugar [U.S. Department of Agriculture (USDA), 1989, 1991]. In 1975, sugar consumption was more than three times that of corn sweeteners. This ratio changed considerably during the late 1970s and early 1980s when HFCS became important as a sugar substitute in soft drinks. As a result, corn sweetener consumption is currently in excess of 50% of the total sweetener market. Per capita consumption of corn sweeteners and refined sugar in 1992 was forecast at 74.5 and 65.1 lb, respectively (USDA, 1991).

II. Starch Production and Application

Corn starch, as manufactured by the corn refining industry, is a 99% pure product containing about 0.65% lipid, 0.25% protein, and 0.1% minerals (Watson, 1977). The starch is composed of amylose and amylopectin polymers in the form of individual granules that are 5–25 μm in size. The amylose fraction represents 25–30% of the starch and is composed of dextrose units arranged in a linear fashion through α-1,4 glucosidic bonds. An amylose molecule consists of 100–1000 dextrose units (Boyer and Shannon, 1987). The amylopectin fraction represents the remaining 70–75% of the granule and is composed of linear starch chains branched through an occasional α-1,6 glucosidic linkage. Amylopectin molecules are very large and range from 50–100 million in molecular weight (Zobel, 1984).

   Corn starch is used for a variety of food and nonfood purposes (Table II). In the unmodified form, starch is used in applications in which the ability to absorb water and swell is important in developing high paste viscosity and strong gels for thickening properties. Modification of starch alters functionality and improves properties for many applications (Rutenberg and Solarek, 1984; Wurzburg, 1986).

Food and nonfood applications account for 35 and 65% of corn starch utilization, respectively. The largest food uses are in brewing, chemicals, and drugs and pharmaceuticals, each of which accounts for about 20% of total usage. Baking, confectionery, and canning each utilize 6–7% of corn starch produced (Orthoefer, 1987). In nonfood applications, over 90% of starch is used in the paper industry to improve paper strength and appearance. Generally, unmodified starch used for this purpose is thinned enzymatically or cooked at high temperature prior to use to improve adhesive quality.

Several different procedures are used to modify starch and alter physical properties (Watson, 1977; Orthoefer, 1987). Methods include acid treatment, oxidation, cross-linking, and pregelatinization. Acid modification is conducted by controlled hydrolysis of an aqueous starch slurry to produce a thin boiling product of reduced viscosity, increased gelling tendency, and improved gel clarity. Acid treatment and roasting of starch under low moisture conditions produces dextrins of various types that exhibit high water

TABLE II
Food, Drug, Cosmetic, and Industrial Uses of Cornstarch[a]

Abrasive paper	Cord polishing	Ore refining
Adhesives	Cork	Paints
Antibiotics	Crayon	Paper
Aspirin	Desserts	Pie filling
Baby foods	Detergents	Pharmaceuticals
Bakery products	Dispersing agents	Photographic films
Baking powder	Dressings, surgical	Plastics
Batteries, dry cell	Drugs	Plywood
Beverages	Dyes	Printing
Binding agents	Fermentation	Precooked meals
Boiler compounds	Fiber board size	Protective colloids
Bookbinding	Fiberglass	Salad dressing
Briquettes	Fireworks	Sauces
Cardboard	Flours	Soaps
Chalk	Gravies	Sugar, powdered
Chemicals	Insecticides	Textiles
Chewing gum	Insulation	Tiles, ceiling
Chocolate drink	Lubricating agents	Tires, rubber
Cleaners	Meat products	Vegetables, canned
Coatings	Mixes, prepared	Wallboard
Color carrier	Mustard, prepared	Wallpaper
Confectionery	Oilcloth	Water recovery
Cosmetics	Oil well drilling	Yeast

[a] Adapted from Corn Refiners Association, 1988.

solubility and low viscosity. Oxidation with sodium hypochlorite converts free hydroxyl groups to carboxyl groups and reduces gelatinization temperature, lowers viscosity, and improves paste clarity by reducing retrogradation. Cross-linking starch with bifunctional reagents restricts swelling and provides a stable paste. Simultaneous pasting and drying of starch produces a pregelatinized product that develops viscosity in cold or warm water.

III. Sweetener and Syrup Production

A. Starch Liquefaction

The first step in the production of corn sweeteners is enzymatic liquefaction, or partial hydrolysis, of corn starch to provide a soluble, low-viscosity, low DE material suitable for subsequent conversion to a variety of products.

Liquefaction is conducted with thermostable bacterial α-amylases that have the ability to withstand the initial high reaction temperatures needed for rapid and complete starch gelatinization. Enzymes that are used for the purpose of starch liquefaction include α-amylases derived from bacterial organisms such as *Bacillus stearothermophilus, B. licheniformis,* and *B. subtilis* (Teague and Brumm, 1992). The *B. stearothermophilus* and *B. licheniformis* enzymes are very thermostable and are used routinely at temperatures in excess of 100°C. The *B. subtilis* enzyme is less thermostable and generally is not used above 90°C, to prevent a rapid loss in activity.

The bacterial α-amylases specifically catalyze the hydrolysis of α-1,4 glucosidic linkages and act in a random but reproducible manner to reduce polysaccharide molecular weight. Saccharide compositions attained with each enzyme at DE levels of 10–40 are shown in Table III. At the normal liquefaction DE of about 10, 85–90% of the oligosaccharides are larger than DP-5 because of the limited degree of hydrolysis. If liquefaction is allowed to proceed to 30 DE, both the *B. stearothermophilus* and the *B. licheniformis* enzyme produce high levels of maltopentaose (23–26%). In contrast, the *B. subtilis* enzyme does not have a DP-5 producing ability, forming only 8–13% pentasaccharide over a 20–40 DE range.

Enzymatic liquefaction requires careful control of reaction parameters such as percentage of solids, temperature, time, pH, and calcium level to insure efficient hydrolysis, minimize enzyme cost, and reduce the potential of downstream processing problems. Typical operational ranges for each of the liquefaction parameters follow.

1. Starch Solids

Starch solids level generally is maintained at 30–40%, preferably at 30–35%. At higher solids concentrations, available water content is not sufficient to achieve complete starch gelatinization, whereas at lower solids, increased cost of water removal impacts negatively on process economics.

TABLE III
α-Amylase Action Pattern[a]

α-Amylase source	DE[c]	\multicolumn{9}{c}{Saccharide distribution (% dry basis) for DP[b]}								
		1	2	3	4	5	6	7	8+
Bacillus stearothermophilus	10	0	2	5	1	3	11	10	68
	20	1	7	9	3	11	20	9	40
	30	4	12	12	5	26	10	3	28
	40	8	23	21	8	15	4	5	16
Bacillus licheniformis	10	0	2	6	3	4	7	6	72
	20	1	7	12	4	12	13	6	45
	30	4	12	15	5	23	7	3	31
	40	11	17	16	6	21	4	4	21
Bacillus subtilis	20	2	7	9	5	8	18	6	45
	30	6	13	12	7	13	10	3	36
	40	13	22	14	7	9	4	4	27

[a] Substrate is 20% w/v acid modified starch at 70°C, pH 6, for *B. stearothermophilus* and *B. licheniformis* enzymes and at pH 7 for *B. subtilis* enzyme.
[b] DP, degree of polymerization, where DP-1 is a monosaccharide, DP-2 is a disaccharide, and so forth.
[c] DE, dextrose equivalent.

2. pH

Starch slurry pH normally is controlled at 5.8–6.5. At a higher pH, by-product formation is increased, resulting in loss of product yield and increased refining costs. As pH is decreased below the optimum range, enzyme becomes more subject to inactivation if upsets in process control occur. The *B. stearothermophilus* amylase has a lower practical pH minimum than the *B. licheniformis* amylase (5.6 vs. 6.0) (Reeve, 1992); a mixture of the two enzymes may offer advantages in low pH operations (Carroll *et al.*, 1990).

3. Calcium

Calcium is added in the chloride or oxide form as an enzyme cofactor to enhance α-amylase thermostability. Generally, a calcium level of 100–200 ppm (starch dry basis) is sufficient for *B. stearothermophilus* and *B. licheniformis* enzymes. A higher calcium level of about 300 ppm (starch dry basis) is often used with *B. subtilis* enzymes.

4. Time/Temperature

Reaction time and temperature are balanced to provide the optimum conditions for enzyme activity and rapid starch gelatinization. Normally, temperature is controlled to be as high as possible while still permitting sufficient enzyme activity to achieve the required degree of hydrolysis.

Typical industrial starch liquefaction processes are diagrammed in Fig. 3.

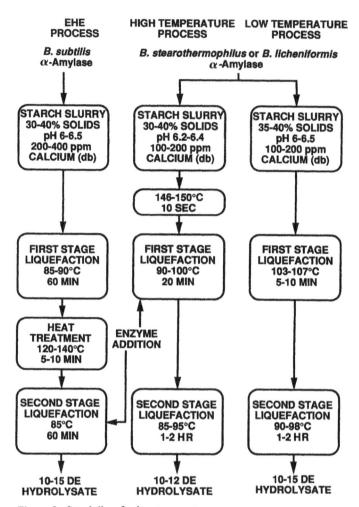

Figure 3 Starch liquefaction processes.

The EHE (Enzyme-Heat-Enzyme) process was the first enzymatic liquefaction process used when *B. subtilis* α-amylases first became commercially available. The HT (high temperature) and LT (low temperature) processes were developed to take advantage of the increased thermostability of the *B. stearothermophilus* and *B. licheniformis* α-amylases.

In the EHE process, a 30–40% starch slurry containing 200–400 ppm calcium (dry basis) is adjusted to pH 6–6.5, dosed with enzyme, and passed through a stream injection heater at a temperature of 85–90°C. The heater is designed to mix the starch slurry and steam thoroughly while rapidly increasing temperature to a level that provides instantaneous gelatinization and hydrolysis. The reaction is continued for about 1 hr to liquefy and solubilize the starch. Under these conditions, however, complete starch solubilization is not attained because of the presence of a small amount of insoluble starch–lipid complex that is not disrupted by enzyme action (Hebeda and Leach, 1974). If the insoluble material is allowed to persist throughout the process, the small starch particles interfere with subsequent filtration and refining operations. Consequently, a 120–140°C, 5–10 min heat treatment step is included to dissociate the complex. The α-amylase is denatured during the heat treatment step, necessitating a second enzyme addition and an additional reaction for 1 hr at about 85°C. The dissociated starch polymers are hydrolyzed sufficiently to prevent reformation of the starch–lipid complex and a 10–15 DE hydrolysate is produced that is suitable for subsequent processing.

In the HT process, the starch slurry is prepared in a manner similar to the one previously described, a thermostable α-amylase is added, and initial liquefaction is conducted at 146–150°C for 10 sec or longer. As in the EHE process, enzyme is inactivated thermally, requiring a second enzyme addition followed by a 95°C reaction for at least 20 min. Further hydrolysis is then conducted at 85–95°C for 1–2 hr to produce a 10–12 DE hydrolysate. The advantage of this process is reported to be enhanced starch gelatinization due to the high initial temperature, resulting in improved filtration characteristics and hydrolysate quality (Shetty and Allen, 1988).

The LT liquefaction technique has become the primary industrial liquefaction process used because of the relatively low temperature needed for efficient hydrolysis. In this process, a starch slurry at pH 6.0–6.5 containing 100–200 ppm (dry basis) calcium is dosed with enzyme, passed through a steam injection heater at 103–107°C, and held at that temperature for 5–10 min. The hydrolysate is flashed to atmospheric pressure and held at 90–98°C for 1–2 hr to produce a 10–15 DE hydrolysate. Compared with the other described processes, the LT process has the advantages of lower temperature (i.e., reduced energy requirements) and only one enzyme addition. The initial temperature is sufficiently high to disrupt the starch–lipid complex but not to cause excessive inactivation of the thermostable enzymes used for liquefaction.

Stability of α-amylase during high temperature liquefaction is important in achieving efficient starch solubilization. In the absence of substrate, *B. stearothermophilus* and *B. licheniformis* α-amylases are 25–135 times more stable than a *B. amyloliquefaciens* (*B. subtilis*) α-amylase because of additional salt bridges that maintain enzyme conformation and minimize thermoinactivation (Tomazic and Klibanov, 1988a,b). Presence of stability factors such as substrate or calcium are effective in increasing enzyme stability by several additional orders of magnitude, as has been shown for α-amylases derived from *B. stearothermophilus* (Brumm *et al.*, 1988) and *B. licheniformis* (Chiang *et al.*, 1979).

Factors that affect α-amylase stability during liquefaction include temperature, time, pH, substrate solids, and calcium level. Half-lives of α-amylases derived from *B. stearothermophilus* and *B. licheniformis* under various combinations of typical process conditions were calculated from the data of

TABLE IV
Stability of *Bacillus stearothermophilus* and *Bacillus lichenlformis* α-Amylases under Process Conditions

Temperature (°C)	pH	Calcium (ppm, as is)	Solids (% w/w)	Half-life (min)	
				BS[a]	BL[b]
102	6	25	37.3	213	70
102	6	25	31.4	126	44
102	6	25	26.5	81	25
105	6.5	40	35	132	50
105	5.5	40	35	43	10
105	6.5	40	25	75	20
105	6.5	10	35	42	15

[a] *B. stearothermophilus* α-amylase (Henderson and Teague, 1988).
[b] *B. licheniformis* α-amylase (Rosendal *et al.*, 1979).

Henderson and Teague (1988) and Rosendal *et al.* (1979) and are listed in Table IV. Lowering solids level from 37.3 to 26.5% or reducing calcium level from 40 to 10 ppm (on an "as is" basis) decreases half-life of either enzyme by a factor of 3. Reducing pH from 6.5 to 5.5 reduces stability of *B. stearothermophilus* and *B. licheniformis* α-amylases by factors of 3 and 5, respectively. Comparing all conditions tested, the *B. stearothermophilus* enzyme is 3 to 4-fold more stable than the *B. licheniformis* enzyme. Consequently, LT process changes have been proposed to take advantage of the highly thermostable nature of the *B. stearothermophilus* α-amylase (Henderson and Teague, 1988). For instance, increasing the initial reaction time in the LT process to 10 min from the typical 6 min allows a reduction in enzyme dosage of 40% while still maintaining efficient liquefaction.

The low DE hydrolysate produced by enzymatic liquefaction is used as substrate for production of corn sweeteners or processed directly to a dry or liquid maltodextrin product.

When the final product is intended to be a maltodextrin, the reaction is conducted to a specific DE and terminated by heat treatment or pH adjustment. The hydrolysate is refined and spray dried to a moisture content of 3–5% or evaporated to a syrup of 75% solids.

B. Dextrose

Glucoamylase is used to convert (saccharify) liquefied starch to a dextrose-containing hydrolysate by cleaving both α-1,4 and α-1,6 glucosidic bonds sequentially from the nonreducing end of the molecule. The enzyme is produced from *Aspergillus* or *Rhizopus* fungal sources. Glucoamylase from *A.*

niger generally is used in commercial processes because of adequate thermostability.

The rate of hydrolysis by glucoamylase depends on the particular linkage, the size of the molecule, and the order in which the α-1,4 and α-1,6 linkages are arranged (Abdullah *et al.*, 1963). For instance, an α-1,4 linkage in maltose is hydrolysed at a rate 100 times faster than an α-1,6 linkage in isomaltose. Increasing the number of α-1,4 linkages in sequence from 1 (DP-2 saccharide) to 4 (DP-5 saccharide) increases the hydrolysis rate of the first glucosidic bond by a factor of 10. In the same manner, increasing the number of α-1,6 linkages from 1 to 4 increases the hydrolysis rate of the nonreducing glucosidic bond by a factor of 23. The hydrolysis rate of an α-1,6 linkage is increased 9-fold when the adjacent linkage is an α-1,4 bond rather than an α-1,6 bond. Conversely, the hydrolysis rate of an α-1,4 linkage is reduced when the adjacent linkage is an α-1,6 bond rather than an α-1,4 bond.

Glucoamylase also catalyzes the reverse reaction (reversion), in which dextrose molecules are combined to form maltose and isomaltose (Hehre *et al.*, 1969; Roels and van Tilburg, 1979; Adachi *et al.*, 1984; Beschkov *et al.*, 1984; Shiraishi *et al.*, 1985). Minimizing this reaction is important in maximizing dextrose yield during commercial operation.

The reversion of dextrose specifically involves the condensation of a β-anomer of D-glucopyranose with either an α- or a β-D-glucose molecule in the presence of glucoamylase (Hehre *et al.*, 1969), as shown here:

$$\text{D-Glucose} + \beta\text{-D-Glucose} \longrightarrow \text{Disaccharide} + H_2O$$

Consequently, the equilibrium constant (K_{eq}) for the reversion of dextrose to maltose or isomaltose can be calculated as:

$$K_{eq} = \frac{C_m \times C_w}{C_d \times 0.63\,C_d} = \frac{C_m \times C_w}{0.63\,C_d^2}$$

where C_m is the molar concentration of maltose, or isomaltose, C_w is the molar concentration of water, C_d is the molar concentration of dextrose, and 0.63 is the decimal fraction of β-D-glucose present at equilibrium.

Rate of maltose formation during saccharification is rapid. The disaccharide reaches an equilibrium level early in the reaction before maximum dextrose is reached. In contrast, formation of isomaltose via reversion is slower than that of maltose, but the equilibrium level is much higher. Therefore, isomaltose level continues to increase during saccharification.

In commercial processes, liquefied hydrolysate at 30–32% w/w solids is adjusted to pH 4–4.5 and 58–61°C. A glucoamylase dosage is added to produce a maximum dextrose level in 1–4 days. Reaction time is inversely related to dose, so if a given dose yields maximum dextrose in 4 days, doubling or quadrupling the dose will reduce reaction time to 2 days or 1 day, respectively (Fig. 4). A high dose, and therefore a short reaction time, requires small saccharification tanks so the final hydrolysate can be processed rapidly to prevent excessive reversion. If large reaction tanks are available, a lower dose generally is used, since a considerable amount of time is required

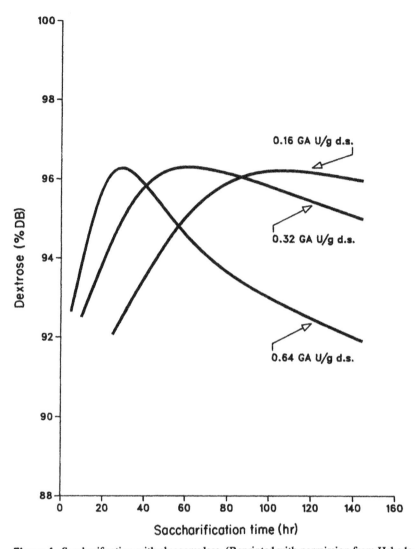

Figure 4 Saccharification with glucoamylase. (Reprinted with permission from Hebeda, 1987.)

to empty the tank. Saccharification tanks range in size from less than 200,000 to several million liters.

Saccharide distribution attained at maximum dextrose is typically about 96.1%, dextrose, 1.5% isomaltose, 1.0% maltose, 0.2% maltulose, 0.3% DP-3, and 0.9% DP-4 and higher saccharides. Both maltose and isomaltose are formed via reversion, that is, the maltose concentration of 1.0% is at the equilibrium level dictated by the final solids level of about 33%, whereas isomaltose continues to increase during saccharification. At maximum dextrose, the isomaltose concentration is about 1.5% but would eventually reach

10–12% if the saccharification reaction was allowed to proceed beyond maximum dextrose. Consequently, overdosing with glucoamylase or extending reaction time beyond the maximum dextrose level will result in a decrease in dextrose yield because of continued formation of isomaltose.

Maltulose concentration is normally 0.1–0.3%, although much higher levels of the disaccharide are produced if the proper pH is not maintained during liquefaction. When starch slurry pH is above 5.8, maltulose precursors are formed by a mechanism in which a portion of the polysaccharide reducing end groups is converted to fructose by alkaline isomerization. Glucoamylase does not hydrolyse the final glucosidic bond of a starch polymer if the end group is fructose. Therefore, the disaccharide maltulose is released during saccharification with an equivalent loss in dextrose yield. As liquefaction pH is increased, maltulose content is increased substantially to a level dependent on time and temperature (Table V). For instance, in a typical LT process conducted with a 3-hr hold time at 98°C, maltulose level in final hydrolysate increases from 0.2% at pH 6 to 2.9% at pH 7.

The DP-3 saccharides generally are present only in trace quantities and are, for the most part, formed via the reversion reaction. DP-4 + saccharides are composed of highly branched material that is hydrolyzed only slowly by glucoamylase (David *et al.*, 1987). This material can be converted completely to dextrose if given sufficient reaction time, but an extended saccharification results in a loss in dextrose yield due to continued formation of isomaltose via reversion.

Maximum dextrose level can be increased by operation at lower solids, since the reverse reaction is less favored as water concentration is increased. For instance, reducing solids from 30 to 20 or 10% increases maximum attainable dextrose level from 96 to 97.5 and nearly 98%, respectively. Problems associated with operating at reduced solids include the requirement for

TABLE V
Effect of Liquefaction Conditions on
Maltulose Level in Dextrose
Hydrolysate

Temperature (°C)	Time (min)	Maltulose (% dry basis) at pH		
		6.0	6.5	7.0
—	0	0.0	0.0	0.1
90	30	0.0	0.0	0.2
90	180	0.1	0.5	1.8
98	30	0.0	0.1	0.4
98	180	0.2	1.1	2.9
105	30	0.1	0.2	0.8
105	180	0.6	1.7	2.9

larger tanks, increased cost of water removal from final product, and increased risk of microbial infection in a dilute system.

Dextrose level also can be increased when certain enzymes are used in combination with glucoamylase. Commercially available enzymes used for this purpose include a debranching enzyme derived from *B. acidopullulyticus* (Jensen and Norman, 1984) and an enzyme exhibiting both amylase and transferase properties that was isolated from *B. megaterium* and transferred genetically into *B. subtilis* (Hebeda *et al.*, 1988).

Debranching enzymes are specific for the hydrolysis of α-1,6 glucosidic bonds and provide additional substrate for the action of glucoamylase. The rate and extent of dextrose production is increased when the enzyme combination is used and an additional 0.5–1.0% dextrose yield is achieved compared with use of glucoamylase alone. The combined action of *B. megaterium* amylase and glucoamylase also provides a 0.5–1.0% increase in dextrose yield. The mechanism for this reaction is shown in Fig. 5. The DP-4+ oligosaccharides that are not easily hydrolyzed by glucoamylase are in the form of a highly branched structure. The *B. megaterium* amylase degrades these oligosaccharides to panosyl units that are then transferred to dextrose, forming a maltotetraose (DP-4) saccharide containing one branch point. The maltotetraose saccharide is hydrolyzed easily to dextrose, so an increase in dextrose yield is achieved. Other advantages of both the debranching and the *B. megaterium* enzyme include reduced reaction time, reduced glucoamylase dose, and the opportunity to operate at higher solids without a loss in dextrose yield.

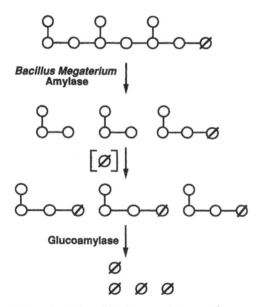

Figure 5   Action of *Bacillus megaterium* amylase.

Saccharified hydrolysate is clarified by vacuum filtration and refined by a combination of carbon and ion-exchange treatments. The refined hydrolysate is processed to a high dextrose corn syrup, crystalline dextrose, or liquid dextrose, or used as feed for HFCS production.

A high dextrose syrup is prepared from the refined hydrolysate by evaporating to 70–75% w/w solids.

Crystalline monohydrate dextrose is produced by concentrating refined hydrolysate to 70–78% w/w, seeding with crystals in a batch crystallizer, and cooling from about 43°C to 20–30°C over several days or until 50–60% of the dextrose has crystallized. The crystals are separated by batch or continuous centrifugation, washed to remove impurities, and dried in hot air rotary dryers to about 8.5% moisture (Hebeda, 1983).

Anhydrous dextrose is produced by dissolving monohydrate dextrose in water and crystallizing at 60–65°C.

Liquid dextrose is prepared by dissolving monohydrate dextrose to 71% solids.

C. High Fructose Syrups

Immobilized glucose isomerase is used to convert a refined 93–96% dextrose hydrolysate to a 42% fructose product. Commercial enzymes are produced intracellularly from a variety of bacterial sources. Selected commercial enzyme products and recommended operating conditions are listed in Table VI.

During the initial stages of HFCS development in the early 1960s, it became obvious that the use of a soluble isomerase for fructose production would require a high enzyme dose or a long reaction time. The high dose was economically unacceptable and the prolonged reaction time caused the formation of undesirable by-products such as mannose, psicose, color, and off-flavors that increased refining costs. Consequently, efforts were directed

TABLE VI
Commercially Available Glucose Isomerase Enzymes

		Recommended operating conditions		
Supplier	Enzyme source	Temperature (°C)	pH	Solids (%)
Enzyme Bio-Systems	*Streptomyces olivochromogenes*	55–60	7.8–8.6	35–50
Genencor International	*Streptomyces rubiginosus*	54–62	7.6	40–50
Gist-Brocades	*Actinoplanes missouriensis*	58–60	7.5	40–45
Solvay Enzyme Products	*Microbacterium arborescens*	≤60	7.5	40–50
Novo Nordisk	*Bacillus coagulans*	59–61	7.8–8.3	40–45
Novo Nordisk	*Streptomyces murinus*	50–60	7.5	35–45

to the development of immobilized enzyme systems that would improve process economics.

An immobilized isomerase system was found to have many of the benefits and few of the disadvantages of bound enzyme systems proposed for other purposes. For instance, the substrate used for HFCS production is a clarified and refined low viscosity stream that is passed easily through a column of bound enzyme. Both the substrate (dextrose) and the product (fructose) are simple low molecular weight saccharides that easily diffuse through porous carriers used for enzyme immobilization. In addition, the bound enzyme exhibits very long half-lives of several months under normal process conditions, resulting in low enzyme cost. Also, reaction time due to the high enzyme concentration is reduced sufficiently in continuous operation to minimize production of by-products. Consequently, with the exception of initial fructose production in the late 1960s, all HFCS has been manufactured using immobilized enzyme.

Two isomerase immobilization processes are used to produce commercial products. These are referred to as "whole cell" and "soluble enzyme" processes.

In the "whole cell" process, the bacterial cells containing the intracellular isomerase are recovered from fermentation broth and treated to immobilize the enzyme within the cells. In one commercial process (Jorgensen et al., 1988), bacterial cells are concentrated by centrifugation, disrupted by homogenization, cross-linked with glutaraldehyde, and flocculated with a cationic flocculent. The resulting material is filtered, extruded, dried, and sieved to produce a 300–1000 μm final product. In another process (Hupkes, 1978), a cell–gelatin mixture is cross-linked with glutaraldehyde, washed, and sieved. Cellular immobilized enzyme products are discarded after use and replaced with new material.

In the "soluble enzyme" process, cells are treated enzymatically or by physical means to break the cell wall and release isomerase. The isomerase is recovered by filtration or centrifugation and concentrated by ultrafiltration. The soluble enzyme is then bound to a carrier. One method (Antrim and Auterinen, 1986) involves recovering a purified soluble isomerase by ultrafiltration and crystallization, and contacting the enzyme for 3–5 hr with a carrier composed of DEAE–cellulose, titanium dioxide, and polystyrene. The material is ground to produce a 400–800 μm final product. In another type of immobilization, polyamine is absorbed on alumina and treated with glutaraldehyde to cross-link the amine and immobilize the isomerase by forming covalent bonds between the enzyme amino groups and the carrier aldehyde groups (Levy and Fusee, 1979; Rohrbach, 1981). In some cases, the carrier can be reused for several cycles by removing residual enzyme and rebinding with fresh enzyme. For instance, the DEAE–cellulose carrier can be regenerated by washing sequentially with water, 2% sodium hydroxide, water, buffer, and rinse water to remove inactivated enzyme and accumulated impurities. Fresh isomerase is then bound by batch or column operation to achieve 95% of the original capacity. Alternatively, additional enzyme can be added periodically via the feed and bound to the carrier during the isomerization process to maintain a constant level of activity over a long period of operation (Antrim *et al.*, 1989).

The immobilized isomerase is loaded in a column and substrate is passed through at a rate that produces an effluent containing 42% fructose. Several parameters must be controlled carefully for maximize productivity. These parameters include feed purity, solids, temperature, pH, oxygen level, magnesium content, and reaction time.

1. Feed Purity

Isomerase feed is refined hydrolysate containing 93–96% dextrose. Efficient refining is required to remove insoluble material that may accumulate in the column over time and cause pressure drop problems and impurities that could cause inactivation of the enzyme. Ion-exchange treatment of feed is required to lower calcium levels to 3 ppm or less, since the metal is inhibitory to isomerase.

2. Solids

Feed solids level is controlled in the 40–50% range, generally at about 45%. If the solids level is too high, increased viscosity may reduce rate of feed diffusion within the carrier pores and cause a reduction in product yield. A solids level that is too low could result in microbial contamination.

3. Temperature

Temperature control is important in maximizing productivity and useful life of the column. Normally, temperature is controlled at 55–61°C. At a lower temperature, the risk of microbial contamination is increased. As temperature is increased beyond the recommended range, activity of the isomerase increases but, at the same time, long-term productivity decreases due to a faster rate of enzyme inactivation. This situation results in a reduction in the useful life of the column. Only when high productivity is required for a short period of time at the expense of long-term productivity is a higher temperature a consideration.

4. pH

Long-term isomerase stability is highly dependent on pH. Depending on the particular enzyme, recommended pH varies from 7.5 to 8.2. Enzyme activity increases at higher pH levels, but productivity decreases due to reduced enzyme stability. For best operation, pH should be controlled as low as possible to minimize by-product formation while maintaining maximum productivity.

5. Oxygen

Presence of dissolved oxygen in dextrose feed is detrimental because of possible oxidation of isomerase cysteine residues, resulting in a loss in activity (Volkin and Klibanov, 1989). Deaeration of feed to remove excess oxygen or addition of SO_2 to reduce oxidation is helpful. Typically, $1-2$ mM SO_2 as sodium bisulfite is added for this purpose.

6. Magnesium

The presence of magnesium is important as an enzyme cofactor for activation and stabilization of glucose isomerase. Generally, a magnesium level of $0.5-5$ mM is recommended. Magnesium counteracts the adverse effect of excess calcium and commonly is added at a molar ratio of 20 times the level of calcium present. Cobalt and iron enhance glucose isomerase activity or stability whereas copper, zinc, nickel, mercury, and silver inhibit activity (van Tilburg, 1985).

7. Reaction Time

Flow rate through a column is controlled to yield an effluent fructose level of $42-45\%$ (dry basis). Under typical operating conditions at a temperature of $60°C$, an equilibrium fructose level of 51.2% is achievable (48.6% fructose with a 95% dextrose feed). However, the very long reaction time needed to achieve the equilibrium level reduces overall productivity significantly.

Control of these parameters is required to achieve maximum productivity. If any of these conditions falls outside the recommended range, a temporary or permanent loss in performance can be expected. Temporary loss in productivity can be caused by a moderate change in pH, low temperature, high solids, insufficient magnesium, or excess calcium. Correction of these parameters will reinstate column operation. More serious problems in productivity are caused by a temperature above $65°C$ or a pH considerably below or above the normal range. The one major disadvantage of column operation is that several months of potential enzyme use may be lost by an uncorrected fluctuation of pH or temperature to an unacceptable level.

If reaction parameters are controlled within the limits recommended by the enzyme manufacturer, efficient operation over several months or longer can be expected. The enzyme normally is used through several half-lives until residual activity is reduced to $10-20\%$ of the original level.

The 42% fructose product containing $51-54\%$ dextrose is refined by carbon and ion-exchange treatments to eliminate color, off-flavors, salts, and other impurities and concentrated to 71% solids. A storage temperature of $27-32°C$ is required to prevent dextrose crystallization.

Products containing higher fructose levels are manufactured by a combination of chromatographic separation and blending. The refined 42% HFCS at about 50% w/w solids is passed through a cationic adsorbent containing calcium groups. The fructose is retained and a stream (raffinate) is collected containing predominantly nonfructose saccharides. The raffinate, composed of 80–90% dextrose, 5–10% fructose, and a higher saccharide fraction, is recycled to the saccharification or isomerization process. Elution of the high fructose fraction with water yields an HFCS product containing 80–90% fructose and 7–19% dextrose. In actual practice, the separation process operates continuously in a simulated moving-bed process, providing a continuous flow of enriched HFCS and raffinate. The primary use of enriched fructose product is for blending with 42% HFCS to produce a 55% fructose product containing about 40% dextrose.

The 55% HFCS is evaporated to 77% solids for storage and shipment. Crystallization is generally not a problem due to the low dextrose level.

A small amount of 80–90% HFCS is evaporated to 77–80% solids and sold as a noncrystallizable syrup.

Crystalline fructose is manufactured using as raw material the 90+% fructose product obtained by chromatographic separation of 42% HFCS. The 90% fructose material is evaporated to a solids level of 90% or higher and the fructose is crystallized by batch or continuous techniques over a period of several hours or days (Hanover, 1992). The crystalline product is washed free of impurities and dried.

D. Traditional Syrups

Traditional corn syrups are manufactured by straight acid, acid–enzyme, or enzyme–enzyme hydrolysis processes (Howling, 1992).

Straight acid syrups have been produced in much the same manner throughout the history of the corn refining industry. Starch at 35–40% solids is acidified with hydrochloric acid to 0.15–0.2N and heated at 140–160°C under pressure until the desired DE is reached; the reaction is stopped by neutralization with sodium carbonate. Hydrolysate is clarified and refined by carbon treatment and, possibly, ion-exchange, and evaporated to 83% solids. Continuous and batch processes are used to produce a standard 42–43 DE product that contains an average saccharide distribution of 19% dextrose, 14% maltose, 12% maltotriose, and 55% DP-4 and higher saccharides.

Lower DE products are produced occasionally. Higher DE products normally are not prepared by acid hydrolysis, since the excessive levels of color and off-flavor formed during the reaction are difficult to remove during refining. Higher DE syrups, however, can be produced by enzymatic hydrolysis of straight acid syrups or by a process in which both liquefaction and saccharification are conducted enzymatically.

The two standard syrups manufactured by acid–enzyme or enzyme–enzyme techniques contain higher levels of dextrose and/or maltose than acid syrups. In one case, a 38–42 DE acid hydrolysate is adjusted to pH 4.8–5.2 and 55–60°C and saccharified with a combination of glucoamylase and *Aspergillus oryzae* fungal α-amylase for several hours or days, depending on enzyme dose. The glucoamylase dose is considerably less than that used in dextrose saccharification so only a low level of dextrose is produced. The fungal α-amylase is a maltogenic enzyme, producing maltose. When saccharification is complete, a 63–65 DE product is produced that exhibits an average composition of 36% dextrose, 30% maltose, and 13% maltotriose. Products of even higher DE can be made by extending the reaction time, although the increased dextrose level may result in crystallization problems during storage or shipment if heating is not used.

Syrups containing high levels of maltose are manufactured using only a maltogenic enzyme for saccharification. Enzymes used for this purpose are fungal α-amylase or barley β-amylase. Saccharification is conducted at 55–60°C, pH 4.8–5.2, and a solids level of 35–45%. Maltose levels of 50–55% are achieved in a reaction time that depends on the dose of enzyme used. If higher levels of maltose are desired, a combination of the maltogenic enzyme and a pullulanase is used for saccharification. The pullulanase hydrolyzes the α-1,6 glucosidic linkages and provides additional substrate for the production of maltose. In this manner, maltose levels of 60–80% or higher are produced. The very high maltose syrups can be prepared only if a low DE substrate is used. If substrate DE is too high, maltotriose will be produced from the oligosaccharides containing an odd number of dextrose units and a high maltose level will not be achieved.

IV. Applications of Sweeteners and Syrups

Corn sweeteners exhibit a variety of properties and are used in many different food and nonfood applications. Functional properties include sweetness, fermentability, viscosity, flavor enhancement, osmotic pressure, humectancy, hygroscopicity, color development, cohesiveness, nutritional quality, texture enhancement, and crystallization prevention. In 1991, total distribution of HFCS, corn syrup, and dextrose to United States food and nonfood industries was 7.7, 2.8, and 0.6 million metric tons, respectively (Table VII).

A. Dextrose

Dextrose is used primarily for sweetening, as a source of fermentable sugars, and for osmotic pressure. Distribution of dextrose for various food applications is shown in Table VII.

The perceived sweetness of dextrose depends on several factors including solids, temperature, and presence of other ingredients. In general, dextrose exhibits about 75% of the sweetness of sucrose and, therefore, is used in applications in which less sweetness is desired. In some formulations, combi-

TABLE VII
Distribution by Industry of Dextrose, HFCS, and Corn Syrup in the United States[a]

Year	Baking	Beverage	Canning	Confectionery	Dairy	Total[b]
Dextrose[c]						
1965	178	8	21	37	6	468
1970	174	8	23	52	7	547
1975	157	18	15	62	7	561
1980	51	66	4	55	2	513
1985	56	81	2	58	1	479
1990	72	102	4	73	NA[d]	595
1991	74	107	4	71	NA	604
HFCS[e]						
1970	18	39	9	0.5	5	99
1975	129	279	64	4	36	715
1980	365	1039	235	15	140	2659
1985	428	4246	288	39	213	6372
1990	397	5253	483	43	259	7368
1991	436	5360	532	36	285	7748
Corn syrup[f]						
1965	209	36	104	410	159	1211
1970	222	93	98	439	213	1449
1975	315	205	174	416	278	2278
1980	196	384	126	446	241	2201
1985	151	377	150	496	290	2469
1990	180	366	164	480	276	2832
1991	180	353	163	484	305	2845

[a] Reprinted with permission from R. E. Hebeda (1983).
[b] Includes other applications not listed.
[c] Monohydrate basis.
[d] NA, Not available.
[e] 71 wt, % solids basis.
[f] 80.3 wt, % solids.

nations of dextrose and sucrose provide a synergistic effect, yielding the same sweetness as sucrose alone.

Dextrose is fermented completely by yeast and is used as a carbohydrate source in applications in which a high degree of fermentability is desired.

In baking, dextrose is used as a fermentable carbohydrate source for yeast to provide the proper degree of carbon dioxide production and product volume. In addition, dextrose is important in improving bread strength for better handling and slicing, for producing desired color and flavor via the Maillard reaction with protein, and for providing good texture and crumb and crust characteristics.

In beverages, dextrose is used to impart sweetness, body, and osmotic pressure. In beverage powders, dextrose is used for flavor enhancement and to reduce excessive sweetness. Dextrose is also used in wine and beer for its fermentable properties. In light beer production, dextrose is used as a completely fermentable additive to reduce residual carbohydrate and caloric level.

In canning, dextrose often is used alone or in combination with sucrose in sauces, soups, gravies, fruit juices, jams, and jellies to control osmotic pressure and to provide flavor, sweetness, and texture.

In confections, dextrose is used in candies to control crystallization and provide sweetness and softness. It is also used for coating, strength, hardness, color, and gloss in candy and gum.

In nonfood applications, dextrose is used in a purified anhydrous form in medical applications such as intravenous solution production and tableting. In addition, it is used as a raw material in fermentations for the production of antibiotics, citric acid, amino acids, enzymes, lactic acid, and ethanol. Dextrose also is used as a raw material for production of sorbitol and polydextrose.

As an additive, dextrose is used in adhesives to control flow, in library paste to increase bonding time, in wallboard to prevent brittle edges, in concrete to retard setting, in resins as a binder, in dyes as a diluent, in metal treatment as a reducing agent, and in leather tanning for pliability.

B. High Fructose Syrups

The most important property of high fructose syrups is sweetness. A 42% HFCS product exhibits about the same sweetness as sucrose, 55% fructose is equivalent to medium sucrose invert, and 80–90% fructose is sweeter than sucrose. In general, HFCS exhibits the greatest sweetness perception under conditions of low temperature and low pH and is less sweet at warm temperatures and neutral pH.

The major use of HFCS is in soft drinks. HFCS first was used in this application in 1974, when HFCS-42 was approved for 25% replacement of sucrose. In 1980, HFCS-55 was used at a 50% sucrose replacement level and, by 1984, 100% replacement of sucrose was approved. Currently, HFCS is used in many cases as a complete replacement for sucrose. In 1991, 69% of HFCS utilization was in the beverage area and amounted to over 5 million metric tons. Specifications to insure continued high quality were designed through a combined effort of the Society of Soft Drink Technologists, the Corn Refiners Association, corn wet-millers, and soft drink bottlers. These specifications include solids, ash, color, pH, fructose level, sediment, SO_2, flavor, and odor (Long, 1986).

HFCS-42 is used in canning to provide sweetness, in dairy products as a bodying agent and to improve texture and mouthfeel, and in confections for sweetness, grain control, and humectancy. HFCS-90 is used in reduced-calorie products to lower caloric levels by 30–50%. Distribution of HFCS for various food applications is shown in Table VII.

Crystalline fructose generally is reported to be 20–80% sweeter than sucrose; a sweetness synergy is often observed when fructose is used in combination with other sugars. (CRA, 1988). The actual degree of sweetness perception obtained depends on several factors including temperature, pH, and concentration (Osberger, 1986). Pure fructose often is used in low calorie and specialty foods to provide sweetness at a reduced caloric level.

C. Traditional Syrups

By definition, corn syrup is any starch hydrolysis product that exhibits a DE between 20 and 99.4. Consequently, a wide range of syrups containing a variety of saccharide distributions is possible. Functional properties can vary considerably depending on composition. For instance, corn syrups are used for sweetness, fermentability, viscosity, humectancy, and hygroscopic properties. In addition, corn syrups contribute osmotic pressure, control crystallization, enhance foam stability, and produce browning reactions.

Sweetness depends on the particular saccharide distribution as well as on pH, solids, and temperature; a synergy often develops when used in combination with sucrose. For instance, a combination of 25% 42 DE corn syrup and 75% sucrose at 45% solids yields the same sweetness as 100% sucrose (Hobbs, 1986). Syrups containing high levels of fermentable saccharides, that is, dextrose and maltose, are used in brewing and baking where fermentability is needed. The high viscosity of the lower DE syrups is used to modify food texture. The humectance and hygroscopic properties of corn syrups allow moisture absorption to maintain product freshness.

Distribution of corn syrup for various food applications is given in Table VII.

In baking, corn syrups are used as a source of fermentables while improving shelf life by maintaining a proper moisture balance and providing good crust color. In beverages, corn syrups are used as an adjunct to provide a source of fermentables and body as well as to enhance flavor. In canning, corn syrups are used to improve flavor, color, and texture while providing body and preventing sucrose crystallization. High osmotic pressure is important for maintaining juices in canned fruit. In confections, corn syrups, provide viscosity, mouthfeel, sweetness, texture, grain inhibition, hygroscopicity, and resistance to discoloration. In some products, corn syrups are instrumental in participating in browning reactions to produce color, give a gloss to hard candies, and provide the proper texture for chewiness in gums. In dairy products, corn syrups are important for texture and smoothness by preventing crystal formation without adding excessive sweetness.

Corn syrups also are used in a wide range of nonfood applications. For instance, syrups are used in colognes and perfumes to control evaporation, in air fresheners as humectants, to improve stability in adhesives, and as a setting retardant in concrete. Corn syrups also are used as carriers and for sweetness in medicinal products such as lozenges. Other uses are in dyes, inks, explosives, metal plating, shoe polish, leather tanning, paper, and textiles.

D. Maltodextrins

Maltodextrins are hydrolysis products of less than 20 DE and, therefore, are bland tasting with essentially no sweetness perception. Maltodextrin often is used in a dry form as a dispersing agent in coffee whiteners and as a bulk carrier for flavor ingredients. Maltodextrin is used in bakery, soup, and sauce mixes, beverage powders, condiments, dehydrated foods, instant tea, low-calorie sweeteners, marshmallows, pan coatings, nougats, and snack foods. Currently, a rapidly growing application area is the use of maltodextrins as fat substitutes in a variety of food products (Alexander, 1992).

References

Abdullah, M., Fleming, I. D., Taylor, M., and Whelan, W. J. (1963). Substrate specificity of the amyloglucosidase of *Aspergillus niger*. *Biochem J.* **89(1)**, 35–36.

Adachi, S., Ueda, Y., and Hashimoto, K. (1984). Kinetics of formation of maltose and isomaltose through condensation of glucose by glucoamylase. *Biotech. Bioeng.* **26,** 121–127.

Alexander, R. J. (1992). Maltodextrins: Production, properties, and applications. *In* "Starch Hydrolysis Products: Worldwide Technology, Production, and Applications" (F. W. Schenck and R. E. Hebeda, eds.), pp. 233–275. VCH Publishers, New York.

Antrim, R. L., and Auterinen, A. L. (1986). A new regenerable immobilized glucose isomerase. *Starch/Stärke* **38(4)**, 132–137.

Antrim, R. L., Lloyd, N. E., and Auterinen, A. L. (1989). New isomerization technology for high fructose syrup production. *Starch/Stärke* **41(4)**, 155–159.

Beschkov, V., Marc, A., and Engasser, J. M. (1984). A kinetic model for the hydrolysis and synthesis of maltose, isomaltose and maltotriose by glucoamylase. *Biotech. Bioeng.* **26,** 22–26.

Boyer, C. D., and Shannon, J. C. (1987). Carbohydrates of the kernel. *In* "Corn: Chemistry and Technology" (S. A. Watson and P. E. Ramstad, eds.), pp. 253–272. American Association of Cereal Chemists, St. Paul, Minnesota.

Brumm, P. J., Hebeda, R. E., and Teague, W. M. (1988). Purification and properties of a new, commercial thermostable *Bacillus stearothermophilus* α-amylase. *Food Biotech.* **2(1)**, 67–80.

Carroll, J. D., Swanson, T. R., and Trackman, P. C. (1990). Starch liquefaction with alpha amylase mixtures. U. S. Patent No. 4,933,279.

Chiang, J. P., Alter, J. E., and Sternberg, M. (1979). Purification and characterization of a thermostable α-amylase from *Bacillus licheniformis*. *Starch/Stärke* **31(3)**, 86–92.

Corn Refiners Association (1988). "1988 Corn Annual." Washington, D.C.

Corn Refiners Association (1991). "1991 Corn Annual." Washington, D.C.

David, M.-H., Gunther, H., and Roper, H. (1987). Catalytic properties of *Bacillus megaterium* amylase. *Starch/Stärke* **39(12)**, 436–440.

Delheye, G., and Moreels, E. (1988). Dextrose equivalent measurements on commercial syrups. *Starch/Stärke* **40(11)**, 430–432.

Hanover, L. M. (1992). Crystalline fructose: Production, properties, and applications. *In* "Starch Hydrolysis Products: Worldwide Technology, Production, and Applications" (F. W. Schenck and R. E. Hebeda, eds.), pp. 201–231. VCH Publishers, New York.

Hebeda, R. E. (1987). Sweeteners. *In* "Corn: Chemistry and Technology" (S. A. Watson and P. E. Ramstad, eds.). P. 512. American Association of Cereal Chemists, St. Paul, Minnesota.

Hebeda, R. E. (1983). Syrups. *In* "Kirk-Othmer: Encyclopedia of Chemical Technology," 3d Ed., Vol. 22, pp. 499–522. John Wiley and Sons, New York.

Hebeda, R. E., and Leach, H. W. (1974). The nature of insoluble starch particles in liquefied cornstarch hydrolysates. *Cereal Chem.* **51**(2), 272–281.

Hebeda, R. E., Styrlund, C. R., and Teague, W. M. (1988). Benefits of *Bacillus megaterium* amylase in dextrose production. *Starch/Stärke* **40**(1), 33–36.

Hehre, E. J., Okada, G., and Genghof, D. S. (1969). Configurational specificity: Unappreciated key to understanding enzymic reversions and *de novo* glycosidic bond syntheses. *Arch. Biochem. Biophys.* **135**, 75–89.

Henderson, W. E., and Teague, W. M. (1988). A kinetic model of *Bacillus stearothermophilus* α-amylase under process conditions. *Starch/Stärke* **40**(11), 412–418.

Hobbs, L. (1986). Corn syrups. *Cereal Foods World* **31**(12), 852–858.

Howling, D. (1992). Glucose syrup: Production, properties, and applications. *In* "Starch Hydrolysis Products: Worldwide Technology, Production, and Applications" (F. W. Schenck and R. E. Hebeda, eds.), pp. 277–318. VCH Publishers, New York.

Hupkes, J. V. (1978). Practical process conditions for the use of immobilized glucose isomerase. *Starch/Stärke* **30**(1), 24–28.

Jensen, B. F., and Norman, B. E. (1984). *Bacillus acidopullulyticus pullulanase:* Application and regulatory aspects for use in the food industry. *Process Biochem* **August,** 129–134.

Jorgensen, O. B., Karlsen, L. G., Nielsen, N. B., Pedersen, S., and Rugh, S. (1988). A new immobilized glucose isomerase with high productivity produced by a strain of *Streptomyces murinus*. *Starch/Stärke* **40**(8), 307–313.

Levy, J., and Fusee, M. C. (1979). Support matrices for immobilized enzymes. U.S. Patent No. 4,141,857.

Orthoefer, F. T. (1987). Corn starch modification and uses. *In* "Corn: Chemistry and Technology" (S. A. Watson and P. E. Ramstad, eds.), pp. 479–499. American Association of Cereal Chemists, St. Paul, Minnesota.

Osberger, T. F. (1986). Pure crystalline fructose. *In* "Alternative Sweeteners" (L. O. Nabors and R. C. Gelardi, eds.), pp. 245–275. Marcel Dekker, New York.

Reeve, A. (1992). Starch hydrolysis: Process and equipment. *In* "Starch Hydrolysis Products: Worldwide Technology, Production, and Applications" (F. W. Schenck and R. E. Hebeda, eds.), pp. 79–120. VCH Publishers, New York.

Roels, J. A., and van Tilburg, R. (1979). Kinetics of reactions with amyloglucosidase and their relevance to industrial applications. *Starch/Stärke* **31**(10), 338–345.

Rohrbach, R. P. (1981). Support matrices for immobilized enzymes. U.S. Patent No. 4,268,419.

Rosendal, P., Nielsen, B. H., and Lange, N. K. (1979). Stability of bacterial α-amylase in the starch liquefaction process. *Starch/Stärke* **31**(11), 368–372.

Rutenberg, M. W., and Solarek, D. (1984). Starch derivatives: Production and uses. *In* "Starch: Chemistry and Technology" (R. L. Whistler, J. N. Bemiller, and E. F. Paschall, eds.), 2d Ed., pp. 311–388. Academic Press, New York.

Shetty, J. K., and Allen, W. G. (1988). An acid-stable, thermostable α-amylase for starch liquefaction. *Cereal Foods World* **33**(11), 929–934.

Shiraishi, F., Kawakami, K., and Kusunoki, K. (1985). Kinetics of condensation of glucose into maltose and isomaltose in hydrolysis of starch by glucoamylase. *Biotech. Bioeng.* **27**, 498–502.

Teague, W. M., and Brumm, P. J. (1992). Commercial enzymes for starch hydrolysis products. *In* "Starch Hydrolysis Products: Worldwide Technology, Production, and Applications" (F. W. Schenck and R. E. Hebeda, eds.), pp. 45–77. VCH Publishers, New York.

Tomazic, S. J., and Klibanov, A. M. (1988a). Mechanisms of irreversible thermal inactivation of *Bacillus* α-amylases. *J. Bio. Chem.* **263**(7), 3086–3091.

Tomazic, S. J., and Klibanov, A. M. (1988b). Why is one *Bacillus* α-amylase more resistant against irreversible thermoinactivation than another? *J. Biol. Chem.* **263**(7), 3092–3096.

U.S. Department of Agriculture (1989). "Sugar and Sweetener Situation and Outlook." Commodity Economics Division, Economic Research Service.

U.S. Department of Agriculture (1991). "Sugar and Sweetener Situation and Outlook." Commodity Economics Division, Economic Research Service.

van Tilburg, R. (1985). Enzymatic isomerization of cornstarch-based glucose syrups. *In* "Starch Conversion Technology" (G. M. A. van Beynum and J. A. Roels, eds.), pp. 175–236. Marcel Dekker, New York.

Volkin, D. B., and Klibanov, A. M. (1989). Mechanism of thermoinactivation of immobilized glucose isomerase. *Biotech. Bioeng.* **33**, 1104–1111.

Watson, S. A. (1977). Industrial utilization of corn. *In* "Corn and Corn Improvement" (G. F. Sprague, ed.), pp. 721–763. American Society of Agronomy, Madison, Wisconsin.

Wurzburg, O. B. (ed.) (1986). "Modified Starches: Properties and Uses." CRC Press, Boca Raton, Florida.

Zobel, H. F. (1984). Gelatinization of starch and mechanical properties of starch pastes. *In* "Starch: Chemistry and Technology" (R. L. Whistler, J. N. Bemiller, and E. F. Paschall, eds.), 2d Ed., pp. 285–309. Academic Press, New York.

CHAPTER 13

Dairy Products

RODNEY J. BROWN

I. Natural Enzymes of Milk

Raw bovine milk contains about 50 native enzymes (Jenness, 1982), at least 27 of which have been classified by the International Union of Biochemistry Enzyme Commission (Walstra and Jenness, 1984). Some, such as catalase and aldolase, may be introduced from blood serum. Most originate in the epithelial cells of the mammary gland (Larson, 1985). Many enzymes found commonly in milk are not intrinsic milk enzymes, but are in milk because of the presence of microorganisms.

Pasteurization eliminates most natural enzyme activity in milk. Some enzymes lose activity after being subjected to treatments as mild as 52°C for 15 sec (Andrews *et al.*, 1987). Other milk enzymes are stable; plasmin, for example, loses only 60% of its activity after 15 sec at 100°C (Alichanidis *et al.*, 1986).

Some native enzymes in milk are employed for analytical or quality control purposes. For instance, destruction of alkaline phosphatase by ordinary pasteurization temperatures is used to monitor adequacy of pasteurization. Microbial phosphatase activity and phosphatase reactivation in high temperature pasteurized milk samples may generate unreliable results in such assays (Druckey *et al.*, 1985; Kosikowski, 1988).

Another example of quality control based on enzymes is measurement of serum enzyme concentrations to detect abnormal milk (Kitchen *et al.*, 1970). Catalase activity in milk is an indicator of elevated leukocyte levels characteristic of mastitic milk. Direct measurements of somatic cells or concentration of deoxyribonucleic acids (DNA) are preferred over the catalase test for this purpose (Richardson, 1985).

Spontaneous development of oxidized flavors is common in dairy products. Several native milk enzymes are implicated in the process. Peroxidase

(EC 1.11.1.7), which is stable at pasteurization temperatures, catalyzes oxidative reactions in milk and dairy products. Xanthine oxidase (EC 1.2.3.2), which catalyzes oxidation of xanthine or hypoxanthine to uric acid and hydrogen peroxide, also causes oxidative flavors (Richardson, 1975).

An important positive use of a native milk enzyme is the activation of lipase by homogenization, which conditions the milk fat substrate in the manufacture of blue cheese (Scott, 1986). Homogenization disrupts the milk fat globule membranes where lipase enzymes in milk are concentrated, making them more available for lipolysis of milk fat.

One milk protein, α-lactalbumin, is more concentrated in milk than is any enzyme. Although not an enzyme, in many respects this protein should be considered with the milk enzymes. α-Lactalbumin is one of two protein moieties essential for optimal activity of lactose synthetase (Larson, 1985); its concentration controls lactose synthesis and, hence, the colligative properties of milk. In addition to its own enzyme activity, α-lactalbumin is closely homologous with the enzyme lysozyme found in egg whites (Jenness, 1982).

II. Milk Clotting

A. Enzymatic Milk Coagulation

The predominant use of enzymes in the dairy industry is in coagulation of milk to make cheese (Brown and Ernstrom, 1988). Most proteolytic enzymes clot milk (Berridge, 1954; Brown, 1988) but not all make acceptable cheese. The Food and Nutrition Board of the United States National Research Council (Anonymous, 1981) uses the term "rennet" to describe all milk-clotting enzyme preparations (except porcine pepsin) used for cheese making. The board defines rennet as "aqueous extracts made from the fourth stomachs of calves, kids, or lambs" and bovine rennet as "aqueous extract made from the fourth stomach of bovine animals, sheep, or goats." Microbial rennet "followed by the name of the organism" is the approved nomenclature for milk-clotting preparations derived from microorganisms.

United States Standards of Identity for Cheddar cheese (Anonymous, 1985a) allow the use of "rennet and/or other clotting enzymes of animal, plant, or microbial origin." Milk-clotting enzymes generally recognized as safe (GRAS) are rennet and bovine rennet (Anonymous, 1984a). Microbial rennets derived from *Cryphonectria parasitica* (formerly *Endothia parasitica*), *Bacillus cereus*, *Mucor pusillus* var. Lindt, and *Mucor miehei* var. Cooney *et* Emerson may be used as secondary direct food additives (Anonymous, 1984b). Porcine pepsin may be used to make cheese also, but is not called rennet (Anonymous, 1981). Chymosin (EC 3.4.23.3) from the abomasa of suckling calves remains the enzyme of choice and the standard against which all others are evaluated. Genetic engineering has been employed to produce chymosin microbially (Teuber, 1990) in *Escherichia coli* (Chen *et al.*, 1984;

Hayenga *et al.*, 1984) and in *Saccharomyces cerevisiae* (Moir *et al.*, 1985). Cheese making trials comparing recombinant chymosin with calf rennet have found no significant differences between the two (Green *et al.*, 1985); recombinant enzymes are now approved for commercial use. Recombinant rennet has the added benefit of meeting some religious dietary requirements not met by calf rennet.

Milk-clotting activity is present as chymosin in the abomasum of a bovine fetus by the sixth month of development and increases in potency as the fetus approaches full term (Pang and Ernstrom, 1986). At birth, chymosin is present in gastric mucosa at $2-3$ mg/g but its production declines after 1 week (Foltmann, 1981). Procedures for extraction of chymosin were described by Ernstrom and Wong (1974). Liquid rennet preparations have a pH between 5.6 and 5.8 to provide the most stable environment. Bacterial content can be reduced by filtration or centrifugation. Crude rennet extract contains active chymosin and an inactive precursor (prochymosin). Removing a peptide from the N terminus of prochymosin (36,000 daltons) converts it to chymosin (31,000 daltons). This transformation is favored below pH 5.0 (Rand and Ernstrom, 1964). Chymosin is most stable between pH 5.3 and 6.3 (Foltmann, 1959b).

Rennet contains bovine pepsin if stomachs of older calves or cows are included with the stomachs of suckling calves. Such rennets are acceptable substitutes in cheese manufacture if their cost is lower than rennets with higher percentages of chymosin. Higher chymosin : pepsin ratios accelerate cheese ripening (Richardson and Chaudhari, 1970). When comparing five commercial milk-clotting preparations, Shaker and Brown (1985a,b) observed that decreasing chymosin : pepsin ratios cause more general proteolysis and loss of curd to the whey.

Porcine pepsin is used as a substitute for part of the rennet in making many varieties of cheese. This enzyme is secreted by hog stomach mucosa as pepsinogen (40,400 daltons). Conversion of pepsinogen to pepsin is catalyzed by pepsin below pH 5. On average, $7-9$ peptide bonds are hydrolyzed during formation of pepsin, which releases about 20% of the molecule. Cleavage of only one of these bonds is likely to be necessary to release the active enzyme (Brown and Ernstrom, 1988). The pH of cheese milk is such that both the activity and the stability of porcine pepsin are far from optimum (pH 2.0); the enzyme is inactivated rapidly (Emmons, 1970). Slow coagulation and a weak set resulting in excessive fat losses and reduced yield occur if insufficient enzyme is used, but porcine pepsin is inexpensive relative to chymosin. Mixtures of porcine pepsin and bovine rennet are used as less expensive coagulants. Such mixtures typically contain $20-25$% chymosin, $40-45$% bovine pepsin, and $30-40$% porcine pepsin activity (McMahon and Brown, 1985; Yiadom-Farkye, 1986).

Fungal proteases are used extensively as substitutes for chymosin in milk clotting. *Mucor miehei* rennet, a protease preparation from *M. miehei* var. Cooney *et* Emerson, is the most common fungal milk-clotting preparation. It is also most heat stable of all the commonly used milk-clotting enzymes

(Sternberg, 1971; Thunell *et al.*, 1979). This enzyme remains active in the whey and is concentrated in condensed whey products, causing problems when whey products containing residual *M. miehei* rennet are mixed with casein. Therefore, nearly all *M. miehei* rennet used by the cheese industry is now modified (Ramet and Weber, 1981) to decrease its heat stability (Branner-Jorgensen *et al.*, 1980; Cornelius, 1982) by treatment with hydrogen peroxide under controlled conditions. Some enzymatic activity is lost, but the modified enzyme has about the same stability as calf rennet (Brown and Ernstrom, 1988).

Rennet from *C. parasitica* has been used with varying success for cheese making (Whitaker, 1970). Among the commercially used fungal rennets, *C. parasitica* rennet is the most proteolytic on α_s- and β-caseins and least proteolytic on κ-casein, but generally causes bitterness in the cheese product (Brown and Ernstrom, 1988). Cheese varieties including Swiss, which are cooked at slightly higher temperatures, can be made successfully with *C. parasitica* rennet (Ernstrom and Wong, 1974). *Mucor pusillus* var. Lindt protease has given satisfactory results as a chymosin substitute in the manufacture of a number of cheese varieties, but not all strains of *M. pusillus* var. Lindt are capable of producing acceptable cheese (Babel and Somkuti, 1968). The clotting activity of *M. pusillus* var. Lindt protease is more sensitive to pH changes between 6.4 and 6.8 than chymosin, but is much less sensitive than porcine pepsin (Richardson *et al.*, 1967).

The search for suitable chymosin substitutes has led to the investigation of a number of proteases produced by plants and bacteria. Occasional reports of success in cheese making reflect the fact that suitable coagulating enzymes may exist as part of crude mixtures containing other highly proteolytic enzymes that are detrimental to cheese making (Shehata *et al.*, 1978; Velcheva and Ghbova, 1978). None of these mixtures is in commercial use. Pepsin from chicken has been used for cheese making when dietary laws restrict use of bovine rennet (Stanley *et al.*, 1980; Green *et al.*, 1984).

B. Milk Coagulation Chemistry

Milk clotting begins with enzymatic cleavage of κ-casein, destroying the ability of κ-casein to stabilize casein micelles (McMahon and Brown, 1984a). This cleavage is followed by a nonenzymatic aggregation of the altered casein micelles; then aggregates of casein micelles form a firm gel structure (Brown, 1981; Green and Morant, 1981; McMahon and Brown, 1990). Possibly, a separate fourth step exists during which the curd structure tightens and syneresis occurs (McMahon and Brown, 1984b).

Proteolytic enzymes are active in milk clotting by cleaving κ-casein to form para-κ-casein and a macropeptide:

$$\kappa\text{-casein} \longrightarrow \text{para-}\kappa\text{-casein} + \text{macropeptide}$$

Although chymosin splits κ-casein at a specific Phe-105–Met-106 bond, other milk-clotting enzymes are less specific (Fox, 1981; Visser, 1981). Splitting of κ-casein destroys the stability of the milk system and casein aggregates. Aggregation of casein micelles begins long before proteolysis of κ-casein is completed (Payens and Wiersma, 1980). The enzymatic and nonenzymatic phases of milk clotting overlap. Under normal cheese making conditions, aggregation of micelles begins at the same time as enzymatic cleavage of κ-casein (Reddy et al., 1986).

Both pH and temperature affect milk coagulation (Ustunol and Brown, 1985). Cheese starter cultures lower the pH before milk-clotting enzyme is added and continue to lower the pH during setting of the curd. Each milk-clotting enzyme has an optimal pH that must be considered in balance with culture activity to attain the correct pH at setting and cutting of the curd (Brown, 1981; Shalabi and Fox, 1982). The temperature coefficient (Q_{10}) for aggregation of casein micelles after rennet cleavage of κ-casein is 5–6 times that of the enzymatic step (Cheryan et al., 1975a,b; McMahon and Brown, 1984b). This fact has been used by many to study the secondary phase of milk clotting after allowing the enzymatic phase to go to completion at a lower temperature.

C. Measurement of Milk Coagulation

Milk-clotting activity is determined by the rapidity with which the enzyme clots milk under a set of specified conditions. This set of conditions includes the whole coagulation process, not just the enzymatic reaction. Measurement of the enzymatic reaction alone must be done with single substrates such as κ-casein or a peptide similar to the cleavage site in κ-casein. It is possible to take advantage of the different temperature coefficients to separate rennet cleavage of κ-casein and aggregation of casein micelles (McMahon and Brown, 1984b), but this separation does not allow separate measurement of enzyme activity.

Apparatuses for visual observation of the formation of a clot and standard substrates for comparing activities of different milk-clotting enzymes have been described by Sommer and Matsen (1935). Berridge (1952) developed a standard substrate to overcome the variability in different milk samples. Berridge substrate is made by reconstituting 60 g low heat nonfat dry milk in 500 ml 0.01 M $CaCl_2$ and holding the solution at 4°C for 20 hr before use.

The Formagraph instrument was designed for following the progress of milk coagulation by recording the movement of small stainless steel loops immersed in milk as the milk is moved (McMahon and Brown, 1982, 1983). In comparison with the visual method of Sommer and Matsen (1935), the Formagraph has no difference in precision, but has the advantage of unattended operation. It can be used over a wide range of chymosin concentrations (0.001 to 0.16 chymosin units/ml).

McMahon *et al.* (1984a,b,c) monitored coagulation of Berridge substrate with a spectrophotometer connected to a computer. An initial dip in the tracing as κ-casein is cleaved is followed by a rapid rise as aggregation begins. The maximum on the curve matches the point of coagulation (τ) described by Payens (1978). More information about the course of milk coagulation is available from this method than from any other devised to date.

Hori (1985) reported an innovative method for following the progress of milk clotting with potential for in-plant use. A 0.1 mm × 106 mm platinum wire is immersed in milk and heated by application of 0.7 amp dc electricity. As the milk surrounding the wire coagulates, dissipation of heat away from the wire decreases and the temperature of the wire increases. This procedure follows the progress of coagulation without disrupting the curd.

Storch and Segelcke (1874) proposed that the product of coagulation time (t_c) and enzyme concentration (E) should be a constant (k) (McMahon and Brown, 1983, 1984b).

$$t_c E = k$$

Holter (1932) added a factor (x) to correct for the time lag between enzymatic cleavage of κ-casein and aggregation (Brown and Collinge, 1986). Foltmann (1959a) rearranged Holter's equation, keeping k and x separate to show that k is a constant but that x varies according to measurement conditions.

$$t_c = \frac{k}{E} + x$$

McMahon *et al.* (1984c) and Brown and Collinge (1986) showed that the x added to Storch and Segelcke's equation by Holter is not necessary if τ [the "actual coagulation time" defined by Payens (1978)] is used instead of t_c.

III. Cheese Ripening

A. Proteolysis

Milk-clotting enzymes are added to milk to cleave κ-casein and begin coagulation of the milk, but they also have general proteolysis capabilities that contribute to cheese aging (Green, 1972; Visser, 1981; Shaker and Brown, 1985a,b; Fox, 1988; Manji and Kakuda, 1988). Chymosin is the least proteolytic of the commonly used milk coagulants (Green, 1977; McMahon and Brown, 1985). Excessive general proteolysis leads to excessive loss of fat and cheese yield and adversely affects flavor and texture (Green, 1977; Sellers, 1982).

Proteolytic activity during the ripening of cheese can be extensive (Fox,

1989). In hard cheese, 25–35% of the insoluble protein of the curd may be converted into soluble protein, whereas in soft varieties such as Brie, Camembert, or Limburger, over 80% of the insoluble protein can be converted to water-soluble compounds such as peptides, amino acids, and ammonia (Foster *et al.*, 1983). Apart from the participation of starter cultures, proteolysis depends in part on the coagulating enzyme system used in the formation of the curd (Lawrence *et al.*, 1972). The amount of residual enzyme in curd after draining the whey depends on the choice of coagulating enzyme (Harper, 1975; Holmes *et al.*, 1977). Production of Cheddar cheese without starter culture results in cheese with much lower levels of free amino acids (Yiadom-Farkye, 1986). Production of bitter flavors generally is attributed to the formation of bitter peptides (Lemieux *et al.*, 1989). Bitter flavor results when these peptides are formed faster than they can be broken down further by proteolytic enzymes of the starter organisms (Cliffe and Law, 1990).

Some rennet substitutes show high rates of proteolysis and frequently produce bitter taste in cheese. However, there are small differences between approved coagulants. When porcine pepsin and calf rennet were used at ratios that gave equivalent clotting times, proteolysis by porcine pepsin was slightly less than that by calf rennet, as determined by formation of nonprotein nitrogen and the total concentration of free amino acids (Richardson, 1975).

B. Lipolysis

Most lipolytic activity during cheese ripening is from enzymes produced by the numerous species of microorganisms in cheese (Bhowmik and Marth, 1988; Kamaly and Marth, 1989). The microflora of cheese is complex and changeable because different organisms predominate at different stages of cheese ripening. Adventitious organisms, including lactobacilli and micrococci, play significant roles in lipolysis during cheese ripening. Most hard cheeses are ripened by bacterial action in the cheese, whereas the soft varieties of cheese are ripened largely by yeasts, molds, or bacteria growing on the cheese surface.

Milk lipases may lead to undesirable rancidity if freshly drawn milk is cooled too rapidly. This effect usually is attributed to the presence of "membrane" lipase, which is adsorbed on the fat globules on rapid cooling and initiates lipolysis. The lipase becomes active if raw milk is homogenized or agitated or if foaming or great temperature fluctuations occur. Because of the fairly high pH range for milk lipase (6–9) and its sensitivity to heat, it does not seem to play an important role in the ripening of cheese made from pasteurized milk.

In some of the mold cheeses, such Roquefort, Gorgonzola, blue, and Stilton, the lipases produced by *Penicillium roqueforti* play an important part in flavor formation.

C. Accelerated Cheese Ripening

Proteases and lipases or the systems producing them are used for manufacture of cheese paste products in which cheese flavor is accelerated significantly (Conner, 1988; Tye *et al.*, 1988). Enzymes often are added as crude preparations made by heat treating or freeze shocking microorganisms (Ardo and Pettersson, 1988; Spangler *et al.*, 1989). Flavor enhancing enzymes have been added as pastes (Kosikowski, 1988), as solutions (Law, 1980), through encapsulation techniques (Braun and Olson, 1986; Alkhalaf *et al.*, 1988, 1989; El Soda *et al.*, 1989), and as soluble enzymes with salt (Green, 1985). Addition of lactase is effective in accelerating the ripening of Manchego cheese (Fernandez-Garcia *et al.*, 1988).

In view of the extensive lipolysis of milk fat during cheese ripening, efforts have been made to supplement lipolytic activity of microorganisms through lipase addition. Addition of pregastric esterases is practiced for production of Italian type cheese. Traditionally, Italian cheeses have been prepared with rennet paste made by drying the entire stomachs of calves, kids, or lambs, including the milk contents of the stomach. The paste made from kids or lambs particularly gives the cheese a characteristic piquant flavor that cannot be obtained with rennet extracts or milk free pastes. The difference is created by enzymes secreted by glands at the base of the tongue of the animals. These enzymes are extracted from the excised glands of calves, lambs, or kids. They are the major source of the characteristic lipolytic activity associated with rennet paste. These pregastric esterase enzymes are sometimes called oral lipases or oral glandular lipases (Richardson, 1975).

IV. Lactase

A. Lactose Intolerance

The inability of some populations to catabolize lactose properly because of inadequate intestinal lactase (β-galactosidase, EC 3.2.1.23) activity is a concern for the dairy industry and for nutritionists (Hourigan, 1984; Houts, 1988). Lactose is the major carbohydrate of milk, and milk is the only source of lactose. Use of dairy foods as dietary sources of specific nutrients such as calcium suggests the need for lactase treatment of milk and milk products.

B. Lactase in Dairy Processing

Use of lactase in the dairy industry is one of the most promising applications of enzymes to food processing. Lactose has poor solubility (about 15% in water at 20°C) and is less than one-sixth as sweet as sucrose. Lactose hydrolysis produces a more soluble, sweeter mixture of equal proportions of glucose and galactose and small amounts of diverse polygalactosides. The hydrolysate has an osmotic pressure about twice that of lactose. Lactose is in abundant supply, particularly in cheese whey, presenting serious disposal problems in the cheese industry (Dziezak, 1991).

The lactose fermenting yeast such as *Kluyveromyces lactis* (formerly *Saccharomyces lactis*), and *Kluyveromyces fragilis*, or fungi such as *Aspergillus niger* and *Aspergillus oryzae* are the most frequently used sources of the enzyme lactase and commonly have been used for its preparation (Holsinger and Klingerman, 1991). Most of the work on dairy applications has been done with the yeast enzyme. Fungal enzymes with lower pH optima have proved advantageous in bread doughs.

Lactose in ice cream mixes can be hydrolyzed by directly incorporating lactase into the mix and holding for 5 days at 7.5°C. Because storage of mix or skim concentrates for the periods required to allow adequate hydrolysis is impractical, such procedures are not useful for adjusting osmotic properties of mixes. Addition of lactase to mix for added sweetness, reduced lactose intolerance potential, and prevention of crystallization in the product is common (Richardson, 1975).

V. Hydrogen Peroxide and Catalase Treatment

Use of hydrogen peroxide (H_2O_2) to reduce the bacterial count of milk has been suggested since about 1900. Two obstacles have prevented extensive use of the process: unavailability of sufficiently pure and stable H_2O_2 solutions and unavailability of catalase, the enzyme used to break down any residual H_2O_2 remaining in milk after treatment (Kosikowski, 1988).

In two situations the H_2O_2/catalase process is preferred over heat pasteurization of dairy products. In many underdeveloped areas where cooling facilities are not available on farms and where heat pasteurization is not feasible, addition of H_2O_2 is a practical means of milk preservation. An effective procedure is addition of 1 ml 33% H_2O_2 per liter of milk at the farm, addition of 1 ml H_2O_2 at the dairy plant, heating to 50°C for 30 min, cooling to 35°C, and adding catalase (Richardson, 1975).

A second reason for the interest in H_2O_2 treatment is the relative advantage over pasteurization in treatment of milk for cheese making. Pasteurization destroys not only milk pathogens but also acid forming microorganisms, and inactivates some of the natural enzymes of the milk. Hydrogen peroxide treatment is effective in reducing the counts of pathogenic microorganisms and inactivates milk catalase and peroxidase, but it permits lactic acid-forming microorganisms to survive, and lipases, proteases, and phosphatases are not affected. The H_2O_2/catalase treatment cannot be substituted legally for pasteurization, since complete elimination of pathogens is not assured (Richardson, 1975).

Hydrogen peroxide is decomposed slowly by the natural catalase of raw milk. However, the amount of catalase in nonpasteurized milk is not sufficiently high to insure complete destruction of H_2O_2. Therefore, addition of catalase is essential to avoid the presence of residual H_2O_2 and a high oxidation/reduction potential, which interferes with the active growth of starter organisms in dairy products. Catalase preparations from liver and bacterial *(Micrococcus lysodeikticus)* sources have been approved for cheese manufacture. Only the liver preparation is sold because of limited demand.

VI. Miscellaneous Application of Enzymes in Dairy Processing

The use of pregastric esterases has been discussed for manufacture of cheese, particularly of the Italian cheese varieties. These esterases also can be used to produced pronounce dairy flavors in milk fat emulsions or in whole milk concentrate substrates. Such flavors are desired for production of milk chocolate and account for the unique superiority of some milk chocolate flavors. The flavors can be produced by treating whole milk or whole milk concentrates with pregastric esterases and drying to whole milk powders. Such powders contain 28.5% milk fat, 2.5% moisture, and otherwise resemble whole milk powders in their gross chemical composition. These powders are remarkably resistant to oxidative rancidity, unlike their whole milk counterparts. They can be used in the production of milk chocolate (0.5–3% based on the weight of the finished chocolate), fudge, caramel fillings, cast creams, and other food applications in which a pronounced dairy flavor is desired. Low levels of lipolyzed concentrates contribute a rich dairy flavor to margarines and baked goods. Basic lipolyzed concentrates have been modified by fortification changes to allow their application in coffee whiteners, snack foods, and imitation dairy foods. Higher levels of these products produce Italian cheese characteristics and can be used in soups, pizza, and similar products where a free fatty acid character is essential (Richardson, 1975).

Another lipolyzed product reported for similar applications is produced by sequential growth of *Lactobacillus bulgaricus* in cream, followed by controlled lipolysis. Bacterial spoilage during cream incubation was prevented using this technique; the lactic culture flavor advantage finds application in some dairy food products (Richardson, 1975).

One of the more uncommon suggestions for the use of enzymes in the dairy industry deals with the use of rennet in ice cream mixes to increase the viscosity of the mix, to eliminate the need for ice cream stabilizers, and to improve the melt characteristics of ice cream. The use of small amounts of pepsin has been suggested to increase the viscosity of evaporated milk so it can be sterilized by high temperature, short time methods. The use of 0.003 ppm crystalline pepsin (1 hr at $0-1°C$) doubles the viscosity of the evaporated milk (Richardson, 1975).

The possibility of future developments of commercial sources of galactose oxidase should be mentioned. The action of this enzyme is specific for galactose. It oxidizes galactose to D-galactohexodialdose and H_2O_2. Although its applications are essentially analytical (Richardson, 1975), the use of such a product could have as wide an application as glucose oxidase in the removal of oxygen.

References

Alichandis, E., Wrathall, J. H. M., and Andrews, A. T. (1986). Heat stability of plasmin (milk proteinase) and plasminogen. *J. Dairy Res.* **53**, 259–269.

Alkhalaf, W., Piard, J.-C., El Soda, M., Gripon, J.-C., Desmazeaud, M., and Vassal, L. (1988). Liposomes as proteinase carriers for the accelerated ripening of Saint-Paulin type cheese. *J Food Sci.* **53**, 1674–1679.

Alkhalaf, W., El Soda, M., Gripon, J.-C., and Vassal, L. (1989). Acceleration of cheese ripening with liposome-entrapped proteinase: Influence of liposome net charge. *J Dairy Sci.* **72**, 2233–2238.

Andrews, A. T., Anderson, M., and Goodenough, P. W. (1987). A study of the heat stabilities of a number of indigenous milk enzymes. *J. Dairy Res.* **54**, 237–246.

Ardo, Y., and Pettersson, H.-E. (1988). Accelerated cheese ripening with heat treated cells of *Lactobacillus helveticus*, and a commercial proteolytic enzyme. *J. Dairy Res.* **55**, 239–245.

Babel, F. J., and Somkuti, G. A. (1968). *Mucor pusillus* protease as a milk coagulant for cheese manufacture. *J. Dairy Sci.* **51**, 937–937.

Berridge, N. J. (1952). Some observations on the determination of the activity of rennet. *Analyst* **77**, 57–62.

Berridge, N. J. (1954). Rennin, and the clotting of milk. *In* "Advances in Enzymology" (E. F. Nord, ed.), Vol. 15, pp. 423–449. Interscience Publishers, New York.

Bhowmilk, T., and Marth, E. H. (1988). Protese, and peptidase activity of *Micrococcus* species. *J. Dairy Sci.* **71**, 2358–2365.

Branner-Jorgensen, S., Schneider, P., and Eigtved, P. (1980). A method of modifying the thermal destabilization of microbial rennet, and a method of cheese making using rennet so modified. U.K. Patent Application 2,045,772A.

Braun, S. D., and Olson, N. F. (1986). Microencapsulation of cell-free extracts to demonstrate the feasibility of heterogenous enzyme systems, and cofactor recycling for development of flavor in cheese. *J. Dairy Sci.* **69**, 1202–1208.

Brown, R. J. (1981). The mechanism of milk clotting. *Proc. 2nd Bienn. Marschall Int. Cheese Conf.* 107–112.

Brown, R. J. (1988). Milk coagulation, and protein denaturation. *In* "Fundamentals of Dairy Chemistry" (N. P. Wong, ed.), 3d Ed., pp 583–607, AVI, Westport, Connecticut.

Brown, R. J., and Collinge, S. K. (1986). Actual milk coagulation time, and inverse of chymosin activity. *J. Diary Sci.* **69**, 956–958.

Brown, R. J., and Ernstrom, C. A. (1988). Milk-clotting enzymes, and cheese chemistry. Part I—Milk-clotting enzymes, and their actions. *In* "Fundamentals of Dairy Chemistry" (N. P. Wong, ed.), 3d Ed., pp. 609–633, Van Nostrand Reinhold, New York.

Chen, M. C. Y., Hayenga, K. J., Lawlis, V. B., and Snedecor, B. R. (1983). Microbially produced rennet, methods for its production, and plasmid used for its production. European Patent Application 0,116,778.

Cheryan, M., Van Wyk, P. J., Olson, N. F., and Richardson, T. (1975a). Continuous coagulation of milk using immobilized enzymes in a fluidized-bed reactor. *Biotechnol. Bioeng.* **17,** 585–598.

Cheryan, M., Van Wyk, P. J., Olson, N. F., and Richardson, T. (1975b). Secondary phase, and mechanism of enzymic milk coagulation. *J. Dairy Sci.* **58,** 477–481.

Cliffe, A. J., and Law, B. A. (1990). Peptide composition of enzyme-treated Cheddar cheese slurries, determined by reverse phase high performance liquid chromatography. *Food Chem.* **36,** 73–80.

Code of Federal Regulations (1984a). 21 CFR 173.150. Food and Drug Administration, Department of Health and Human Services, Washington, D.C.

Code of Federal Regulations (1984b). 21 CFR 184.1685. Food and Drug Administration, Department of Health and Human Services, Washington, D.C.

Code of Federal Regulations (1985). 21 CFR 133.113. Food and Drug Administration, Department of Health and Human Services, Washington, D.C.

Conner, T. (1988). Advances in accelerated ripening of cheese. *Cult. Dairy Prod. J.* **23,** 21–25.

Cornelius, D. A. (1982). Process for decreasing the thermal stability of microbial rennet. U.S. Patent No. 4,348,482.

Druckrey, I., Kleyn, D. H., Murthy, G. K., and Wehr, M. W. (1985). Phosphatase methods. In "Standard Methods for the Examination of Dairy Products." (G. H. Richardson, ed.), 15th Ed., pp. 311–326. American Public Health Association, Washington, D.C.

Dziezak, J. D. (1991). Enzymes: Catalysts of food processes. *Food Technol.* **45,** 78–85.

El Soda, M., Johnson, M., and Olson, N. F. (1989). Temperature sensitive liposomes: A controlled release system for the acceleration of cheese ripening. *Milchwissensch.* **44,** 213–214.

Emmons, D. B. (1970). Inactivation of pepsin in hard water. *J. Dairy Sci,* **53,** 1177–1182.

Ernstrom, C. A., and Wong, N. P. (1974). Milk-clotting enzymes, and cheese chemistry. In "Fundamentals of Dairy Chemistry" (B. H. Webb, A. H. Johnson, and J. A. Alford, ed.), 2d Ed., pp. 662–771. AVI Publishing, Westport, Connecticut.

Fernandez-Garcia, E., Ramos, M., Polo, C., Juarez, M., and Olano, A. (1988). Enzyme accelerated ripening of Spanish hard cheese. *Food Chem.* **28,** 63–80.

Foltmann, B. (1959a). On the enzymatic, and coagulation stages of the renneting process. *XV Int. Dairy Congr. Proc.* **2,** 555–655.

Foltmann, B. (1959b). Studies on rennin. II. On the crystallisation, stability, and proteolytic activity of rennin. *Acta Chem. Scand.* **13,** 1927–1935.

Foltmann, B. (1981). Mammalian milk-clotting proteases: Structure, function, evolution, and development. *Neth. Milk Dairy J.* **35,** 223–366.

Foster, E. M., Nelson, F. E., Speck, M. L., Doetsch, R. N., and Olson, J. C. (1983). "Dairy Microbiology." Ridgeview Publishing, Atascadero, California.

Fox, P. F. (1981). Proteinases in dairy technology. *Neth. Milk Dairy J.* **35,** 233–253.

Fox, P. F. (1989). Proteolysis during cheese manufacture, and ripening. *J. Dairy Sci.* **72,** 1379–1400.

Green, M. L. (1977). Review of the progress of dairy science: Milk coagulants. *J. Dairy Res.* **44,** 159–188.

Green, M. L. (1985). Effect of milk pretreatment, and making conditions on the properties of Cheddar cheese from milk concentrated by ultrafiltration. *J. Dairy Res.* **52,** 555–564.

Green, M. L., and Morant, S. V. (1981). Mechanism of aggregation of casein micelles in rennet-treated milk. *J. Dairy Res.* **48,** 57–63.

Green, M. L., Valler, M. J., and Kay, J. (1984). Assessment of the suitability for Cheddar cheesemaking of purified, and commercial chicken pepsin preparations. *J. Dairy Res.* **51,** 331–340.

Green, M. L., Angal, S., Lowe, P. A., and Marston, F. A. O. (1985). Cheddar cheesemaking with recombinant calf chymosin synthesized in *Escherichia coli. J. Dairy Res.* **52,** 281–286.

Harper, W. J., and Lee, C. R. (1975). Residual coagulants in whey. *J. Food Sci.* **40,** 282–284.

Hayenga, K. J., Lawlis, V. B., and Snedecor, B. R. (1984). Microbially produced rennet, methods for its production, and reactivation, plasmids used for its production, and its use in cheesemaking. European Patent Application 0,114,507,A1.

Holmes, D. G., Duersch, J. W., and Ernstrom, C. A. (1977). Distribution of milk clotting enzymes between curd, and whey, and their survival during Cheddar cheese making. *J. Dairy Sci.* **60**, 862–869.

Holsinger, V. H., and Klingerman, A. E. (1991). Applications of lactase in dairy foods, and other foods containing lactose. *Food Technol.* **45**, 92–95.

Holter, H. (1932). Über die Labwirkung. *Biochem. Z.* **255**, 160.

Hori, T. (1985). Objective measurement of the process of curd formation during rennet treatment of milks by the hot wire method. *J. Food Sci.* **50**, 911–917.

Hourigan, J. A. (1984). Nutritional implications of lactose. *Aust. J. Dairy Technol.* **39**, 114–120.

Houts, S. S. (1988). Lactose intolerance. *Food Technol.* **42**, 110–113.

Jenness, R. (1982). Inter-species comparison of milk proteins. In "Developments in Dairy Chemistry-1, Proteins" (P. F. Fox, ed.), pp. 87–114. Applied Science Publishers, London.

Kamaly, K. M., and Marth, E. H. (1989). Enzyme activities of lactic streptococci, and their role in maturation of cheese: A review. *J. Dairy Sci.* **72**, 1945–1966.

Kitchen, B. J., Taylor, G. C., and White, I. C. (1970). Milk enzymes—Their distribution, and activity. *J. Dairy Res.* **37**, 279–288.

Kosikowski, F. V. (1988). Enzyme behavior, and utilization in dairy technology. *J. Dairy Sci* **71**, 557–573.

Larson, B. L. (1985). Biosynthesis, and cellular secretion of milk. In "Lactation" (B. Larson, ed.), pp. 129–163. Iowa State University Press, Ames.

Law, B. (1980). Accelerated ripening of cheese. *Dairy Ind. Int.* **45**, 15–48.

Lawrence, R. C., Creamer, L. K., Gilles, J., and Martley, F. G. (1972). Cheddar cheese flavour. I. The role of starters, and rennets. *N. Z. J. Dairy Sci. Technol.* **7**, 32–37.

Lemieux, L., Puchades, R., and Simard, R. E. (1989). Size-exclusion HPLC separation of bitter, and astringent fractions from Cheddar cheese made with added *Lactobacillus* strains to accelerate ripening. *J. Food Sci.* **54**, 1234–1237.

McMahon, D. J., and Brown, R. J. (1982). Evaluation of Formagraph for comparing rennet solutions. *J. Dairy Sci.* **65**, 1639–1642.

McMahon, D. J., and Brown, R. J. (1983). Milk coagulation time: Linear relationship with inverse of rennet activity. *J. Dairy Sci.* **66**, 341–344.

McMahon, D. J., and Brown, R. J. (1984a). Composition, structure, and integrity of casein micelles: A review. *J. Dairy Sci.* **67**, 499–512.

McMahon, D. J., and Brown, R. J. (1984b). Enzymic coagulation of casein micelles: A review. *J. Dairy Sci.* **67**, 919–929.

McMahon, D. J., and Brown, R. J. (1985). Effects of enzyme type on milk coagulation. *J. Dairy Sci.* **68**, 628–632.

McMahon, D. J., and Brown, R. J. (1990). Development of surface functionality of casein particles as the controlling parameter of enzymic milk coagulation. *Coll. Surfaces* **44**, 263–279.

McMahon, D. J., Brown, R. J., and Ernstrom, C. A. (1984a). Enzymic coagulation of milk casein micelles. *J. Dairy Sci.* **67**, 745–748.

McMahon, D. J., Brown, R. J., Richardson, G. H., and Ernstrom, C. A. (1984b). Effects of calcium, phosphate, and bulk culture media on milk coagulation properties. *J. Dairy Sci.* **67**, 930–938.

McMahon, D. J., Richardson, G. H., and Brown, R. J. (1984c). Enzymic milk coagulation: Role of equations involving coagulation time, and curd firmness in describing coagulation. *J. Dairy Sci.* **67**, 1185–1193.

Manji, B., and Kakuda, Y. (1988). The role of protein denaturation, extent of proteolysis, and storage temperature on the mechanism of age gelation in a model system. *J. Dairy Sci.* **71**, 1455–1463.

Moir, D. T., Mao, J. I., Ducan, M. J., Smith, R. A., and Kohno, T. (1985). Production of calf chymosin by the yeast *S. cerevisiae*. *Dev. Ind. Microbiol.* **26**, 75–85.

National Research Council, Food, and Nutrition Board (1981). "Food Chemicals Codex." National Academy Press, Washington, D.C.

Pang, S. H., and Ernstrom, C. A. (1986). Milk clotting activity in bovine fetal abomasa. *J. Dairy Sci.* **69**, 3005–3007.

Payens, T. A. J. (1978). On different models of casein clotting: The kinetics of enzymatic, and non-enzymatic clotting compared. *Neth. Milk Dairy J.* **32**, 170–183.

Payens, T. A. J., and Wiersma, A. K. (1980). On enzymatic clotting processes. V. Rate equations for the case of arbitrary rate of production of the clotting species. *Biophys. Chem.* **11**, 137–146.

Ramet, J. P., and Weber, F. (1981). Cheesemaking properties of a thermolabile milk-clotting enzyme from *Mucor miehei. Lait* **61**, 458–464.

Rand, A. G., and Ernstrom, C. A. (1964). Effect of pH, and sodium chloride on activation of prorennin. *J. Dairy Sci.* **47**, 1181–1187.

Reddy, D., Payens, T. A., and Brown, R. J. (1986). Effect of pepstatin on the chymosin-triggered coagulation of casein micelles. *J. Dairy Sci.* **69 (Suppl. 1)**, 72.

Richardson, G. H. (1975). Dairy industry. In "Enzymes in Food Processing" (G. Reed, ed.), 2d Ed., pp. 361–395. Academic Press, New York.

Richardson, G. H. (ed.) (1985). "Standard Methods for the Examination of Dairy Products," 15th Ed. American Public Health Association, Washington, D.C.

Richardson, G. H., and Chaudhari, R. V. (1970). Differences between calf, and adult rennet. *J. Dairy Sci.* **53**, 1367–1372.

Richardson, G. H., Nelson, J. H., Lubnow, R. E., and Schwarberg, R. L. (1967). Rennin-like enzyme from *Mucor pusillus* for cheese manufacture. *J. Dairy Sci.* **50**, 1066–1072.

Scott, R. (1986). "Cheesemaking Practice," 2nd Ed. Elsevier Applied Science, New York.

Sellers, R. L. (1982). Effect of milk-clotting enzymes on cheese yield. Presented at the 5th Biennial Cheese Industry Conference, Utah State University, Logan.

Shaker, K. A., and Brown, R. J. (1985a). Effects of enzyme choice, and fractionation of commercial enzyme preparations on protein recovery in curd. *J. Dairy Sci.* **68**, 1074–1076.

Shaker, K. A., and Brown, R. J. (1985b). Proteolytic, and milk clotting fractions in milk clotting preparations. *J. Dairy Sci.* **68**, 1939–1942.

Shalabi, S. I., and Fox, P. F. (1982). Influence of pH on the rennet coagulation of milk. *J. Dairy Res.* **49**, 153–157.

Shehata, A. E., Ismail, A. A., Hegazi, A., and Hamdy, A. M. (1978). Fractionation of commercial rennet enzymes on Sephadex G-100. *Milchwissensch.* **33**, 693–695.

Sommer, H. H., and Matsen, H. (1935). The relation of mastitis to rennet coagulability, and curd strength of milk. *J. Dairy Sci.* **18**, 741–749.

Spangler, P. L., El Soda, M., Johnson, M. E., Olson, N. F., Amunson, C. H., and Hill, C. G. (1989). Accelerated ripening of Gouda cheese made form ultrafiltered milk using liposome-entrapped enzyme, and freeze shocked lactobacilli. *Milchwissensch.* **44**, 199–203.

Stanley, D. W., Emmons, D. B., Modler, H. W., and Irvine, D. M. (1980). Cheddar cheese made with chicken pepsin. *Can. Inst. Food Sci. Technol. J.* **13**, 97–102.

Sternberg, M. Z. (1971). Crystalline milk clotting protease from *Mucor miehei,* and some of its properties. *J. Dairy Sci.* **54**, 159–167.

Storch, V., and Segelcke, T. (1874). *Milchforsch. Milchprax.* **3**, 997.

Teuber, M. (1990). Production of chymosin (EC 3.4.23.4) by microorganisms, and its use for cheesemaking. *Bull. Int. Dairy Fed.* **251**, 3–15.

Thunell, R. K., Duersch, J. W., and Ernstrom, C. A. (1979). Thermal inactivation of residual milk clotting enzymes in whey. *J. Dairy Sci.* **62**, 373–377.

Tye, T. M., Haard, N. F., and Patel, T. R. (1988). Effects of bacterial protease on the quality of Cheddar cheese. *Can. Inst. Food Sci Technol. J.* **21**, 373–377.

Ustunol, Z., and Brown, R. J. (1985). Effects of heat treatment, and post-treatment holding time on rennet clotting of milk. *J. Dairy Sci.* **68**, 526–530.

Velcheva, P., and Ghbova, D. (1978). Proteolytic activity of a milk-coagulating bacterial enzyme isolated from *Bacillus mesentericus* strain 90. *Acta Microbiol. Bulg.* **1**, 12–20.

Visser, S. (1981). Proteolytic enzymes, and their action on milk proteins. A review. *Neth. Milk Dairy J.* **35**, 65–88.

Walstra, P., and Jenness, R. (1984). "Dairy Chemistry and Physics." John Wiley and Sons, New York.

Whitaker, J. R. (1970). Protease of *Endothia parasitica. In* "Methods in Enzymology" (G. E. Perlman and L. Lorand, eds.), Vol. 19, pp. 436–445. Academic Press, New York.

Yiadom-Farkye, N. (1986). "Role of Chymosin and Porcine Pepsin in Cheddar Cheese Ripening." Ph.D. Thesis. Utah State University, Logan.

CHAPTER 14

Pectic Enzymes in Fruit and Vegetable Juice Manufacture

WALTER PILNIK and ALPHONS G. J. VORAGEN

I. Introduction

Pectic enzymes occur in higher plants and are produced by microorganisms. Their substrates are the pectic substances that occur as structural polysaccharides in the middle lamella and the primary cell walls of higher plants and are prominent in parenchymous tissues. Endogenous pectic enzymes therefore can produce important textural changes in fruits and vegetables during ripening and storage. The necessity to activate or inactivate them often has a decisive influence on processing steps in the manufacture of products derived from fruits and vegetables. Microbial pectic enzymes serve functions in plant pathology and fermented foods but also are produced industrially as processing aids in the food industry. Pectic enzymes purified to well defined activities are used in pectin analysis, structure research of pectins, and cell wall studies. This subject has been reviewed extensively in the last decade by Rombouts and Pilnik (1979, 1980, 1984), and Pilnik and Rombouts (1979a, b, 1981), Gierschner (1981), Rombouts (1981), Pilnik (1982), Kilara (1982), Voragen and Pilnik (1989), and Pilnik and Voragen (1989).

II. Pectic Substances and Pectic Enzymes

Pectic substances (pectins) are chain molecules with a rhamnogalacturonan backbone. This backbone consists of "smooth" α-D-1,4-galacturonan regions that are interrupted to a small extent by insertion of 1,2-linked α-L-rhamnosyl residues and highly branched regions with an almost alternating rhamno-

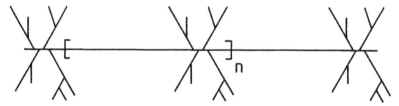

Figure I Schematic structure of apple pectin; rhamnogalacturonan backbone with regions rich in neutral sugar side chains (hairy regions). (Reprinted with permission from De Vries *et al.*, 1986.)

galacturonan chain (Aspinall, 1980; Neukom *et al.*, 1980; McNeil *et al.*, 1984; Selvendran, 1985). Side chains composed of neutral sugars are attached by glycosidic linkage to carbon atoms 3 and 4 of the rhamnose units and carbon atoms 2 and 3 of the galacturonic acid units, giving the rhamnogalacturonan portion of the pectin backbone a "hairy" character (de Vries *et al.*, 1986; Fig. 1). The predominant sugars D-galactose and L-arabinose are present in complex chains of considerable length (McNeil *et al.*, 1984), whereas some xylose occurs in monomeric or oligomeric side chains. Commercial apple or citrus pectins, extracted for use as gelling agents, have a degree of polymerization that usually does not exceed 500, but pectin fractions from fruits may have much higher molecular weights. Pectins extracted from most fruits under mild conditions usually have a degree of esterification (DE, percentage of galacturonic acid monomers that are methyl esterified) of over 70 (Does-burg, 1965). Acetic acid may occur as substitute on the hydroxyl groups of

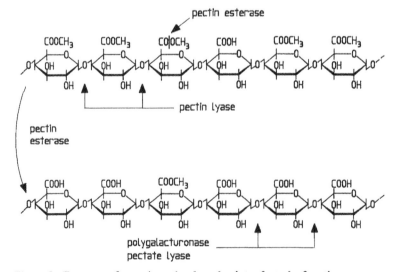

Figure 2 Fragment of a pectin molecule and points of attack of pectic enzymes.

galacturonic acid at carbon atom 2 or 3. Acetyl groups are prominent in sugar beet pectins, which also carry feruloyl groups linked as esters to terminal galactose or arabinose units of side chains (Fry, 1982; Rombouts and Thibault, 1986a).

Pectic enzymes are classified according to their mode of attack on the galacturonan part of the pectin molecule. Referring to the scheme in Fig. 2, one can distinguish between pectin methylesterases (PE; EC 3.1.11.1), which deesterify pectins to low methoxyl pectins or pectic acid, and pectin depolymerases, which split the glycosidic linkages between galacturonosyl (methylester) residues. Polygalacturonases (PG; EC 3.2.2.15 and 3.2.1.67) split glycosidic linkages next to free carboxyl groups by hydrolysis; pectate lyases (pectic acid lyases, PAL; EC 4.2.2.2 and 4.2.2.9) split glycosidic linkages next to free carboxyl groups by β-elimination (Fig. 3). Both endo and exo types of PGs and PALs are known. The endo types (EC 3.2.1.15, EC 4.2.2.2) split the pectin chain at random. Exo-PGs (EC 3.2.1.67) release monomers or dimers from the nonreducing end of the chain, whereas exo-PALs (EC 4.2.2.9) release unsaturated dimers from the reducing end. Pectates and low-methoxyl pectins are, therefore, the preferred substrates for PGs and PALs; the enzymes have very low activity on highly methylated pectins. Such pectins are degraded by pectin lyases (PL; EC 4.2.2.10), for which only the endo type has been described. Highly esterified pectins are, of course, also degraded by a combination of PE with PG or PAL (Neukom, 1969; Fogarty and Ward,

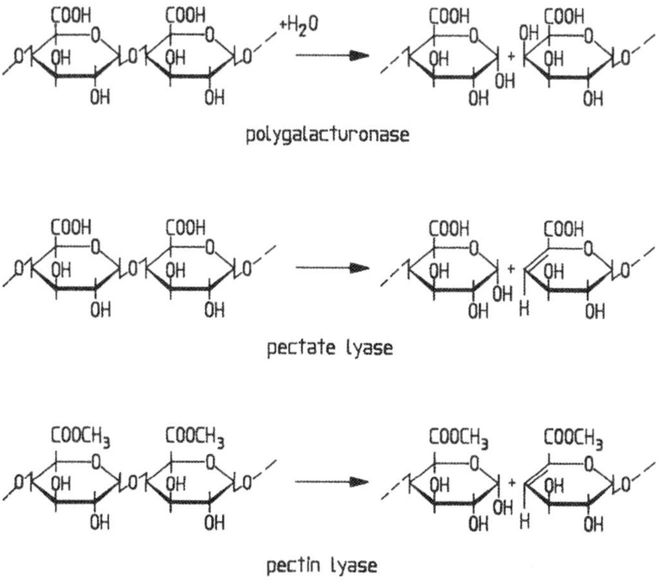

Figure 3 Splitting of glycosidic bonds in pectin by hydrolysis (polygalacturonase) and by β-elimination (pectate lyase and pectin lyase).

TABLE I
Occurrence of Pectin Esterase and Polygalacturonase in Higher Plants[a]

Plant	PE	PG Endo	PG Exo	PG NS	Plant	PE	PG Endo	PG Exo	PG NS
Apple	+		+		Passion fruit	+	+	+	
Apricot	(+)			+	Peach		+	+	
Avocado		+			Pear	+	+	+	
Banana	+	+			Plums	+			
Berries	+			+	Beans	+			
Citrus					Carrots	+		+	
lime	+				Cauliflower	+			
orange	+				Cucumber	+	+		
grapefruit	+		+		Leek	+			
mandarin	+				Onions	+			
Cherries	+	+			Pea	+			
Current				+	Potato	+		+	
Grapes	+			+	Radish	+			+
Mango	+	+			Tomato	+	+	+	
Papaya	+	+	+						

[a]Abbreviations: PE, pectin esterase; PG, polygalacturonase; ns, not specified.

1974; McMillan and Sheiman, 1974; Kulp, 1975; Rexová-Benková and Markovic, 1976; Rombouts and Pilnik, 1980). The existence of a polymethylgalacturonase for hydrolytic degradation of high ester pectin has never been proved and has not been found by careful screening of commercial preparations and microorganisms in the authors' laboratory.

III. Occurrence and Microbial Production of Pectic Enzymes

Table I shows the presence of PEs and PGs in higher plants. These enzymes also are produced by microorganisms, which produce PAL and PL as well (Table II; Rombouts and Pilnik, 1980).

IV. Food-Processing-Related Properties of Pectic Enzymes

Some molecular and kinetic properties are given in Table III.

A. Pectinesterases

Plant PEs (Solms and Deuel, 1955; Miller and Macmillan, 1971) are thought to attack either at the nonreducing end or next to a free carboxyl group and to proceed along the molecule by a single chain mechanism, creating blocks of unesterified galacturonic acid that are extremely calcium sensitive (Kohn *et al.*, 1968). Irregularities in the galacturonan chain, such as acetylated monomers, ester groups transformed into amides or reduced to primary alcohol, and occurrence of hairy regions, inhibit PE activity (Solms and Deuel, 1955). Pectinesterase is highly specific for the methylester of polygalacturonic acid. Other esters are attacked only very slowly (MacDonnell *et al.*, 1950; Manabe, 1973). The methylester of polymannuronic acid is not attacked at all (MacDonnell *et al.*, 1950). The rate of pectin deesterification depends on chain length; trimethyl trigalacturonate is not attacked at all (McCready and Seegmiller, 1954). Fungal PEs differ from plant PEs by obeying a multichain mechanism, removing methoxyl groups at random (Ishii *et al.*, 1979b). Further properties of PEs will be discussed in the section on technological functions.

TABLE II
Occurrence of Pectic Enzymes in Microorganisms[a]

		PG		PAL		PL
Organism	PE[b]	Endo	Exo	Endo	Exo	Endo
Fungi						
Aspergillus (niger)	++[c]	++	++			+
Penicillium	+	+				+
Fusarium	+	+		+		
Rhizopus	+	++				
Sclerotinia	+	+				+
Collectotrichum		+	+			
Yeasts						
Kluyveromyces		++				
Bacteria						
Bacillus (polymyxa)				+++		
Clostridium	+				+	
Erwinia		+	+	+++	+	+
Pseudomonas				+++		
Arthrobacter				+		

[a]Reprinted with permission from Rombouts and Pilnik, 1980.
[b]Abbreviations: PE, PG, see Table I. PAL, pectate lyase; PL, pectin lyase.
[c]+++ means produced in high levels; no data means activity absent.

TABLE III
Properties of Pectic Enzymes from Some Plants and Food-Grade Microorganisms

Enzyme type/source	Molecular weight	Isoelectric point	Specific activity (units/mg)	Optimum pH	K_m value (mg/ml)	Reference[a]
Pectin esterase						
Orange I[b]	36,000	10.0	694	7.6	0.083	Versteeg et al., 1980
II[b]	36,200	11.0	762	8.8	0.0046	Versteeg et al., 1980
III[b]	54,000	10.2	≥ 180	8	0.041	Versteeg et al., 1980
Aspergillus niger	39,000	3.9	406	4.5	3	Baron et al., 1980
Polygalacturonase						
Tomato I[b]	100,000	8.6		4.5		Pressey, 1986
II[b]	44,000	9.4		4.5		Pressey, 1986
Kluyveromyces fragilis	46,000 48,000	6.1	89	4.0		
Aspergillus aculeatus	47,000	4.7–4.8	67	4.0–5.0		
A. niger[b]	42,000 49,000	6.0–6.4	275	4.0		
Pectin lyase						
A. niger I	35,400	3.65	17	6	5.0	van Houdenhoven, 1975
A. niger II	33,100	3.75	44	6	0.9	van Houdenhoven, 1975
Pectate lyase						
Bacillus subtilis	33,000	9.85		8.5		Chesson and Codner, 1978

[a]Unpublished results of our laboratory unless otherwise indicated.
[b]Multiple molecular forms (isoenzymes).

B. Polygalacturonases

Activity measurements by viscosity on methyl and glycol esters of pectic acid show a rapid decrease in the rate and degree of hydrolysis with increasing esterification (Pilnik et al., 1973). Rexova-Benkova et al. (1977a) have shown that acetyl groups decrease the extent of degradation by lowering the affinity to the substrate molecules through the blocking of binding sites. The limitation of degradation by the presence of acetyl groups was confirmed by Rombouts and Thibault (1986b) using sugar beet pectin as substrate (Table IV). The limiting effect of methoxyl groups is shown in the same experiment. However, Koller and Neukom (1969) have isolated a fungal PG that is fully active on 70% acetylated pectin. Also, the mode of attack is known to differ for PGs of different origins. An endo enzyme may hydrolyze one bond randomly in a single enzyme–substrate encounter, followed by complete

TABLE IV
Degradation Limits of Sugar Beet Pectins before and after
Enzymatic Demethylation and Saponification[a]

Pretreatment and mode of degradation	Pectins			
	Water soluble	Oxalate soluble	Acid soluble	Alkali soluble
No pretreatment				
β-Elimination	7.2[b]	4.9	4.6	—
Endopectin lyase	8.6	4.3	4.1	—
Endopolygalacturonase	1.2	3.7	1.7	26.9
Endopectate lyase	≤1.0	4.8	1.3	23.7
After pectin esterase treatment				
Endopolygalacturonase	19.1	22.4	16.3	—
Endopectate lyase	11.8	13.0	9.9	—
After saponification				
Endopolygalacturonase	37.5	36.7	30.8	29.3
Endopectate lyase	33.4	34.5	31.0	26.4

[a]Reprinted with permission from Rombouts and Thibault, 1986b.
[b]Percentages of galactosyluronic acid bonds broken.

dissociation of enzyme and products. Additional scissions of this type follow, resulting in a gradual appearance of oligomers from the depolymerization of higher oligogalacturonides. This pattern is observed with endo-PG from *Kluyveromyces fragilis* (Phaff, 1966). It is characterized as a multichain attack. In the case of single-chain multiple attack, for example, with PG from *Colletotrichum lindemuthianum* (English *et al.*, 1972), a single random hydrolytic scission is followed by a number of nonrandom attacks on one of the products, resulting in rapid liberation of oligogalacturonates. Many multiple forms and isoenzymes of PG have been described (Pilnik and Rombouts, 1981). Points of difference are dependence on cations and mechanism of action on oligomers. Particularly relevant to their industrial application is possible inhibition by vegetable extracts, which is noted for fungal PGs (Albersheim and Anderson, 1971) but not for PG from tomatoes (Bock *et al.*, 1970a).

C. Endopectin Lyase

The best substrate for endopectin lyases, measured at pH ≥ 7, is completely esterified pectin, as indicated by degradation limits and affinity (M^{-1}). At lower pH values, affinity for less highly esterified pectins increases and a marked stimulation by calcium and other cations is observed. This behavior makes PL a useful enzyme for fruit processing (Voragen, 1972); Ishii and Yokotsuka, 1975). These enzymes need the methylester groups; they are inactive on glycol esters and on amidated pectates (Pilnik *et al.*, 1973, 1974).

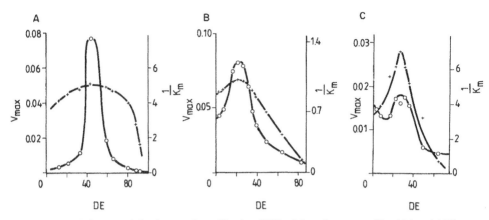

Figure 4 Influence of the degree of esterification (DE) of the substrate on V_{max} (+) and $1/K_m$ (O) of pectate lyases of *Arthrobacter* 370 (A), of *Arthrobacter* 547 (B), and of *Bacillus polymyxa* (C) measured by spectrophotometry. (Reprinted with permission from Pilnik *et al.*, 1973.)

D. Endopectate Lyase

Endopectate lyases, on the other hand, do not differentiate between the methyl ester and the glycol ester of pectic acids. Surprisingly, pectate is not the best substrate for bacterial PALs. These enzymes show maximal activity (initial velocity and degree of degradation) on low methoxyl pectins. Figure 4 shows that the DE optima for PALs from *Arthrobacter* 547, from *Arthrobacter* 370, and from *Bacillus polymixa* are 21, 44, and 26% respectively (Pilnik *et al.*, 1973). This feature and their high pH optimum make them interesting for vegetable processing. Changes produced by PALs are studied easily since enzyme action can be stopped by the addition of a chelating agent because of an absolute requirement for calcium ions (Rombouts, 1972).

E. Rhamnogalacturonase

An enzyme has been described (Schols *et al.*, 1990) that splits the galacturonic acid–rhamnose glycosidic linkage in hairy regions of apple pectins with strongly enhanced activity when the ester groups have been deesterified and arabinose has been removed by acid hydrolysis (modified hairy regions). The enzyme was found in a commercial pectinase preparation and surely must be grouped with pectic enzymes. The end products are oligomers with alternating galacturonic acid and rhamnose units, rhamnose forming the nonreducing end. The availability of these compounds make it possible to determine the anomeric form of the rhamnose units as α using advanced NMR techniques (Colquhoun *et al.*, 1990).

F. Commercial Pectinases

Commercial pectinases are fungal preparations, mainly from *Aspergillus* species, produced by a number of companies (Gist-Brocades, NOVO-Nordisk, Röhm, Biocon, Genencor International, Miles Kalichemie, ABM Sturge, Amano, A.T.P., Shin-Nihon). These preparations are usually mixtures of PE, PG, and PL activities, hemicellulases, and endo-β-glucanases (C_x-cellulases). The activities of three different commercial preparations are shown in Table V (Voragen *et al.*, 1986a). The activities found are all from the mold used for the production of the pectic enzymes, with the exception of the C_1-cellulase (cellobiohydrolase), which is added for technological purposes. The arabinanases play a special role in fruit juice processing (see subsequent text). Preparations containing polygalacturonase as the sole pectic activity are on the market and experimental quantities of fungal PE are available. Only one pectinase to date has been seen to contain rhamnogalacturonase activity. Gene technology surely will be used to produce desirable single-activity preparations that will open new possibilities for use alone or in optimized combinations.

TABLE V
Specific Enzyme Activities of Multienzyme Complexes in Technical Fungal Pectinase Preparations[a,b]

Enzyme activity	Substrate	A	B	C
Polygalacturonase	Polygalacturonic acid	1982	3314	1878
Pectin lyase	Pectin DE 90	43	53	74
Pectin esterase	Pectin DE 65	548	448	227
Combined pectolytic	Pectin DE 75	198	290	274
C_x-Cellulase	CMC	998	180	1228
C_1-Cellulase	Avicel	99	1	22
Arabanase				
linear arabinan	1-5-α-L-Arabinan	9	10	16
branched arabinan	1,3;1,2;1,5-α-L-Arabinan	14	14	16
α-L-Arabinofuranosidase	PNP-Arabinofuranoside	35	37	333
Galactomannanase	Galactomannan	3	4	9
Mannanase	1,4-β-D-Mannan	7	0	0
Galactanase	1,4-β-D-Galactan	11	58	91
Xylanase	1,4-β-D-Xylan	4	0	2

[a]Reprinted with permission from Voragen *et al.*, 1986c.
[b]Activity is expressed in international units per gram enzyme preparation.

V. Methods Used in Technological Pectic Enzyme Research

A. Assay Methods

An extended discussion by many authors on methods using pectic enzymes from specific microorganisms and plant cell cultures has been compiled by Wood and Kellog (1988). This section is a more general treatment of the subject.

1. Pectinesterase Activities

Pectinesterase activities produce free carboxyl groups and free methanol. The increase of free carboxyl groups is monitored easily by automatic titration. At pH values below 4.5, or when assessing PE action in a natural substrate not suited for automatic titration, the increase in free methanol can be measured by gas liquid chromatography (GLC) (McFeeters and Armstrong, 1984) or high performance liquid chromatography (HPLC). Headspace GLC after conversion of methanol to methyl nitrite is convenient and sensitive (Bartolome and Hoff, 1972; Litchman and Upton, 1972). Even more convenient, although less sensitive, is HPLC executed, for example, on an alkaline isopropanol extract with an ion-exchange column (Voragen et al., 1986b). One unit of PE is defined as the amount of enzyme that liberates 1 micromole free carboxyl groups or methanol per minute under specified assay conditions. Random or blockwise distribution of free carboxyl groups obtained by various PEs (discussed subsequently) can be distinguished by high performance ion-exchange chromatography of the substrate (Schols et al., 1989) or by measuring the affinity for calcium ions, which is much stronger in the case of blockwise distribution (Kohn et al., 1968). Semiquantitative tests are sensitive and conveniently done by observing pH shift in a neutralized pectin solution, or gelling when calcium ions have been added (Pilnik and Rothschild, 1960). Such tests are useful to check for (undesirable) residual PE activity in citrus juices after pasteurization (International Fruit Juice Union FFU, 1985).

2. Lyase Activities

Lyase activities can be measured by monitoring the increase in light absorbance at 232 nm caused by the formation of the double bond (Fig. 3) at the new nonreducing ends of the molecules. Differentiation between PALs and PLs is possible by choosing low methoxyl or high methoxyl pectins, respectively, as substrates. Care must be taken to use clear pectin solutions, since any change in turbidity by pectin degradation would affect the values measured. Exo and endo activities, initial velocity, and extent of depolymerization are measured in this way. One unit lyase liberates 1 micromole unsaturated products per min under specified assay conditions. The specific extinctions at 232 nm are 4600 $M^{-1}cm^{-1}$ for PAL (Macmillan and Phaff, 1966) and 5500 $M^{-1}cm^{-1}$ for PL (Edstrom and Phaff, 1964).

3. Polygalacturonase Activities

Polygalacturonase activities can be determined by measuring the increase in reducing end groups. The tests of Nelson-Somogyi (Spiro, 1966) and Milner and Avigad (1967) are used preferentially, using galacturonic acid as standard. These tests measure endo and exo activities as well as the activity of oligopolygalacturonases. One unit of activity is defined as the quantity of enzyme that liberates 1 micromole reducing groups per min under specified conditions. The substrate used is low-methoxyl pectin or pectate, which also responds to PAL, the activity of which also results in an increase in reducing end groups. PG, however, does not form unsaturated products and, thus, can be differentiated. Further, the pH optima differ widely (Table III); PAL has an absolute requirement for calcium ions. It is recommended that viscosity measurements be combined with end-group (PG) and light-absorbance (lyase) measurements. This allows the calculation of glycosidic linkages split when the specific viscosity has reached half the original value. For endo activities, only a small percentage must be split whereas values as high as 40% are necessary for exo activities. Information about the presence of exo activities and the breakdown patterns in general can be obtained by gel chromatography of oligomers and by peak shifts in elution diagrams from size exclusion chromatography for which versatile HPLC methods are now available (Schols et al., 1990). A reliable qualitative test for complete pectin degradation, for example, in fruit juice clarification, is the absence of flocculation on addition of two volumes of alcohol.

B. Preparation of Pectic Enzymes for the Laboratory

Pectinesterase with only weak exopolygalacturonase activity is obtained from orange pulp by extraction with a borax–acetate buffer (Krop, 1974). Polygalacturonase is found as the only pectic activity in liquid growth medium of *Kluyveromyces* (Phaff, 1966). Pectate lyases as single pectic activities are isolated from liquid cultures from certain strains of *Pseudomonas fluorescence* (Pilnik et al., 1973; Rombouts et al., 1978). There is no simple source of pectin lyase. Separation and purification of these activities is achieved by chromatographic methods. The usual steps are:

- desalting by gel filtration (Biogel P100)
- protein separation by anion exchange (DEAE Biogel A) or cation exchange (CM Biogel A)
- separation of pectic enzymes by cross-linked alginate
- final cleaning-up step by fast protein liquid chromatography (FPLC)

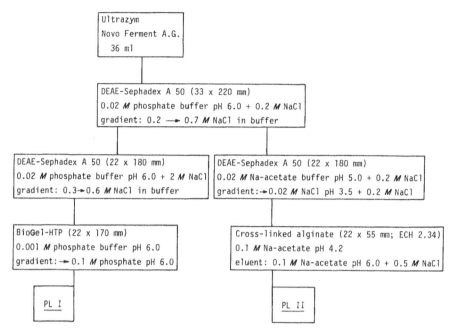

Figure 5   Purification scheme for pectin lyase from a technical pectinase preparation of fungal origin (Modified from van Houdenhoven, 1975).

Cross-linked alginate works by a combination of affinity and electrostatic effects and replaces cross-linked pectate (Versteeg *et al.*, 1978; Rombouts *et al.*, 1982a). It is less problematic to prepare and has a greater capacity. In the authors' laboratory, Biogel is used because of its stability in the presence of glycosidases. Detailed procedures have been published for polygalacturonase (Phaff, 1966; Versteeg, 1979), pectate lyase (Rombouts *et al.*, 1978), pectin lyase (van Houdenhoven, 1975), and pectinesterase (Rexova-Benkova *et al.*, 1977; Versteeg *et al.*, 1978; Rombouts *et al.*, 1979; Versteeg 1979; Baron *et al.*, 1980). Figure 5 gives the scheme for the purification of pectin lyases I and II from a commercial preparation (van Houdenhoven, 1975).

VI. Use of Commercial Fungal Enzymes as Processing Aids

Figure 6 shows a general flow diagram for the processing of fruit and vegetable juices. The arrows indicate the point in the process at which enzymes are applied usefully.

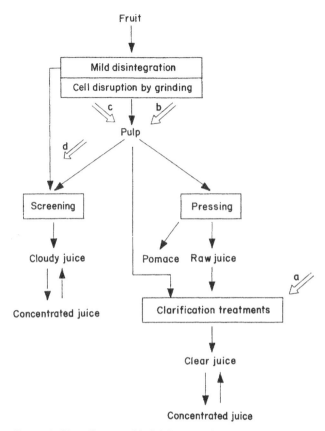

Figure 6 Flow diagram of fruit juice manufacture. Arrows indicate eventual enzyme treatments by (a) pectinases for clarification; (b) pectinases for pulp enzyming; (c) pectinases and C_1 cellulases for liquefaction; and, (d) polygalacturonase, pectin lyase, or pectate lyase for maceration. (Reprinted with permission from Pilnik and Voragen, 1989.)

A. Fruit Juice Clarification

Fruit juice clarification (Fig. 6, arrow a) is the oldest and still the largest use of pectinases, applied mainly to deciduous juices and grape juice (Kertesz, 1930; Mehlitz, 1930). The traditional way to prepare such juices is by crushing and pressing the pulp. The raw press juice is a viscous liquid with a persistent cloud of cell wall fragments and complexes of such fragments with cytoplasmic protein. Addition of pectinase lowers the viscosity and causes cloud particles to aggregate to larger units ("break"), which sediment and are removed easily by centrifugation or (ultra)filtration. Indeed, pectinase preparations once were known as filtration enzymes. Figure 7 (Yamasaki *et al.*, 1964) shows a schematic presentation of cloud particles: a protein nucleus

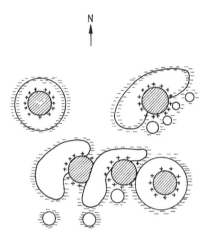

Figure 7 Aggregation of cloud particles. (Reprinted with permission from Yamasaki *et al.*, 1964.)

with a positive surface charge is coated by negatively charged pectin molecules. Partial degradation of this pectin by pectinase results in aggregation of oppositely charged particles. Also, the reduction of the viscosity of the raw juice is due to pectinase action (Endo, 1965a; Yamasaki *et al.*, 1967). Note that the mechanism of enzymatic clarification was not elucidated until 35 years after its introduction into the fruit juice industry. Careful experiments with purified enzymes have shown that this effect is reached either by a combination of PE and PG (Endo, 1965b) or by PL alone in the case of apple juice, which contains highly esterified pectin ($\geq 80\%$; Ishii and Yokotsuka, 1972), whereas in grape juice, which contains pectins with a lower DE (44– 65), PL alone does not perform as well (Ishii and Yokotsuka, 1973). A hot clarification process saving enzyme and time (Grampp, 1977) and continuous clarification processes have been described (Richard, 1974; Wucherpfenning and Otto, 1988). The advent of membrane processes has created the possibility for ultrafiltration of enzyme-treated juices, a process that is being introduced increasingly in the juice industry (Milner, 1984; Moslang, 1984). The clarification of juices by pectin degradation is also important in the manufacture of high Brix concentrates to avoid gelling and the development of haze.

B. Enzyme Treatment of Pulp for Juice Extraction

In the early period of use of pectinase for clarification, it was found, first for black currants (Charley, 1969), that enzyme treatment of the pulp (Fig. 6, arrow b) before pressing improved juice and color yield. Pulping of such soft fruits with much soluble pectin results in a semigelled mass that is very difficult to press. Enzymatic pectin degradation yields thin free-run juice and a pulp with good pressing characteristics (Beltman and Pilnik, 1971). In the early 1970s this treatment was applied to apples with poor pressing characteristics created by cultivar nature (Golden Delicious) or the necessity for storage (de Vos and Pilnik, 1973). Prevention of the inactivation of added enzymes by polyphenols is an important aspect of the process, and is achieved by aeration of the pulp. Endogenous polyphenol oxidases oxidize the polyphenols, which subsequently polymerize to high molecular weight compounds that are unable to inhibit the added enzymes. At the same time, the danger of a polyphenol haze in the finished product is diminished (Van Buren and Way, 1978). Addition of microbial polyphenol oxidases has been suggested (Maier et al., 1990). An alternative is the addition of polyvinylpolypyrrolidone, which binds the phenolic substances. Enzyme preparations performing well in juice clarification are also suitable for the enzyme treatment of pulp. In the case of apples, it has been shown that any combination of enzymes that depolymerizes highly esterified pectin (DE ≥ 90) can be used successfully (Voordouw et al., 1974). The good pressability of enzyme-treated pulp is shown in Table VI (Bielig et al., 1971; Hak et al., 1973). High yields are always connected with enzyme treatment. The better release of anthocyanins of red fruits into the juice, achieved by cell wall destruction, is another advantage of the pulp enzyme process, which has received attention from red wine manufacturers also. The process permits good yields of red

TABLE VI
Effect of Enzyme Treatment and Storage on Apple Juice Yields

Variety	Storage	Enzyme treatment	Press yield (%)
Treatment with 0.04% Pectinol flüssig® at 45°C for 10 min[a]			
Horneburger	+	−	66
Horneburger	+	+	86
Golden Delicious	+	−	59
Golden Delicious	+	+	73
Golden Delicious	−	−	71
Golden Delicious	−	+	83
Jonathan	−	−	61
Jonathan	−	+	88
Treatment with 0.004% Ultrazym 100® at 40°C for 20 min after preoxidation[b]			
Golden Delicious	+	+	78
Golden Delicious	+	−	52
Golden Delicious	+	+	75
Golden Delicious	+	−	62
Karmijn	+	+	82
Karmijn	+	+	74

[a]Data from Bielig et al., 1971.
[b]Data from Hak et al., 1973.

grape juice, especially in combination with a mild heat treatment, which then can be fermented directly, as is the case for white wine. This technique is simpler than the classical fermentation "on the skins" as required for red wine to obtain the desired color. In the United States, extraction of red grape juice from the pectin-rich Concord grape *(Vitis labrusca)* by the enzymatic process is common (Neubeck, 1975). Enzyme treatment of pulp of olives, palm fruit, and coconut to increase oil yield has been described also (Neubeck, 1975).

C. Liquefaction

In Fig. 6, arrow c indicates a process in which pulp is liquefied enzymatically so pressing is not necessary. The enzymes needed are pectinase and exo-β-glucanase (C_1-cellulase, cellobiohydrolase). The latter enzyme is derived from *Trichoderma* spp.; the so-called liquefaction process was developed when the enzyme became commercially available. Exo-β-glucanase works in conjunction with endo-β-glucanases (C_X-cellulases) which are present as *Aspergillus*-derived enzymes in most preparations. Figure 8 (Pilnik *et al.*, 1975) shows the viscosity decrease of stirred apple pulp during treatment with pectinase, C_1-cellulase, and a mixture of the two enzyme preparations. For the mixture, a synergistic effect is seen. The low viscosity values reached correspond to complete liquefaction; at that stage, microscopic examination shows that cell walls have disappeared. The liquefied juices are almost clear (papaya, cucumber), cloudy (apples, peaches), or pulpy (carrots), depending on the accessibility of the cell wall compounds to the enzymes (lignin). The liquefaction products can be clarified further by usual techniques; increased preference is given to ultrafiltration. The action of the enzymes can be followed chemically in enzyme-treated cell wall material by comparing the amounts of galacturonic acid and neutral sugars in alcohol-insoluble residues with the amounts produced in the control sample (Voragen *et al.*, 1980). Table VII gives results

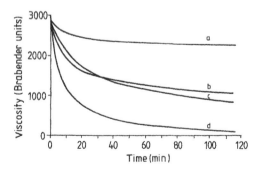

Figure 8   Viscosity decrease of stirred apple pulp at 25°C. (a) No enzyme added; (b) Cellulase added ($C_1 + C_X$); (c) Pectinase added; and (d) Cellulase and pectinase added. (Reprinted with permission from Pilnik *et al.*, 1975.)

TABLE VII
Neutral Sugars and Galacturonides Released from Apple AIS by Pure Enzymes[a]

Enzyme treatment	Neutral sugars[b,c]							AGA	Total
	Rha/Fuc	Ara	Gal	Man	Xyl	Glc	Total		
PE (citrus)	—	—	—	—	—	—	—	—	0.1
PG	4	9	5	—	4	—	3	21	8
PL	19	39	27	—	20	4	17	57	28
PE + PG	25	52	34	—	13	—	18	75	33
C_1	9	3	12	26	20	22	16	5	13
C_1 + PE + PG	58	89	69	84	71	79	78	82	80

[a]Reprinted with permission from Voragen et al., 1980.
[b]Abbreviations: AGA, anhydrogalacturonic acid; Rha, rhamnose; Fuc, fucose; Ara, arabinose; Gal, galactose; Man, mannose; Xyl, xylose; Glc, glucose.
[c]Data presented as percentage recovery, corrected for control.

of experiments on apple cell wall material with purified enzymes, showing that pectinase activity (PE + PG) alone releases 75% of the pectic material, including the pectin sugars arabinose, galactose, xylose, and rhamnose. No glucose is released from cellulose. Cellulase alone had little effect on pectin and solubilized only 22% of cellulose. Combined cellulase–pectinase activities released 80% of the polysaccharides. This synergistic effect is noticed in pulp viscosity experiments (Fig. 8). A similar effect has been found for grapefruit segment membranes (Ben-Shalom, 1986) when treated enzymatically. The breakdown products increase the soluble solids content of the juices (Voragen et al., 1980). In this way, high yields of juice and solids are obtained, which is of special interest to manufacturers of concentrated juices. Table VIII compares pressing and liquefying apple juice (Pilnik, 1988). Liquefaction is also interesting for processing of vegetables and fruits that yield no juice on pressing (Dongowski and Bock, 1977; Steinbuch and van Deelen, 1986; Pilnik, 1988) or for which no presses have been developed

TABLE VIII
**Comparison of Yields from 1000 kg Apples with
Good Pressing Characteristics[a]**

Pressing
    786 kg juice 12°Bx (750 liters)
    131 kg concentrate 72°Bx
    1000 kg 72°Bx concentrated juice from 7633 kg apples
Enzymatic liquefaction
    950 kg juice 13.5°Bx (900 liters)
    178 kg concentrate 72°Bx
    1000 kg 72°Bx concentrated juice from 5618 kg apples

[a]Reprinted with permission from Pilnik, 1988.

(mango, guava, bananas) (Schreier et al., 1985). Therefore much interest is shown in this new, simple, low capital, easy maintenance technology in developing countries. Enzymatic liquefaction is attractive to all countries because costly waste disposal is virtually eliminated. Further, socioeconomic developments in industrial countries make seasonal operations undesirable if not impossible. More attention will be paid to the transformation of stored apples, which are very difficult to press, so enzymatic liquefaction will have to be used (Pilnik, 1988).

D. Maceration

Maceration (Fig. 6, arrow d) is the process by which organized tissue is transformed into a suspension of intact cells, resulting in pulpy products used as base material for pulpy juices and nectars, as baby foods, and as ingredients for dairy products such as puddings and yogurts. The enzymatic process for the manufacture of such products must be contrasted to the mechanical process in which many cells inevitably are disrupted, with negative consequences for quality because of action of endogenous enzymes that damages flavor, color, and ascorbic acid. To limit the action of these enzymes, heat must be applied with the concomitant danger of heat damage. In any case, finisher operations to remove coarse tissue result in mechanical losses of nutritive constituents and a smooth consistency is not always achieved. Enzymatic degradation of pectin after a mild mechanical treatment often improves product properties if the process is carried out to leave as many cells as possible intact. Since the aim of the enzyme treatment is the transformation of tissue into a suspension of intact cells (Grampp, 1972; Bock et al., 1983), pectin degradation should affect only the middle lamella pectin. This process is called enzymatic maceration. The so-called macerases are enzyme preparations with only PG (Zetelaki-Horvath and Gatai, 1977a,b; Struebi et al., 1978; Zetelaki-Horvath and Vas, 1980) or PL activity (Ishii and Yokotsuka, 1971, 1975). The effectiveness of these macerases on various substrates has been compared (Ishii, 1976). For maceration of vegetables, bacterial PAL is interesting because of its high pH optimum (Rombouts et al., 1978; Bock et al., 1983; Wegener and Henniger, 1987). The restricted pectin degradation solubilizes rather than degrades the middle lamella pectin, which may improve the creamy mouthfeel. The process is often used for carrots (Bock et al., 1984). Whether maceration or cell wall breakdown has occurred can be checked by microscopy and by the absence of enzymatic browning. A very interesting use of enzymatic maceration is pulp for the production of dried instant potato mash (Bock et al., 1979). Enzymatic maceration prevents starch leaking out of the cells, avoiding the quality defect of gluiness of the reconstituted product. Potatoes are blanched to gelatinize starch and to inactivate endogenous PE, which in conjunction with the added PG would extend the possibilities for pectin degradation and turn the maceration process into pulp enzyming technology (Bock et al., 1979). Inactivation of endogenous PE is important for the maceration of many products (Sulc and Vujicic, 1973; Dongowski and Bock, 1980).

Most uses of exogenous pectinases for fruit juice extraction and clarification are based on the destruction of pectin and other desirable dietary fiber components of fruits. Macerated products with only limited pectin degradation therefore may gain appeal for the consumer in future and promote the production of purified PGs and lyases. The use for pulpy products also means that immobilization of these enzymes is of little interest.

E. Product-Specific Uses of Pectinases

1. Pulp Wash

In the citrus industry, pulp wash is the product obtained by counter-current water extraction of citrus pulp after dejuicing. Pulp wash is added back to the juice before concentration, but is often too viscous for this purpose because of dissolved pectin. The viscosity is reduced easily and efficiently by addition of pectinase. There are also claims of an increased solids yield (Braddock and Kesterson, 1979). Also a process in citrus production has been patented (Bruemmer, 1981) to facilitate the manufacture and improve the quality of citrus fruit segments. Citrus fruit is heated to a core temperature of $20-40°C$; then peel surface is pierced to penetrate the albedo barely without penetrating the juice sections. Gas bubbles in the albedo are replaced with pectinase solution and the fruit is incubated to permit removal of peel and other membranes from intact juice segments. The use of enzymes to obtain citrus carotenoid suspensions as natural coloring matter will probably gain interest because of current consumer attitudes (Emmam et al., 1990).

2. Apricot Nectar Stabilization

Apricot nectar exhibits a yet unexplained phenomenon of cloud loss. The nectar is made by mixing finely screened pasteurized apricot pulp with water, sugar, and acid to contain $30-50\%$ apricot ingredient. On standing, the slowly sedimenting pulp particles form a gel that contracts, leaving clear synersis fluid as supernatant (Siliha and Pilnik, 1985). This separation can be avoided by grinding the pulp so finely that at least one-third of the cells is broken open (Weiss and Saemann, 1972). This process is difficult to control during production, but cells can be broken open by the action of pectolytic enzymes (Siliha and Pilnik, 1985). For this purpose, an arabinanan-rich pectin fraction must be degraded (Siliha, 1985). Pure PG or commercial preparations containing mainly PG are not suitable. The cell walls become thinner but do not break. If PE or an exoarabinanase is supplemented, the enzyme preparation becomes a cloud-stabilizing enzyme, capable of degrading the extracted arabinan-rich pectin and of breaking open cells in the apricot pulp from which a cloud-stable product can be made. Figure 9 shows the elution

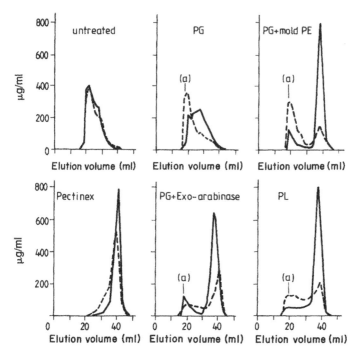

Figure 9 Gel filtration chromatography of the HCl-soluble apricot pectin fraction degraded by various enzyme systems on Sephacryl S-300 column. Anhydrogalacturonic acid (AUA; μg/ml) indicated by solid line. Sugars (μg/ml) indicated by dashed line. Pectinex is a wide spectrum commercial pectinase. PG activity alone hardly degrades the pectin and does not prevent cloud loss. (Reprinted with permission from Siliha, 1985.)

patterns of gel filtration chromatography of the arabinan-rich pectin after treatment with various enzyme systems. Only treatments with cloud-stabilizing enzymes move the uronide peak to an elution volume approximately equal to the bed volume, indicating depolymerization (Siliha and Pilnik, 1985).

VII. Quality Aspects of Clear Fruit Juices Obtained by Enzyme Treatment of Juices or Pulp

A. Sparkling Clarity

The primary quality requirement of the consumer for sparkling clarity is achieved in all processes (juice clarification, pulp enzyming, liquefaction), provided pectin is degraded sufficiently (alcohol flocculation test). Nevertheless, haze has appeared in concentrated apple and pear liquefaction juices and sometimes in concentrated juices from enzyme-treated pulp. Analysis of the haze material shows it to be 90% arabinose, of which 88% is α-1,5 linked (Table IX; Voragen *et al.*, 1982; Churms *et al.*, 1983). Structure work on nonhaze apple juice arabinans (Voragen *et al.*, 1986) and the study of mold arabinanases (Voragen *et al.*, 1987) indicate the haze-forming mechanism shown in Fig. 10. The enzyme treatment of pulp releases a branched arabinan. The low uronide content indicates that the haze may come from the hairy regions of the pectin molecule. Its main chain is α-1,5 and the side chains are α-1,3. The authors and their co-workers have identified and purified various arabinanases from commercial pectinase preparations. An exo-α-L-arabinofuranosidase II preferentially splits the α-1,2 linkages of the side chains (Voragen *et al.*, 1987), which results in debranching. Remaining chains then can retrograde under conditions such as low water activity and low temperature and crystallize, generating haze. Addition of sufficient endoarabinanase activity to depolymerize the α-1,5 chain inhibits haze formation. Enzyme manufacturers are now providing for this effect in their preparations to minimize the problem. Table X shows the monomer analysis of freeze-dried dialysis retentate of apple juices made in the laboratory using the three technologies. The table provides information on the degree of polymerization of the solubilized cell wall polysaccharides and on the activities of the enzyme preparations used. Thus, liquefied juices with a high arabinan content of a high total carbohydrate retentate may be susceptible to haze formation (In't Veld, 1986). A simple test such as the alcohol flocculation test for pectin or the iodine test for starch must still be developed. Detection of the

TABLE IX
Sugar Composition of Haze Material and Glycosidic
Linkage Composition for Arabinosyl Residues[a]

Sugar	Amount (g/100 g haze)	Glycosidic linkages[b] arabinosyl units (%)
Rhamnose	1.2	4.2 1-linked (terminal
Arabinose	90	furanose unit)
Xylose	0.3	0.2 1-linked (terminal
Mannose	0.7	pyranose units)
Galactose	1.6	2.3 1,3-linked (unbranched
Glucose	0.1	(furanose units)
Galacturonic acid	3.9	88.4 1,5-linked (unbranched
		furanose units)
Total	97.8	3 1,3,5-linked (branched
		furanose units)
		1 1,2,5-linked (branched
		furanose units)

[a]Reprinted with permission from Voragen *et al.*, 1982.
[b]Derived from methylation studies.

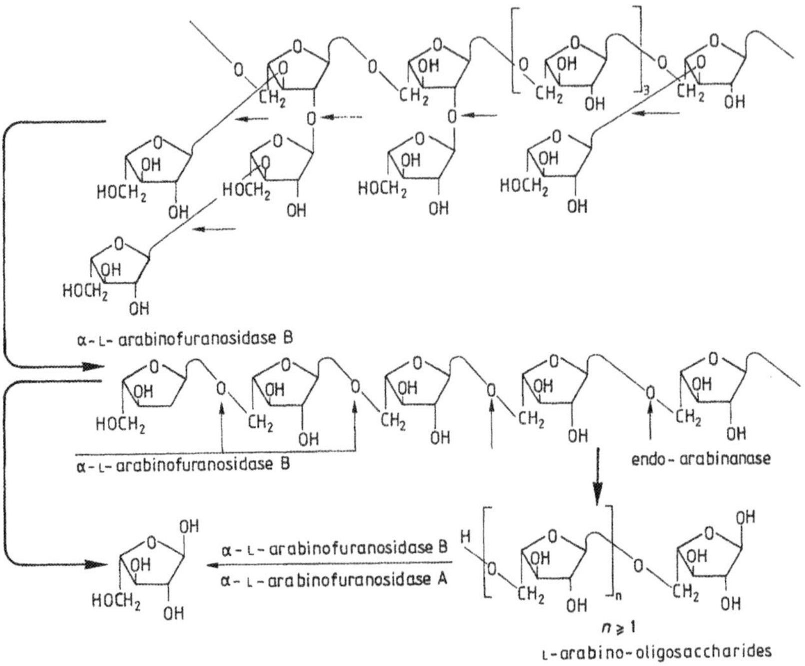

Figure 10 Structure of arabinan and points of attack of arabinan-degrading enzymes. (Reprinted with permission from Voragen *et al.*, 1987).

presence of the (partly) debranched arabinan chain involves separation by gel filtration and degradation by pure endoarabinanase (Endresz *et al.*, 1986; Ducroo, 1987).

B. Uronide Content of Apple Juices

Table XI shows the total uronide content of juices obtained in a laboratory study on the composition of juices prepared with the use of enzymes (Kolkman, 1983). The uronide content is seen to increase with intensity of pectic enzyme treatment. Free methanol from PE action is also present in quantities corresponding to about 10% of the uronide content, that is, 60–450 mg/ liter. Concern about free methanol is expressed recurrently but does not consider the fact that ester-bound methanol also enters the metabolism. The higher figure would be about 100 mg/kg above the range found in all kinds of beverages (Wucherpfenning *et al.*, 1989). Of course, in modern fruit juice technology, volatiles are stripped from juice, kept, and traded separately or mixed for use. Thus methanol content in reconstituted juices can be controlled. Depending on the PL action of the enzyme preparations used, 20– 40% of the uronides are present as unsaturated oligomers (Fig. 3). These

TABLE X
**Amounts and Characteristics of Polysaccharides Isolated from Apple Juice
Prepared by Different Processes and Different Enzymes**[a]

Analysis[b]	Pressing	Pulp enzyme treatment[c]			Liquefaction[c]	
		A	B	C	C	D
Polysaccharide content (mg/kg pulp)	155	1075	1860	1075	3000	4300
Sugar composition (mole %)						
Rhamnose	3	5	3	5	4	4
Arabinose	21	38	6	52	43	54
Xylose	2	5	2	5	6	8
Galactose	19	9	4	7	6	8
Glucose	2	1	1	1	1	0
Anhydrogalacturonic acid	53	42	84	30	40	26
DM[d]	52	72	76	31	59	39
DA[e]	15	18	6	13	28	29

[a]Reprinted with permission from In't Veld, 1986.
[b]Analysis was of dialysis retentates.
[c]Different enzyme preparations were used: A, B, and D were experimental preparations from Gist-Brocades, Delft, The Netherlands: A was a pectinase, B a pectinase rich in hemicellulase activity, and D a pectinase–cellulase mix. C was ultra SP-L from Novo Ferment, Dittingen, Switzerland.
[d]Degree of methylation.
[e]Degree of acetylation calculated from anhydrogalacturonic acid content.

enzyme-treated juices undergo more rapid browning than traditional juices. Figure 11 presents the results of model experiments from which it is seen that the unsaturated oligomeric uronides are powerful browning precursors (Voragen *et al.*, 1988). The models for concentrated apple juices were: Indicated quantities of oligouronides and fructose were brought to 70°Brix with sorbitol and to pH 3.5 with malic acid. In some experiments, 0.5% amino

TABLE XI
Uronide Content of Apple Juice (12° Brix)[a]

Methods of preparation	Uronide content (mg/ml)
Pressing and clarification	0.58
Water extraction	1.31
Methanol extraction	0.35
Enzymatic pulp treatment	2.51
Enzymatic pulp liquefaction	4.32

[a]Reprinted with permission from Kolkman, 1983.

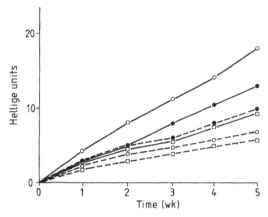

Figure 11 Unsaturated oligogalacturonides (○, 0.5%) as precursors for nonenzymatic browning reactions. Saturated oligogalacturonides (0.5%), □; fructose (43%), ●; with amino acids (0.5%), ——— ; without amino acids, ---.

acid mixture (asparagine, aspartate, and glutamate) were also added. Storage temperature was 50°C. At high temperatures, unsaturated oligomers were seen to split into saturated oligomers with one unit less than the original compound. The unsaturated monomer reacts further to give a number of compounds, among which are the known browning precursors 2-furoic acid and 2-formyl-2-furoic acid. These findings are surely of interest in view of the large quantities of concentrated apple and pear juice shipped under nonrefrigerated conditions.

VIII. Technological Functions of Endogenous Pectic Enzymes

Commercial pectinases used as processing aids make certain processes possible; the presence of endogenous pectic enzymes makes certain processes necessary. PE action during fermentation of fruit is responsible for methanol in the distillates from the fruit. Distilleries pasteurize fruit pulps before fermentation to insure that methanol content of end products remains within legal limits (Tanner, 1970). This release of methanol was observed by von Fellenberg (1913) and led to the conclusion that pectin is a methyl ester.

The tomato processing industry differentiates between a hot break and a cold break process. The hot break process is necessary to obtain the highly viscous juice the consumer prefers and the high consistency concentrated juice (tomato paste) used for sauces, soups, ketchup, and similar products. In this process, ingenious systems are used to heat-inactivate the abundant PE and PG activities of tomatoes (McColloch and Kertesz, 1949; Gould, 1974), for example, by crushing the tomatoes directly into circulating hot tomato pulp. If tomatoes are used for color and flavor only, and consistency is provided by other ingredients such as starch, a more easily handled thin "cold break" juice is the starting material for paste production. A holding time is introduced between crushing and heat treatment to insure breakdown of the pectins by combined PE/PG action. A process has been patented to macerate vegetables in fresh unpasteurized tomato pulp for the preparation of vegetable juices (Bock *et al.*, 1973). A tomato cultivar has been developed in which most PG activity has been removed by gene technology; a firmer consistency of the fruit and a higher viscosity of the juice is claimed.

Citrus fruits are very active in PE; only a weak exo-PG activity as an additional pectic enzyme can be detected. Citrus PE causes one of the most intensively studied problems in food technology: the cloud loss of citrus juices. A juice protected against microbial spoilage can be seen to separate on standing into a clear supernatant and a layer of sedimented pulp. In concentrated juice, a gel is formed from which no juice can be reconstituted (Rouse and Atkins, 1953; Kew and Veldhuis, 1961). These serious quality defects are caused by the action of PE, which deesterifies the native pectin which, at a certain point of deesterification, coagulates with native calcium ions just as a solution of commercial low methoxyl pectin is coagulated by the addition of calcium ions, forming flocs in diluted or a gel in a more concentrated solution (Pilnik, 1958; Joslyn and Pilnik, 1961; Krop, 1974; Versteeg, 1979). The manufacture of single strength juice with a stable cloud, or of concentrated juice that can be reconstituted to a stable cloud juice, therefore involves inactivation of PE, either by heat or by freezing. Both possibilities have disadvantages. The heat-sensitive orange juice flavor is easily damaged by heat treatments; freezing, on the other hand, is an expensive technology involving an uninterrupted line for the frozen concentrate from the juice extraction plant to the (overseas) customer. Alternatives are, therefore, investigated:

1. Inhibition of PE. Addition of polyphenols has been patented (Kew and Veldhuis, 1961). A protein extract from the kiwi fruit (*Actinidia chinensis*) has been shown to inhibit PEs from various sources and to maintain cloud in unpasteurized orange juices (Castaldo *et al.*, 1990). Both additions would need legal clearance and may create flavor problems. PE is known to display end-product inhibition, but the addition of pectic acid clarifies the juice, thus creating the effect one wants to prevent (Krop and Pilnik, 1974a; Baker, 1976). However, we have been able to establish that, by degrading pectic acid, a degree of polymerization can be found (8–10) that has the inhibitory effect of the high-polymer preparations but does not coagulate with calcium ions (Termote *et al.*, 1977). The addition of such preparations will not prevent but will delay self-clarification, probably sufficiently long to be useful for short-term distribution of fresh chilled juice. No taste problems were noted and food clearance may be obtained more easily for a natural citrus juice constituent.

2. Binding of calcium ions. Figure 12a shows the cloud-preserving effect of adding ammonium oxalate to PE active orange juice (Krop and Pilnik, 1974a). Similar results are obtained with chelating agents such as EDTA. This phenomenon is more interesting as an explanation of cloud loss than as an industrial practice.

3. Removal of PE substrate by addition of exogenous enzymes. Figure 12b shows that PG prevents cloud loss by depolymerizing the low ester pectins since they are formed by PE before calcium coagulation occurs (Krop and Pilnik, 1974b). The successful use of PL (not readily available) also has been described; in this case, the high ester juice pectin is depolymerized to low molecular weight products that are not calcium sensitive (Baker, 1980), even if deesterified, and that, in fact, might serve the function of PE inhibitor also. Such depolymerizing action by exogenous enzymes also has been considered for lowering juice viscosity and, thus, facilitating concentration of orange juice to higher Brix values than ordinarily possible (Crandall *et al.*, 1986). Multistage evaporators that allow optimized heat treatment and aroma recovery, combined with subsequent frozen storage and shipment of the concentrated juice, largely have overcome the PE inactivation–heat damage problem. This problem, however, still exists in developing countries for the manufacture of other cloudy juices that have strong PE activity and a heat-sensitive flavor, for example, guava and mango. The general trend for mini-

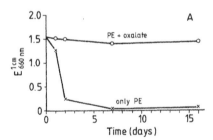

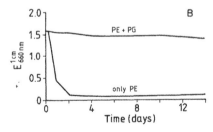

Figure 12 Cloud stabilization of pasteurized orange juice with added PE measured as extinction of supernatant after centrifugation. (A) Cloud stabilized by calcium binding as oxalate. (Reprinted with permission from Krop and Pilnik, 1974a.) (B) Cloud stabilized by PG. (Reprinted with permission from Krop and Pilnik, 1974b.) Not shown: Almost equal release of methanol with PE alone and PE + oxalate, respectively PE + PG. This proves that PE is not inhibited and cloud is stabilized as described in text.

TABLE XII
General Properties of Pectinesterases Isolated from Navel Oranges[a]

Characteristic	PE I	HMW-PE	PE II
Specific activity (u/mg protein)	200	≥ 180	762
Molecular weight	36200	54000	36200
Isoelectric point	10.05	10.2	≥ 11
pH optimum	7.6	8.0	8.0
K_m value (mg pectin/ml)	0.083	0.041	0.0046
K_i value (mg pectate/ml)	0.416	0.010	0.0016
V_{max} (mmol/mg min)	0.564	—	0.561
Turnover number (mmol/mmol sec)	340	—	338

[a]Data from Rombouts *et al.*, 1982b; Versteeg *et al.*, 1978, 1980.

mally heat processed products may bring the PE problem back, even to the modern citrus industry.

The intensive research on citrus juice cloud and gelling problems has resulted in better knowledge of citrus PEs. Versteeg *et al.* (1978; Versteeg, 1979) have isolated various multiple forms of PEs in oranges and have shown that they differ in affinity to pectins and pectates (Table XII; Rombouts *et al.*, 1982b). They also have different heat stabilities (Fig. 13; Versteeg *et al.*, 1980). It is understandable that they play different roles in cloud loss phenomena. One form, characterized by its high molecular weight, represents

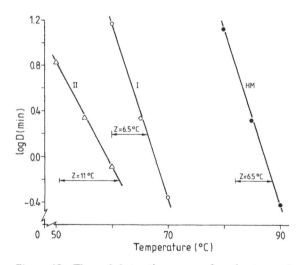

Figure 13 Thermal destruction curves of pectinesterases I and II and of the high molecular weight pectinesterase (HM) in orange juice at pH 4.0. (Reprinted with permission from Versteeg *et al.*, 1980.)

about 5% of the total activity and has various outstanding properties. It is heat resistant and therefore constitutes the main activity responsible for cloud loss after insufficient heat treatment of pasteurized juice. This enzyme is also active at low temperature and low pH, and must have been the important activity in the now almost extinct process of self-clarification for clear lemon and lime juices that were obtained in holding tanks under SO_2 preservation. Such clear juices are still important commercial products and are now produced with commercial pectinases. Only form I of the two other isoenzymes in Table XII has clarifying power. PE II has no clarifying power; the amount of free methanol generated is less than that generated by the clarifying enzymes. Table XII shows the extremely high inhibition (K_i) by pectate that probably stops the activity at a certain point of saponification. This effect surely is enhanced by a blockwise distribution of free acid groups. Figure 14 shows that PE reactivation occurs when PG is added. Versteeg

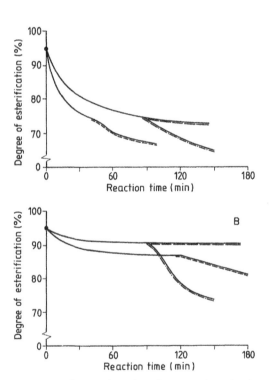

Figure 14   Stimulation of pectinesterase by polygalacturonase $(- \cdot - \cdot - \cdot)$ and by pectate lyase (– – –). The stimulation is stronger for polygalacturonase and stronger in experiment B, in which the esterase used was inhibited more strongly by end product than in experiment A. Obviously, pectin lyase (———), which cannot degrade the deesterified regions, has no effect. (Reprinted with permission from Versteeg, 1979.)

(1979) has noted also that the rate at which free methanol is formed in PE active orange juice is enhanced by PG. Such kinetic differences between PEs also can explain incomplete inhibition by oligogalacturonates. The different affinities to pectates can be used for separating PEs also (Rombouts *et al.*, 1979). One PE in orange juice does not cause self-clarification, even after liberating the same amount of methanol as do the clarifying PEs (Versteeg, 1979). A different mode of attack must be supposed that produces low-ester pectins with reduced calcium sensitivity, probably by random distribution of free carboxyl groups. This mechanism, usually connected with fungal PEs, also has been postulated for a PE from *Aspergillus japonicus* (Ishii *et al.*, 1979b); its production has been patented for the manufacture of low ester pectin (Ishii *et al.*, 1979a) for the preparation of low sugar jams. Such pectins are usually manufactured by an acid process (May, 1989). Fungal PE also has been developed by some enzyme manufacturers for clarification in the French cider industry (Baron and Drilleau, 1982). The two isoenzymes with orange juice clarifying properties are found in all component parts of orange fruit. All orange cultivars and citrus species tested so far contain the 12 PEs found in our laboratory (Rombouts *et al.*, 1982b).

PE activity must be destroyed when peel is processed to so-called pectin pomace, the raw material for industrial pectin extraction. This modification must be done immediately after juice extraction by a blanching process, since even slight PE action creates pectins that are too calcium sensitive for normal use as gelling agents (May, 1989). In contrast to this application, PE is activated in the much larger operation of dried citrus pomace production as cattle feed. In the so-called liming process, a slurry of calcium hydroxide is added to the peels while grinding them, bringing the pH to neutral and to alkaline levels. The high pH and the calcium ions activate the PE; rapid deesterification and calcium pectate coagulation occur. A bound liquid phase is released and easily pressed out, so only a fraction of the original water content must be removed by expensive thermal drying. The press liquor is similar in composition to the juice and is concentrated to molasses, which is also used as cattle feed (Kesterson and Braddock, 1976). This application is probably the second largest use of a native enzyme in the food and feed industry, the largest being the use of native amylases in the cereal processing industries (malt, beer, bread). Endogenous PE is exploited also to protect and improve texture and firmness of several processed fruits and vegetables, including apple slices (Wiley and Lee, 1970), canned tomatoes (Hsu *et al.*, 1965), cauliflower (Hoogzand and Doesburg, 1961) carrots (Lee *et al.*, 1979), potatoes (Bartolome and Hoff, 1972), and peas (Steinbuch, 1976). A low temperature–long time blanching process will activate PE causing a partial deesterification of the pectins, which then react with (added) calcium ions, resulting in stronger intercellular cohesion.

Pectolytic activities from microbial infections are, of course, of exogenous nature, but the consequences of their presence make them similar to endogenous activities. Yeast PG and bacterial PAL are involved in the softening of cucumbers and olives in brine (Bock *et al.*, 1970). This quality defect can be avoided by inhibitors such as polyphenols leaching out from wine leaves into the brine (Bell *et al.*, 1968), as practiced in some Mediterranean regions. Recirculated brine must, in any case, be pasteurized. *Byssochlamys fulva*, a heat-resistant mold, produces enzymes that cause decay of strawberries in syrup or jam (Put and Kruiswijk, 1964; Eckhardt and Ahrens, 1977). A highly resistant PG from *Rhizopus* spp. is responsible for texture breakdown in canned apricots (Luh *et al.*, 1978). Mold infection has been described to cause clarification of aseptically filled orange juice. It is not clear whether this effect is caused by PE, PG, or both, but an increased amount of methanol was found in the clarified juices (Nussinovitch and Rosen, 1989). Yeast PG can cause pectin breakdown in fermenting apple pomace, which then loses its value as raw material for the pectin industry (Sulc and Ristic, 1954). Desirable aspects of pectolysis by microorganisms are seen in coffee (Castelein and Pilnik, 1976) and cocoa fermentation. During fermentation, the mucilage layer surrounding these beans is decomposed and washing of the beans before drying is facilitated greatly. Noble rot or *Botrytis cinerea* rot of overripe grapes is often considered desirable. The skin of the grapes is pierced. As a result of water evaporation, higher sugar concentrations are obtained, yielding a wine with higher alcohol content and a special flavor.

References

Albersheim, P., and Anderson, A. J. (1971). Proteins from plant cell walls inhibit polygalacturonases secreted by plant pathogens. *Proc. Natl. Acad. Sci. U.S.A.* **68**, 1815–1819.

Aspinall, G. O. (1980). Chemistry of cell wall polysaccharides. *In* "The Biochemistry of Plants" (J. Preis, ed.) Vol. III, pp. 473–500. Academic Press, New York.

Baker, R. A. (1976). Clarification of citrus juices with polygalacturonic acid. *J. Food Sci.* **41**, 1198–1200.

Baker, R. A. (1980). The role of pectin in citrus quality. *In* "Citrus Nutrition and Quality" (S. N. Nagy and J. A. Attaway, eds.) pp. 109–128. American Chemical Society, Washington, D.C.

Baron, A., and Drilleau, J. F. (1982). Utilisation de la pectinestérase dans les industries cidricoles. *In* "Use of Enzymes in Food Technology" (P. Dupuy, ed.) pp. 471–476. Technique et Documentation Lavoisier, Paris.

Baron, A., Rombouts, F. M., Drilleau, J. F., and Pilnik, W. (1980). Purification et propriétés de la pectinestérase produite par *Aspergillus niger*. *Lebensm. Wiss. Technol.* **13**, 330–333.

Bartolome, L. G., and Hoff, J. E. (1972). Gas chromatographic methods for the assay of pectin methylesterase, free methanol, and methoxyl groups in plant tissues. *J. Agric. Food Chem.* **20**, 262–266.

Bell, T. A., Etchells, J. L., Williams, C. F., and Porter, W. L. (1962). Inhibition of pectinase and cellulase by certain plants. *Bot. Gaz.* **123(3)**, 220–223.

Beltman, H., and Pilnik, W. (1971). Die Kramer'sche Scherpresse als Laboratoriums-Pressvorrichtung und Ergebnisse von Versuchen mit Aepfeln. *Confructa* **16(1)**, 4–9.

Ben-Shalom, N. (1986). Hindrance of hemicellulose and cellulosic hydrolysis by pectic substances. *J. Food Sci.* **51(3)**, 720–721, 730.

Bielig, H. J., Wolff, J., and Balcke, K. J. (1971). Die Fermentation von Apfelmaische zur Entsaftung durch Packpressen oder Dekanter. *Flüss. Obst* **38**, 408, 410–412, 414.

Bock, W., Krause, M., and Dongowski, G. (1970a). Charakterisierung qualitätsverändernder Prozesse bei der Herstellung von Salzgurken. *Ernährungsforsch.* **15**, 403–415.

Bock, W., Krause, M., and Dongowski, G. (1970b). Wirkungsunterschiede pflanzeneigener Polygalakturonase-Hemmstoffe. *Nahrung* **14**, 375–381.

Bock, W., Dongowski, G., and Krause, M. (1973). Nützung des Tomaten-Enzymkomplexes zur Herstellung von trinkfähigen Mischmazeraten aus Gemüse und Obst. *Nahrung* **17**, 757–767.

Bock, W., Dongowski, G., Maischack, H., and Ruttloff, H. (1979). Verfahren zur Herstellung von Trockenprodukten aus Speisekartoffeln. DDR Patent 135 322.

Bock, W., Krause, M., and Dongowski, G. (1983). Verfahren zur Herstellung von Pflanzenmazeraten. DDR Patent 84317.

Bock, W., Krause, M., Henniger, H., Andeerson, M., and Molnar, I. (1984). Verfahren zur enzymatischen Verflüssigung von Gemüse und Speisekartoffeln. DDR Patent 216 617.

Braddock, R. J., and Kesterson, J. W. (1979). Use of enzymes in citrus processing. *Food Technol.* **33**(11), 78–83.

Bruemmer, J. H. (1981). Method of preparing citrus fruit sections with fresh fruit flavor and appearance. U.S. Patent No. 4,284,651.

Castaldo, D., Lovoi, A., Balestrieri, C., Giovane, A., Quagliuolo, L., and Servillo, L. (1990). Inhibition of pectinesterase activity in fruit juices. Presented at the XX International Symposium of the International Federation of Fruit Juice Producers, Paris, 16–18 May.

Castelein, J. M., and Pilnik, W. (1976). The properties of the pectate-lyase produced by *Erwinia dissolvens*, a coffee fermenting organism. *Lebensm. Wiss. Technol.* **9**(5), 277–283.

Charley, V. L. S. (1969). Some advances in food processing using pectic and other enzymes. *Chem. Ind.* 635–641.

Chesson, A., and Codner, R. C. (1978). The maceration of vegetable tissue by a strain of *Bacillus subtilis. J. Appl. Bacteriol.* **44**, 347–364.

Churms, S. C., Merrifield, E. H., Stephen, A. M., Walwylor, D. R., Poson, A., van der Merke, K. J., Spies, H. S. C., and Losta, N. (1983). An arabinan from apple juice concentrate. *Carbo. Res.* **113**, 339–344.

Colquhoun, I. J., de Ruiter, G. A., Schols, H. A., and Voragen, A. G. J. (1990). NMR identification of oligosaccharides obtained by treatment of the pectic hairy regions of apple with rhamnogalacturonase. *Carbo. Res.* **206**, 131–144.

Crandall, P. G., Chen, C. S., Marcy, J. E., and Martin, F. G. (1986). Quality of enzymatically treated 72°Brix orange juice stored at refrigerated temperatures. *J. Food Sci.* **51**, 1017–1020.

de Vos, L., and Pilnik, W. (1973). Pectolytic enzymes in apple juice extraction. *Process Biochem.* **8**(8), 18–19.

de Vries, J. A., Voragen, A. G. J., Rombouts, F. M., and Pilnik, W. (1986). Structural studies of apple pectin with pectolytic enzymes. *In* "Chemistry and Function of Pectins" (M. L. Fishman and J. J. Jen, eds.), pp. 38–48. American Chemical Society, Washington, D.C.

Doesburg, J. J. (1965). "Pectic Substances in Fresh and Preserved Fruits and Vegetables. Institut voor Bewaring en Verwerking van Tuinbouwprodukten. Communication Nr. 25. Wageningen, The Netherlands.

Dongowski, G., and Bock, W. (1977). Enzymatische Verflüssigung von Speisemöhren. *Lebensmittelind.* **24**, 33–39.

Dongowski, G., and Bock, W. (1980). Eine Prüfmethode zur Ermittlung der mazerierenden und depektinisierenden Eigenschaften von pektolytischen Enzympräparaten. *Ber. Wiss. Technol. Komm. IFU* **XVI**, 275–288.

Ducroo, P. (1987). Research on a prediction method for araban haze formation in apple juice. *Flüss. Obst* **5**, 265–269.

Eckhardt, C., and Ahrens, E. (1977). Untersuchungen über *Byssochlamis fulva* Olliver und Smith als potentiellem Verderbniserreger in Erdbeerkonserven. *Chem. Microb. Technol. Lebensm.* **5**, 71–75.

Edstrom, R. D., and Phaff, H. J. (1964). Purification and certain properties of pectin-transeliminase. *J. Biol. Chem.* **239**, 2403–2407.

Emmamn, S. S., Ibrahim, S. S., Ashour, M. M. S., and Askar, A. (1990). Gewinnung von Carotinoiden aus Orangenschale durch Enzymbehandlung. *Flüss. Obst* **57**, 295–297.

Endo, A. (1965a). Studies on pectolytic enzymes of molds. Part XVI: Mechanism of enzymatic clarification of apple juice. *Agric. Biol. Chem.* **29,** 229–233.

Endo, A. (1965b). Studies on pectolytic enzymes of molds. Part XIII: Clarification of apple juice by the joint action of purified pectolytic enzymes. *Agric. Biol. Chem.* **29,** 129–136.

Endresz, H.-U., Tableros, M., Omran, H., and Gierschner, K. (1986). A test for arabinan-caused cloudiness. *Intern. Fruit Juice Symp.* Vol. XIX, 293–306.

English, P. D., Maglothin, A., Keegstra, K., and Albersheim, P. (1972). A cell wall-degrading endopolygalacturonase secreted by *Colletotrichum lindemuthianum*. *Plant Physiol.* **49,** 293–297.

Fogarty, W. M., and Ward, O. P. (1974). Pectinases and pectic polysaccharides. *In* "Progress in Industrial Microbiology" (D. J. D. Hockenhull, ed.), Vol. 13, pp. 59–119. Churchill Livingstone, Edinburgh.

Fry, S. C. (1982). Phenolic components of the primary cell wall. *Biochem. J.* **203,** 493–504.

Gierschner, K. (1981). Pectin and pectic enzymes in fruit and vegetable technology. *Gordian* **81,** 171–210.

Gould, W. A. (1974). "Tomato Production, Processing, and Quality Evaluation." AVI, Westport, Connecticut.

Grampp, E. (1972). Die Veränderung der Technologie am Beispiel der Obst- und Gemüseverarbeitung mit einem speziellen Pektinasepräparat. *Dechema-Monogr.* **70,** 175–186.

Grampp, E. (1977). Hot clarification process improves production of apple juice concentrates. *Food Technol* **31(11),** 38–41.

Hak, J., Pilnik, W., Voragen, A. G. J., and de Vos, L. (1973). "De Maische Fermentering van Golden Delicious." Master's Thesis. Agricultural University, Wageningen, The Netherlands.

Hoogzand, C., and Doesburg, J. J. (1961). Effect of blanching on texture and pectin of canned cauliflower. *Food Technol.* **15,** 160–163.

Hsu, C. P., Deshpande, S. N., and Desrosier, N. W. (1965). Role of pectin methylesterase in firmness of canned tomatoes. *J. Food Sci.* **30,** 583–588.

Internationale Fruchtsaft Union (1985). "Detection of Pectinesterase Activity in Citrus Juices and Their Concentrates." Analysis Method No. 46. International Federation of Fruit Juice Producers.

In't Veld, P. (1986). "Invloeden van Enzymen en Ultrafiltratie op een Aantal Kenmerken van Appelsap." Master's Thesis. Agricultural University, Wageningen, The Netherlands.

Ishii, S. (1976). Enzymatic maceration of plant tissues by endo-pectin lyase and endo-polygalacturonase from *Asp. japonicus*. *Phytopathol.* **66,** 281–289.

Ishii, S., and Yokotsuka, T. (1971). Maceration of plant tissues by pectin transeliminase. *Agric. Biol. Chem.* **35,** 1157–1159.

Ishii, S., and Yokotsuka, T. (1972). Clarification of fruit juice by pectin transeliminase. *J. Agric. Food Chem.* **20,** 787–791.

Ishii, S., and Yokotsuka, T. (1973). Susceptibility of fruit juice to enzymatic clarification by pectin lyase and its relation to pectin in fruit juice. *J. Agric. Food Chem.* **21,** 269–272.

Ishii, S., and Yokotsuka, T. (1975). Purification and properties of pectin lyase from *Asp. japonicus*. *Agric. Biol. Chem.* **39,** 313–321.

Ishii, S., Kiho, K., Sugiyama, S., and Sugimoto, H. (1979a). Verfahren zur Herstellung einer neuen Pektinesterase und Verfahren zur Herstellung von demethoxyliertem Pektin unter Verwendung der neuen Pektinesterase. BRD Patent Offenlegungsschrift 2843351.

Ishii, S., Kiho, K., Sugiyama, S., and Sugimoto, H. (1979b). Low-methoxyl pectin prepared by pectinesterase from *Aspergillus japonicus*. *J. Food Sci.* **44,** 611–614.

Joslyn, M. A., and Pilnik, W. (1961). Enzymes and enzyme activity. *In* "The Orange: Its Biochemistry and Physiology" (W. B. Sinclair, ed.), pp. 373–435. University of California. Riverside.

Kertesz, Z. I. (1930). "A New Method for Enzymic Clarification of Unfermented Apple Juice." New York State Agricultural Experimental Station (Geneva) Bull. No. 689. U.S. Patent No. 1.932.833.

Kesterson, J. W., and Braddock, R. J. (1976). By-products and speciality products of Florida citrus. Technical Bull. No. 784. Agricultural Experimental Station, University of Florida, Gainesville.

Kew, T. J., and Veldhuis, M. K. (1961). Cloud stabilization in citrus juice. U.S. Patent No. 2,995,448.

Kilara, A. (1982). Enzymes and their use in the processed apple industry: A review. *Process Biochem.* **17,** 35–41.

Kohn, R., Furda, I., and Kopec, Z. (1968). Distribution of free carboxyl groups in the pectin molecule after treatment with pectin esterase. *Coll. Czech. Chem. Commun.* **33,** 264–269.

Kolkman, J. R. (1983). "Oligo- en Polysacchariden in Appelsappen, als Karakteristiek voor de Bereidingswijze." Master's Thesis. Agricultural University, Wageningen, The Netherlands.

Koller, A., and Neukom, H. (1969). Untersuchungen über den Abbaumechanismus einer gereinigten Polygalakturonase aus *Aspergillus niger. Eur. J. Biochem.* **7,** 485–489.

Krop, J. J. P. (1974). "The Mechanism of Cloud Loss Phenomena in Orange Juice." Ph.D. Thesis. Agricultural University, Wageningen, The Netherlands.

Krop, J. J. P., and Pilnik, W. (1974a). Effect of pectic acid and bivalent cations on cloud loss of citrus juice. *Lebensm. Wiss. Technol.* **7**(1), 62–63.

Krop, J. J. P., and Pilnik, W. (1974b). Cloud loss studies in citrus juices: Cloud stabilization by a yeast-polygalacturonase. *Lebensm. Wiss. Technol.* **7**(2), 121–124.

Kulp, K. (1975). Carbohydrases. *In* "Enzymes in Food Processing" (G. Reed, ed.), pp. 53–122. Academic Press, London.

Lee, C. Y., Bourne, M. C., and van Buren, J. P. (1979). Effect of blanching treatments on the firmness of carrots. *J. Food Sci.* **44,** 615–616.

Litchman, M. A., and Upton, R. P. (1972). Gas chromatographic determination of residual methanol in food additives. *Anal. Chem.* **44,** 1495–1497.

Luh, B. S., Ozbilgin, S., and Liu, Y. K. (1978). Textural changes in canned apricots in the presence of mold polygalacturonase. *J. Food Sci.* **43,** 713–720.

McColloch, R. J., and Kertesz, Z. I. (1949). Recent developments of practical significance in the field of pectic enzymes. *Food Technol.* **3**(3), 94–96.

McCready, R. M., and Seegmiller, C. G. (1954). Action of pectic enzymes on oligogalacturonic acids and some of their derivatives. *Arch Biochem. Biophys.* **50,** 440–450.

MacDonnell, L. R., Jang, R., Jansen, E. F., and Lineweaver, H. (1950). The specificity of pectinesterases from several sources with some notes on the purification of orange pectinesterase. *Arch. Biochem.* **28,** 260–273.

McFeeters, R. F., and Armstrong, S. A. (1984). Measurement of pectin methylation in plant cell walls. *Anal. Biochem.* **139,** 212–217.

Macmillan, J. D., and Phaff, H. J. (1966). Exopolygalacturonate lyase from *Clostridium multifermentans. Meth. Enzymol.* **8,** 632–635.

McMillan, J. D., and Sheiman, M. I. (1974). *In* "Food Related Enzymes" (J. R. Whitaker, ed.), pp. 101–130. American Chemical Society, Washington, D.C.

McNeil, M., Darvill, A. G., Frey, S. C., and Albersheim, P. (1984). Structure and function of the primary cell walls of plants. *Ann. Rev. Biochem.* **53,** 625–663.

Manabe, M. (1973). Saponification of ester derivatives of pectic acid by *Matsudaidai* pectinesterase (Studies on the derivatives of pectic substances, Part V). *J. Agric. Chem. Soc. Japan* **47,** 385–390.

May, D. (1989). Industrial pectins: Sources, production and application. *Carboh. Polymers* **12,** 79–99.

Mehlitz, A. (1930). Ein neues Verfahren zur Herstellung von Fruchtsäften, *Süssmosten Konservenind.* **17,** 306, 321.

Miller, L., and MacMillan, J. D. (1971). Purification and pattern of action of pectinesterase from *Fusarium oxysporum* f.sp. *vasinfectum. Biochem.* **10,** 570–576.

Milner, B. A. (1984). "The Application of Ultrafiltration in Apple Juice Processing." N.Y. State Agricultrual Experimental Station (Geneva) Special Report No. 54.

Milner, Y., and Avigad, G. (1967). A copper reagent for the determination of hexuronic acids and certain ketohexoses. *Carboh. Res.* **4,** 359–361.

Moslang, H. (1984). Ultrafiltration in the fruit juice industry. *Confructa* **28,** 219–224.

Neubeck, C. E. (1975). *In* "Enzymes in Food Processing" (G. Reed, ed.) 2d Ed. pp. 397–442. Academic Press, New York.

Neukom, H. (1969). Pektinspaltende Enzyme. *Z. Ernährungs wissensch. (Suppl.)* **8,** 91–96.

Neukom, H., Amado, R., and Pfister, M. (1980). Neuere Erkenntnisse auf dem Gebiete der Pektinstoffe. *Lebensm. Wiss. Technol.* **13,** 1–6.

Nussinovitch, A., and Rosen, B. (1989). Cloud destruction in aseptically filled citrus juice. *Lebensm. Wiss. Technol.* **22,** 60–64.

Phaff, H. J. (1966). α-1,4-Polygalacturonide glycanohydrolase (endopolygalacturonase) from *Saccharomyces fragilis. Meth. Enzymol.* **8,** 636–641.

Pilnik, W. (1958). Die Trubstabilität von Citrus-Konzentraten. *In* "Berichte der Wissenschaft Technolog. Kommission der Internationalen Fruchtsaft Union," pp. 203–228. Juris Verlag, Zürich.

Pilnik, W. (1982). Enzymes in the beverage industry. *In* "Use of Enzymes in Food Technology" (P. Dupuy, ed.), pp. 425–450. Technique et Documentation Lavoisier, Paris.

Pilnik, W. (1988). From traditional ideas to modern fruit juice winning processes. *X. Intern. Congr. Fruit Juices,* Vol. X, 159–179.

Pilnik, W., and Rombouts, F. M. (1979a). Pectic enzymes. *In* "Polysaccharides in Food." (J. M. V. Blanshard and J. R. Mitchell, eds.), pp. 109–126. Butterworths, London.

Pilnik, W., and Rombouts, F. M. (1979b). Utilization of pectic enzymes in food production. *In* "Proceedings of the 5th International Congress of Food Science and Technology." (H. Chiba, M. Fujimaki, K. Iwai, H. Mitsuda, and Y. Morita, eds.), pp. 269–277. Elsevier Scientific, Amsterdam.

Pilnik, W., and Rombouts, F. M. (1981). Pectic enzymes. *In* "Enzymes and Food Processing" (G. G. Birch, N. Blakebrough, and K. J. Parker, eds.), pp. 105–128. Applied Science, London.

Pilnik, W., and Rothschild, G. (1960). Trubstabilität und Pektinesterase—Restaktivität in pasteurisiertem Orangenkonzentrat. *Fruchtsaftind.* **5,** 131–138.

Pilnik, W., and Voragen, A. G. J. (1989). Effect of enzyme treatment on the quality of processed fruits and vegetables. *In* "Quality Factors of Fruits and Vegetables, Chemistry and Technology" (J. J. Jen, ed.), pp. 250–269. American Chemical Society, Washington, D.C.

Pilnik, W., Rombouts, F. M., and Voragen, A. G. J. (1973). On the classification of pectin depolymerases: Activity of pectin depolymerases on glycol esters of pectate as new classification criterion. *Chem. Microbiol. Technol.* **2,** 122–128.

Pilnik, W., Voragen, A. G. J., and Rombouts, F. M. (1974). Specificity of pectic lyases on pectin and pectate amides. *Lebensm. Wiss. Technol.* **7(6),** 353–355.

Pilnik, W., Voragen, A. G. J., and de Vos, L. (1975). Enzymatische Verflüssigung von Obst und Gemüse. *Flüss. Obst* **42(11),** 448–451.

Pressey, R. (1986). Polygalacturonases in higher plants. *In* "Recent Advances in the Chemistry and Function of Pectins." (M. Fishman and J. J. Jen, eds.), pp. 157–174. American Chemical Society, Washington, D.C.

Put, H. M. C., and Kruiswijk, J. T. (1964). Disintegration and organoleptic deterioration of processed strawberries caused by the mould *Byssochlamys nivea. J. Appl. Bacteriol.* **27,** 53–58.

Rexová-Benková, L., and Marcovic, O. (1976). Pectic enzymes. *Adv. Carbo. Chem. Biochem.* **33,** 323–385.

Rexová-Benková, L., Mracková, M., Luknár, D., and Kohn, R. (1977a). The role of secondary alcoholic groups of D-galacturonan in its degradation by endo-D-galacturonase. *Coll. Czech. Chem. Commun.* **42,** 3204–3213.

Rexová-Benková, L., Markovic, O., and Foglietti, M. J. (1977b). Separation of pectic enzymes from tomatoes by affinity chromatography on cross-linked pectic acid. *Coll. Czech. Chem. Commun.* **42,** 1736–1741.

Richard, J. P. (1974). Clarification en continu du jus de pommes. U.S. Patent No. 3,795,521.

Rombouts, F. M. (1972). "Occurrence and Properties of Bacterial Pectate Lyases." Ph.D. Thesis. Agricultural University, Wageningen, The Netherlands.

Rombouts, F. M. (1981). Pectic enzymes, their biosynthesis and roles in fermentation and spoilage. *In* "Advances in Biotechnology." (M. Moo-Young, ed.). Vol. 3, pp. 585–592. Pergamon Canada, Willowdale, Ontario.

Rombouts, F. M., and Pilnik, W. (1979). Enzymes in the fruit juice industry. Presented at the Congres International Microbiologie et Industrie Alimentaire, October 1979, Paris.

Rombouts, F. M., and Pilnik, W. (1980). Pectic enzymes. *In* "Economic Microbiology" (A. H. Rose, ed.), Vol. 5, pp. 227–282. Academic Press, London.

Rombouts, F. M., and Pilnik, W. (1984). Enzymes for structural analysis of plant polysaccharides. Presented at the International Workshop on Plant Polysaccharides, Structure, and Function, Nantes, July 9–11.

Rombouts, F. M., and Thibault, J. F. (1986a). Feruloylated pectic substances from sugar-beet pulp. *Carboh. Res.* **154,** 177–187.

Rombouts, F. M., and Thibault, J. F. (1986b). Enzymic and chemical degradation and the fine structure of pectins from sugar-beet pulp. *Carboh. Res.* **154,** 189–203.

Rombouts, F. M., Spaansen, C. H., Visser, J., and Pilnik, W. (1978). Purification and some characteristics of pectate lyase from *Pseudomonas fluorescens* GK-5. *J. Food Biochem.* **2,** 1–22.

Rombouts, F. M., Wissenburg, A. K., and Pilnik, W. (1979). Chromatographic separation of orange pectinesterase isoenzymes on pectates with different degrees of crosslinking. *J. Chromatogr.* **168,** 151–161.

Rombouts, F. M., Geraeds, C. C. J. M., Visser, J., and Pilnik, W. (1982a). Purification of various pectic enzymes on crosslinked polyuronides. *In* "Affinity Chromatography and Related Techniques" (T. C. J. Gribnau, J. Visser, and R. J. V. Nivard, eds.), pp. 255–260. Elsevier, Amsterdam.

Rombouts, F. M., Versteeg, C., Karman, A. M., and Pilnik, W. (1982b). Pectinesterase in component parts of citrus fruits related to problems of cloud loss and gelation in citrus products. *In* "Use of Enzymes in Food Technology" (P. Dupuy, ed.), pp. 483–487. Technique et Documentation Lavoisier, Paris.

Rombouts, F. M., Voragen, A. G. J., Searle-van Leeuwen, M. F., Geraeds, C. C. J. M., Schols, H. A., and Pilnik, W. (1988). The arabinanases of *Aspergillus niger:* Purification and characterization of two α-L-arabinofuranosidases and an endo-1,5-α-L-arabinanase. *Carboh. Polymers* **9,** 25–47.

Rouse, A. H., and Atkins, C. D. (1953). Further results from a study on heat inactivation of pectinesterase in citrus juices. *Food Technol.* **7,** 221–223.

Schols, H. A., Reitsma, J. C. E., Voragen, A. G. J., and Pilnik, W. (1989). High performance ion exchange chromatography of pectins. *Food Hydrocoll.* **6(2),** 115–121.

Schols, H. A., Geraeds, C. C. J. M., Searle-van Leeuwen, M. F., Kormelink, F. J. M., and Voragen, A. G. J. (1990). Hairy (ramified) regions of pectins. II. Rhamnogalacturonase: A novel enzyme that degrades the hairy regions. *Carboh. Res.* **206,** 105–115.

Schreier, R., Kiltsteiner-Eberle, R., and Idstein, H. (1985). Untersuchungen zur enzymatischen Verflüssigung tropischer Fruchtpülpen: Guava, Papaya, und Mango. *Flüss. Obst* **52,** 365–370.

Selvendran, R. R. (1985). Developments in the chemistry of pectic and hemicellulosic polymers. *J. Cell. Sci. Suppl.* **2,** 51–88.

Siliha, H. A. I. (1985). "Studies on Cloud Stability of Apricot Nectar." Ph.D. Thesis. Agricultural University, Wageningen, The Netherlands.

Siliha, H. A. I., and Pilnik, W. (1985). Cloud stability of apricot nectars. *Ber. Wiss. Techn. Komm. IFU* **XVIII,** 325–334.

Solms, J., and Deuel, H. (1955). Über den Mechanismus der enzymatischen Verseifung von Pektinstoffen. *Helv. Chim. Acta* **38,** 321–329.

Spiro, R. G. (1966). Analysis of sugars found in glycoproteins. *Meth. Enzymol.* **8,** 3–26.

Steinbuch, E. (1976). Improvement of texture of frozen vegetables by stepwise blanching treatment. *J. Food Technol.* **11,** 313–316.

Steinbuch, E., and van Deelen, W. (1986). Zur enzymatischen Saftgewinnung aus Gemüse und Früchten. *Confructa* **30,** 15–23.

Struebi, P., Escher, F. E., and Neukom, H. (1978). Use of a macerating pectic enzyme in apple nectar processing. *J. Food Sci.* **43,** 260–263.

Sulc, D., and Ristic, M. (1954). Über den Pektinabbau in gärenden Apfeltrestern. *Z. Lebensm. Forsch.* **98**, 430–434.

Sulc, D., and Vujicic, B. (1973). Untersuchungen der Wirksamkeit von Enzympräparaten auf Pektinsubstraten und Frucht- und Gemüsemaischen. *Flüss. Obst* **40**, 79–83, 130–137.

Tanner, H. (1970). Ueber die Herstellung von methanolarmen Fruchtbranntweinen. *Schweiz. Z. Obst Weinbau* **106**, 625–629.

Termote, F., Rombouts, F. M., and Pilnik, W. (1977). Stabilization of cloud in pectinesterase active orange juice by pectic acid hydrolysates. *J. Food Biochem.* **1**(1), 15–34.

van Buren, J. P., and Way, R. D. (1978). Tannin hazes in deproteinized apple juice. *J. Food Sci.* **43**, 1235–1237.

van Houdenhoven, F. E. A. (1975). "Studies on Pectin Lyase." Ph.D. Thesis. Agricultural University, Wageningen, The Netherlands.

Versteeg, C. (1979). Pectinesterases from the Orange Fruit—Their Purification, General Characteristics and Juice Cloud Destabilizing Properties." Ph.D. Thesis. Agricultural University, Wageningen, The Netherlands.

Versteeg, C., Rombouts, F. M., and Pilnik, W. (1978). Purification and some characteristics of two pectinesterase isoenzymes from orange. *Lebensm. Wiss. Technol.* **11**, 267–274.

Versteeg, C., Rombouts, F. M., Spaansen, C. H., and Pilnik, W. (1980). Thermostability and orange juice cloud destabilizing properties of multiple pectinesterases from orange. *J. Food Sci.* **45**, 969–971; 998.

von Fellenberg, T. (1913). Über den Ursprung des Methylalkohols in Trinkbranntweinen. *Mitt. Lebensm. Hyg.* **4**, 122, 273.

Voordouw, G., Voragen, A. G. J., and Pilnik, W. (1974). Apfelsaftgewinnung durch Maischefermentierung. III. Der Einfluss spezifischer Enzymaktivitäten. *Flüss. Obst.* **41**(7), 282–284.

Voragen, A. G. J. (1972). "Characterization of Pectin Lyases on Pectins and Methyl-Oligogalacturonates." Ph.D. Thesis. Agricultural University, Wageningen, The Netherlands.

Voragen, A. G. J., and Pilnik, W. (1989). Pectin-degrading enzymes in fruit and vegetable processing. *In* "Biocatalysis in Agricultural Biotechnology" (J. R. Whitaker and P. E. Sonnet, eds.), pp. 93–115. American Chemical Society, Washington, D.C.

Voragen, A. G. J., Heutink, R., and Pilnik, W. (1980). Solubilization of apple cell walls with polysaccharide-degrading enzymes. *J. Appl. Biochem.* **2**, 452–468.

Voragen, A. G. J., Geerst, F., and Pilnik, W. (1982). Hemi-cellulases in enzymic fruit processing. *In* "Use of Enzymes in Food Technology" (P. Dupuy, ed.), pp. 497–502. Technique et Documentation Lavoisier, Paris.

Voragen, A. G. J., Schols, H. A., Siliha, H. A. I., and Pilnik, W. (1986a). Enzymic lysis of pectic substances in cell walls: Some implications for fruit juice technology. *In* "Chemistry and Function of Pectins" (M. L. Fishman and J. J. Jen, eds.), pp. 230–247. American Chemical Society, Washington, D.C.

Voragen, A. G. J., Schols, H. A., and Pilnik, W. (1986b). Determination of the degree of methylation and acetylation of pectins by HPLC. *Food Hydrocoll.* **1**, 65–70.

Voragen, A. G. J., Wolters, H., Verdonschot-Kroef, T., Rombouts, F. M., and Pilnik, W. (1986). Effect of juice-releasing enzymes on juice quality. *Int. Fruit Juice Symp.* Vol. XIX. 453–462.

Voragen, A. G. J., Rombouts, F. M., Searle-van Leeuwen, M. F., Schols, H. A., and Pilnik, W. (1987). The degradation of arabinans by endo-arabinanase and arabinofuranosidases purified from *Aspergillus niger*. *Food Hydrocoll.* **1**(5/6), 423–437.

Voragen, A. G. J., Schols, H. A., and Pilnik, W. (1988). Non-enzymatic browning of oligosaccharides in apple juice models. *Z. Lebensm. Unters. Forsch.* **187**, 315–320.

Wegener, C., and Henninger, H. (1987). Charakterisierung und Einsatzmöglichkeiten einer Pektatlyase aus *Erwinia carotovora*. *Nahrung* **31**, 297–303.

Weiss, J., and Sämann, H. (1972). Turbidity stability of apricot nectar. *Mitt. Rebe Wein Obstbau Früchteverwert.* **22**, 177–184.

Wiley, R. C., and Lee, Y. S. (1970). Modifying texture of processed apple slices. *Food Technol.* **24**, 1168–1170.

Wood, W. A., and Kellogg, S. T. (eds.) (1988). "Biomass: Lignin, Pectin and Chitin," Methods in Enzymology Vol. 161, pp. 329–399. Academic Press, New York.

Wucherpfennig, K., and Otto, K. (1988). Beitrag zum kontinuierlichen Abbau von Pektinstoffen durch Enzyme bei der Klärung von Apfelsaft. *Flüss. Obst* **55(3)**, 112–116.

Wucherpfennig, K., Dietrich, H., and Bechtel, J. (1989). Alcohol: Total, potential and actual methanol content of fruit juices. *Flüss. Obst* **83**, 348–354.

Yamasaki, M., Yasui, T., and Arima, K. (1964). Pectic enzymes in the clarification of apple juice. Part I. Study on the clarification reaction in a simplified mode. *Agric. Biol. Chem.* **28**, 779–787.

Yamasaki, M., Kato, A., Chu, S. Y. and Arima, K. (1967). Pectic enzymes in the clarification of apple juice. II. The mechanism of clarification. *Agric. Biol. Chem.* **31(5)**, 552–560.

Zetelaki-Horváth, K., and Gátai, K. (1977a). Disintegration of vegetable tissues by endo-polygalacturonase. *Acta Aliment.* **6(3)**, 227–240.

Zetelaki-Horváth, K., and Gátai, K. (1977b). Application of endo-polygalacturonase in vegetables and fruit. *Acta Aliment.* **6**, 335–376.

Zetelaki-Horváth, K., and Vas, K. (1980). Solubilization of nutrients of vegetables by endo-polygalacturonase treatment. *Ber. Wiss. Technol. Komm. IFU* **XVI**, 291–308.

CHAPTER 15

Enzymes Associated with Savory Flavor Enhancement

TILAK NAGODAWITHANA

I. Introduction

Enzymes have long been exploited to improve the palatability of foods, even before their existence was known and understood. The manufacture of cheese using calf stomach as catalyst was probably the first example of enzyme use; this practice is so old that its discovery predates recorded history. With the understanding of the nature of enzymes and their catalytic potential, the use of enzymes gradually has been extended to a variety of fields. Among these fields, enzymes have found the most ready usage in the rapid evolution of the food industry. Some of the beneficial effects are flavor development, more desirable physical characteristics, greater nutritional value, and cost effectiveness. Although this entire field is of interest to many, this chapter will deal only with the production of flavor potentiating substances commonly found in savory foods and the enzymes associated with their production.

Flavor enhancers or flavor potentiators are known to be chemical substances that have little flavor of their own but, when mixed with food products, have the ability to enhance the flavor of the food. The best known flavor enhancers in use extensively worldwide are monosodium glutamate, disodium 5'-inosinate, and disodium 5'-guanylate. These compounds accentuate meaty flavor and have found applications as flavor potentiators in soups, sauces, gravies, and many other savory products.

Traditionally, the Japanese culture has used naturally occurring sources of flavor potentiators in their food preparations. Typical sources of these flavor potentiators were sea tangle (*Laminaria japonica*), dried bonito tuna, and the black mushroom (*Lentinus edodus*). These products have been reported to exhibit a distinct savory and delicious taste commonly referred to in Japanese as "umami." The important chemical compounds responsible for flavor potentiation action later were identified as monosodium glutamate (MSG), inosine 5'-monophosphate (5'-IMP; Kodama, 1913), and guanosine-5'-monophosphate (5'-GMP; Kuninaka, 1960). In the 1960s, Kuninaka (1966,1967) observed that 5'-nucleotides have a remarkable synergistic effect with MSG in flavor enhancement. All three potentiators that have been produced commercially are now used extensively in the food industry worldwide.

Among the more important developments in the use of enzymes is their application in the production of 5'-IMP and 5'-GMP. Of the 5'-nucleotides that exist in nature, these two show the most acceptable flavor enhancement properties. Although xanthosine 5'-monophosphate (5'-XMP) has some flavor enhancement properties, it is significantly less effective than 5'-GMP or 5'-IMP; hence, it has not been produced commercially.

Several researchers have been able to elucidate further the relationship between flavor potentiating properties and the chemical structure of the nucleotides. The nucleotides first recognized as having umami taste were those compounds that had a purine base with a hydroxyl group (or keto group, its tautomeric form) at the 6 position and a ribose moiety esterified with phosphoric acid in the 5 position. Additionally, it was implied that the hydrogen at the 2 position could be replaced by other groups such as a hydroxyl or an amino group without adversely affecting the flavor-enhancing properties. Thus, it is interesting to note that the three major purine-based 5'-nucleotides (5'-GMP, 5'-IMP, and 5'-XMP) have these common structural features which seem to be important in flavor potentiation. However, the intensity of flavor perception and enhancement was defined clearly by the specific nature of the substituents at position 2 of the purine moiety. From the standpoint of flavor enhancement, 5'-GMP has been judged to be the most effective, followed by 5'-IMP and 5'-XMP.

Both 5'-IMP and 5'-GMP occur naturally in foods. Animal muscles are generally rich in 5'-IMP. Several fresh seafoods are rich in 5'-AMP, which may be converted to 5'-IMP due to the action of endogenous enzymes released during processing. Raw vegetables have neither 5'-IMP nor 5'-GMP, yet these compounds were detected after the vegetables were cooked (Nguyen *et al.*, 1988). Thus, it seems probable that the 5'-nucleotides found in cooked foods are the products formed following enzymatic degradation of intracellular RNA. Several important enzymes that have applications in flavor-nucleotide production have been identified in naturally occurring products of plant, animal, or microbial origin. Best known within this group are the two enzymes 5'-phosphodiesterase and adenylic deaminase, derived from *Penicillium citrinum* and *Asperqillus* strains, respectively. This chapter primarily includes information relevant to the biochemical formation of the flavor enhancer nucleotides through the degradation of RNA by the aforementioned enzymes. Some effort has been made also to describe the characteristics of a few competing enzymes that must be minimized or eliminated for the desirable enzymes to be most effective in the flavor-nucleotide production process.

A. Substrate Specificity

Before addressing how different types of enzymes degrade the nuclear material in all biological systems, we must understand the structure of these macromolecules and the specific bonds that the enzymes break down during the process of hydrolysis. Like proteins, nucleic acids are informational macromolecules and therefore contain nonidentical monomeric units in a specific sequence. The two major types of nucleic acids are DNA (deoxyribonucleic acid) and RNA (ribonculeic acid). Both RNA and DNA are long chain polymers made up of mononucleotides that contain three characteristic components: (1) a nitrogenous purine or pyrimidine base, (2) a five-carbon sugar, and (3) a phosphate group. The nitrogenous bases of DNA include adenine, guanine, cytosine, and thymine. The composition of RNA is very similar to that of DNA with one exception. Thymine is replaced by uracil in the polymer chain. Further, RNA contains D-ribose as the 5-carbon sugar component. DNA contains 2-deoxy-D-ribose sugar. Functionally, DNA serves primarily as a repository of genetic information, whereas RNA plays several different roles in the expression of that information.

In recent years, yeast has found application as a substrate and a source of nuclear material for the production of flavor nucleotides. Although yeast contains both DNA and RNA, the latter typically makes up nearly 8–12%, with less than 1% DNA on a dry solids basis. Yeast extracts produced from baker's or brewer's yeast have been cleared for use in the United States by the Food and Drug Administration (Code of Federal Regulations 199010). Clearly, yeasts appeared particularly attractive as a source of nucleic acid for the production of flavor nucleotides because these microorganisms are well known to be edible with little risk of toxicity.

Three basic classes of ribonucleic acids are distinguished by their molecular weight and the way they function during protein synthesis. Messenger RNA (mRNA) provides the information that dictates amino acid sequence during polypeptide synthesis, transfer RNA (tRNA) directs the correct amino acids to the correct site in an elongating polypeptide chain, and ribosomal RNA (rRNA) provides the site for protein synthesis. Structurally, these polymers are formed by the linking of each nucleotide to the next through a phosphate group. Specifically, the phosphate group already attached by a phosphoester bond to the 5' carbon of one nucleotide becomes linked by a second phosphoester bond to the 3' carbon of the adjacent nucleoside phosphate (Fig. 1). The resulting linkage is called a 3',5'-phosphodiester linkage.

One member of a nucleotide pair must be a purine and the other a pyrimidine in order to meet steric requirements. Pairing in RNA always occurs between guanine and cytosine, and adenine and uracil. The phosphate

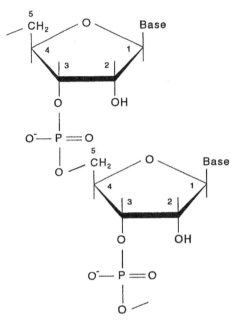

Figure 1   3',5'-Phosphodiester bond.

and sugar skeleton of the model is regular, but any sequence of base pairs will fit the overall structure. The four nucleotides can give rise to a variety of permutations; molecular masses of different RNA molecules range from 25,000 daltons for tRNA to over 500,000 daltons for rRNA. It is difficult to prepare particular RNA in a undegraded form. These high values obtained for freshly isolated RNA therefore would be possible only if RNA escaped the action of nucleolytic enzymes during the purification process. Ribosomal RNA is by far the most abundant and most stable form of RNA in most cells. Usually, this RNA is referred to when speaking of "yeast RNA" for flavor-nucleotide production.

II. Enzymatic Hydrolysis of Nuclear Material

Several enzymes, predominantly belonging to the general group of hydrolases, are capable of digesting the nucleic acid chain, resulting in the formation of fragments of nucleotides, nucleosides, and free bases. Such a breakdown is determined primarily by the specificity of the enzymes. The enzymes that attack nucleic acid are distributed widely in yeast and in other biological materials in the microbial, animal, and plant kingdoms. The kinetics of enzymatic hydrolysis of various synthetic substrates by these nucleolytic enzymes has been the subject of a great deal of work in the past few decades.

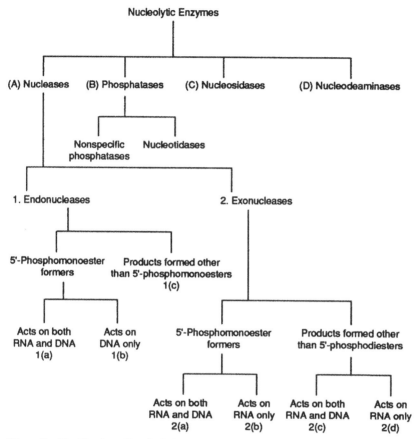

Figure 2 Classification of nucleolytic enzymes.

The distribution of the four major classes of nucleolytic enzymes, that is, nucleases, phosphatases, nucleosidases, and nucleodeaminases, as found in all biological systems is outlined in Fig. 2. It is important for living cells to carry such an array of enzymes, both to break down the nuclear material that the cell may absorb from the environment during the growth phase and to break down the specific substrates inside that the cell no longer requires for sustenance. Under normal conditions, these hydrolytic enzymes must be sequestered carefully until actually needed by the cell, lest they digest vital cellular components that are not designed for destruction. Most of these enzymes are present in yeast and other biological materials at varying concentrations. Thus it becomes critical for the enzyme preparation used in the production of 5'-nucleotide rich extracts to have the maximum activity for the formation of the expected product, with minimal activity of the other competing enzymes. Needless to say, this characteristic is an absolute requirement for achieving acceptable flavor and optimum product yields in the process.

A. Nucleases

Of the four major groups of nucleolytic enzymes presented in Fig. 2, several enzymes that are nucleases have found use in the production of flavors and flavor enhancers. All nucleases are known to catalyze the breakdown of internucleotide bridges without liberating the inorganic phosphates. Hence, these enzymes are capable of producing a wide variety of nucleotides when present in the reaction mixture. Therefore, a careful selection of the desirable enzymes is critical for the development of the desirable nucleotides for the flavor industry.

Three main features of nuclease activity can be used as the basis for classification. First is the mode of attack: polynucleotides can be attacked at points within the polymer chain (endolytically) or stepwise from one end of the chain (exolytically; Fig. 3). The second feature is the substrate specificity, which is based on whether the action is on RNA, DNA, or both. The third feature is the mode of cleavage of the phosphodiesterase bond. This feature depends on whether the hydrolysis of the bond occurs between 3'-OH and the phosphate group of the phosphodiester bond, to give 5'-phosphoryl end groups, or whether the attack is specific for the bond between the 5'-OH and the phosphate, which results in 3'-phosphoryl end groups. Based on these criteria, the industrially useful nucleases have been described using the classification scheme shown in Fig. 2.

1. Endonucleases

Endonucleases are enzymes capable of hydrolyzing phosphodiester linkages within the nucleic acid chain at several points, with the simultaneous

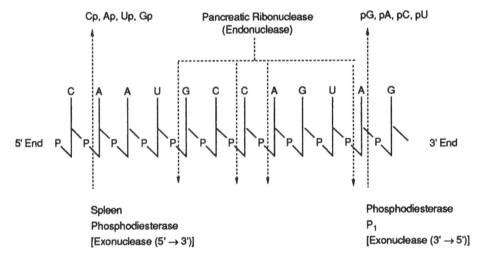

Figure 3 Substrate specificity of nuclease enzymes.

liberation of several mono- and oligonucleotide fragments. This class of enzymes can again be subdivided into two groups, namely, those enzymes that produce 5′-phosphomonoesters, some of which are capable of enhancing savory flavor, and others that produce nucleotides other than the 5′-phosphomonoesters. The very common and widely recognized nucleotides belonging to the latter category do not enhance flavor and include 3′- and 2′-phosphomonoesters and, at times, 2′,3′ cyclic nucleotides.

a. Nonspecific endonucleases forming 5′-phosphomonoesters (EC 3.1.30) The endonuclease enzymes that produce 5′-phosphomonoester compounds as a result of their activity on RNA or DNA commonly are derived from sources of plant or microbial origin. The best known enzyme belonging to this category that has found commercial use is nuclease P_1 derived from *Penicillium citrinum*. This enzyme shows a preference for cleaving phosphodiester bonds in single-stranded fragments of nucleic acids yielding 5′-nucleotides. Other less well known but nevertheless important enzymes that exhibit similar enzyme specificities are those from *Aspergillus* (*Aspergillus* nuclease S_1), *Neurospora crassa* (*N. crassa* nuclease), and Mung beans (*Phoseolus aureus* nuclease). An enzyme of similar properties also has been isolated from malt rootlets, and now has become of commercial importance in the production of flavor nucleotides. All these enzymes are capable of cleaving phosphodiester bonds of polynucleotides, regardless of the type of sugar and the base. The reaction proceeds with either denatured DNA or RNA; the extent of formation of 5′-phosphomononucleotides and oligonucleotides depends largely on the pH and the temperature of the reaction medium.

Nuclease P_1 shows a preference for certain specific linkages over others in the nucleic acid chain. For example, in the case of dinucleotide monophosphates, X-Y, nuclease P_1 showed a higher preference when X is adenosine or deoxyadenosine. Further, the base preference seems to depend largely on X rather than Y when nuclease P_1 acts on substrates that are typically of X-Y type. It has been claimed that, for the substrate to be recognized by nuclease P_1, the critical structural requirements are the nucleoside 3-phosphate portion, with preferential breakdown of the diester bond taking place between the 3′-hydroxyl group of adenosine or deoxyadenosine and the 5′-phosphoryl group of the succeeding nucleotide (pA-Y) in the chain.

Although nuclease P_1 is placed among the endonucleases according to the most recent classification, it is also known for its simultaneous exonucleolytic activity, with the rapid generation of 5′-mononucleotides (Fujimoto *et al.*, 1974). Accordingly, the rate of appearance of mononucleotides from RNA was shown to be

$$AMP > GMP > UMP > CMP$$

whereas the rate for denatured DNA was

$$dAMP > dTMP > dCMP > dGMP$$

(see Table I).

TABLE I
**Molar Ratio of 5'-Nucleotides Released during
Partial Hydrolysis of RNA and Heat Denatured
DNA[a]**

	Molar ratio (%)	
Nucleotide type	RNA	DNA
5'-AMP	37.7	39.5
5'-GMP	24.6	17.8
5'-CMP	17.1	20.4
5'-UMP or 5'-dTMP	20.6	22.3

[a]Reprinted with permission from Fujimoto et al., 1974.

Nuclease P_1 from *Penicillium citrinum* thus possesses some unique properties that differ from certain other well-studied nucleases. For example, as mentioned earlier, the enzyme is capable of hydrolyzing both RNA and denatured DNA in an endo- and exonucleolytic manner, yielding predominantly 5'-mononucleotides. Although similar, the mode of action of snake venom phosphodiesterase is somewhat different because it functions only as an exonuclease, yielding only 5'-phosphomononucleotides. On the other hand, although the structural requirements necessary for nuclease P_1 and spleen phosphodiesterase are similar, the latter enzyme is known for its action at the diester bond between the phosphate and the adjacent nonspecific hydroxyl component of the diester bond, thereby yielding 5'-phosphomononucleotides and 3'-phosphooligonucleotides as end products. Subsequent studies by Kuninaka (1971) have demonstrated that the 2',5'-phosphodiester linkage was resistant to both nuclease P_1 and spleen phosphodiesterase enzymes, although these bonds were highly susceptible to snake venom diesterases. *Penicillium* nuclease P_1 also was noted for its ability to attack a variety of nucleic acids, regardless of the type of sugar or base, the size of the molecule, or the presence or absence of the terminal phosphate group in the nucleic acid chain. Thus it is clear that the type and the extent of the formation of phosphomononucleotides depend largely on the origin of the enzyme. Nevertheless, nuclease P_1 can be regarded as one of the most useful enzymes available on a commercial scale for the production of 5'-mononucleotide-rich flavor extracts from substrates such as yeast that contain substantial levels of nucleic acids.

Among the nucleases known to date, an endonuclease enzyme of plant origin that has been studied rather extensively is that of potato tuber. According to Suno et al. (1973), the structural requirements of the substrate and the cleavage sites on the nucleic acid chain for the nuclease enzyme from potato have been found to be similar to those of nuclease P_1. This enzyme is regarded as an endonuclease active on either RNA or denatured DNA to yield, almost exclusively, 5'-phosphomononucleotides and 3'-phosphooligonucleotides.

The purification of cell wall-bound nucleases from potato tuber was realized by Nguyen et al. (1988) in small amounts that were, nevertheless, adequate for the systematic study of the nucleotides produced by its action. The results of this study indicated an estimated molecular mass of 37.6 kDa for this enzyme. The enzyme showed a pH optimum in the range 6.5–7.5 and a temperature optimum at or around 70°C. However, there appeared to be a basic difference between the cell wall-bound nuclease and the cytoplasmic nuclease derived from potato with respect to the optimum temperature of activity. Thus, according to the experimental data, the cytoplasmic nuclease performed best at 80°C, 10°C higher than the cell wall nuclease. The cell wall nuclease differed from the cytoplasmic nuclease in potato tuber with respect to its relative molecular size (37.7 kDa for cell wall; 34.2 kDa for the cytoplasmic nuclease) and its substrate specificity (the Michaelis–Menten constants, K_m, of the cell wall and cytoplasmic nucleases were 60 μg and 40 μg, respectively, when yeast RNA was used as substrate).

Table II shows the relative activity of the purified potato cell wall nuclease enzyme in the presence of various activators and inhibitors. These results have indicated a strong inhibition of enzyme activity in the presence of Zn^{2+}, Cu^{2+}, Mn^{2+}, pyrophosphates, citrate, phosphate, and EDTA. Nitrate, Mg, glutamate, and sulfite showed minimal or no inhibition of the enzyme.

TABLE II
Effect of Various Ions and EDTA on Nuclease Activity[a]

Additive	Concentration	Nuclease relative activity (%)
Control (none)	—	100
Na-Glutamate	10	82
Na-Citrate	10	21
$NaNO_3$	10	104
$NaHSO_3$	10	95
NaH_2PO_4	10	51
Na-Pyrophosphate	10	0
$CuSO_4$	1	0
$ZnCl_2$	1	0
$MnCl_2$	1	20
$MgCl_2$	1	92
EDTA	1	17

[a]Reprinted with permission from Nguyen et al., 1988.

b. **Endoribonucleases forming 5'-phosphomonoesters (EC 3.1.26)** For most of the ribonucleases, the position of cleavage is such that the fragments formed predominantly terminate in 3'-phosphate groups. However, at least seven endoribonucleases have been identified that are capable of degrading ribonucleic acid chains to form fragments that terminate in 5'-phosphate groups. The best known examples of enzymes in category are ribonuclease P, ribonuclease III, and ribonuclease IV. Although these enzymes can yield 5'-nucleotides, they are of lesser importance in flavor-nucleotide production research because of their lack of availability on a commercial scale. Further, their presence in yeast has not been reported to date in the literature. Hence, they will not be discussed here in much detail.

c. **Endoribonucleases forming products other than 5'-phosphomonoesters (EC 3.1.27)** These endonucleases preferentially hydrolyze RNA with the formation of mononucleotides and oligonucleotides that terminate in 3'-phosphate groups. Also it is now uncommon to find 2',3' cyclic phosphate intermediates in the reaction mixture. The best known enzymes belonging to this class of enzymes are RNase T_1, RNase T_2, and pancreatic ribonuclease. Enzymes similar in activity to these RNases have been identified in several materials of microbial and plant origin. Hydrolysis of RNA by these RNases leads almost exclusively to the formation of the four ribonucleoside 3'-phosphates.

These enzymes catalyze the hydrolysis of RNA in a two-step process. In the first step, the enzyme catalyzes the cleavage of RNA to form 2',3' cyclic phosphate. In the second step, it catalyzes the hydrolysis of the cyclic phosphate to ribonucleoside 3'-phosphates following the regeneration of the hydroxyl group at the C-2 position of the ribose unit (Fig. 4). The presence of these contaminating enzymes, either in the enzyme preparations or in the substrates for the production of 5'-nucleotide rich extracts, can increase the risk of possible deteriorative changes to potentially important flavor enhancing compounds and their precursors. Such contamination seriously decreases the overall quality of the final product.

The comparative properties of RNase T_1 and RNase T_2 have been reviewed by Egami *et al.* (1964). According to these studies, the molecular weights of these two enzymes were found to be 11,000 and 30,500, respectively. Their optimum pH values were determined to be 7.5 and 4.5. Both enzymes were inhibited by Cu^{2+} and Zn^{2+} ions. Studies have indicated that RNase T_1 cleaves phosphodiester bonds adjacent to the 3'-phosphate attached to a guanosine residue, yielding oligonucleotides terminating in guanosine 3'-phosphate. In contrast, pancreatic ribonucleases are pyrimidine specific. They cleave bonds adjacent to the 3'-phosphate linked to cytidine and uridine, resulting in a mixture of oligonucleotides terminating in either cytidine 3'-phosphate or uridine 3'-phosphate. The specificity of RNase T_2 does not appear to be as strict as that of RNase T_1. However, it is capable of producing the two-stage endonucleolytic cleavage liberating 3'-phosphomon-

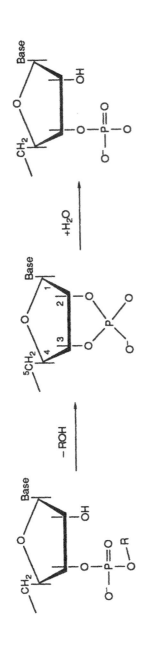

Figure 4 Mechanism of hydrolysis of RNA by endoribonucleases.

onucleotides and 3'-phosphooligonucleotides with 2',3'-cyclic phosphates as intermediates. Similar types of enzymes that are likely to be present in 5'-phosphodiesterase preparations from either potato or malt rootlets or from substrates such as yeast must, however, be minimized or eliminated, probably by heat inactivation, to achieving best results in the production of extracts rich in purine-based 5'-nucleotides.

2. Exonucleases

So far we have considered the typical endonucleases that are capable of attacking the nucleic acid chain to result in the rupture of phosphodiester bonds at several points with simultaneous liberation of mono- and oligonucleotide fragments. With this information in hand, we are ready to examine the exonucleases that are common to all biological systems. These enzymes require a free 3'-hydroxyl group or a free 5'-hydroxyl group and proceed stepwise from the ends of the polynucleotide chain. These enzymatic reactions progressively increase the level of 3'- or 5'-nucleotides in the reaction medium, depending on the enzyme used in the reaction.

a. **Nonspecific exonucleases forming 5'-phosphomonoesters (EC 3.1.15)** The best known example in this category is venom exonuclease (EC 3.1.15.1), the exonuclease cleavage reaction of which occurs in the 3' to 5' direction to yield the 5'-phosphomonoesters. In the purified state, this enzyme is active on both RNA and DNA. It exhibits no strict specificity with respect to the nucleotide sequence, but always yields 5'-mononucleotides. In this respect, snake venom phosphodiesterase appears to have properties similar to those of nuclease P_1 described previously. Its activity was, however, found most prominent at alkaline pH, between 7 and 10, with an optimum at pH 8.6. Although snake venom phosphodiesterase enzyme has the required properties for the production of flavor nucleotides, it is not certified as a food-grade enzyme for obvious reasons. A similar type of exonuclease enzyme has been observed in *Lactobacillus casei* cells; this enzyme degrades RNA to nucleoside-5'-monophosphates. The activity of the enzyme is stimulated by the presence of K^+ ions (Keir *et al.*, 1964). Its optimum activity is at pH 8.0–8.2. Very little work has been published on this enzyme.

b. **Exoribonucleases forming 5'-phosphomonoesters (EC 3.1.13)** These enzymes attack only RNA, providing exonucleolytic cleavage either in the 3' to 5' or 5' to 3' directions, yielding 5'-phosphomononucleotides. The best known examples of enzymes in this category are exoribonuclease II and exoribonuclease H. Although these enzymes are of interest to a flavor-nucleotide manufacturer, suitable food-grade sources have not been identified to date to produce this enzyme in a cost effective manner commercially.

c. Nonspecific exonucleases forming products other than 5'-phosphomono-esters (EC 3.1.16) These enzymes react with RNA and DNA, resulting in the exonucleolytic cleavage on the nucleic acid chain in the 5' to 3' direction to yield 3'-phosphomononucleotides. An example of an enzyme with these characteristics is spleen exonuclease. However, similar properties have been found in *Lactobacillus acidophilus* nuclease and *Bacillus subtilis* nuclease enzymes. The presence of enzymes of similar properties can lead to lower yields of 5'-nucleotides in a flavor-nucleotide production process. Typically, these enzymes have lower temperature optima and can be inactivated by simple heat treatment without destroying the more temperature-tolerant 5'-phosphodiesterase enzymes.

d. Exoribonuclease forming products other than 5'-phosphomonoesters (EC 3.1.14) These enzymes have the characteristic property of hydrolyzing the polynucleotide chain of RNA in a stepwise manner by successively detaching mononucleotides in sequence, starting from one end of the chain, to liberate primarily 3'-phosphomononucleotides. The best known exonuclease in this category is yeast ribonuclease. This enzyme is of particular interest to the producers of 5'-nucleotide rich extracts from yeast because of its negative effect on the yields of 5'-nucleotides if not removed completely from yeast prior to 5'-phosphodiesterase enzyme treatment. It is also important to inactivate similar enzymes that could possibly be present in the commercial grade 5'-phosphodiesterase preparations used in the production of 5'-nucleotide rich extracts. Typically, this inactivation is achieved by differential heat treatment.

B. Phosphatases

Phosphatases are phosphoric monoester hydrolase enzymes found extensively in almost all tissues and microorganisms. These enzymes are able to hydrolyze the terminal phosphoryl group of nucleotides to cause the formation of free inorganic phosphates and nucleosides.

These enzymes have a relatively broad specificity, so they are able to act on many compounds with common structural features. They thus may hydrolyze many different esters of phosphoric acid and are not entirely limited to nucleotides.

Placed in this group (EC 3.1.3) are 51 enzymes according to the Enzyme Nomenclature (1984). Again these enzymes are grouped broadly under nonspecific phosphatases and specific phosphatases. As the term implies, the enzymes included in the first category are known to hydrolyze all phosphomonesters without distinguishing between the groups esterifying the phosphoryl residue. These enzymes would, therefore, hydrolyze both 3'-phosphates and 5'-phosphates. Included in this group are two subgroups that have been known for a long time under the familiar names alkaline and acid phosphatases. Since they are nonspecific phosphatases, they are capable of hydrolyzing other phosphomonoesters such as glycerophosphates and phenyl phosphates, in addition to catalyzing the familiar reaction with nucleic acids and their corresponding mononucleotides.

The enzymes that belong to the second category are different phosphatases with varying specificities, each being very specific in the nature of the substituent on the phosphoryl group during hydrolysis. At least 49 enzymes are grouped in this category (Enzyme Nomenclature, 1984). Among the enzymes in this series, those that are considered destructive to the processing of flavor-enhancing nucleotides are the nucleotidases, the best known of which are 5'-nucleotidases that are characterized by the position at which the phosphate group is hydrolyzed on the respective nucleotides. For example, the 5'-nucleotidases are known for their activity in splitting flavor-enhancing 5'-nucleotides to their corresponding nucleosides and orthophosphates, causing a decline in the flavor-enhancing ability of the final product.

Nonspecific phosphatases and nucleotidases are distributed widely in gramineous seeds, potatoes, bacteria, and fungi. Use of any such ingredients or their extracts for the production of flavor nucleotides requires the prior elimination of the aforementioned degradative enzymes to insure optimal flavor-nucleotide production in the process.

C. Nucleosidases

The nucleosidase subgroup of enzymes under hydrolases (EC.3) is likely to be of interest to the flavor industry, particularly because of the problems these enzymes can cause by destroying the desirable flavor-enhancing nucleotides formed in the manufacturing process. These contaminating enzymes have the ability to hydrolyze N-glycosyl compounds thereby liberating free bases as a result of the hydrolysis of the β-glycosidic bond. These reactions are generally single step conversions, for example, the conversion of 5'-inosinate to hypoxanthine and D-ribose-5'-phosphate in the presence of inosinate nucleosidase (EC 3.2 .2.12).

$$5'\text{inosinate} + H_2O \xrightarrow[\text{nucleosidase}]{\text{Inosinate}} \text{hypoxanthine} + \text{D-ribose-phosphate}$$

These enzymes are likely to be present in high concentrations in many of the materials employed in the production of 5'-phosphodiesterase enzymes used in flavor-nucleotide production. Further, the presence of these contaminating enzymes in yeast can be a serious concern to those who use this substrate as a source of RNA for nucleotide production. Every precaution must be taken to eliminate or minimize such competing enzymes prior to the use of yeast in the process.

D. Nucleodeaminase

The nucleodeaminases preferentially act on C–N bonds other than peptide bonds in cyclic amidines, liberating the free ammonia by deamination. This reaction can occur in compounds that contain purine or pyrimidine bases. In a typical reaction, adenine deaminase (EC 3.5.4.2) converts adenine to hypoxanthine and ammonia:

$$\text{Adenine} + H_2O \xrightarrow[\text{deaminase}]{\text{Adenine}} \text{hypoxanthine} + NH_3$$

Adenosine phosphate deaminase (EC 5.4.17) is the best known enzyme in this category that has the greatest influence on the production of flavor nucleotides. Like the 5′-phosphodiesterase enzyme, adenosine phosphate deaminase, sometimes referred to as adenylic deaminase, has become an important enzyme for the production of 5′-nucleotide rich extracts. Its primary function is to convert 5′-adenosine monophosphate (5′-AMP), which offers no flavor-enhancing properties, to 5′-inosine monophosphate (5′-IMP), a product that has marked flavor-enhancing properties in food systems (Fig. 5).

Yeast is used as a common source of RNA for the commercial production of flavor-nucleotide rich extracts. Of the four 5′-nucleotides formed during the hydrolysis of RNA by the action of 5′-phosphodiesterase enzyme, only 5′-GMP is known to provide the flavor-potentiating properties. Nonetheless, the 5′-AMP produced, which has no flavor-enhancing properties, can be converted to 5′-IMP, which has flavor-enhancing properties like those of 5′-GMP, by applying a second enzyme, adenosine phosphate deaminase, as described. Both compounds then have flavor-enhancing properties, although guanylate is approximately twice as effective as inosinate.

Although adenylic deaminase is highly specific to 5′-AMP, it also can act on ADP, ATP, NAD$^+$, and adenosine, in decreasing order of activity (Su *et al.*,

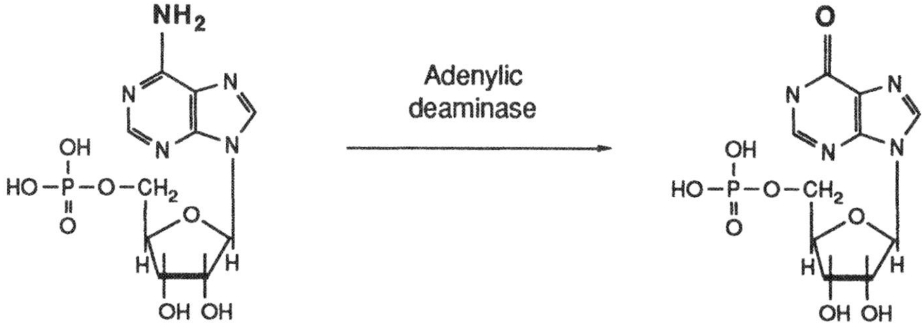

Figure 5 Conversion of 5′-AMP to 5′-IMP by reaction of the enzyme adenylic deaminase.

1966; Yates, 1969). Adenylic deaminase has been produced commercially from *Aspergillus oryzae.*

III. Commercial Production of Extracts Rich in Flavor Enhancers

Traditionally, the most commonly used flavor potentiators other than table salt are the naturally occurring food preparations rich in monosodium glutamate, as exemplified by Kombu or sea tangle *(Laminiarie japonica)* used by oriental cultures for centuries. This flavor potentiator, like the 5'-nucleotides 5'-GMP and 5'-IMP, generally is known to stimulate the corresponding taste buds to bring about the full savory character in foods for maximum enjoyment.

A. Yeast Extracts

Yeast extracts and hydrolyzed vegetable proteins are commercially available products rich in glutamic acid. They have been used extensively by the food industry as flavoring agents. Yeast extract is a concentrate of the soluble fraction of yeast generally made following autolysis, which is essentially a degradation process carried out by activating the yeast's own degradative enzymes to solubilize the cell components found in the cell.

Autolysis or self-digestion of viable yeast cells takes place at an acceptable pace when the yeast slurry is maintained at 45–55°C for 24–36 hr at pH 5–6. Lysis of these live whole cells is primarily brought about the action of β-1,3 gluconase and protease enzymes present in the cells. The β-1,6 gluconase and mannanase enzymes participate in the further solubilization of the cell wall matrix.

More than 40 proteolytic enzymes have been identified to date in *Saccharomyces cerevisiae,* of which only a few can be described as vital to the autolytic process. A few of the important proteolytic enzymes that play a major role during lysis are proteinase ysc A, proteinase ysc B, carboxypeptidase ysc Y, and carboxypeptidase ysc S. All these soluble proteases appear to be localized in the vacuole, whereas their inhibitors are in the cytoplasm outside the vacuole. Such compartmentalization segregates biologically compatible components from the corresponding enzymes to allow the control and integration of the intracellular activities for the orderly functioning of the viable cell (Table III).

Under autolytic conditions, the gradual breakdown of barriers in the highly compartmentalized cell matrix releases proteolytic enzymes from the vacuole into the degenerating cell matrix. These enzymes are, at first, inactivated because of the formation of complexes with corresponding inhibitors present in the immediate vicinity. This inactivation is evident when proteinase

TABLE III
Properties of Yeast Proteases[a]

Property	Proteinase ysc A	Proteinase ysc B	Carboxypeptidase ysc Y	Carboxypeptidase ysc S
Type	Acid endopeptidase	Serine endopeptidase	Serine exopeptidase	Metallo exopeptidase (Zn^{2+})
Optimum pH	2–6	6–7	4–7	7
Optimum T (°C)	35–40	45–55	45–55	60
Cell location	Vacuole	Vacuole	Vacuole	Vacuole
Solubility	Soluble	Soluble	Soluble	Soluble
Molecular weight	60,000	32,000–44,000	61,000	Not determined
Isoelectric point	3.8	5.8	3.6	—
Inhibitors	Pepstatin, etc.	Chymostatin, etc.	Hg^{2+}	EDTA
Cellular role	Protein degradation	Protein degradation	Protein degradation	Protein degradation

[a]Reprinted with permission from Reed and Nagodawithana, 1991.

ysc A and proteinase ysc B are extracted by autolysis at pH 7. However, incubation of the yeast at pH 5 caused reactivation of these proteinase enzymes (Hata *et al.*, 1969) by the proteolytic inactivation of the firmly bound inhibitors (Lenney and Dalbec, 1969).

The vacuolar proteinases are involved in highly nonspecific protein and peptide degradations. Likewise, nucleases present in the cell begin to act on the RNA and DNA, reducing them to polynucleotides, mononucleotides, and nucleosides or free bases. These processes can, however, be enhanced by initially subjecting the yeast to an external protease attack, for example, by papain, followed by treatment with a lytic enzyme such as glucanase. These enzymes are currently available commercially to improve the extract yield during yeast autolysis.

Glucanase enzymes have been isolated from yeast, and are known to hydrolyze the β-1,3 and β-1,6 linkages of the glycan building blocks of the cell wall. The glucanase enzymes are known to participate in the budding process. Nonetheless, under autolytic conditions, β-1,3 glucanase enzymes, with support from the proteinases, are capable of disrupting the cell wall, thereby facilitating the release of the soluble compounds such as amino acids, peptides, mononucleotides, and other degraded low molecular weight materials from within the cell matrix.

The yeast proteases have a broad specificity for their substrates. They are known to exhibit different pH and temperature optima. Therefore, the process conditions used by extract manufacturers are not optimal for all proteolytic enzymes. Consequently, the variation of flavor profiles of extracts made by autolysis under different pH and temperature conditions is understandable. Thus, the control of conditions of autolysis is vital to maintaining product quality and consistency.

Glutamic acid is the most abundant amino acid in a vast majority of proteins found in foods. Proteolytic degradation during autolysis results in the release of much glutamic acid into the medium. When protein hydrolysis is incomplete, it is common to find some glutamic acid units as parts of peptides. Such bound glutamic acid generally exhibits minimal influence on the flavor. The free glutamic acid can be in equilibrium with a number of ionic forms that are dependent on the environmental pH. The predominant ionic form of glutamic acid at a pH range of 5.5–8, which is the common pH of most food systems, has a net negative charge. Under these conditions, both glutamic acid and MSG appear to be present as the glutamate ion, thus enhancing the sensory properties of the food.

Yeast extracts rich in flavor-enhancing nucleotides such as 5'-GMP and 5'-IMP also are being produced from *Candida utilis* by heating a cell suspension for 30 min at 80–100°C, cooling the medium to 40°C, and treating with proteinase, 5'-phosphodiesterase, and deaminase enzymes (Harada *et al.*, 1988). The suspension is then centrifuged and the supernatant rich in 5'-nucleotides is concentrated and spray dried. The resulting product is claimed to have properties particularly effective in improving the flavor of savory meats, gravies, sauces, and soups.

B. Hydrolyzed Vegetable Protein

Another important development in the food industry is the production of amino or peptide savory flavorings from soy beans, wheat, and other plant substrates by acid hydrolysis. The products made by this procedure are commonly referred to as hydrolyzed vegetable protein (HVP). However, acid hydrolysis is known to bring about the destruction of certain amino acids and vitamins, resulting in a decline in the nutritive value. Additionally, HVPs made using HCl have been found to contain monochloro- and dichloropropanols (MCPs and DCPs) that are generally known as carcinogens.

One way to minimize the formation of such undesirable compounds is to provide a partial proteolysis of the protein substrate initially after which the content is subjected to a much milder acid hydrolysis. Some of the vegetable proteins selected for this process are generally high in glutamic acid. Processes that are designed to use acid hydrolysis exclusively or those that use the combination of partial proteolysis followed by acid hydrolysis are known for their higher degree of hydrolysis of proteins than the autolysis process described previously. Hence, these products are likely to have higher levels of free glutamic acid than yeast extracts. Under physiological conditions, these hydrolysates have a higher level of glutamate ions with a higher efficacy for flavor enhancement. These products are known worldwide for their versatility and ease of the application in a variety of savory food systems.

IV. Conclusion

Trends in consumer preference for all-natural foods and a desire on the part of the food industry for a "clear label" has resulted in an increased demand for substitutes such as yeast extracts to serve the function of the commonly used flavor enhancers in processed foods. As has been described previously, the procedure for the production of such yeast extracts requires the use of two enzymes, 5'-phosphodiesterase and adenylic deaminase.

One of the problems facing extract manufacturers is the escalating price of the enzymes. Because the current usage levels of these enzymes in the food industry are relatively low, the total amounts produced are correspondingly small. As expected, the cost of such low volume products is high and will continue to increase. A substantial improvement in the market share for these yeast extracts in the flavor industry should result in an increase in the usage volumes for these enzymes. More competition would result in more reasonably priced enzymes available for production of extracts.

However, it is not critical to operate such processes with highly purified enzymes. Indeed, one easy and certainly viable alternative is the use of crude enzyme preparations since, by adopting this procedure, the purification and stability problems could be circumvented. Further, this method is highly cost effective. However, one must bear in mind that, when operating with crude enzyme preparations, there is the risk of side reactions competing for the same substrate, thereby causing significant reduction in flavor-nucleotide yields. There is currently a shortage of crude enzymes and the yeast extract manufacturers are not comfortable with the quality and price of those crude preparations. Accordingly, there is a definite need to make these enzymes available in commercial quantities with higher purity at affordable prices.

The enzyme 5'-phosphodiesterase is known to be highly unstable after it is extracted from its native environment. The factors that influence stability most are pH, temperature, moisture, ionic strength, shear, pressure, and any combination of these effects. A more complete understanding of these effects should enable enzyme producers to produce and commercialize more active and stable 5'-phosphodiesterase enzyme for use in extract production.

Opportunities still exist to reduce the cost of enzyme use further without affecting the quality of the extracts. Use of immobilized enzymes to carry out the precise degradation of the RNA lowers costs substantially. An important feature of this method is that immobilization enhances enzyme stability and biocatalytic recovery. The economic benefit of using this approach is based on the fact that it allows for decreased fixed costs, including those of labor and construction. Moreover, the immobilized enzymes are ideally suited for long-term continuous operation, which can bring about economic benefits to the processor. Studies have also shown that, when the two enzymes critical for production of flavor enhancers are bound to an alkylamine ceramic support using glutaraldehyde, the pH optima of both enzymes tend to shift toward the more acidic side, thereby providing improved heat stability for both enzymes. A change of this nature could allow the reaction to be carried out at a higher temperature (60°C) and a lower pH (pH 5). Although the technology is available and economically attractive, the extract manufacturers must grapple with this problem to decide when and how the immobilization technology will be applied to commercial processing.

Current trends in biotechnology and, in particular, genetic engineering should widen significantly the scope of research in areas that directly relate to flavor enhancement. Indeed, this technology will be most useful to introduce specific genes, for example, those encoding the enzymes 5'-phosphodiesterase and adenylic deaminase, into yeast in high copy number without affecting the RNA production capacity or its growth characteristics. A successful transformation would allow extract manufacturers to produce yeast extracts high in 5'-nucleotides without relying on external sources of enzyme. The economic advantages of a change of this nature would be overwhelming.

Current commercial strains do not yield more than 10–12% RNA that is of value to extract manufacture. To realize higher yields of 5'-nucleotides, the RNA content of the yeast must be increased, either by genetic means or by control of the processing of yeast. Unfortunately, genetic alterations of yeast to enhance its RNA content have received very little attention to date.

Although protein engineering is still in its infancy with respect to application to food technology, there seems to be an almost limitless potential to develop more active or more stable enzymes by making targeted alterations to the native enzyme. Such changes would permit the selected tailored enzymes to perform certain specific process tasks more efficiently. However, the future prospects for development of such superactive, stable 5′-phosphodiesterase and adenylic deaminase enzymes are not clear.

Acknowledgments

The author thanks Gerald Reed and Thanh Nguyen for helpful suggestions provided during the preparation of this manuscript. The invaluable clerical help of Virginia Teat in the preparation of this chapter is also gratefully appreciated.

References

Egami, F., Takahashi, K., and Uchida, T. (1964). Ribonucleases in taka-diastase: Properties, chemical nature, and applications. *In* "Progress in Nucleic Acid Research" (J. N. Davidson and W. E. Cohn, eds.), Vol. 3, 59–101. Academic Press, New York.

Enzyme Nomenclature (1984). Academic Press, New York.

Fujimoto, M., Fujiyama, K., Kuninaka, A., and Yoshino, H. (1974). Mode of action of nuclease P on nucleic acids and its specificity for synthetic phosphodiesters. *Agric. Biol. Chem.* **38(11)**, 2141–2147.

Harada, S., Itou, J., Yano, M., Aoyagi, Y., and Maekawa, H. (1988). Manufacture of yeast extracts as food seasonings. Patent Corporation Treaty International Patent Application W-08,805,267.

Hata, T., Hayashi, R., and Doi, E. (1969). Purification of yeast proteinases. Part I. Fractionation and some properties of the proteinases. *Agric. Biol. Chem.* **31**, 150–159.

Keir, H. M., Mathog, R. H., and Carter, C. E. (1964). Purification of a K^+ ion-activated RNA 5′-phosphodiesterase from *Lactobacillus cesei. Biochemistry* **3**, 1188–1193.

Kodama, S. (1913). Isolation of inosinic acid. *J. Tokyo Chem. Soc.* **34**, 751–757 *(in Japanese).*

Kuninaka, A. (1960). Studies on taste of ribonucleic acid derivatives. *J. Agric. Chem. Soc. Jpn.* **34**, 489–492.

Kuninaka, A. (1966). Recent studies of 5′-nucleotides as new flavor enhancers. In "Flavor Chemistry" (R. F. Gould, ed.), pp. 261–274. American Chemical Society, Washington, D.C.

Kuninaka, A. (1967). Flavor potentiators. *In* "Symposium on Foods: the Chemistry and Physiology of Flavors" (H. W. Schultz, E. A. Day, and L. M. Libbey, eds.), pp. 515–535. AVI Publishing, Westport, Connecticut.

Kuninaka, A. (1971). Proceedings of the International Symposium on Conversion and Manufacture of Foodstuffs by Microorganisms. Kyoto, Japan.

Lenny, J. F., and Dalbec, M. (1969). Yeast proteinase B. Identification of the inactive form as an enzyme inhibitor complex. *Arch. Biochem. Biophys.* **129**, 407–409.

Nguyen, T. T., Palcic, M. M., and Hadziyev, D. (1988). Characterization of cell-wall bound nucleases and ribonucleases from potato tuber. *Agric. Biol. Chem.* **52**, 957–965.

Reed, G., and Nagodawithana, T. W. (1991). "Yeast Technology," 2nd Ed. Van Nostrand Reinhold, New York.

Su, J. C., Li, C.-C. and Ting, C. C. (1966). A new adenylate deaminase from red marine alga *porphyrarispata. Biochemistry* **5**, 536–543.

Suno, M., Nomura, A., and Mizuno, Y. (1973). Studies on 3′-nucleotidase-nuclease from potato tuber. *J. Biochem.* **73**, 1291–1297.

Yates, M. G. (1969). Non-specific adenine nucleotide deaminase from *Desulfovibrio desulfuricans. Biochim. Biophys. Acta* **171**, 299–310.

CHAPTER 16

Wine

Jean-Claude Villettaz

I. Introduction

The production of high quality wines requires carefully optimized parameters such as the growing environment for a given grape variety and adapted wine technology. In the wine making process, the enzymes play a very important role. In Pasteur's time, it was claimed, "Ce sont les microorganismes qui font le vin;" today we could claim, "ce sont les enzymes qui font le vin." In fact, more than 10 different enzymes are involved in alcoholic fermentation. Without these enzymes, grape juice would never become wine.

In addition to these enzymes are others that act during the wine making process. Some of them are beneficial, others are detrimental to wine quality. Therefore, it is extremely important to the enologist to know precisely the nature and the behavior of these enzymes. The enologist must create the optimal conditions for the desirable enzymes to work and, at the same time, inhibit those enzymes detrimental to wine quality.

To accelerate certain enzymatic reactions to improve the economy and the quality of the wines, Cruess and Besone (1941) suggested the use of commercial enzyme preparations developed for the fruit juice industry. The positive results obtained led to the use of these biotechnological products in wine making. Research on grape juice and wine colloids have initiated development of new enzymatic preparations in wine making.

The first part of this chapter discusses endogenous enzymes, while the second part is devoted entirely to the discussion of the use of commercial enzyme preparations.

II. Endogenous Enzymes

This section is a brief survey of a few endogenous enzymes that are of technological relevance to the winemaker. Among the desirable enzymes, the pectolytic enzymes and the proteases are described. The enzymes involved in the formation of volatile C_6 alcohols and the polyphenol oxidases will be mentioned as detrimental to wine quality. A more complete survey has been published by Villettaz (1984). A more recent survey has been published by Villettaz and Dubourdieu (1991).

A. Pectolytic Enzymes

The pectolytic enzymes present in juice or wine have several origins: the grapes, the yeasts, and other microorganisms such as *Botrytis cinerea*. All types of pectolytic enzymes are found, including pectinesterases, polygalacturonases, pectin lyases, and pectate lyases.

1. Technological Relevance of Pectolytic Enzymes

Some pectolytic enzymes (pectinesterases) deesterify the methylated pectin and release methanol. Others, called depolymerases (polygalacturonases, pectin and pectate lyases), hydrolyze the pectin chain into small fragments.

The depolymerases, especially the endo- type, are able to reduce the viscosity of a pectin solution in a very short period of time. The hydrolysis of a small percentage of the linkages in the main chain is sufficient to reduce the viscosity of a pectin solution by more than 50%. This rapid drop of viscosity allows a faster sedimentation of the cloud particles. The benefits of such a reaction are a better juice extraction, a faster and more complete juice clarification, and, finally, an easier wine filtration.

2. Sources of Endogenous Pectolytic Enzymes

a. **Grapes** Grapes contain mainly pectinesterase (PE) and polygalacturonase (PG) activities. These activities increase during the ripening of the grapes and decrease toward the end of maturation (Grassin, 1987). The activities are thermolabile; in the case of PG, an exposure at 60°C for 20 min reduces the activity by more than 50% (Olivieri, 1975). PE is located mainly in the skin of the grape berries. Its activity decreases during alcohol fermentation (Montedoro and Bertucioli, 1975; Marteau *et al.*, 1961).

b. **Yeasts** Very little has been published on the pectolytic activities of yeast (Demain and Phaff, 1954; Patel and Phaff, 1959). Yeast has been reported to produce PG and, to a lesser extent, PE activities. The hydrolysis mechanisms of yeast PG have been described by Demain and Phaff (1954).

c. Botrytis cinerea   Work performed by Grassin (1987) indicates that *Botrytis cinerea* produces PE and PG mainly during the growth phase. The PG level of a "botrytized" juice is about 200-fold higher than that of a healthy juice, explaining why such juices contain much less pectin than juices extracted from healthy grapes.

B. Proteases

1. Technological Relevance of Proteases

Grape proteins are considered responsible for a certain type of haze formed during the storage of wine. If this precipitation occurs after bottling, the wine becomes unsalable. The molecular mass of these "haze forming" proteins ranges between 12,600 and 30,000 daltons (Hsu and Heatherbell, 1987; Dubourdieu *et al.*, 1988b). To prevent this precipitation, bentonite has been used as an adsorbing agent. Although this technology is successful in removing the proteins, it is not selective and removes other desirable compounds from the wine. This lack of selectivity often affects the organoleptic qualities of the wine, especially when a high dose of bentonite is applied. Therefore, the use of a more selective technology (e.g., microbial protease) is more suitable (Blade and Boulton, 1988). However, the hydrolysis products of grape protein serve as nutrient for the bacteria involved in the second or malolactic fermentation (Feuillat *et al.*, 1980).

2. Sources of Endogenous Proteases

a. Grapes   A protease activity has been detected in grapes, but this activity is weak and is inhibited rapidly by the alcohol formed during fermentation, as well as by juice clarification (Cordonnier and Dugal, 1968; Feuillat *et al.*, 1980; Canonica and Ferrari, 1987).

b. Yeasts and other microorganisms   The proteolytic activity produced by yeast apparently is more stable than the one produced by the grapes. This enzyme is still active during alcohol fermentation. The enzyme activity varies greatly from one type of yeast to another. Positive correlation has been observed between the rapid start of malolactic fermentation and the ability of the yeast to hydrolyze protein and produce peptides that serve as nutrients for *Lactobacillus oenos*.

According to Lurton (1988), the major enzymes involved in autolysis of yeast cells are proteases, whereas liberation of mannoproteins from yeast cell walls is caused by a β-1,3-glucanase (Llaubères *et al.*, 1987; Llaubères, 1988). The diffusion of these macromolecular compounds into the wine has a positive effect on the organoleptic quality of the wine (Feuillat and Charpentier, 1982; Silva *et al.*, 1987).

A protease produced by *Botrytis cinerea* has been identified by Heale and Movahedi (1989).

C. Enzymes Involved in the Formation of C₆ Alcohols

1. Technological Relevance

Volatile C_6 alcohol compounds such as hexanol are responsible for the grassy taste of juices and wines. Crushing of the grapes leads to a greater formation of C_6 compounds than the pressing of uncrushed grapes. Juice clarification reduces the amount of C_6 compounds in the juice (Dubourdieu *et al.*, 1986; Ollivier, 1987).

Work done by Rapp *et al.* (1976), Cayrel *et al.* (1985), and Crouzet *et al.* (1985) has shown that the formation of C_6 compounds in juice requires the sequential action of four enzymes, namely, acyl hydrolases, lipoxygenase, a hydroperoxide-cleaving enzyme, and an alcohol dehydrogenase.

2. Enzyme Sources

The C_6 compounds are formed in the presence of oxygen during the prefermentation phase, after the mechanical treatment of the grapes. During alcohol fermentation, hexanal and hexenals are transformed almost entirely into hexanol and hexenols by the yeasts. Grape and yeast enzymes therefore are involved in the formation of these unpleasant grassy-tasting compounds.

D. Polyphenol Oxidases

The oxidation of phenolic compounds in the juice is catalyzed by polyphenol oxidases. One must distinguish between the grape polyphenol oxidase and polyphenol oxidase produced by *Botrytis cinerea* (Dubernet and Ribéreau-Gayon, 1973,1974; Dubernet, 1974; Dubernet *et al.*, 1977).

1. Polyphenol Oxidase of Grapes

The polyphenol oxidase of the grapes, also called tyrosinase, is mainly bound to insoluble cellular compounds such as chloroplasts. The enzyme can be removed from the juice by a clarification (natural sedimentation of the cloud or centrifugation). This enzyme is not stable under wine making conditions (pH, SO_2, alcohol). Therefore, this polyphenol oxidase can be controlled easily by clarification and SO_2 addition.

In hyperoxygenation technology, this enzyme is used advantageously to reduce the level of oxidizable polyphenols in white wines (Müller-Späth, 1989). The influence of this technology on the varietal flavors of white wines remains to be investigated more closely.

2. *Polyphenol Oxidase of* Botrytis cinerea

The polyphenol oxidase produced by *Botrytis cinerea* is called laccase. This enzyme is about 30 times more active than the tyrosinase. However, it is pH and SO_2 resistant and stable under wine making conditions. Because of its characteristics, this enzyme is considered detrimental by winemakers. Laccase is responsible for the loss of color and the browning of red wines. Therefore it is of greatest importance to understand and control this polyphenol oxidase. Inactivation usually is performed by heat treatment and/or SO_2 addition. Laccase can survive for some time in the wine. The presence and activity level of this enzyme therefore must be controlled from time to time, especially before the wine is exposed to air (racking). A selective and simple colorimetric method to determine laccase activity has been developed by the Wine Research Institute of Bordeaux (Dubourdieu *et al.*, 1984; Grassin and Dubourdieu, 1989). A test kit based on this method is available commercially (NOVO Test Botrytis® NOVO NORDISK FERMENT).

III. Use of Commercial Enzyme Preparations in Wine Making

Commercial enzymes have been developed to reinforce the endogenous enzymes present under wine making conditions. In some cases, the suitable enzymes are missing; exogenous enzymes are used to fill that gap.

Two types of commercial enzymes are used: the pectolytic enzymes and the beta glucanases. The pectolytic enzymes were introduced into wine making in the early 1970s. The betaglucanases were introduced in 1984. Pectolytic enzymes are used to improve juice extraction or the extraction of other valuable compounds, as well as to improve clarification and filtration. The beta glucanases were developed to solve specific clarification and filtration problems caused by the presence of Botrytis beta glucan.

A. Use of Pectolytic Enzymes

The potential uses of pectolytic enzymes in wine making are illustrated in this section, using a few examples.

1. White Wine Technology

a. Juice extraction Pectin is a structural compound of the cell wall. It is responsible for the firmness of the grape berries. A low free-run juice yield and a long pressing time usually are correlated with the pectin level of the grapes. The influence of enzymatic pectin hydrolysis on the juice yield is shown in two examples.

TABLE I
Influence of Enzyme Treatment (Rapidase CX)
on Juice Yield

	Control	Enzyme treated
Batch size	4000 kg	4000 kg
Free run juice and first press fraction	2305 liter	2650 liter
Second and third press fraction	630 liter	500 liter
Fourth and fifth press fraction	230 liter	150 liter
Total yield	3165 liter	3300 liter

The first example illustrates the results obtained on Kadarka, a Hungarian grape variety that has a high pectin content (Canal-Llaubères, 1989). The enzyme preparation (Vinozym® NOVO NORDISK FERMENT) is added continuously to the crushed mash while filling the press. The contact time of enzyme and mash was 1–2 hr. The mash temperature was 14°C. The amount of juice obtained in the different fractions — free-run juice including the first and second press and press juice (third to fifth press) — is indicated in Fig. 1. As we can see from the figure, the first quality juice fraction (free-run juice and first and second press) is much higher for the enzyme-treated batch than for the control.

Another trial, performed on grapes containing less pectin than Kadarka (Table I), shows similar results. This trial was performed on Spanish grapes (Rnairen, Verdejo) with Rapidase CX® (Gist brocades). Table I indicates the yield obtained for the different press fractions, as well as the total juice yield. The juice yield obtained for the first quality juice (freejuice and first press fraction) is about 66.3% for the enzyme-treated batch and 57.6% for the control. The overall juice yield is 82.5% for the enzyme-treated batch and 79.1% for the control.

These two examples demonstrate very clearly the positive influence of an enzyme treatment. In addition to the positive effect on juice extraction, the enzymatic mash treatment also improved the extraction of aromatic compounds such as terpenic substances. Table II indicates the results obtained by a mash treatment (Semillon) using Ultrazym 100® NOVO NORDISK FERMENT. The enzymatic maceration of the mash took place at 15°C over 20 hr (Ollivier, 1987).

b. Juice clarification The effect of an enzyme treatment on the juice clarification and especially on the compactness of the cloud sediment is indicated in Fig. 2. To illustrate this effect, a trial on Chasselas grape juice (low pectin level) was chosen (Canal-Llaubères, 1989). The differences obtained between treated and untreated samples are considerable. The effect would have been greater in a juice of high pectin content. As we can see in Fig. 2, the sedimentation of the cloud is much faster after enzyme treatment.

Control

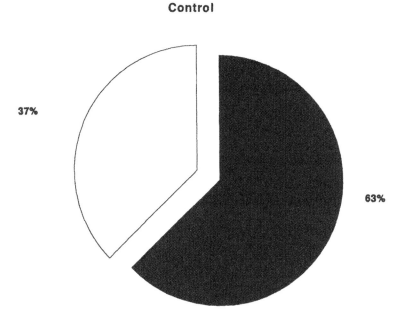

Enzyme treated (vinozym 2g/hl)

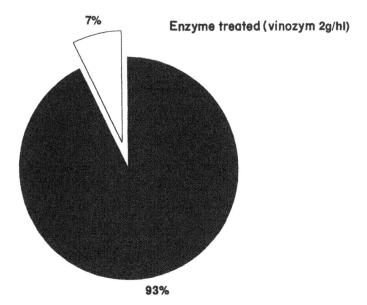

Figure 1 Influence of an enzymatic mash treatment on the free run juice (including first and second press; ■) and press fractions (third to fifth press; □).

TABLE II
Influence of Enzyme Treatment (Ultrazym 100) during Maceration on Level of Terpenic Substances in Wine (Semillon)[a]

Terpenic substance	Control	Enzyme treated (5 g/hl)
Linalol	2.5	3.1
Terpineol	0.5	0.9
Citronellol	5.3	8.3
Geraniol	6.7	7.5
Total	15.0	20.0

[a] Reprinted with permission from Olivieri, 1987.

Sedimentation of the control required 4–5 times more time without obtaining the same compactness of sediment.

In another trial performed on Chasselas, the degree of clarification of the juices was followed for 20 hr. The turbidity of the juices (supernatant) was determined after centrifugation (10 ml juice centrifuged at 3000 g for 5 min). The results obtained are indicated in Fig. 3. The clarification effect obtained after enzyme treatment is quite spectacular. Even after 20 hr, the control samples do not reach the same degree of clarification obtained by the enzyme-treated juice after 1 hr.

c. **Wine filtration** To illustrate the effect of enzyme treatment on wine filtration, we have chosen a trial performed with Sylvaner grapes (medium pectin level). The enzyme treatment was done on the mash before pressing. No further enzyme treatment took place between pressing of the grapes and filtration of the wine. Filtration was carried out with a 10-m² Kieselgur filter. The filtration behavior of both wines is indicated in Table III. As we can see from the table, the enzyme treatment allowed filtration of about 2.5 times more wine per filtration run. The economy realized by such an enzyme treatment (filtration material, filtration time, wine losses) is considerable.

2. Red Wine Technology

The classical application of pectolytic enzymes in red wine technology is the treatment of press wines to improve clarification and filtration. The results obtained are generally on the same order as or better than those reached for white wines.

Another enzyme application in red wine making is color extraction. Experimental trials on a laboratory scale (Villettaz, 1986) indicate that pectolytic enzyme preparations with a broad spectrum of activities (hemicellulases, cellulases) give better results than standard pectinases. Also it has been observed that the contact time during maceration must be about 1 week to extract valuable polyphenols that subsequently stabilize the red color com-

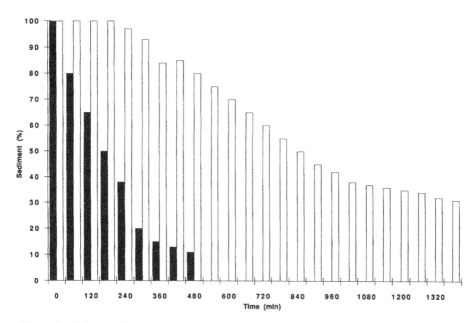

Figure 2 Influence of enzymatic treatment of the juice on the compactness of the sediment. Control, open bars; enzyme treated, shaded bars.

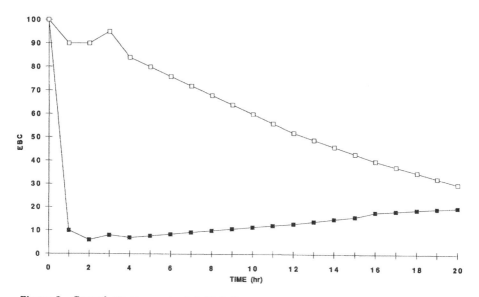

Figure 3 Control, □; enzyme treated, ■. Influence on enzyme treatment of clarification of juice. (Reprinted with permission from Llaubères, 1989.)

TABLE III
Effect of Enzymatic Treatment (Vinozym, NOVO NORDISK FERMENT) of a Sylvaner Grape Mash on Filterability of the Wine[a]

Filtration time (hr)	Filtration behavior of the wine			
	Enzyme treated		Control	
	Differential pressure (bar)	Volume filtered (liter)	Differential pressure (bar)	Volume filtered (liter)
0.5	3.0	3,500	3.0	3500
1.0	3.0	7,000	6.0[b]	6500
1.5	3.0	10,500		
2.0	3.0	13,500		
2.5	6.0[b]	15,500		

[a] Filtration over Kieselgur (10 m²).
[b] Filtration surface clogged; filtration run stopped.

pounds. Thus, color intensity is not correlated with color stability. Well-colored wines can be obtained after 2–3 days maceration with enzyme treatment, but experience has shown that such wines generally lose a great amount of color after malolactic fermentation. Therefore it is necessary to perform the maceration for 8–10 days to achieve good color stability. More work is necessary to understand and master the color stability problems in red wine technology.

B. Use of β-Glucanases

Wines obtained from botrytized grapes (grey mold or noble mold) are always difficult to clarify and to filter because of the presence of a high molecular weight β-glucan (in the past, erroneously called dextran) produced by *Botrytis cinerea* (Dubourdieu, 1981).

The presence of Botrytis glucan can be detected by the following specific and very simple test. Of the wine to be tested, 10 ml is poured into a test tube. Then, 5 ml 96% alcohol (industrial or denatured alcohol) are added to the wine. The solution is shaken lightly just to insure thorough mixing of the two liquids. If more than 15 mg glucan per liter are present in the wine, a filamentous precipitate will become visible within minutes.

To detect lower quantities (3 mg/liter and above), the following test has been developed. In a conical centrifuge tube, 5 ml 96% alcohol are added to 5 ml wine. The contents of the tube are mixed and left to stand at room temperature for 30 min. The mixture is centrifuged at 3000 RPM (700–800 *g*) for 20 min. The supernatant is decanted carefully. The sediment is dissolved in 1.0 ml water, 0.5 ml 96% alcohol is added. The formation of a filamentous precipitate indicates the presence of small amounts of Botrytis glucan. This test is somewhat more laborious than the first one but is, nevertheless, highly recommended, since filtration problems already start with the presence of 3–5 mg glucan per liter.

Specific clarification and filtration problems cannot be mastered by centrifugation, fining, or addition of pectolytic enzymes. The development of a specific β-glucanase has solved all glucan-related problems in a spectacular way. The effects obtained by this kind of preparation are shown in the next two examples.

A Bordeaux sweet white wine containing 22 mg glucan per liter was divided into two portions. The first portion was treated enzymatically (Glucanex,® NOVO NORDISK FERMENT) whereas the second portion was kept untreated as control. The filtration took place over Kieselgur (filtration surface, 4 m²). The filtration conditions were the same for both wines. The results obtained are summarized in Fig. 4. As we can see from this trial, the enzyme treatment allowed the filtration of almost 10 times more wine per filtration run (Villettaz *et al.*, 1982).

The second example shows the filtration behavior over filter pads. In this case, both wines (the enzyme-treated and control wines) were centrifuged prior to filtration. The results obtained are indicated in Figs. 5 and 6. This trial demonstrates that centrifugation does not remove the glucan and, consequently, does not solve the filtration problems (Villettaz *et al.*, 1984).

Glucanase treatment solves glucan filtration problems in a spectacular way. It permits better filtration at lower pressure and reduces the filtration

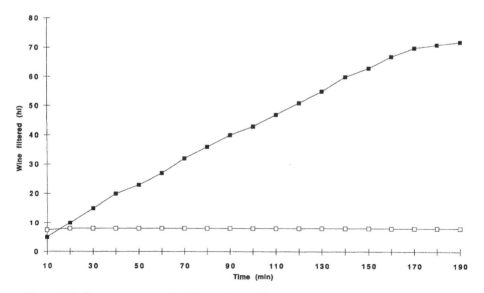

Figure 4 Influence on enzymatic β-glucan degradation on wine filterability. Control, □; enzyme treated, ■.

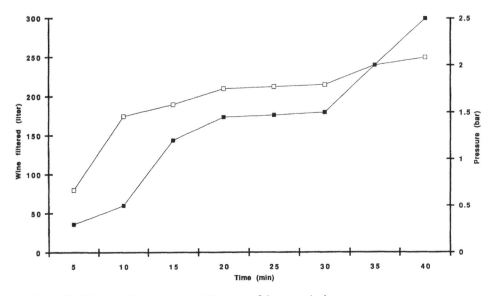

Figure 5 Filtration (□) and pressure (■) curves of the control wine.

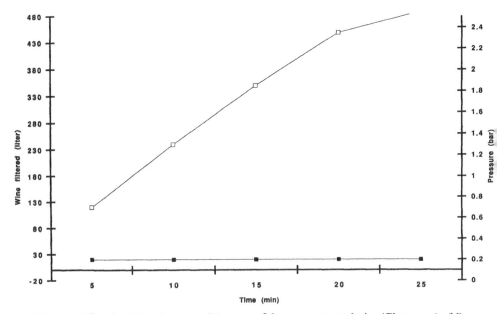

Figure 6 Filtration (□) and pressure (■) curves of the enzyme treated wine (Glucanex, 1 g/hl).

costs drastically. The organoleptic properties of the wine are not affected by such a treatment.

IV. Future Developments

The development of pectolytic enzymes with a broad spectrum of activities (hemicellulases, cellulases) for color extraction in red wine technology is certainly one of the projects that will be carried out in the near future.

As already mentioned, the replacement of bentonite by a microbial protease for the stabilization of white wines against protein precipitation remains a desirable goal. Considerable research work is presently being done in this field.

Another interesting subject is the hydrolysis of bound terpenes (terpenes bound to sugar) to reinforce the aromatic profile of certain wines. This treatment can be done using different glucosidases in a sequential order (Gunata et al., 1990). All the microbial glucosidases tested to date are inhibited by the presence of glucose. Therefore, such enzymes cannot be used for the treatment of grape juices. In the wine treatment, the majority of enzyme reactions take place after alcohol fermentation. Several research groups are presently working on this project (Dubourdieu et al., 1988a; Grossmann and Rapp, 1988; Gunata et al., 1988; Cordonnier et al., 1989; Cuenat et al., 1992).

It is possible, by means of glucose oxidase/catalase treatment of grape juice, to produce a high amount of gluconic acid in the wine. Since gluconic acid is not metabolized by wine yeasts into alcohol, the fermentation of such juices can lead to the production of low alcohol wines. In this case, the excess of gluconic acid must be removed. If gluconic acid is left in the wine, the wine will be used as an "acid reserve" to blend with acid deficient wines, allowing production of better balanced wines. Glucose oxidase/catalase-related technologies are not optimized to date and must be developed further (Villettaz, 1986).

References

Blade, H., and Boulton, R. (1988). Adsorption of protein by bentonite in two model wine solution. *Am J. Enol. Vitic.* **39,** 193–199.

Canal-Llaubères, R. M. (1989). Les enzymes industrielles dans la biotechnologie du vin. *Rev Oenolog.* **53,** 17–22.

Canonica, B., and Ferrari, G. (1987). "Mémoire Fin D'étude." Enita, Dijon, France.

Cayrel, A., Crouzet, J., Chan, H. W. S., and Price, K. R. (1983). Evidence for the occurence of lipoxygenase activity in grapes (variety carigrare). *Am. J. Enol. Vitic.* **34,** 64–77.

Cordonnier, R., and Dugal, A. (1968). Les activités protéolytiques du raisin. *Ann. Technol. Agric.* **17,** 189–206.

Cordonnier, R., Gunata, Y. Z., Baumes, R., and Bayonove, C. (1989). Recherche d'un matériel enzymatique adapté à. l'hydrolyse des précurseurs d'arôme de nature glycosidique du raisin. *Conn. Vigne Vin* **23**(1), 7–23.

Crouzet, J., Nicolas, M., Molina, J., and Valentine, G. (1985). *In* "Progress in Flavour Research" (J. Adda, ed.), pp. 401–408. Elsevier Science, Amsterdam.

Cruess, W. W., and Besone, J. (1941). *Fruit Prod. J. Am. Vin. Ind.* **20**, 365.

Cuenat, P., Canal-Llaubères, R. M., and Leyat, C. (1992). *Rev. Suisse Vitic. Arb. Hortic.* **24**, 73–74.

Demain, A. L., and Phaff, H. J. (1954). Hydrolysis of the oligogalacturonides and pectic acid by yeast polygalacturonase. *J. Biol. Chem.* **210**, 381–393.

Dubernet, M. (1974). "Recherches sur la Tyrosinase de Vitis Vinifera et la Laccase de *Botrytis cinerea*." Applications technologiques, Thèse 3ème cycle, Université Bordeaux II.

Dubernet, M., and Ribéreau-Gayon, P. (1973). Les polyplenoloxydases du raisin sain et du raisin parasité par Botrytis cinerea. *R. Acad. Sci.* **277D**, 245–470.

Dubernet, M., and Ribéreau-Gayon, P. (1974). Isoelectric point changes in vitis vinifera catechol oxidase. *Phytochem.* **13**, I-085–I-087.

Dubernet, M., and Ribéreau-Gayon, P. (1975). Etude de quelques propriétés caractéristiques de la laccase de Botrytis cinerea. *C. R. Acad. Sci.* **280D**, I-313.

Dubernet, M., Ribéreau-Gayon, P., Lerner H. R., Harel, R., and Mayer, A. M. (1977). Purification and properties of laccase from Botrytis cinerea. *Phytochem.* **16**, 191–193.

Dubourdieu, D., Fournet, B., Ribéreau-Gayon, P. (1981). Identification d'un glucane sécrété dans la baie de raisin par *Botrytis cinerea. Carbohydr. Res.* **93**, 294–299.

Dubourdieu, D., Grassin, C., Deruche, C., and Ribéreau-Gayon, P. (1984). Mise au point d'une mesure rapide de l'activité laccase dans les moûts et dans les vins par la méthode à la syringaldazine. Application à l'appréciation de l'état sanitaire des vendanges. *Conn. Vigne Vin* **18**, 237–252.

Dubourdieu, D., Ollivier, C., and Boidron, J. N. (1986). Incidence des opérations préfermentaires sur la composition chimique et les qualités organoleptiques des vins blancs secs. *Conn. Vigne Vin* **20**, 53–76.

Dubourdieu, D., Darriet, P., Ollivier, C., Boidron, J. N., and Ribéreau-Gayon, P. (1988a). Rôle de la levure S. cerevisiae dans l'hydrolyse enzymatique des hétérosides terpéniques du jus de raisin. *C. R. Acad. Sci. Paris* **306**, 489–493.

Dubourdieu, D., Serrano, M., Vannier, A. C., and Ribéreau-Gayon, P. (1988). Etude comparée des tests de stabilité protéique. *Conn Vigne Vin* **22**, 261–273.

Dubourdieu, D., Darriet, P., and Chatonnet, P., and Boidron, J. N. (1989). Intervention de systèmes enzymatiques de Saccharomyces cerevisiae sur certains précurseurs d'arômes du raisin. *In* "Actualités Oenologiques," IV Symposium International d'Oenologie de Bordeaux. pp. 151–159. Dunod, Paris.

Feuillat, M., and Charpentier, C. (1982). Autolysis of yeast in champagne. *Am. J. Enol. Vitic.* **33**, 5–13.

Feuillat, M., Brillant, G., and Rochard, J. (1980). Mise en évidence d'une production de protéases exacellulaires par les levures au cours de la fermenttion alcoolique du moût de raisin. *Conn. Vigne Vin* **14**, 37–52.

Grassin, C. (1987). "Recherche sur les Enzymes Extracellulaires Secrétées par *Botrytis cinerea* dans la Baie de Raisin. Applications Oenologiques et Phytopathologiques." Ph.D. Thesis. Université de Bordeaux II.

Grassin, C., Dubourdieu, D. (1989). Quantitative determination of *Botrytis* laccase in musts and wines by the syringaldazine test. *J. Sci. Food Agric.* **48**, 369–376.

Grossmann, M., and Rapp, A. (1988). Steigerung des sortentypischen Weinbuketts nach Enzymbehandlung. *Dtsch. Lebensm. Rdsch.* **84** (2), 35–37.

Gunata, Y. Z., Bitteur, S., Brillouet, J. M., Bayonove, C., and Cordonnier, R. (1988). Sequential enzymic hydrolysis of potentially aromatic glycosides from grapes. *Carboh. Res.* **184**, 139–149.

Gunata, Y. Z., Dugelay, I., Sapis, J. C., Baumes, R., and Bayonove, C. (1990). Action des gluycosidases exogènes au cours de la vinification: libération de l'arôme à partir de précurseurs glycosidiques. *Conn. Vigne Vin* **24**, 133–144.

Heale, J., and Movahedi S. (1989). Primary role of an aspartic proteinase enzyme produced by Botrytis cinerea in causing cell death in plant tissues. *In* "Actualités Oenologiques," IV Symposium International d'Oenologie de Bordeaux. pp. 127–132, Dunod, Paris.

Hsu, J. C., and Heatherbell, D. A. (1987). Heat unstable proteins in wine. I. Characterization and removal by bentonite fining and heat treatment. *Am. J. Enol. Vitic.* **38**, 11–16.

Llaubères, R. M. (1988). "Les Polysaccharides Sécrétés dans les Vins par *Saccharomyces cerevisiae* et *Pédiococcus* species." Ph.D. Thesis. Université de Bordeaux II.

Llaubères, R. M., Dubourdieu, D., and Villettaz, J. C. (1987). Exacellular polysaccharides from S cerevisiae in wine. *J. Sci. Food Agric.* **41**, 277–286.

Lurton, L. (1988). "Etude de la Proteolyse Intervenant au Cours du Processus d'Autolyse Chez *Saccharomyces cerevisiae*. Applications Oenologiques." Ph.D. Thesis. Université de Dijon.

Marteau, G., Scheur, J., and Olivieri, C. (1961). *Ann. Technol. Agric.* **10**, 161–183.

Montedoro, G., and Bertucioli, M. (1985). IV Symposium International d'Oenologie, Valence.

Müller-Späth, H. (1989). Vinification en blanc. Influence de l'oxygène et de la température avant la fermentation. *In* "Actualités Oenologiques," IV Symposium d'Oenologie de Bordeaux. pp. 139–145. Dunod, Paris.

Olivieri, C. (1975). Considération sur l'évolution des activités enzymatiques lors du traitement thermique de la vendange à différents pH. *Prog. Agric. Vitic.* **7**, 225–230.

Ollivier, C. (1987). "Recherche sur la Vinification des Vins Blancs Secs." Master's Thesis. Université de Bordeaux II.

Patel, D. S., and Phaff, H. J. (1959). *J. Biol. Chem.* **234**, 237–241.

Rapp, A., Hastrich, H., and Engel, L. (1976). Gaschromatographische Untersuchungen über die Aromastoffe von Weinbeeren. *Vitis* **15**, 183–192.

Silva, A., Fumi, M., Montesissa, G., Colombi, M. G., and Cologrande, O. (1987). Effect of storage in the presence of yeasts on the composition of sparkling wines. *Conn. Vigne Vin* **3**, 141–162.

Villettaz, J. C. (1984). Les enzymes en oenologie. *Bull Office International de la Vigne et du Vin* **57**, 19–29.

Villettaz, J. C. (1986). A new method for the production of low alcohol wines and better balanced wines. Proceedings of the Sixth Australian Wine Industry Technical Conference, Adelaide (T. H. Lee, ed.). pp. 125–128. Australian Industrial Publisher.

Villettaz, J. C., and Dubourdieu, D. (1991). Enzymes in wine-making. *In* "Food Enzymology" (P. F. Fox, ed.), Vol. 1. pp. 427–453. Elsevier Applied Science, Amsterdam.

Villettaz, J. C., Dubourdieu, D., Lefèbvre, A. (1982). L'emploi des beta glucanases en oenologie. *In* "Utilisation des enzymes en technologie alimentaire" (P. Dupuy, ed.). pp. 457–462, Lavoisier, Paris

Villettaz, J. C., Steiner, D., Trogus, H. (1984). The use of a beta glucanase as an enzyme in wine clarification and filtration *Am. J. Enol. Vitic.* **35**, 253–256.

CHAPTER 17

Enzymes In Brewing

JOSEPH POWER

I. Introduction

Beer is a beverage made by alcoholic fermentation of a sugar solution derived from grain. Most carbohydrates in grain is present in the form of starch, which cannot be fermented by brewer's yeast. Before beer can be produced, the starch of the grain must be transformed into fermentable sugar by enzymatic digestion.

The brewing of beer is an ancient enzymatic process that was developed, as were other traditional processes, without knowledge of the existence and nature of the enzymes involved. This chapter begins with an examination of the conventional brewing processes established long ago and reviews the enzymatic reactions involved.

By the beginning of the twentieth century, enzymes were recognized as part of yeast, malt, and other living organisms. Many enzymes became available in commercial quantities. The application of enzymes from nontraditional sources made possible many changes in the brewing industry. In 1911, Leo Wallerstein demonstrated that, by adding papain from papaya to beer, the beer could be kept clear during shipment over long distances and during storage for long periods of time. This discovery was critical to the establishment of larger brewing companies that market their packaged beer over expanded geographical areas.

One of the most significant recent developments in the brewing industry is the use of microbial enzymes for the production of lower calorie or "light" beers. Over the last 20 years, these beers have become an important part of the market, $\sim 30\%$ of the total volume in the United States (Reid, 1992). The success of enzymatically produced "light" beer has led to changes in marketing strategy throughout the industry and has encouraged research into the development of other new products.

Other applications of enzymes not native to the brewing process have been developed and have found industrial application. However, none have had as dramatic an effect on the industry as the applications just described. The use of enzymes to process cheaper or more readily available raw materials has allowed their incorporation into beer in place of more expensive traditional barley malt. Other enzymes have been found that facilitate filtration of the cloudy beer liquid immediately after fermentation to yield a clear final product. Enzymes that react with the small amount of oxygen found in packaged beer have been used to extend the shelf life.

II. Malt and Its Reactions during Brewing

Malt is grain that has been moistened and allowed to germinate under controlled conditions and is then dried to a stable form that is rich in enzymes but similar to the original grain in external appearance. Malt is stable; it can be shipped over long distances and stored for long periods of time. The most commonly malted grain, and the primary grain used in brewing, is barley. Barley is preferred for many reasons, two of the most important of which are that barley has an adhering husk that protects the malt and acts as a filter aid during separation of the grain from the liquid after brewing and that barley has a good ability to form enzymes during germination.

During germination of the barley, several important hydrolytic enzymes are produced or released. Some enzyme hydrolysis of the grain occurs during the several days of germination. Phytase digests much of the phytin of the grain, releasing phosphates and associated metallic minerals that are important mineral nutrients for yeast during fermentation. Hemicellulases are produced that degrade much of the cell wall structure of the endosperm tissue, which holds most of the starch in the grain. The chief component of the barley endosperm cell wall is $(1\rightarrow3)$ and $(1\rightarrow4)$-β-D-glucan. A number of enzymes generally referred to as β-glucanases are involved in degradation of cell wall hemicellulose. Usually, degradation of the cell walls is extensive, leading to the production of a malt that is called "modified", that is, less rigid in structure than the original grain and capable of releasing soluble components relatively easily when used for brewing. A variety of proteases is produced also that begins breaking down protein in the grain. Insoluble storage protein such as hordein is converted partially to large soluble peptides and to smaller peptides and amino acids. The larger peptides are important in brewing as a source of protein, an important component of the foam formed when a beer is poured. The smaller digestion products are an important source of nitrogen for growth of the yeast during beer fermentation.

Starch digestion is very limited during germination. Barley α-amylases are formed and digest the starch to a limited degree. The bound β-amylase found in ungerminated barley is released by the action of proteases during germination, but the β-amylase is not active on the native starch granules found in the germinating barley.

During germination, enzyme hydrolysis is minimized to obtain sufficient modification of the malt. The definition of sufficient modification varies depending on the type of malt required. Ale malts of the type classically used in England are very thoroughly modified because the traditional ale brewing equipment offers little of the flexibility required for extensive enzyme hydrolysis when the malt is used for brewing. Extended hydrolysis during germination results in higher levels of sugars and amino compounds that are converted to colored products, melanoidins, by browning reactions when the malt is dried in a kiln. Traditional ale malt gives a relatively dark-colored beer. The popular Pilsner style of beer, which is as pale as possible, requires much less color development in the malt. The malts used to make pale lager and Pilsner style beers are, therefore, less modified during germination. Thus, they require a more intense enzymatic hydrolysis during brewing, which is brought about by a more complicated time–temperature schedule.

The extent of germination also influences the amount of enzyme present in malt. Pale malts with low modification intended for Pilsner production have lower enzyme levels than a darker distiller's malt. Many of the enzymes in the germinated barley or green malt, particularly the α- and β-amylases, survive the kilning process and are activatable in the brewing process. Several publications (Palmer, 1989; Bamforth and Quain, 1989) discuss the biochemistry of malt in detail.

Brewing beer begins with making a mixture of hot water and grain, called a mash. The mash usually contains malt as the major grain; the malt acts as a source of enzymes for brewing. The mash undergoes a program of varied time and temperature to accomplish the desired hydrolysis and solubilization of the grain components. At the end of the mash program, the liquid contains dissolved sugars and other compounds derived from hydrolysis of starch and other components of the grain. The liquid, called wort, is separated from the remaining undigested or spent grain by filtration, using the hulls of the malt to form the filter bed. A gravity filter, called a Lauter bed, or a special plate-and-frame wort filter that operates with pressure provided by a pump may be used. The wort is boiled with flavoring hops, cooled, and fermented after the addition of yeast and oxygen. The mash program determines some of the basic properties of the beer.

To produce a Pilsner style beer from pale malt alone, at least three temperature stages are used (Fig. 1). The mash begins at the lowest temperature, often called the "mashing in" or "proteolytic" temperature. As the name suggests, the most important function of this step is proteolysis of the malt, to break down insoluble protein in the malt to soluble peptides of varied molecular weights and even to free amino acids. The temperature used may vary from 40°C to as high as 52°C. The time may be 15 min to 1 h. Several enzymes from malt, both endopeptidases and carboxypeptidases, are involved. The temperature stability of the enzymes determines the range in

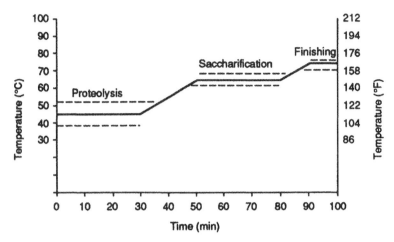

Figure I Typical time and temperature mash program for the production of a lager beer. Dotted lines indicate maximum and minimum temperature for different stages.

which they are effective. The endopeptidases tend to be stable up to ~50°C, whereas the carboxypeptidases are stable to ~55°C (Moll *et al.*, 1981). The aminopeptidases of barley are thermolabile; little activity for use in the mash remains in malt after kilning. Soluble nitrogenous products of these enzymes are important in supporting yeast growth during fermentation as well as in forming beer foam. Considerable amounts of soluble nitrogenous compounds will have been formed already during germination of the barley; only about one-third of the total nitrogen in the wort is released during this stage of the mash. For beers made with malt as the only grain, or made with a high percentage of malt, this stage of the mash is not very critical and may be abbreviated or run at the high end of the temperature range given. Occasionally, malts with a low degree of modification may require additional action of β-glucanase or phytase during the mashing process. These malt enzymes are more heat labile than the proteases, and their reactions are best carried out at the low end of the temperature range shown (40°C; Fig. 1). Normally malt will have completed most of the reactions that occur at this temperature to a sufficient degree earlier during germination of the barley. In the early twentieth century, barley varieties did not have as great an enzyme forming capacity as present day varieties, so low temperatures and long mashing times were common. In some breweries, the practices have not changed although the malt has.

After completion of the proteolytic stage, the mash is raised to a higher temperature. Increase in temperature can be accomplished by removing part of the mash, bringing it to a boil in another vessel, and returning it to the original vessel. The resulting rise in temperature can be controlled accurately using rudimentary equipment. The method is called decoction mashing and is a very old approach, predating the availability of thermometers for temperature control. More commonly, the temperature of the entire mash is raised at a controlled rate using steam heat or even direct flame on the vessel in a process called programmed infusion mashing. Regardless of the method used to increase the temperature, the mash is brought to the next stage, commonly referred to as the saccharification stage (Fig. 1), at a temperature of 60– 70°C.

Prior to the saccharification stage, the starch in the malt remains largely intact because it has not yet been gelatinized and native starch is not hydrolyzed readily by enzymes. The starch of barley malt gelatinizes at a temperature of 60°C. Once gelatinized, the starch is digested rapidly by the α- and β-amylases present in the malt. In the saccharification temperature range, both types of amylase are active. Their combined action results in production of large quantities of fermentable sugar, especially maltose, which is the only sugar produced by β-amylase; hence, this stage receives the name saccharification stage. Smaller amounts of glucose and maltotriose, which are also fermentable by yeast, are formed from starch by α-amylase, in addition to larger nonfermentable dextrins.

The saccharification stage seldom lasts longer than 30 min. At temperatures above 60°C, which are required to obtain gelatinization of malt starch, the malt amylases, especially β-amylase, are not very stable. During saccharification at 65°C, β-amylase is inactivated almost completely (Narziss, 1976; Moll *et al.*, 1981); within 30 min (Fig. 2). The α-amylase also is deactivated significantly (Fig. 2), but some of this enzyme is still present at the end of saccharification.

Some of the malt starch usually remains ungelatinized at the saccharification temperature. The saccharification stage represents a compromise between the need for higher temperatures to obtain gelatinization and lower temperatures to preserve enzyme activity. The temperature chosen results in optimal sugar formation but incomplete starch degradation. Starch degrada-

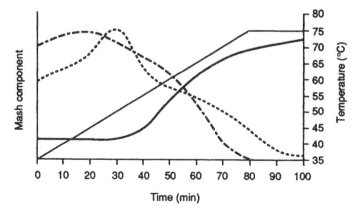

Figure 2 Concentration of mash carbohydrate (——) and carbohydrases (α-amylase, - - -; β-amylase, — · —), in the liquid phase during a laboratory mash. Data from Moll *et al.*, 1981.

tion must be finished by raising the mash temperature again so gelatinization can be completed. The finishing temperature is at least 70°C, but below 80°C. Again, the exact temperature is a compromise selected to be as high as possible while still permitting significant activity of the malt α-amylase (Fig. 2). The small amount of α-amylase still present is sufficient to hydrolyze starch to soluble dextrins, but not to form much more fermentable sugar.

The basic program for an infusion mash outlined here is modified at different breweries to provide different types of beers. By adjusting the saccharification temperature, smaller or larger proportions of fermentable sugar will be formed in the wort, giving beers with different ratios of alcohol to nonfermented dextrins. Intermediate steps may be used also. A complicating factor in understanding and predicting the rates of enzyme reactions in a mash is, as Narziss (1976) points out, that the enzymes are present inside the structure of the malt and time is required for them to become soluble and able to act on their substrates. Soaking at a lower temperature will affect reaction rates at a higher temperature. Brewers' experience suggests that 63°C is optimum for fermentable sugar formation and that a temperature as high as 73°C can be used for simultaneous saccharification and finishing to give a wort with the lowest percentage of fermentable sugar. By changing the temperature schedule of a mash, the percentage of fermentable sugar in wort can be varied from about 50% to 75% of total dissolved solids using the malt enzymes. This range will give beers of different alcohol contents. Most beers are fermented from worts with 65–70% fermentable sugar. The presence of some preformed sugars in the malt makes it difficult to reduce the percentage of fermentable sugar in a wort and, therefore, to make a very low alcohol beer.

The traditional ale brews of England were made with a very well-modified malt. The reactions of the proteolytic stage of the mash are completed sufficiently during germination of the malt. Therefore, the lower temperature stage can be omitted. A traditional ale mash begins at a saccharification temperature of about 65°C. The mash can be warmed to finishing temperature by using hotter water to sparge or rinse wort from the grains. Therefore, there is no need for equipment to heat the mash, which means that ale breweries can be simpler and less expensive to build.

III. Use of Brewer's Adjuncts

The preceding discussion of brewing reactions describe reactions that occur in a brew made with malt alone. Historically, beers were made using only malted grain, but brewers found that at least small amounts of unmalted grain could be used also. By the end of the nineteenth century, brewers determined that malt was the source of whatever converted the starch of other grain to fermentable sugar. The malt sugar-producing enzyme diastase was one of the first enzyme systems recognized. The ability of malt to convert unmalted grains, called adjuncts, was recognized clearly in the United States. The varieties of six-row malting barley grown in most of the country have high protein levels and produce malts with enzyme capacities much higher than those necessary to process the malt starch alone. The malts are well suited to use with adjuncts because of this excess enzyme. The process of making whiskey from corn using a small amount of malt as an enzyme source, commonly used in the United States, suggested that corn could be used to make beer also. Experimentation and publications, especially by Anton Schwarz and John E. Siebel (Max Henius Memoir Committee, 1937) popularized and promoted the use of adjuncts, especially corn and rice, in brewing beer. Handke (1896) described production of a beer in Milwaukee for which corn was used to provide 45% of the extract or soluble material. This amount is essentially the same as the practical limit on the percentage of adjunct material used with malt today. By the end of the nineteenth century, the use of adjuncts digested by malt enzymes in making beer had been perfected.

An important aspect of the use of unmalted grain as an adjunct for brewing is that the most commonly used grains, corn and rice, have a higher gelatinization temperature than barley malt. From the preceding discussion of malt reactions in brewing, it is obvious that malt starch gelatinizes at a temperature just low enough to allow the amylases of malt to digest the starch. The gelatinization temperatures of corn, about 70°C, and rice, 70–80°C, are so high that malt β-amylase cannot provide sufficient activity for normal sugar formation as the starch gelatinizes. Malt α-amylase will, however, be active for a short time when corn and even rice starch gelatinizes. The activity of malt α-amylase permits the use of these grains as adjuncts. Because of their higher gelatinization temperatures, corn, rice, and sorghum must be boiled to be used completely and efficiently. Boiling is carried out in a separate vessel, a cooker. A mash containing a small amount of malt as a source of enzyme and the ground unmalted grain, preferably degerminated, is made and heated to a temperature at which the adjunct starch just begins to gelatinize. Malt α-amylase hydrolyzes the starch so excess thickening does not occur; enzyme action is facilitated by holding this temperature for a short period of time, perhaps 10 min. The temperature is then increased to boiling to complete gelatinization of the enzymatically thinned mixture. After thorough boiling, the cooked adjunct mash is added to and mixed with a separate mash containing most of the malt which is proteolyzing at a lower temperature. The combined mashes are sized most conveniently so the complete mixture is at saccharification temperature. Otherwise temperature adjustment will be necessary. The amylases present in the enzyme-rich six-row barley malt are sufficient to complete digestion of the starches of both malt and adjunct grains at the same temperature as an all-malt brew.

Malt α-amylase is not the ideal enzyme for thinning the starch solution that develops as adjuncts are cooked. The amylase is stable enough to work fairly well with corn starch. However, when using rice, which usually has a higher starch gelatinization temperature, sufficient thinning becomes difficult to obtain; production problems are not uncommon. Rice for brewing is milled rather finely before it is cooked to make the starch more accessible to the amylase, but the enzyme is barely stable enough to digest the gelatinizing starch sufficiently to keep the adjunct mash from being too viscous to handle. The availability of amylases from the bacteria *Bacillus subtilis* and *Bacillus licheniformis* offers the brewer an alternative source of amylase for thinning the cooking adjunct. These enzymes, especially the thermostable *Bacillus licheniformis* enzyme, are more stable than the malt enzyme and are more effective in reducing viscosity build-up. Some, but not most, breweries have adopted one of the bacterial enzymes for use in adjunct cooking. Commercial literature on these enzymes suggests use of 0.5–1 kg standard enzyme solutions for 1000 kg adjunct grain. At this rate, these enzymes are both economical and effective.

Cooked adjuncts such as corn and rice are not subjected to digestion by the malt at temperatures suitable for protease activity, so they have insignificant amounts of soluble nitrogen compounds. As a result, these starchy adjuncts contribute no soluble nitrogen compounds to the wort. Wort and beers made with these adjuncts will have lower nitrogen content than beers made from malt only. This characteristic can be an advantage in preserving clarity of the beer after it is packaged because higher molecular weight beer peptides tend to form hazes as the beer ages, and haze formation becomes a problem. If the malt provides sufficient nitrogen for the yeast to grow during fermentation, a good tasting beer can be made with adjunct.

A few grains, such as barley and wheat, contain starch with a gelatinization temperature low enough that the grains can be converted by malt enzymes without prior boiling. The grains are mixed with malt in the mash and can be converted to sugar by the malt enzymes using the temperature that would be used for an all-malt mash. The problem with using unmalted wheat and barley arises in filtration to separate wort from grains or yeast from beer. The hemicelluloses of the unmalted grains tend to obstruct filtration. If more than 10% of these grains are used as adjuncts, an additional enzyme such as glucarase must be added to digest hemicellulose (Ducroo and Delecourt, 1972).

IV. Brewing with Very High Adjunct Levels

The practical limit for the amount of starchy adjunct that can be used to make a beer is determined by the failure of common adjuncts to contribute nitrogen to the wort. Experimental work has shown that the amylase content of malt is sufficient to convert the starch of adjuncts to normal amounts of sugar, even when only one-third of the sugar comes from malt (Denault *et al.*, 1981). However, the practical minimum is 50–55% from malt. Malt is a source of the nitrogen and minerals that the yeast requires for growth during a normal fermentation. Because the starchy adjuncts contribute only sugars and dextrin to the wort, the nitrogen and minerals become diluted by the adjunct. The nutrient that first becomes critical is the nitrogen. The flavor that a yeast imparts to a beer depends, among other factors, on the type of amino acids the yeast uses for growth. When crucial amino acids are in short supply, the yeast begins to form higher quantities of off-flavors caused by compounds such as diacetyl or hydrogen sulfide. At some level, the flavor change in the beer becomes unacceptable. The critical point and the degree of flavor change depends on the yeast strain used, but a level of free amino nitrogen of about 120 mg/liter in wort before it is fermented is necessary to produce an acceptable beer flavor. To provide a margin of safety, a level of 150 mg/liter is prudent (Button and Palmer, 1974). This level is reached when 50% of the solids in wort come from a high-nitrogen, high-enzyme six-row barley malt.

To use more than 50% adjunct, more nitrogen than can be provided by malt and its enzymes must be supplied. The malt contains more protein than is released in the mash; only ~40% of the total protein becomes soluble. Addition of protease from another source is one method that has been used to provide more nitrogen from the protein present in the malt. Plant sulfhydryl proteases, ficin and papain, have been added to the mash successfully to increase the levels of soluble nitrogen in the wort. Although the major flavor problems of making a very high adjunct beer are solved by this approach, the application has never become common commercial practice. Extreme dilution of some minor flavor components that come from the malt seems to be a problem.

Another procedure for making beer that contains only a small percentage of malt was worked out in detail by Nielsen (1971) and Weig (1973). Unmalted barley is used to replace a major portion of the malted barley. The barley provides some of the flavor normally contributed by malt. The unmalted barley must undergo many of the enzymatic changes that occur during several days of germination in a short period of time in a mash for this approach to be successful. The major changes that occur during formation of the malt are the degradation of the endosperm cell walls, the digestion of protein to form lower molecular weight nitrogen compounds, and the formation of α-amylase and release of β-amylase, which are needed for sugar formation. The major and most efficient enzyme involved in cell wall degradation is endo-β-glucanase. Alone, this enzyme significantly digests barley cell walls without the full complement of glucanases found in germinating barley. Bacterial α-amylase from *Bacillus subtilis* reacts with starch at the same temperatures, or slightly higher, as the malt α-amylase and is a suitable substitute for this malt enzyme. Barley possesses a high level of β-amylase in bound form. The bound enzyme can be released by proteolytic action. Any of a number of proteases are suitable for this reaction. Since *Bacillus subtilis* produces both endo-β-glucanase and α-amylase, a protease also produced by this organism is the most efficient choice. *Bacillus subtilis* actually produces two different proteases, an alkaline protease and a neutral protease. The alkaline protease is susceptible to an inhibitor present in barley and is not useful in brewing (Munck *et al.*, 1985). The neutral protease is very effective in digesting barley protein and is suitable for use in brewing. Thus, all the necessary enzymes can be produced by the same organism; they can be produced simultaneously in the same bacterial fermentation.

The final mixture of enzymes contains all the basic enzyme activities for making a malt substitute from barley. The mixed enzymes have been successful in producing beer from a mash containing as much as 90% unmalted barley (Nielsen, 1971). Further experience has shown that some varieties of barley are more amenable to enzymatic digestion than others. The enzymatic process has been accepted more in areas of the world where these barley varieties are grown. Beers made with barley as an adjunct have many of the minor constituents found in malt, for example, polyphenols, but objective evaluation still finds that their flavor is not quite the same as that of beers made with malt as the major component. Brewing with barley and enzymes is still a rather rare procedure.

V. Keeping Beer Clear with Chillproofing Enzymes

The major use of non-malt enzyme in the brewing industry is as enzyme that is added to fermented beer to "chillproof" the beer. Beer is fermented and aged under chilled conditions. Almost all packaged beer is filtered while cold to achieve clarity. In spite of this filtration, the beer becomes cloudy after it is packaged, distributed to customers, and chilled again for serving. The cloudiness that develops is caused by formation of haze particles, termed "chill haze". The haze that develops results largely from the interaction of beer peptides with polyphenolic procyanidins (Hough *et al.*, 1982). Other material, such as carbohydrate or metal ions, also may be incorporated into the haze. The amount of haze varies with factors such as the type of beer and increases with age, exposure to oxygen, and agitation during shipping. Eventually the beer becomes hazy even at warm temperature and a sediment may develop. The haze is not desirable for aesthetic reasons: it resembles cloudiness produced by microbial spoilage.

In a series of patents, Leo Wallerstein (1911) proposed the use of proteolytic enzymes to prevent development of chill haze in beer. In a highly informative publication, Wallerstein (1961) looked back at the response to and development of his very successful patented process exactly 50 years after the patents had been awarded. By the early 1900s, advances in bottling and the introduction of pasteurization created the potential for greatly increased sales of bottled beer. Packaged foods in general were becoming popular. Unfortunately, the beer could not be shipped great distances or stored for long periods of time because of problems with chill haze. The brewing industry needed a solution to this problem. In 1909 and 1910, the United States Brewmasters Association offered cash awards for the best papers submitted on the causes of haze formation in beer. In 1911, Wallerstein's patents and a paper presented at the Second International Brewers' Conference in Chicago announced that addition to the beer of "a proportion of proteolytic enzymes active in slightly acid media sufficient to modify the proteids contained in the beer in such a manner that they will not be precipitated upon chilling subsequent to pasteurization, the beer being rendered chillproof in the sense that it is capable of remaining brilliant even when kept upon ice for a considerable time" (Wallerstein, 1911).

After some initial skepticism, the treatment with protease suggested by Wallerstein was tried with great success. Belgian scientist Effront (1977) reported of United State's beer that the Wallerstein treatment was applied to "the greater part of the beer made in that country." The original patents described the use of bromelin, papain, and pepsin for chillproofing. Further experience with the process (Wallerstein, 1961) led to improvement in commercial preparations. Although data on the composition of proprietary products normally are not published, papain (the extract of papaya that contains mixed proteases) is far superior to the other enzymes used for chillproofing. Similar plant sulfhydryl proteases such as bromelin and ficin are not as effective. Purified papain and chymopapain components of commercially available papain have been shown to be effective in chillproofing beer by Cayle *et al.*, (1964). Virtually all the enzyme used for chill stabilization has been and still is papain. As new proteases were discovered, they were tested for effectiveness as chillproofing agents for beer. Only the proteases produced by some *Streptomyces* species have been effective enough to have some commercial use (Posada *et al.*, 1981). However, adding a small amount of α-amylase from *Aspergillus oryzae* to commercial papain-based chillproofing preparations has become common. The amylase is thought to help keep the beer clear during storage by hydrolyzing any large dextrins that may be present in the beer and that would tend to become insoluble with age. Experience has shown that the amylase often does help keep beers clear.

Wallerstein's original patents described the action of proteases in chillproofing as modifying the proteins contained in the beer "in such a manner that they will not be precipitated upon chilling subsequent to pasteurization." He stated that the enzymes became active during the heating of treated beer to pasteurize it. Despite almost a century of industrial application of the enzymatic chillproofing process, the exact mechanism has never been determined. The amount of proteolytic hydrolysis in the beer is very limited, so no significant increase in amino nitrogen is found in treated beer. A decrease in trichloroacetic acid-precipitable protein is detected. Whether the chillproofing activity of papain is one of protein digestion alone is uncertain (Horie, 1964). Hebert *et al.*, (1975) also have detected an increase in coagulable protein after papain treatment and have suggested that papain-treated haze precursors coagulate very quickly, preventing further reaction with polyphenolics to form larger amounts of haze during chilling.

Whatever the reaction of papain with beer haze precursors, the affinity of papain for the precursor protein must be extremely specific because so little hydrolysis is involved in effective treatment and no other enzyme is as effective as papain.

VI. Making New and Special Types of Beer with Enzymes

The amount of alcohol in a beer is limited by the amount of solids, or extract, dissolved in the wort and by the percentage of that extract that is fermentable sugar. The percentage of fermentable sugar in wort is controlled by the extent of digestion of starch by amylases in the mash. The upper limit of fermentable sugar is $\sim 75\%$. The amount of fermentable sugar that can be formed by α- and β-amylase is limited by the temperature stability of these enzymes and by the fact that both enzymes hydrolyze only $\alpha(1\rightarrow4)$ bonds of starch, leaving the $\alpha(1\rightarrow6)$ branch points intact.

Over the years, brewers often have attempted to make special beers with a higher alcohol content than standard beer. Alcohol can be increased either by increasing the total percentage of solids or extract in the wort or by increasing the percentage of the extract that can be fermented by yeast. Increasing the fermentability requires the application of enzymes to hydrolyze more of the nonfermentable dextrins.

The term "malt liquor" has no strict definition and can be applied to any beer. In practice, however, the name usually is applied to a beer that is higher in alcohol than a regular beer, not very bitter, and fermented to give a relatively high ester content. Generally, the higher alcohol level is achieved by increasing the percentage of fermentable sugar in the wort. The average percentage of extract fermented in beers labeled malt liquor in the United States market is 75%. In making a malt liquor, therefore, the percentage of fermentable sugar usually is increased by a combination of increased enzyme action and addition of a completely fermentable sugar, such as pure glucose, as an adjunct.

A simple traditional method of increasing fermentable sugar during fermentation is adding wort that was separated from the mash at an early stage while the temperature was still low to cooled, fermenting wort. Some of the first "malt liquors" produced were made using this technique. The malt extract has a full complement of malt α-amylase and β-amylase as well as a small amount of a thermolabile starch debranching enzyme found in malt (Manners and Rowe, 1971). Even at the low temperature of fermentation, 15°C or less for a lager beer, the enzymes are capable of significant hydrolysis of dextrins during the several days of fermentation. The percentage of fermentable sugar in the wort can be increased by $\sim 10\%$ using this method, elevating total fermentable extract from 65% to 75%. Unfortunately, a variety of bacteria can be introduced into the fermenting wort with the enzyme when using this method, so it is not very popular. A more recent version of this method is adding commercially available barley β-amylase to the fermenting wort (Norris and Lewis, 1985). Malt α-amylase is not very active in fermenting wort because the pH drops during fermentation, usually to pH 4.5 or less. The malt debranching enzyme is present only in small quantity, so at the pH of fermentation the only malt starch-degrading enzyme that is very active is β-amylase, which is quite acid resistant. Addition of barley β-amylase gives essentially the same results as addition of a crude malt extract.

Commercially prepared α-amylase from the mold *Aspergillus oryzae* was developed originally by Takamine as a substitute for malt diastatic enzyme. This enzyme has been added to brewer's wort during fermentation to increase the level of fermentable sugar by hydrolyzing dextrins. Although the enzyme is an α-amylase, it produces a large amount of fermentable sugar from dextrin, and it is stable at the low pH of fermentation. The enzyme is capable of increasing the percentage of fermentable sugar in the wort to $\sim 75\%$. The amount of enzyme that must be added is about 1500 SKB units (American Association of Cereal Chemists, 1985) per liter of wort.

In the late 1960s, a new type of beer was developed using enzyme to increase the percentage of fermentable extract in the wort. The beer type is lower calorie beer, commonly called "light" beer. This new type has become tremendously successful. In 1991, the three most popular brands of light beer constituted 23% of total United States beer sales. Most of this beer is made using microbial enzymes.

In a regular beer, about one-third of the wort extract material is soluble dextrins derived from starch. These dextrins include the $\alpha(1\rightarrow6)$ branch linkages of starch. Dextrins are not fermented by the yeast, so they are present in the finished beer, to which they contribute about one-third of the caloric content. By converting these dextrins into fermentable sugar, a beer with significantly higher alcohol content and very little residual carbohydrate can be produced. If this high-alcohol beer is diluted so the alcohol content is reduced to that of a regular beer, the resulting beer has fewer calories than regular beer and very little residual carbohydrate. The diluted beer is sold as "light" beer or, in some countries, as low carbohydrate beer.

The enzyme originally applied to beers to give near complete breakdown of the dextrins normally found in beer is glucoamylase from *Aspergillus niger*. This enzyme is an exoamylase capable of hydrolyzing both $\alpha(1\rightarrow4)$ and $\alpha(1\rightarrow6)$ bonds in the dextrins. The application of this enzyme to increase the amount of fermentable sugar in wort was investigated by Saletan (1966) soon after the enzyme became commercially available. Under optimum conditions, glucoamylase is capable of transforming at least 95% of dextrins in beer to glucose that is fermented by the yeast. Since compounds other than carbohydrate, such as peptides of various sizes, are present in beer, the percentage of the total wort extract fermented is $\sim 85\%$. The enzyme is added to fermenting wort. Different units are used by different manufacturers to measure glucoamylase activity in commercial preparations, but the most common units are comparable in magnitude. The amount of enzyme added to fermenting wort for optimal reaction is 5–10 Units per liter. A Unit may be gm glucose produced from soluble starch per hr at pH 4.2 and 60°C. The assay conditions presented are quite different from conditions in fermenting wort. The pH is close to 4.2, the optimum for the enzyme, but the temperature will not be above 15°C. Therefore, the enzyme is used quite inefficiently during fermentation. However, the marketability of the beer produced justifies the inefficiency. Usually, the enzyme is not used at such a high level that excess glucose accumulates during fermentation. Conversion of dextrin to glucose is the rate limiting step. After the sugar that is formed during brewing of the wort is used by the yeast, fermentation continues for an additional 1 – 2 days as the dextrins are digested by the enzyme and the products almost immediately fermented by the yeast. The original procedure for producing light beer was developed by Owades at Narragansett Brewery and gradually spread to other breweries, and has been improved over subsequent years.

The increase in business that occurred as a result of the popularity of light beers made coping with some of the production problems that arise from the use of glucoamylase worthwhile. The biggest problem is that, if beer containing the enzyme is mixed with regular beer with a full complement of dextrin, the enzyme will begin digesting the dextrin in the regular beer, resulting in production of large amounts of glucose and an abnormally sweet tasting beer. Most beer is pasteurized at a temperature of 60°C to prevent spoilage. This temperature is optimum for glucoamylase activity; beer pH is the optimum pH for the enzyme. Pasteurization of regular beer mixed with as little as 5% enzyme-containing light beer will change the taste of the beer significantly by making it sweet. Extreme vigilance and control is required to insure that this does not occur, especially since it is common practice to filter two different types of beer in sequence through the same filter or to fill tanks with new beer when they are nearly but not completely empty.

Because of the potential problem of accidentally mixing glucoamylase-containing beer with regular beer, the prospect of using glucoamylase early in the brewing process, in the mash where starch from grains is made soluble and hydrolyzed by malt enzymes, is of great interest. All enzymes, from malt or other sources, present in the mash are inactivated completely during the subsequent boiling of the extracted wort. Conditions in the mash during starch gelatinization and digestion include temperatures above 60°C and a pH in the range of 5.2 to 6.0. The glucoamylase of *Aspergillus niger* begins to be inactivated at temperatures above 60°C. It is most stable and most active in a pH range of 4.0 – 4.5. Conditions in a brewers' mash, therefore, are not favorable for activity of this enzyme. Nevertheless, by using several-fold more enzyme than would be needed to obtain sufficient hydrolysis during fermentation, by adjusting the pH of the mash to the low end of the practical range, and by holding the mash at relatively low starch saccharification temperature (~63°C) for some time, almost as much conversion of starch to fermentable sugar, mainly glucose, occurs as occurs when enzyme is added during fermentation. Because of the advantages of absence of glucoamylase from the beer during later stages of the process, some brewers prefer to use the enzyme in this manner.

Production of light beer with glucoamylase is a good source of examples of the importance of side activities in commercial enzymes. *Aspergillus niger* is capable of producing many enzymes other than glucoamylase that can have an effect on beer. The acid protease produced by this organism is a good example. In beer, the ability to form foam when poured is an important characteristic. Foam formation depends on the presence of high molecular weight peptides, often called proteins, in beer. The acid protease of *Aspergillus niger* is ideally suited for hydrolysis of peptide bonds at the acid pH of beer. If too much protease is included in glucoamylase preparations, beer treated with the enzyme is likely to be rejected because of poor foam formation.

However, the protease side activity can have a positive effect in fermentation of light beer. Many yeast strains have the ability to form flocs after they finish growing during fermentation. The yeast cell surface changes so the cells adhere to each other on contact. Hundreds or thousands of cells will adhere to each other to form a floc, which allows the yeast to settle to the bottom of the fermenting tank much more quickly than single yeast cells would. The change at the cell surface must involve protein because a variety of proteases are able to change yeast cells from the floc-forming state to the non-floc-forming state. Fermenting beer with a floc-forming yeast is often advantageous because the yeast settle out of the beer quickly after fermentation is over, making the beer easier to clarify by filtration. In production of light beer, the yeast must stay suspended for a longer period of time to ferment all the glucose released by the glucoamylase. A small amount of protease present in glucoamylase can be just enough to prevent the formation of flocs for 1–2 days. The extra time is enough to allow fermentation to be completed. When batches of glucoamylase containing no protease are used in brewing with a floc-forming yeast strain, the yeast is able to settle out of the beer too quickly and the beer does not ferment to the extent that it should. These brewers must use a glucoamylase that contains some protease, but not so much that it reduces foam quality.

Other problems occurred when glucoamylase was first applied to production of light beer. A side activity, possibly β-galactosidase, which can hydrolyze galactosyldiglycerides found in malt, caused severe problems in foam production, especially in breweries in which the equipment was known to allow higher than normal amounts of lipid to be carried into the beer. Glucoamylase produced by certain enzyme-producing plants tended to induce haze formation in beer treated with that enzyme. Beer is expected to remain clear after it is filtered. The tendency to form haze was linked to protease levels, but the exact cause of the haze was never determined. As the techniques of enzyme production evolved, purer and more concentrated glucoamylase became available and many of the earlier problems of beer production disappeared.

The glucoamylase produced by *Rhizopus* species has attracted some interest for use in brewing because it is inactivated at ordinary beer pasteurization temperatures (60°C) and should not cause problems of excess sweetness if enzyme-treated and untreated beers become mixed. This enzyme is capable of essentially complete degradation of dextrins to fermentable glucose during fermentation of a wort in a manner similar to that of the *Aspergillus* enzyme. However, side activities or impurities in commercially available preparations lead to problems of haze formation and poor foam characteristics in beer produced with this enzyme. Therefore, this enzyme has never been used for commercial production of a light beer.

VII. Improving Filtration with Enzymes

Filtrations are critical to the brewing process. Before fermentation can begin, the liquid wort must be filtered from the grains; after fermentation, almost all beers are filtered to a high degree of clarity. In addition, some beers are filtered almost to the point of sterility if they are to be packaged and sold without pasteurization. The $(1\rightarrow3)$ and $(1\rightarrow4)$-linked β-D-glucan of the barley cell wall has been identified as a particular source of problems in filtration. Enzymes that digest this polysaccharide, identified simply as β-glucanases, have been employed as an aid to filtration.

The separation of liquid wort containing the soluble products of hydrolysis of the grains from the insoluble portion of the grain, or spent grain, is critical from the standpoints of time required and of the amount of material, called extract, recovered from the grains used. Ordinarily a gravity filtration, called lautering, requires about 2 hr whereas a pressure filtration requires a little less time. A difference in time of only a few minutes can affect significantly the throughput and efficiency of equipment operating on a 24-hr basis. The amount of extract recovered from the grains obviously affects production cost. The two factors are interrelated. If wort filtration is slow, time can be saved by washing the grains with less "sparge" water, but this adjustment leads to less efficient extraction. Therefore, factors that affect the rate of wort filtration raise considerable interest. Incompletely digested cell wall fragments or undigested small starch granules in the grain are important elements that can impede passage of liquid through the grain (Barrett *et al.* 1973). Malt that has not undergone adequate digestion of cell walls during the germination process, referred to as undermodified malt, often gives poor wort filtration when it is used for brewing. Addition of β-glucanase enzyme from *Bacillus, Aspergillus, Penicillium,* or *Trichoderma* sources has been found to improve filtration when undermodified malt or even unmalted barley is present in the mash (Letters *et al.,* 1985; Oksanen *et al.,* 1985). This effect is sometimes attributed to viscosity reduction by the β-glucanase but, at the high temperature of the mash, the soluble β-glucans in the wort do not increase the viscosity of the wort significantly. Careful study has determined that blockage by small insoluble pieces of grain is more likely to be the source of filtration difficulty (Barrett *et al.,* 1973). Addition of enzyme also may increase the amount of extract recovered from the grain, but the mechanism for the increase has not been determined carefully. Increased filterability due to the enzyme may allow more thorough washing of already soluble material from the grain. Alternatively, the enzyme actually may allow more components of the grain to become soluble.

The soluble β-glucan that is extracted from malt into the wort remains in solution during fermentation and is still present in the fermented beer. Although poorly modified malts will contribute more β-glucan, even the highest quality well-modified malts still have some soluble β-glucan content. This β-glucan becomes part of the beer that needs to be filtered to achieve clarity and, perhaps, near-sterility. For many years it was known that poorly modified malts produce beers that are difficult to filter; it was assumed that much of the problem was due to the β-glucan in the beer. Wagner *et al.* (1988) rigorously have correlated high molecular weight β-glucan content of beer with filtration characteristics of the beer. Letters *et al.,* (1985) have shown that filtration difficulties are caused by gel formation by the β-glucans, which depends on factors such as temperature and the shear developed as cold beer passes through a filter.

Addition of β-glucanase to beer during wort production or during storage after fermentation is effective in reducing the content of high molecular weight β-glucan and in making beer easier to filter, especially when the untreated beer is difficult to filter. Enzyme from *Trichoderma* is especially effective in making beers easier to filter (Oksanen *et al.,* 1985). The preparations contain a cellulase that hydrolyzes the $\beta(1{\rightarrow}4)$ bonds of the glucan as well as hemicellulases that may aid in glucan breakdown.

VIII. Extension of Shelf Life of Packaged Beer

Beer is extremely sensitive to oxygen and oxidative staling. Oxygen disappears during fermentation because it is taken up by the yeast, and remains below 0.2 mg/liter during processing in a well-operated plant in which the atmosphere in tanks and equipment is pure carbon dioxide. Levels of air in packaged beer are usually no more than 1 ml per package; some breweries operate below 0.5 ml. Once the development of enzymatic chillproofing made it practical to package beer and ship it over long distances while maintaining its appearance, further developments in packaging technology that reduced the air content in the package helped make beer flavor more stable as well as improved the clarity of the beer. Further improvement in packaged beer stability came with the development of antioxidants that are added to beer to retard oxidation, a development also credited to Wallerstein (1961) and his company. Antioxidants commonly used in beer include sulfite salts, sodium hydrosulfite ($Na_2S_2O_4$), and sodium erythorbate.

A continuing interest in the application of enzymes that react with oxygen to reduce oxygen in beer is the result of the importance of oxidation in beer flavor maintenance. The most readily available enzyme for this purpose is glucose oxidase from *Aspergillus niger*. A combination of glucose oxidase, which produces peroxide from glucose and oxygen, and catalase from the same organism, which produces oxygen and water from the peroxide, will after a number of cycles through peroxide intermediates eventually remove all oxygen from a water solution containing the enzymes, glucose, and water. Blockmans *et al.*, (1987) applied the glucose oxidase–catalase system to beer. Although glucose is present in unfermented wort, it is used completely by yeast during fermentation so insufficient glucose is present to support complete removal of the oxygen trapped in a sealed beer package. These investigators found that addition of 0.1 g/liter glucose to the beer allows complete reaction of the oxygen normally present in the package with glucose oxidase. Unexpectedly, however, beer treated with glucose and glucose oxidase–catalase enzyme tasted more stale after aging than did untreated beer. The next discovery was that, in the presence of ethanol, catalase does not convert peroxide to oxygen and water but instead acts like a peroxidase, catalyzing the oxidation of beer components by the peroxide intermediate. To eliminate the peroxide, it is necessary to add free sulfite to the beer. The sulfite is oxidized preferentially by the peroxide, providing a pathway for oxygen removal that does not adversely affect beer flavor. By adding glucose, enzyme, and sulfite to beer these individuals finally were able to achieve an improvement in flavor after aging.

Subsequent experimentation with glucose oxidase flavor stabilization by Goossens *et al.*, (1989) used glucose oxidase immobilized with glucose in the bottle crown inlay. These investigators were able to remove 90% of the oxygen in the headspace gas above the beer at a rate dependent on design parameters of the crown inlay. Further experimentation with oxygen removal by enzymes is necessary to determine whether removal of the low level of oxygen present in a commercial package results in significant flavor stabilization and whether the system is economically feasible.

References

American Association of Cereal Chemists (1985). α-Amylase activity of malt. In "Approved Methods of the American Association of Cereal Chemists," Vol. I. American Association of Cereal Chemists, St. Paul, Minnesota.

Bamforth, C. W., and Quain, D. E. (1989). Enzymes in brewing and distilling. *In* "Cereal Science and Technology" (G. H. Palmer, ed.). Aberdeen University Press, Aberdeen, Ireland.

Barrett, J., Clapperton, J. F., Divers, D. F., and Rennie, H. (1973). Factors affecting wort separation. *J. Inst. Brew.* **79**, 407–413.

Blockmans, C., Heilporn, M., and Masschelein, C. A. (1987). Scope and limitations of enzymatic deoxygenating methods to improve flavor stability of beer. *J. Am. Soc. Brew. Chem.* **45**, 85–90.

Button, A. H., and Palmer, J. R. (1974). Production scale brewing using high proportions of barley. *J. Inst. Brew.* **80**, 206–213.

Cayle, T., Saletan, L. T., and Lopes-Ramos, B. (1964). Some papain fractions and their characteristics. *Am. Soc. Brew. Chem. Proc.* 142–151.

Denault, L. J., Glenister, P. R., and Chau, S. (1981). Enzymology of the mashing step during beer production. *J. Am. Soc. Brew. Chem.* **39**, 46–52.

Ducroo, P., and Delecourt, R. (1972). Enzymatic hydrolysis of barley beta-glucans. *Wallerstein Lab. Commun.* **35**, 219–226.

Effront, J. (1917). "Biochemical Catalysts in Life and Industry." (English translation by S. C. Prescott). John Wiley and Sons, New York.

Goossens, E., Dillemans, M., and Masschelein, C. A. (1989). Headspace oxygen removal from packaged beer using crown cork inlays coated with glucose oxidase. *Proc. Eur. Brew. Conv. (Zurich)* **22**, 625–632.

Handke, E. (1896). "Handbuch für den Amerikanischen Brauer und Mälzer," Vol. 1. Wetzel, Milwaukee, Wisconsin. (German).

Hebert, J. P., Scriban, R., and Strobbel, B. (1975). Action of papain on beer proteins during pasteurization. *Proc. Eur. Brew. Conv. (Nice)* **15**, 405–421. (French).

Horie, Y. (1964). Protein-clotting activity of papain. *Am. Soc. Brew. Chem. Proc.* 174–182.

Hough, J. S., Briggs, D. E., Stevens, R., and Young, T. W. (1982). "Malting and Brewing Science," Vol. II, pp. 826–828. Chapman and Hall, London.

Letters, R., Byrne, H., and Doherty, M. (1985). The complexity of beer β-glucans. *Proc. Eur. Brew. Conv. (Helsinki)* **20**, 395–402.

Manners, D. J., and Rowe, K. L. (1971). Studies on carbohydrate-metabolizing enzymes. XXV. The debranching enzyme system in germinated barley. *J. Inst. Brew.* **77**, 358–365.

Max Henius Memoir Committee (1937). "Max Henius." Kingsport Press, Kingsport, Tennessee.

Moll, M., Flayeux, R., Lipus, G., and Marc, A. (1981). Biochemistry of mashing. *Master Brewer's Assoc. Am. Tech. Quart.* **18**, 166–170.

Munck, L., Mundy, J., and Vaag, P. (1985). Characterization of enzyme inhibitors in barley and their tentative role in malting and brewing. *J. Am. Soc. Brew. Chem.* **43**, 35–38.

Narziss, L. (1976). The influence of mashing procedure on the activity and effect of some enzymes—A survey. *Master Brewer's Assoc. Am. Tech. Quart.* **13**, 11–20.

Nielsen, E. B. (1971). Brewing with barley and enzymes—A review. *Proc. Eur. Brew. Conv. (Estoril)* **15**, 149–170.

Norris, K., and Lewis, M. J. (1985). Application of a commercial barley beta-amylase in brewing. *J. Am. Soc. Brew. Chem.* **43**, 96–101.

Oksanen, J., Ahvenainen, J., and Home, S. (1985). Microbial cellulase for improving filtrability of wort and beer. *Proc. Eur. Brew. Conv. (Helsinki)* **20**, 419–425.

Palmer, G. H. (1989). Cereals in malting and brewing. *In* "Cereal Science and Technology" (G.H. Palmer, ed.). Aberdeen University Press, Aberdeen, Ireland.

Posada, J., Segura, R., Perez, J. B., and Martinez, J. L. (1981). Colloidal stabilization of beer by addition of proteases. *Proc. Eur. Brew. Conv. (Copenhagen)* **18**, 443–450.

Reid, P. V. K. (ed.) (1992). The year in review. *Modern Brewery Age* **43**, 6–9.

Saletan, L. T. (1966). Carbohydrases of interest in brewing with particular reference to amyloglucosidase. *Proceedings of the Ninth Convention, the Institute of Brewing (Australian Section)*, pp. 127–140. Auckland, New Zealand.

Wagner, N., Esser, K. D., and Krüger, E. (1988). Analysis and importance of high molecular weight β-glucan in beer. *Monatsschr. Brauwiss.* **10**, 384–395. (German).

Wallerstein, L. (1911). United States Patents No. 995,820; 995,823; 995,824; 995,825; 995,826.

Wallerstein, L. (1961). Chillproofing and stabilization of beer. *Wallerstein Lab. Commun.* **24**, 158–167.

Weig, A. J. (1973). Brewing adjuncts and industrial enzymes. *Master Brewer's Assoc. Am. Tech. Quart.* **10**, 79–86.

CHAPTER 18

Fish Processing

Gudmundur Stefansson

I. Introduction

The importance of enzymes in fish processing has, until recently, mainly been restricted to endogenous enzymes naturally present in the fish tissues. The deliberate use of added enzymes in fish processing is not common. However, in the last several years, interest in controlling or aiding traditional processes that rely on endogenous enzymes by using added enzymes has increased. The use of enzymes as "specific" tools for fish processing operations has appeared also. Deskinning of fish, membrane removal, and roe purification are examples of such processes.

II. Traditional Processes

A. Production of Fish Sauce

Fermented fish sauce is a traditional product of the Far East. The products are important protein sources and are used mainly as a condiment for rice dishes (Amano, 1962). The products are consumed by over 250 million people in Southeast Asia (Saisithi *et al.*, 1966) and annual production has been estimated at about 250,000 tons (Campbell-Platt, 1987). The production procedure consists of mixing small uneviscerated fish with high concentrations of salt and storing the mixture in sealed vessels at ambient temperatures for a period of 6–18 months (van Veen, 1965; Saisithi *et al.*, 1966; Beddows, 1985). During the process, hydrolytic enzymes slowly degrade the fish tissues. Massive proteolysis occurs, with concomitant formation of peptides and amino acids. Other nitrogenous compounds such as ammonia

(Amano, 1962) and trimethylamine (Beddows *et al.*, 1979) are formed also. The final product is a clear liquid that is separated by filtration or is decanted off. The liquid has a high content of soluble nitrogenous compounds and a salt concentration in the range of 20–30% (Amano, 1962; Beddows, 1985). The remaining residue consists mainly of bones and scales and commonly has been used as a fertilizer (van Veen, 1965).

The hydrolysis of the tissues appears to be mainly an autolytic process by endogenous fish enzymes (Uyenco *et al.*, 1953; Amano, 1962; Beddows *et al.*, 1979). Orejana and Liston (1981) report that trypsin or trypsin-like enzymes are the principal agents of proteolysis in patis (a fish sauce) production. The pH during fish sauce fermentation is usually in the range of 5–6 (Uyenco *et al.*, 1953; Orejana and Liston, 1981). Bacteria or their enzymes are only of minor importance during the digestion, because of the inhibitory effects of the high salt concentration (Uyenco *et al.*, 1953; Orejana and Liston, 1981). It appears, however, that bacteria play a role in forming the distinct flavor and aroma that, in turn, determine the quality of the fish sauces (Amano, 1962; van Veen, 1965; Saisithi *et al.*, 1966; Beddows *et al.*, 1980). Fish enzymes and/or atmospheric oxidation also have been reported as possible contributors to the aroma of the fermented products (Dougan and Howard, 1975).

The subject of fish sauce and other fermented fish products has been reviewed by Amano (1962), van Veen (1965), Adams *et al.* (1985), and Beddows (1985).

In recent years, there has been some interest in accelerating fish sauce processing to reduce production cost by shortening the digestion or fermentation period. Amano (1962) and van Veen (1965) state that pineapple juice is sometimes used as a source of proteolytic enzymes to hasten the process. Beddows and Ardeshir (1979a) found that the plant enzymes bromelain, ficin, and papain could digest fish tissues in a short period, thus producing hydrolysates with distribution and concentration of nitrogenous compounds similar to that of fish sauce, but lacking in aroma. Ooshiro *et al.* (1981) came to a similar conclusion when using the enzyme papain; the added enzyme accelerated proteolysis but the aroma typical of traditionally produced fish sauce was not observed. Fish hydrolysates manufactured by the use of plant proteolytic enzymes can be added to traditionally produced fish sauce without affecting the nutritional quality (Beddows and Ardeshir, 1979a). However, on their own, the artificially produced fish hydrolysates are inadequate since they lack aroma, and aroma and flavor are the most important factors in consumer acceptability of the traditional fermented fish sauces (Beddows, 1985).

It has been reported that a good quality fish sauce can be prepared from capelin *(Mallotus villosus)* with the help of squid hepatopancreas (Raksakulthai *et al.*, 1986). These investigators conclude that the amount and type of amino acids and other non-amino-acid nitrogen compounds are important for the acceptability of capelin fish sauces. Addition of proteinases such as fungal, pronase, trypsin, and chymotrypsin to minced capelin increased the initial rate of protein hydrolysis, but did not yield a product with a much higher content of free amino acids in comparison with sauces without added enzymes (Raksakulthai *et al.*, 1986).

It may be possible to accelerate fish sauce processing using halophilic bacteria. Ok *et al.* (1982) found that the addition of halophilic *Bacillus* strains accelerated the production of fish sauce considerably. No great differences in free amino acid profile were observed between samples with or without the added bacteria.

Acid digestion has been suggested as a possible means to accelerate the processing of fish sauce (Uyenco *et al.*, 1953; Amano, 1962; Beddows and Ardeshir, 1979b). The disadvantage with hydrolysates thus produced is the lack of aroma and flavor (Uyenco *et al.*, 1953; Beddows and Ardeshir, 1979b). The acid hydrolysates can, however, be neutralized and added to traditionally prepared fish sauces, to increase the volume and thus reduce production cost (Beddows and Ardeshir, 1979b). Gildberg *et al.* (1984) found that fish sauce hydrolysis could be accelerated considerably by autolysis at pH 4 and low salt concentrations, followed by subsequent neutralization and salt addition. The final product, although initially lacking in flavor, developed flavor characteristics similar to those of traditionally produced fish sauce during storage (Gildberg *et al.*, 1984).

In this context, the production of fish silage can be mentioned. The process has been used commercially for some years, mainly in Denmark, to produce animal feed (Raa *et al.*, 1983). The raw material is mostly fish waste and trash fish, which is homogenized and acidified by inorganic and/or organic acids, thus creating a suitable environment for the proteolytic enzymes in the fish to degrade the tissues and form a liquefied mass (Raa and Gildberg, 1982; Gildberg and Almås, 1986). During fish silage production, no salt is used since the low pH prevents microbial spoilage. The fish silage thus produced is not suitable for human consumption because of the bitter tasting hydrophobic peptides that are formed during the enzymatic hydrolysis (Raa and Gildberg, 1982). These bitter peptides apparently are not observed in the production of traditional fish sauce, possibly because of the high salt concentration (Raa and Gildberg, 1982).

Many aspects of traditional fish sauce manufacture are still unclear. Better understanding is needed of the nature of the enzymatic hydrolysis and especially of the development of the flavor and aroma to obtain the best results with added commercial enzymes.

B. Salting of Herring

Endogenous fish enzymes are also of importance in traditional salt-curing of herring *(Clupea harengus)*. The process probably originated in Scotland in the eighth century. Over the years, it has been changed and improved by the various countries that produce salted herring (Voskresensky, 1965). The salting of herring is carried out by all countries that fish in the North Atlantic and the North Sea. The salting or processing of other species such as sprat *(Sprattus sprattus)* or anchovies *(Engraulis encrasicholus)* will not be addressed in this discussion.

Salting of herring is carried out by mixing salt and fresh herring (whole or gutted) and packing it in barrels. In some countries, a salt brine is added before the barrels are closed. The barrels are stored at low temperatures ($-5-10°C$) for a few weeks or up to several months until the herring is ripe. The final salt concentration of the products varies (approximately $4-18\%$), depending on the recipes used. Sugar, spices, and preservatives (potassium nitrate, sodium nitrite, potassium sorbate, and sodium benzoate) are some-times used to produce certain varieties of ripened herring. If the recipes are followed closely during the manufacture of salted herring and the raw mate-rial is of good quality, the final product obtained will be characteristic of a well-ripened herring with a tender consistency and a pleasant taste and odor.

The process consists of two stages, that is, salting and ripening but the latter stage is caused mainly by enzymes naturally present in the fish. The salting stage is characterized by salt penetration into the fish muscle and ends when the concentration of salt in the tissues is equal to that of the surround-ing brine (Voskresensky, 1965). During the ripening stage, biochemical and chemical processes occur that change the characteristics of the fish tissues. Proteolysis causes an increase in soluble nitrogenous compounds, such as peptides and amino acids (Kiesvaara, 1975), with concomitant change in the tissue structure. Both endo- and exopeptidases are considered to be of importance during ripening. Luijpen (1959) found that trypsin was the major contributor to endopeptidase activity at the pH normally found in herring brines (pH $5.5-6.5$). Various exopeptidases are believed to be important, especially during development of the sensory characteristics (Kiesvaara, 1977).

Although the salting of herring has been practiced for centuries, the nature of the ripening process is not known in any detail. Experience shows that herring that has been thoroughly cleaned of intestines ripens slowly and does not acquire the characteristic taste and odor. Luijpen (1959), investigat-ing the salt curing of matjes herring, concluded that pyloric caeca were responsible for the sensory properties of the finished product. Enzymes from muscle tissue also may be of importance (Voskresensky, 1965), although it has been shown that some muscle enzymes, for example, cathepsins, may lose activity at relatively low salt concentrations (Reddi *et al.*, 1972). Microorga-nisms and their enzymes are believed to be of minor importance during the ripening of herring (Knøchel and Huss, 1984).

Proteolytic enzymes from the internal organs, particularly the digestive enzymes, are of primary importance during the ripening of herring. How-ever, the activity of the endogenous digestive enzymes vary with the season and is at its highest before spawning (Kiesvaara, 1975). Spent herring is considered unsuitable for salting because of the bad appearance and the lack of the characteristic odor and taste in the finished salted product (Ruiter, 1972). In effect, the best products are obtained from herring salted during the feeding stage, which lasts for only a few months, when the endogenous proteolytic activity is high and the fat content is high as well. Considerable effort has been put into developing a manufacturing process that relies more on added enzymes than on endogenous enzymes, so salt-ripened herring can be produced throughout the year. Acceleration of the ripening process also has been considered (Opshaug, 1982).

It has been reported (Ruiter, 1972) that an acceptable ripened product can be obtained from eviscerated lean herring by the addition of a homogenate prepared from the appendices pyloricae of well-fed herring. A process for the manufacture of ripened herring has been patented (Eriksson, 1975) that is based on an enzyme preparation from herring viscera. The mixture is reported to be suitable for ripening herring caught during nonfeeding periods (both fresh and frozen) as well as herring fillets. According to Eriksson (1975), enzyme concentrates from cod fish also can be used to control the ripening of herring. Opshaug (1982) has described a process based on the addition of homogenized herring viscera to accelerate the ripening of herring or herring fillets (fresh or frozen). The ripening takes place during a period of 1–5 days and the resulting product is described as having a smooth consistency and a good aroma. However, the use of added herring viscera enzymes to control the ripening of herring does not appear to be widely used at the industrial level; the enzyme preparations are not readily available as commercial products.

Ritskes (1971) and Ruiter (1972) found that a mixture of commercial proteolytic preparations (from beef organs) containing low amounts of lipase activity had a favorable effect on both the flavor and the texture of salted matjes herring. A process for the manufacture of salted matjes herring based on a proteolytic mixture, containing mainly trypsin and chymotrypsin, has been patented (Anonymous, 1975). The enzyme mixture should, according to the patent, be substantially free of pepsin and papain, but also preferably free of lipolytic and amylolytic activities. The addition of cod trypsin to eviscerated brined herring has been shown to cause a more rapid accumulation of soluble protein than addition of bovine trypsin (Simpson and Haard, 1984). The accumulation of free amino acids increased with trypsin supplementation of eviscerated herring samples, although the amino acid profile was different from that of conventionally prepared samples (Simpson and Haard, 1987). Moreover, trypsin-supplemented herring samples were ranked lower than conventionally produced herring samples by sensory panelists (Simpson and Haard, 1987). Kiesvaara (1975) found that an incubation of herring fillets in a solution of pancreatic trypsin increases the release of peptides whereas, in traditional barrel ripening, peptides do not accumulate since the natural fish enzymes seem to cause further proteolysis. However, it has been reported that cod trypsin and an enzyme preparation from squid liver containing catepsin C (dipeptidyl peptidase I) activity accelerated the formation of free amino acids when supplemented to a brined eviscerated squid. An informal taste evaluation indicated that the product prepared with squid liver catepsin C had a better taste than other samples (Lee *et al.*, 1982).

Today, products are available on the European market that are ripened by adding commercially available enzymes or enzyme mixtures (Børresen, 1992). The quality, especially the taste quality, of herring products ripened this way is likely to be inferior to that of traditional salt ripened herring. The commercially available enzymes or enzyme mixtures that have been used to affect the ripening of herring seem to contain mainly endopeptidase activity, whereas the natural herring enzymes contain both exo- and endopeptidase activity (Ritskes, 1971; Ruiter, 1972; Kiesvaara, 1975). The difference between the activity of added enzymes and that of the endogenous enzymes may affect the quality of the final product, because the amount and type of free amino acids formed appears to be important in development of the flavor of ripened herring (Kiesvaara, 1975).

Although some success has been obtained with artificial ripening of herring, more work is needed, especially to understand in detail the nature of the traditional enzymatic ripening process and to select the enzymes that have properties and specifications similar to those of the endogenous enzymes.

III. New Applications of Enzymes in Fish Processing

Some new interesting applications of added enzymes for fish processing operations have emerged, partly because of increased knowledge of enzymes and their uses but also because many enzymes including some fish enzymes have become commercially available. In this section, some uses of enzymes in fish processing are discussed. Note, however, that not all the examples are used on an industrial level at this time.

A. Deskinning

The removal of skin from various fish species has been investigated by several workers. Fehmerling (1973) described a process for deskinning of fish species, such as tuna, that are difficult to skin by manual or mechanical means. Briefly, the process consists of preheating tuna fish (e.g., skipjack tuna) to 60°C with steam and placing the fish in a warm water bath (50–60°C) containing mainly proteolytic enzymes for approximately 10–90 min, following the immersion, most of the skin can be removed by water jets.

A process has been reported for the deskinning of skate *(Raja radiata)* or, more specifically, skate wings (Stefansson, 1988). This skate species has proven to be difficult to deskin manually or by skinning machines because of the spiked skin. The main structural component of the skin is collagen, which gives the skin physical strength. The skinning is accomplished by a gentle heat denaturation of the skin collagen in warm water, followed by immersing the skate wings at low temperatures (0–10°C) in an enzyme bath for several hours. The final step of the process consists of rinsing the dissolved skin from the skate wings and trimming the wings. A suitable enzyme solution should contain not only proteinases but also carbohydrases. The carbohydrases are apparently important during the enzymatic skinning, although not essential. Although the process is promising, it is as yet not used commercially.

Joakimsson (1984) described a process for the deskinning of whole herring *(Clupea harengus)*. The skin collagen is first denatured and the scales removed by a treatment with 5% acetic acid. The skin is solubilized by a cod pepsin preparation at low pH. Solubilization is reported to require 30–120 minutes at 15–20°C. The final step is removal of skin and enzymes by water jets (Joakimsson, 1984). However, this process has not yet been applied at an industrial level.

A method for deskinning and tenderizing squid using enzymes has been patented by Raa and Nilsen (1984). The skinning of squid is necessary to improve the appearance and to prevent discoloration, off-flavor, and odor of products for human consumption (Buisson *et al.*, 1985). However, mechanical or manual skinning is normally the bottleneck during processing. The enzymatic method can be accomplished by immersing the squid at low temperatures in a salt solution (5% NaCl) containing the enzyme papain (Raa and Nilson, 1984). The squid also can be skinned by immersing the tube and tentacles in a salt solution at 45°C for 10 min, whereby the endogenous enzymes of the squid are activated (Raa and Nilsen, 1984). According to the patent, the outer and the inner skins of the squid tube and the skin from the tentacles are all removed during the process. The enzymatic process is being used at the industrial level and has proved successful (Hempel, 1983). The outline of the industrial process, described by Wray (1988), indicates that the outer pigmented skin of the squid is removed by skinning machines prior to enzymatic treatment.

Deskinning of shark fin with enzymes has been described (Anonymous, 1992). Shark fin is a delicacy in Chinese cookery, used either for soups or as an appetizer. The preparation of shark fin is a traditional process that is both time consuming and labor intensive. Using a bacterial proteinase, the fin can be prepared in a shorter time. In the enzymatic process, the shark fin is immersed in a warm solution containing the enzyme until the skin can be loosened by rubbing. The enzyme activity is destroyed by transferring the fin to a solution with 1% hydrogen peroxide (Anonymous, 1992).

B. Removal of Membranes

Membrane removal can be a problem during fish processing. The production of canned cod liver is an example. The raw material for canning, cod *(Gadus morhua)* liver, is sometimes infested with sealworm *(Phocanema decipiens)*, which must be removed during processing. Removal is accomplished by removing a thin membrane that surrounds the liver since the sealworms infest the membrane. Suitable machinery is not available. Therefore, the membrane is removed by hand, but only with moderate success and at considerable labor cost. Steingrimsdottir (1987) developed a method in which this membrane is dissolved by proteolytic fish enzymes or bromelain. The method was patented in Iceland (Steingrimsdottir and Stefansson, 1990).

Membrane removal is also a problem during the production of salted cod swim bladders. The product is manufactured in small quantities (30–50 tons) for markets in Southern Europe. The fresh swim bladder used for the process is enclosed in a black membrane which must be removed before salting. The membrane is tightly bound to the bladder itself and is virtually impossible to remove by manual or mechanical means. However, a simple process has been developed (Steingrimsdottir and Stefansson, 1988) to dissolve this membrane using enzymes. The enzymatic process increases the rate of handling of swim bladders during membrane removal at least eightfold. The method is now used by all producers of salted cod swim bladder with a high degree of success.

C. Fish Roe Purification

A method of fish roe purification has been patented based on the use of proteinase to hydrolyze both the supportive and the connective tissues that envelop individual salmon and trout roe eggs and the roe sack (Sugihara *et al.*, 1973). The product yield is claimed to be more than 85%, whereas it is 50–60% using conventional methods. According to the patent, any proteinase may be used, for example, proteinases active under acid, neutral, or alkaline conditions (pH 3–10). Fish pepsins are reported to be useful during the production of salmon and trout roe caviar (Gildberg and Almås, 1986; Gildberg, 1988). According to Gildberg (1988), the fish pepsins split the linkages that adhere the egg cells to the connective tissue of the roe sack without affecting the egg cells. The enzymatic process is used in Scandinavia and Canada. In 1986, 40–50 tons of caviar were produced by this method (Wray, 1988). An enzymatic process is also reported to be used in Scotland during the production of trout roe caviar (Anonymous, 1989).

D. Miscellaneous Applications

Fehmerling (1973) described processes to remove intestines from shellfish such as clams, scallops, and mussels by the use of enzymes. The process consists of immersing the shellfish, after removing the shells, in a warm solution containing hydrolytic enzymes (mainly proteinases, lipases, and glycosidase hydrolases) until the intestines can be removed by spraying with water jets. Enzymes also may be of use in loosening or removing the shells from shellfish (Fehmerling, 1970).

The production of fish protein hydrolysate by proteinase treatment is a means to transform cheap pelagic fish and fish offal into protein concentrates with better functionality than fish meal. The manufacture consists of mixing ground fish with proteinase (e.g., ficin, papain, or microbial proteinases) that hydrolyzes part of the fish protein into peptides and amino acids. After hydrolysis, the enzyme is inactivated, the suspension is sieved, and usually it is concentrated and spray dried (Mohr, 1980; Mackie, 1982). Fish protein hydrolysates are soluble and can be used to supplement the nutritional value of liquid foods. They have been produced commercially in France as animal feed, mostly for use in milk replacers (Mackie, 1974). However, fish protein hydrolysates are not suitable for human consumption because of the bitter peptides that are formed during hydrolysis and because of their fishy flavor (Mohr, 1980).

A method has been described in which an emulsified protein mixture is produced from fish using enzyme hydrolysis (Shoji, 1990). The raw material is fish or shellfish. The fish is eviscerated and skinned. In some cases, bones and shells are removed before further processing. The raw material is then ground into particles of 100μm or smaller. A proteinase is added and the proteins are broken down so that coagulation will be minimal when heated. After hydrolysis, the mixture is heated to inactivate the enzyme and heat sterilized to insure good storage life. The paste-like product thus produced is a creamy textured fish protein that has a moderate fish taste. It is claimed that the product can be used for various purposes as a food ingredient, feed (after drying), or fertilizer (Shoji, 1990; Uchida et al., 1990).

Seafood flavorants from various sources of raw material can be produced by enzymatic hydrolysis (In, 1990a,b). For the process, mainly by-products from fish and shellfish are used. The process consists of liquefaction of the raw material by enzymatic hydrolysis, thermal inactivation of enzymes, separation of bones and shells, filtration or centrifugation, and, finally, concentration of the flavors naturally occurring in the raw materials. The purpose of the enzymatic hydrolysis is to liquefy but not produce any flavor or taste. Exogenous enzymes are used but endogenous enzymes are also of importance for the hydrolysis (In, 1990a,b).

A process has been patented based on the addition of an extract of crude papaya latex to salmon during canning (Yamamoto, 1983). The enzymatic activity is claimed to reduce the amount of heat-coagulated soluble proteins ("curd") formed during the canning process. According to the patent, other proteinases are also useful, for example, pepsin, pancreatic proteinases, ficin, and bromelain.

A bacterial proteinase has been used to hydrolyze protein in stickwater during fish meal processing (Jacobsen and Lykke-Rasmussen, 1984). The enzyme is claimed to reduce the viscosity of the stickwater and therefore generate considerable energy savings.

The combined action of the enzymes glucose oxidase and catalase in a solution of glucose has been suggested to extend the storage life of fresh fish (Wesley, 1982). The formation of gluconic acid with the concomitant drop in pH at the fish surface is claimed to inhibit spoilage microorganisms. Field et al. (1986) found that the enzymes could extend the storage life of both fillets and whole winter flounder. However, Shaw *et al.* (1986) found that the enzymes glucose oxidase and catalase or gluconic acid alone had little potential in extending the storage life of Atlantic cod fillets.

References

Adams, M. R., Cooke, R. D., and Rattagool, P. (1985). Fermented fish products of South East Asia. *Trop. Sci.* **25**, 61–73.

Amano, K. (1962). The influence of fermentation on the nutritive value of fish with special reference to fermented fish products of Southeast Asia. *In* "Fish in Nutrition" (E. Heen and R. Kreuzer, eds.), pp. 180–200. Fishing News, London.

Anonymous (1975). Production process for salted herrings (Matjes). British Patent No. 1,403,221.

Anonymous (1989). How to handle caviar with care. *Biotimes* **IV (4)**,12–13.

Anonymous (1992). Sharks are no match for enzymes. *Biotimes* **VII(1)**, 6–7.

Beddows, C. G. (1985). Fermented fish and fish products. *In* "Microbiology of Fermented Foods" (B. J. B. Wood, ed.), Vol. 2, pp. 1–39. Elsevier Applied Science, London.

Beddows, C. G., and Ardeshir, A. G. (1979a). The production of soluble fish protein solution for use in fish sauce manufacture. I. The use of added enzymes. *J. Food Technol.* **14**, 603–612.

Beddows, C. G., and Ardeshir, A. G. (1979b). The production of soluble fish protein solution for use in fish sauce manufacture. II. The use of acids at ambient temperature. *J. Food Technol.* **14**, 613–623.

Beddows, C. G., Ardeshir, A. G., and bin Daud, W. J. (1979). Biochemical changes occurring during the manufacture of budu. *J. Sci. Food Agric.* **30**, 1097–1103.

Beddows, C. G., Ardeshir, A. G., and bin Daud, W. J. (1980). Development and origin of the volatile fatty acids in budu. *J. Sci. Food Agric.* **31**, 86–92.

Børresen, T. (1992). Biotechnology, by-products, and aquaculture. *In* "Seafood Science and Technology" (E. G. Bligh, ed.), pp. 278–287. Fishing News Books, Oxford, England.

Buisson, D. H., O'Donnell, D. K., Scott, D. N., and Ting, S. C. (1985). "Squid Processing Options for New Zealand." DSIR Fish Processing Bulletin No. 6. Division of Horticulture and Processing, Auckland, New Zealand.

Campbell-Platt, G. (1987). "Fermented Foods of the World—A Dictionary and Guide." Butterworths, London.

Dougan, J., and Howard, G. E. (1975). Some flavouring constituents of fermented fish sauces. *J. Sci. Food Agric.* **26**, 887–894.

Eriksson, C. (1975). Method of controlling the ripening process of herring. Canadian Patent No. 969419.

Fehmerling, G. B. (1970). Process and composition for loosening and removing edible tissue from shells of marine creatures. U.S. Patent No. 3,513,071.

Fehmerling, G. B. (1973). Separation of edible tissue from edible flesh of marine creatures. U.S. Patent No. 3,729,324.

Field, C. E., Pivarnik, L. F., Barnett, S. M., and Rand, A. G., Jr. (1986). Utilization of glucose oxidase for extending the shelf-life of fish. *J. Food Sci.* **51** (1), 66–70.

Gildberg, A. (1988). Recovery of digestive enzymes from fish waste and application of such enzymes in fish and food processing. Presented at the International Symposium on Application of Biotechnology for Small Industries Development in Developing Countries. Bangkok.

Gildberg, A., and Almås, K. A. (1986). Utilization of fish viscera. *In* "Food Engineering and Process Applications" (M. Le Maguer and P. Jelen, eds.), Vol. 2, pp. 383–393. Elsevier Applied Science, London.

Gildberg, A., Espejo-Hermes, J., and Magno-Orejana, F. (1984). Acceleration of autolysis during fish sauce fermentation by adding acid and reducing the salt content. *J. Sci. Food Agric.* **35,** 1363–1369.

Hempel, E. (1983). Taking a short-cut from the laboratory to industrial-scale production. *Infofish Marketing Dig.* **4,** 30–31.

In, T. (1990a). Seafood flavorants produced by enzymatic hydrolysis. *In* "Advances in Fisheries Technology and Biotechnology for Increased Profitability" (M. N. Voigt and J. R. Botta, eds.), pp. 425–436. Technomic, Lancaster, Pennsylvania.

In, T. (1990b). Seafood flavorants produced by enzymatic hydrolysis. *In* "Making Profits out of Seafood Wastes" (S. Keller, ed.), pp. 197–201. Proceedings of the International Conference on Fish By-Products, Anchorage, Alaska, April 25–27.

Jacobsen, F., and Lykke-Rasmussen, O. (1984). Energy-savings through enzymatic treatment of stickwater in the fish meal industry. *Process Biochem.* **19**(5), 165–169.

Joakimsson, K. G. (1984). "Enzymatic Deskinning of Herring *(Clupea harengus).*" Thesis. Institute of Fisheries, University of Tromsø, Norway *(in Norwegian).*

Kiesvaara, M. (1975). "On the Soluble Nitrogen Fraction of Barrel Salted Herring and Semi-Preserves during Ripening." Thesis. Technical Research Center of Finland, Helsinki, Finland.

Kiesvaara, M. (1977). Ripening of salted fish. *Livsmeddelsteknik* **7,** 351–353 *(in Swedish).*

Knøchel, S., and Huss, H. H. (1984). Ripening and spoilage of sugar salted herring with and without nitrate. I. Microbiological and related chemical changes. *J. Food Technol.* **19,** 203–213.

Lee, Y. Z., Simpson, B. K., and Haard, N. F. (1982). Supplementation of squid fermentation with proteolytic enzymes. *J. Food Biochem.* **6,** 127–134.

Luijpen, A. F. M. G. (1959). "The Influence of Gibbing on the Ripening of Maatjes Cured Herring." Thesis. University of Utrecht, Holland *(in Dutch).*

Mackie, I. M. (1974). Proteolytic enzymes in recovery of proteins from fish waste. *Process Biochem.* **9**(10), 12–14.

Mackie, I. M. (1982). Fish protein hydrolysates. *Process Biochem.* **17**(1), 26–23, 31.

Mohr, V. (1980). Enzymes technology in the meat and fish industries. *Process Biochem.* **15**(6), 18–21, 32.

Ok, T., Matsukura, T., Ooshiro, Z., Hayashi, S., and Itakura, T. (1982). Study on the use of halophilic bacteria in production of fish sauce. *Nippon Shokuhin Kogyo Gakkaishi* **29**(10), 623–627.

Ooshiro, Z., Ok, T., Une, H., Hayashi, S., and Itakura, T. (1981). Study on use of commercial proteolytic enzymes in production of fish sauce. *Mem. Fac. Fish. Kagoshima Univ.* **30,** 383–394.

Opshaug, K. R. (1982). Method for enzyme maturing of fish. PCT Patent No. 82/03533.

Orejana, F. M., and Liston, J. (1981). Agents of proteolysis and its inhibition in patis (fish sauce) fermentation. *J. Food Sci.* **47,** 198–203, 209.

Raa J., and Gildberg, A. (1982). Fish silage: A review. *CRC Crit. Rev. Food. Sci. Nutr.* **16,** 383–419.

Raa, J., and Nilsen, K. (1984). A method for the removal of connective skins from squid. Norwegian Patent No. 150304 *(in Norwegian).*

Raa, J., Gildberg, A., and Strøm, T. (1983). Silage production—Theory and Practice. *In* "Upgrading Waste for Feed and Food" (D. A. Ledward, A. J. Taylor, and R. A. Lawrie, eds.), pp. 117–132. Butterworths, Boston.

Raksakulthai, N., Lee, Y. Z., and Haard, N. F. (1986). Effect of enzyme supplements on the production of fish sauce from male capelin *(Mallotus villosus). Can. Inst. Food Sci. Technol. J.* **19**(1), 28–33.

Reddi, P. K., Constantinides, S. M., and Dymsza, H. A. (1972). Catheptic activity of fish muscle. *J. Food Sci.* **37,** 643–648.

Ritskes, T. M. (1971). Artificial ripening of maatjes-cured herring with the aid of proteolytic enzyme preparations. *Fish. Bull.* **69**(3), 647–654.

Ruiter, A. (1972). Substitution of proteases in the enzymic ripening of herring. *Ann. Technol. Agric.* **21**(4), 597–605.

Saisithi, P., Kasemsarn, B-O., Liston, J., and Dollar, A. M. (1966). Microbiology and chemistry of fermented fish. *J. Food Sci.* **31,** 105–110.

Shaw, S. J., Bligh, E. G., and Woyewoda, A. D. (1986). Spoilage pattern of Atlantic cod fillets treated with glucose oxidase/gluconic acid. *Can. Inst. Food Sci. Technol. J.* **19**(1), 3–6.

Shoji, Y. (1990). Creamy fish protein. *In* "Making Profits out of Seafood Wastes" (S. Keller, ed.), pp. 87–93. Proceedings of the International Conference on Fish By-Products, Anchorage, Alaska, April 25–27.

Simpson, B. K., and Haard, N. F. (1984). Trypsin from Greenland cod as a food-processing aid. *J. Appl. Biochem.* **6,** 135–143.

Simpson, B. K., and Haard, N. F. (1987). Cold-adapted enzymes from fish. *In* "Food Biotechnology" (D. Knorr, ed.), pp. 495–527. Marcel Dekker, New York.

Stefansson, G. (1988). Enzymes in the fishing industry. *Food Technol.* **42**(3), 64–65.

Steingrimsdottir, U. (1987). Enzymatic removal of cod liver membrane. Internal Report. Icelandic Fisheries Laboratories, Reykjavik, Iceland *(in Icelandic).*

Steingrimsdottir, U., and Stefansson, G. (1990). A method to remove membranes, skin or scales from marine animals. Icelandic Patent No. 1447 *(in Danish).*

Steingrimsdottir, U., and Stefansson, H. (1988). Enzymatic removal of swimbladder membrane. Internal Report, Icelandic Fisheries Laboratories, Reykjavik, Iceland *(in Icelandic).*

Sugihara, T., Yashima, C., Tamura, H., Kawasaki, M., and Shimizu, S. (1973). Process for preparation of Ikura (salmon egg). U.S. Patent No. 3,759,718.

Uchida, Y., Hukuhara, H., Shirakawa, Y., and Shoji, Y. (1990). Bio-fish flour. *In* "Making Profits out of Seafood Wastes" (S. Keller, ed.), pp. 95–99. Proceedings of the International Conference on Fish By-Products. Anchorage, Alaska, April 25–27.

Uyenco, V., Lawas, I., Briones, P. R., and Taruc, R. S. (1953). Mechanics of bagoong (fish paste) and patis (fish sauce) processing. *In* "Proceedings of the 4th Meeting of the Indo Pacific Fisheries Council, Bangkok," pp. 210–222. Food and Agriculture Organization, Bankok, Thailand.

van Veen, A. G. (1965). Fermented and dried seafood products in Southeast Asia. *In* "Fish as Food" (G. Borgstrom, ed.), Vol. III, pp. 223–250. Academic Press, New York.

Voskresensky, N. A. (1965). Salting of herring. *In* "Fish as Food" (G. Borgstrom, ed.), Vol. III, pp. 107–131. Academic Press, New York.

Wesley, P. (1982). Glucose oxidase treatment prolongs shelf life of fresh seafood. *Food Dev.* 36–38.

Wray, T. (1988). Fish processing: New uses for enzymes. *Food Manufac.* **63**(7), 48–49.

Yamamoto, M. (1983). Fish canning process. U.S. Patent No. 4,419,370.

Index

FOOD SCIENCE AND TECHNOLOGY

International Series

S. M. Herschdoerfer (ed.), *Quality Control in the Food Industry*, second edition. Volume 1—1985. Volume 2—1985. Volume 3—1986. Volume 4—1987.

Walter M. Urbain, *Food Irradiation*. 1986.

Peter J. Bechtel (ed.), *Muscle as Food*. 1986.

H. W.-S. Chan (ed.), *Autoxidation of Unsaturated Lipids*. 1986.

F. E. Cunningham and N. A. Cox (eds.), *Microbiology of Poultry Meat Products*. 1987.

Chester O. McCorkle, Jr. (ed.), *Economics of Food Processing in the United States*. 1987.

J. Solms, D. A. Booth, R. M. Dangborn, and O. Raunhardt (eds.), *Food Acceptance and Nutrition*. 1987.

Jethro Jagtiani, Harvey T. Chan, Jr., and Williams S. Sakai (eds.), *Tropical Fruit Processing*. 1988.

R. Macrae (ed.), *HPLC in Food Analysis*, second edition. 1988.

A. M. Pearson and R. B. Young, *Muscle and Meat Biochemistry*. 1989.

Dean O. Cliver (ed.), *Foodborne Diseases*. 1990.

Majorie P. Penfield and Ada Marie Campbell, *Experimental Food Science*, third edition. 1990.

Leroy C. Blankenship (ed.), *Colonization Control of Human Bacterial Enteropathogens in Poultry*. 1991.

Yeshajahu Pomeranz, *Functional Properties of Food Components*, second edition. 1991.

Reginald H. Walter (ed.), *The Chemistry and Technology of Pectin*. 1991.

Herbert Stone and Joel L. Sidel, *Sensory Evaluation Practices*, second edition. 1993.

Robert L. Shewfelt and Stanley E. Prussia (eds.), *Postharvest Handling: A Systems Approach*. 1993.

R. Paul Singh and Dennis R. Heldman, *Introduction to Food Engineering*, second edition. 1993.

Tilak Nagodawithana and Gerald Reed (eds.), *Enzymes in Food Processing*, third edition. 1993.

Takayaki Shibamoto and Leonard Bjeldanes, *Introduction to Food Toxicology*. 1993.

Dallas G. Hoover and Larry R. Steenson (eds.), *Bacteriocins of Lactic Acid Bacteria*. 1993.

John A. Troller, *Sanitation in Food Processing*, second edition. 1993.

Ronald S. Jackson, *Wine Science: Principles and Applications*. In Preparation.

Robert G. Jensen, and Marvin P. Thompson (eds.), *Handbook of Milk Composition*. In Preparation.

Tom Brody, *Nutritional Biochemistry*. In Preparation.

Printed and bound by CPI Group (UK) Ltd, Croydon, CR0 4YY

03/10/2024

01040422-0014